Cable Stayed Bridges

From Concept to Performance-based Design

Ayman Shama
Program Manager Parsons
New York, NY, USA

CRC Press
Taylor & Francis Group
Boca Raton London New York

CRC Press is an imprint of the
Taylor & Francis Group, an **informa** business
A SCIENCE PUBLISHERS BOOK

First edition published 2025
by CRC Press
2385 NW Executive Center Drive, Suite 320, Boca Raton FL 33431

and by CRC Press
4 Park Square, Milton Park, Abingdon, Oxon, OX14 4RN

CRC Press is an imprint of Taylor & Francis Group, LLC

Library of Congress Cataloging-in-Publication Data (applied for)

ISBN: 978-1-138-55789-5 (hbk)
ISBN: 978-1-041-02655-6 (pbk)
ISBN: 978-1-315-15113-7 (ebk)

DOI: 10.1201/b22457

Typeset in Times New Roman
by Prime Publishing Services

To
My wife Inas and our children,
Kareem, Kariman, and Myra
for support, patience and endurance

Preface

Cable-stayed bridges are elegant structures favorable for long spans. They are characterized by high-strength steel cables as main structural elements transferring the superstructure weight to the foundations through pylons. They are both visually pleasing and economically advantageous due to their structural redundancy, light decking, and ease of design, detail, and construction.

Less than 120 major cable-stayed bridges were constructed in the early 1980s, but currently there may be over 1000 open and planned cable-stayed bridges. Their spans have greatly expanded, rising from about 400 meters in the 1970s to 1104 meters now. They are superior to suspension bridges because they eliminate the need for large anchorage systems and have better aerodynamic stability. Nowadays, the cable-stayed system is positioned as a substitute for the suspension system for spans between 700 meters and 1000 meters due to advancements in the fabrication of high-strength steel cables.

The objective of this book is to compile the most recent developments in cable-stayed bridge design, analysis, and building techniques into a single volume. Practicing engineers, bridge owners, and stake holders looking to stay current on construction and design techniques of cable-stayed bridges can find in this book a convenient resource of up-to date construction and design methods. The book can also be regarded as a mini-encyclopedia on notable cable-stayed bridges worldwide, since it includes design data of more than 150 significant bridges from North America, Central and South America, Europe, Asia, and Africa. The book explores the history of the cable-stayed bridge, from its initial use to widespread in Germany following World War II and its expansion into other countries. A detailed description is given of the main characteristics of contemporary bridges from different countries, such as their geometrical arrangements, the kinds and designs of their pylons, and the numerous types of decks constructed using diverse materials and techniques. Photographs and detailed illustrations of each bridge's unique qualities are also included with concise discussion of each bridge design.

Performance-based design is the concept of designing the structure for different levels of hazard by selecting for each hazard level its overall performance, which is tied to specific damage levels of its components. The concept has been applied in the past to numerous cable-supported bridges. Compared to prescriptive design, performance-based design has several benefits. Targeted performance can be achieved with revealed confidence and dependability when PBD is carried out correctly. Through PBD, stakeholders can choose performance objectives that meet the bridge's cost, serviceability, and sustainability needs. This design philosophy is further discussed and its applicability to cable-stayed bridges is explained in this book. The book consists of fourteen chapters organized as follows:

The history of the initial attempts to create a stay cable system to facilitate crossings over various obstacles is presented in Chapter 1. Early systems are included such as the cable stayed bridge by Fausto Veranzio in 1595, the Ordish–Lefeuvre bridge system, 1858 in England, the Runyon-William Cable-stayed system (1883-1893) in Texas, the John Roebling stayed-suspended bridges in Ohio and New York (1867-1883), and the Arnodin-Gisclard Bridge System in France (1879).

Chapter 2 discusses fundamental aspects of cable-stayed bridges such as the structural system and support conditions, the longitudinal and transverse arrangement of cables, and the behavior of cable-stayed bridges under live loads.

Chapter 3 provides a detailed description of the stay cable system and its constituent parts. A description of cable types is given, emphasizing parallel strand cables. Various anchorage solutions for concrete and steel decks are described. The cable saddle system at concrete pylons is also presented. Design requirements of stay cables are also explained. Finally, protection of cables against fire and blast is covered.

Chapter 4 covers the main parameters that govern the selection and sizing of different components of cable-stayed bridges. factors considered in the selection of sizes and types of major components, including the foundations, pylons, superstructure, span lengths, and their ratios are outlined. Before diving into chapters 5 through 9, it is important to learn and comprehend the various concepts in chapters 2 through 4.

The development of modern cable-stayed bridges in Europe is covered in Chapter 5, beginning with Dischinger's 1956 Stromsund Bridge in Sweden and continuing through the post-World War II reconstruction of Germany's bridges. Notable cable-stayed bridges in Europe are covered including the Russky Bridge in Russia, which became the world's longest cable-stayed bridge in 2012. Design data and illustrations of more than 65 European cable-stayed bridges are presented in this chapter.

Chapter 6 outlines major cable-stayed bridges built in the US starting from the John O'Connell Memorial Bridge in Alaska to the upcoming Gordie Howe International Bridge, Michigan. Major Cable-stayed bridges in Canada are also outlined in this chapter including the Samuel De Champlain Bridge.

Chapters 7 and 8 include major cable-stayed bridges in Asia and Africa. Bridges in China and Japan are covered in Chapter 7. Chapter 8 includes bridges in South Korea, Vietnam, Thailand, Malaysia, Indonesia, India, Turkey, Egypt, Libya, Morocco, and more.

Chapter 9 includes significant signature cable-stayed bridges in Mexico, Central and South America. Included in this chapter bridges in Puerto Rico, Dominican Republic, Panama, Ecuador, Columbia, Peru, Venezuela, Argentina, Chile, Uruguay, and Brazil.

Advanced topics in cable-stayed bridges are included in chapters 10 through 14. Chapter 10 covers analysis, design and construction techniques of cable-stayed bridges. This chapter covers first, non-linear static and dynamic analyses of cable-stayed bridges. Aspects associated with modeling and analysis of composite and reinforced concrete segmental bridges are outlined. Approximate design method of cables is explained. Design of different types of superstructures such as composite, orthotropic steel, and precast box girders are outlined. Discussions of construction methods of superstructure, pylons and different kinds of foundations are also included.

Chapter 11 discusses the aerodynamic stability of cable-stayed bridges. In this chapter, phenomena related to cable-stayed bridge vibration are described. These include buffeting, flutter, galloping, and vortex shedding. Additionally, the steps involved in the aerodynamic investigation of cable-stayed bridges are covered. Wind-induced vibration of stay cables and mitigation strategies are also covered.

The performance-based design concept is discussed in Chapter 12 along with how it applies to seismic and fire dangers, two threats that concern cable-stayed bridges. Seismic performance-based design is discussed including determination of the seismic hazard levels, establishment of general performance requirements and local damage levels. Additionally modeling techniques for seismic PBD are covered. Performance-based fire safety design of cable-stayed bridges is also covered, including of definition of performance objectives and identification of bridge fire scenarios.

Chapter 13 covers structural health monitoring (SHM) techniques for cable-stayed bridges. The Sunshine Skyway Bridge in Florida, the Tatara Bridge in Japan, the Stonecutters Bridge in Hong Kong, the Jindo Bridge in South Korea, and the Charles W. Cullen Bridge are few examples of SHM application on existing cable-stayed bridges. Inspection of cable-stayed bridges is covered in Chapter 14. Techniques for cable maintenance and inspection are also described.

I am grateful to many engineers with whom I have collaborated on numerous projects over the past few decades, especially Tom Spoth, Seth Condell, Jamey Barbas, Kenneth Serzan, Michael Jones, and Greg Orsolini.

Ayman Shama
Princeton, New Jersey
November 2024

Contents

CHAPTER 1

History of Stay Cable Systems

1.1 THE FIRST STAYED STRUCTURES

The concept of supporting a pole or a beam by firm cables, has its roots that extend to ancient times. The system's first known use dates back thousands of years. In 1943 B.C., Egypt's Queen Hatshepsut issued an expeditionary decree to Punt (today known as Somalia). As seen in Fig. 1.1, the ancient Egyptians utilized ships equipped with inclined ropes to support a beam from the ship mast on this voyage. Additionally, Borneo and Laos were home to archaic bamboo pedestrian stay bridges (Troitsky, 1988). Despite the fact that these prehistoric constructions provided evidence for an early comprehension of the stayed system, stayed bridges were not documented again until the seventeenth century.

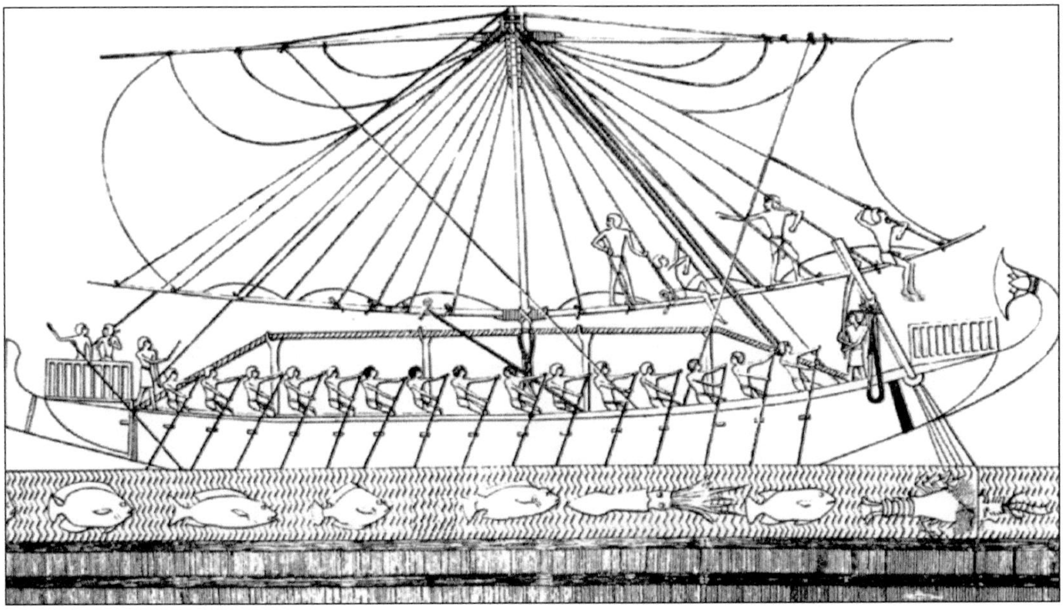

Fig. 1.1 Egyptian Queen's Hatshepsut expedition ship 1493 B.C.

The first stayed bridge designs in Europe can be found in Fausto Veranzio's book *Machinae Novae* (New Machines), published in 1595 (Fig. 1.2). The drawings show the bridge supported by many chain stays. The earliest stayed bridge described in European literature was this one. Eventually, in 1784, Immanuel Loscher, a German carpenter, created an all-timber bridge with a

44-meter span made of wooden stays fastened to a wooden pylon (Fig. 1.3). It was this bridge that originally gave birth to the idea of a bridge supported solely by inclined stays. The use of steel bar stays hanging from tall pylons was proposed by French architect Poyet in 1787 (Fig. 1.4). Redpath and Brown, two British engineers, constructed the King's Meadow Bridge in England in 1817 (Fig. 1.5). This footbridge had a span of approximately 34 m and was built using sloping stayed cables made of chain links consisting of looped wires attached to cast iron pylons (Ponaldy and Scalzi, 1986).

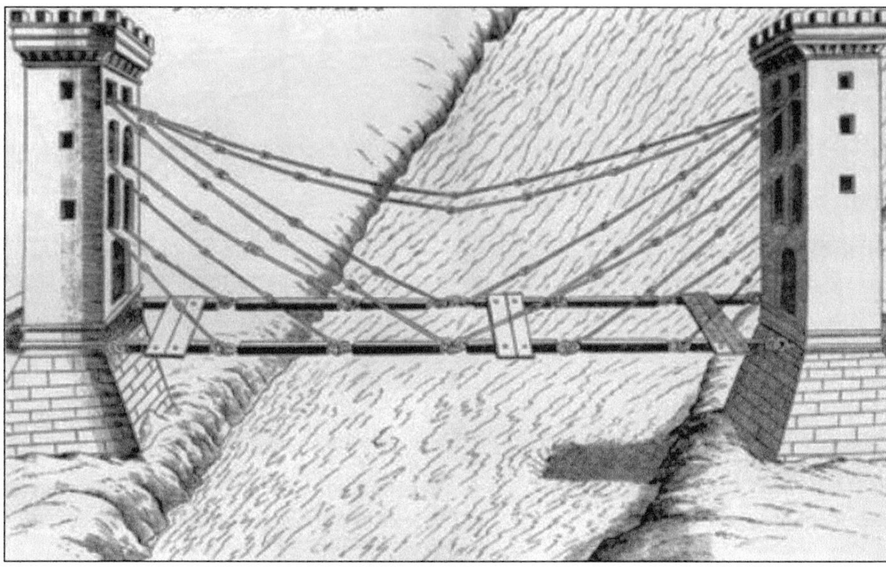

Fig. 1.2 The first stayed bridge in Europe, Italy, 1595

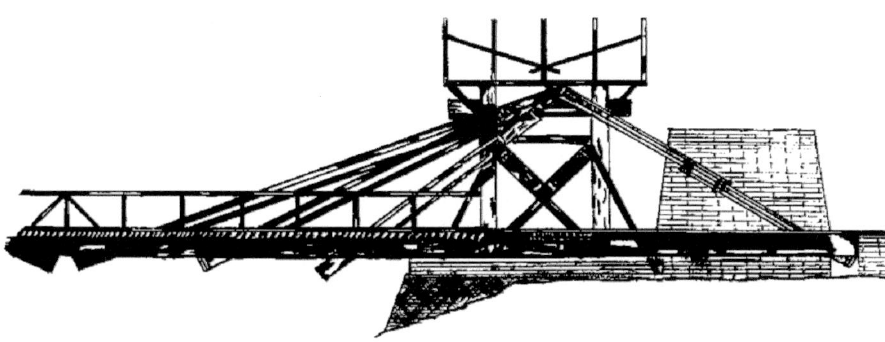

Fig. 1.3 The all-timber stayed bridge designed by Loscher in Germany, 1784

There are two recorded incidents of collapse of such constructions at the beginning of the eighteenth century. In 1817, the system of inclined stay wires previously employed for the King's Meadow Bridge was used in the design of the first bridge across the Tweed River in Scotland near Dryburgh Abbey (Drewry, 1832), which had a 79 m main span and a width of 1.2 m. Soon after its completion, the bridge suffered significant vibrations under wind and pedestrian loads. Eventually, it was destroyed six months after its opening when it was exposed to severe gust winds due to fractures in the inclined stays. It was soon restored in 1818 with added auxiliary stay cables. This

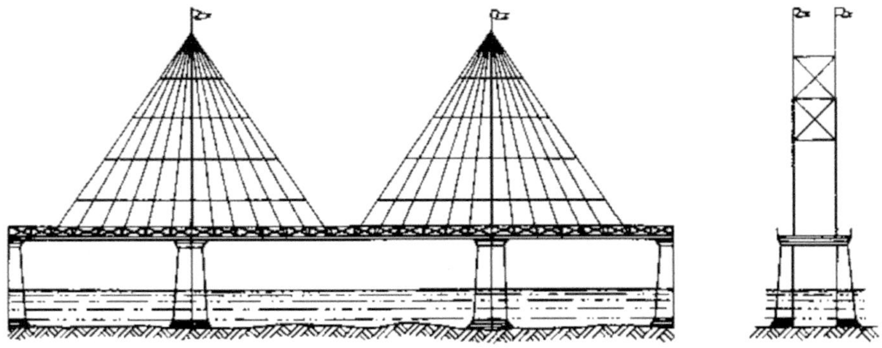

Fig. 1.4 Poyet's proposal of a cable stayed bridge, France, 1787

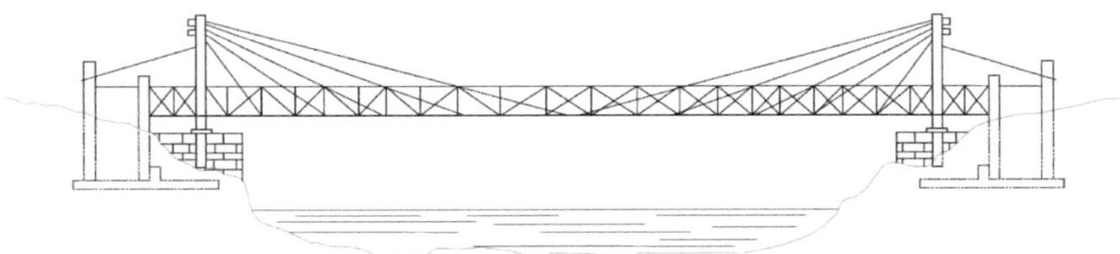

Fig. 1.5 Illustration of the King's Meadow Bridge, England, 1817

Fig. 1.6 Dryburg Bridge, Scotland, 1817

was the first application of the stayed suspension bridge, wherein the stays were added with an aim to provide additional stiffness and stability against winds as shown in Fig. 1.6. Another bridge was constructed in Germany over the River Saale near Nienburg in 1824. This bridge, which had a 78 m span, collapsed in the following year under overloading by a crowd of people (Feige, 1966). This accident was the main reason that the development of stayed bridges was delayed in Germany for about 125 years.

These early failures were due to the engineers' lack of knowledge of the actual structural behavior of these bridges such as the dependence of the overall structure stiffness on the stiffness of the stays. Moreover, the procedures used during this period for design could not address the complex indeterminate stay system of cable-stayed bridges. The materials used for the stays in these early designs, such as steel bars and chains, lacked the strength required for these structures to support large loads. Also, the stays were usually attached to the superstructure and pylons by pinned connections, which prevented proper tensioning of the stays. Consequently, these members performed only after considerable deformation and did not contribute to the behavior of the bridge to the degree they were intended for.

In 1830 the famous engineer and French scientist Claude-Louis Navier, published his report *"Rapport et Memoire sur les Ponts Suspendus"* (Navier, 1830) after two visits to England to study the art of building suspension bridges. His report presented a mathematical analysis of suspension bridges, and also devoted a brief treatment of stay bridges. Navier acknowledged the complexity of the cable-stayed bridges as statically indeterminate structures and realized the impossibility to evaluate accurately their response to loads by the available methods of his time. His final conclusion was that the suspension system should be used instead of the stayed system (Billington and Nazmy, 1990). This conclusion was significantly influenced by the failure accidents of the cable-stayed bridges in both England and Germany. Navier clearly ruled out cable-stayed bridges and his recommendations resulted in a very limited number of stayed bridges being built up to the 1950s. Nevertheless, systems that combined the suspension system with stays continued to be used in many major bridges built in the second half of the 19th century.

Curtis published a concept namely, the inflexible suspension bridge in 1837 (Curtis, 1837). He used radial rods (inclined stays) as well as a stiff beam to carry the loads as illustrated in Fig. (1.7). He claimed that with this arrangement, the bridge possesses a degree of stiffness and solidity wholly unattainable by traditional methods of constructing suspension bridges. This approach is clearly equivalent to the nowadays extradosed bridge concept.

Fig. 1.7 Curtis inflexible suspension bridge concept, 1837

Thomas Motley adopted Curtis's approach in the design of a new bridge across river Avon at Twerton, England (Fig. 1.8). The bridge was erected in 1837 with a span of 37 m and two side spans each of 17 m. The superstructure consists of two 0.18 m bars 10 mm thick that are connected by brace plates. This bridge, when erected, was described as the first of this kind of construction (Motley, 1838). The inflexible suspension bridge approach was not widely accepted by engineers

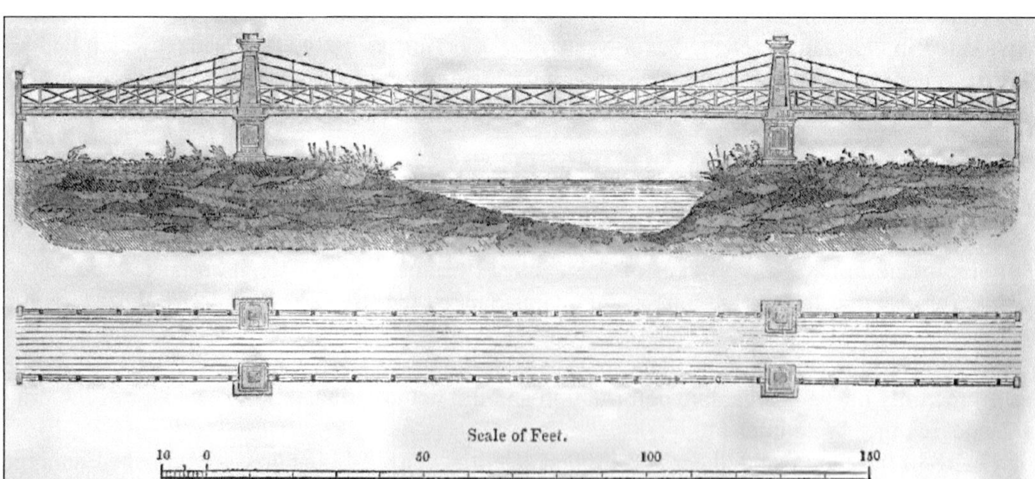

Fig. 1.8 Bridge across River Avon at Twerton, England, 1837

during this time, as they preferred to use suspension bridges with stiffening stays for relatively long spans. The most notable bridges of this type were designed by John A. Roebling and constructed in the United States.

1.2 THE JOHN ROEBLING STAYED-SUSPENSION BRIDGES

John Roebling was born in June 1806 in the little town of Muhlhausen in Thuringen Germany. He graduated in 1826 with honors with a civil engineering degree from the Royal Polytechnique School at Berlin (Steinmann and Watson, 1941). He immigrated to the United States in 1831, where he settled in western Pennsylvania and established with his elder brother the borough of Saxonburg in Butler County, about 40 km from Pittsburgh. His wire rope cable found its first application when the Pennsylvania canal company authorized him to build several suspension wooden aqueducts. One of these crosses the Delaware River at Lackawaxen, Pennsylvania and has been converted to a suspension bridge and is now carrying modern automobile traffic after being renovated. Roebling later in 1849 moved to Trenton, New Jersey, where he established his new wire mills. Perhaps Roebling's motivation with stayed-suspension bridges came following the tragic collapses of the suspension bridge across the Ohio River at Wheeling, West Virginia in 1854 and the Father Louis Hennepin Suspension Bridge, Minnesota in 1855, both due to windstorms. The wheeling bridge was designed by Charles Ellet in 1847 and was completed in 1849 with a main span of 310 m. In 1854 the bridge's deck failed due to flutter during a strong windstorm, and then it was rebuilt again by Ellet in 1859. Another round of improvements followed in 1874 by William Hildenbrand and William Roebling who designed auxiliary stay cables that were added to the bridge to provide stiffness and stability against wind loads. The bridge remains today in active service (Fig. 1.9). The Father Louis Hennepin Suspension Bridge was designed by Thomas Griffith in 1854 with a main span of 190 m. Soon after its opening in January 1855, the deck was destroyed by a strong wind storm. It was then restored in July 1855. Nevertheless, safety and capacity concerns persisted through its lifetime that eventually led to its replacement in 1876.

Fig. 1.9 The wheeling bridge, West Virginia, 1849-1874

John Roebling got involved in the design and construction of the Niagara Suspension Bridge in 1851 after Charles Ellet's engagement with the project was terminated. This bridge was a great challenge since it was the first suspension bridge in the world to carry a railroad and many engineers of the time proclaimed its impossibility. In order to achieve his goals, Roebling developed a hybrid structural system consisting of three primary elements: parabolic suspension cables, inclined (or diagonal) cable stays and stiffening trusses. The stiff deep girder timber truss was employed to reduce the deflection under rail load and the stiffening stays were introduced to create additional support for the bridge and to stiffen the floor against cumulative undulation that may be started by the wind reaction. To distribute the total load between the parabolic suspension cable and the inclined stays, Roebling divided the load between the cable and stays; typically, 1/4 to 1/3 of the total load was assigned to the stays and the remainder to the parabolic suspension cable. This ratio was based on his intuitive grasp rather than a theoretical basis. The span of this bridge was 280 m (Fig. 1.10) and was suspended from four cables, which rested on top of masonry pylons at each end of the central span. The ends of the cables were carried to anchorage in the solid rock. The superstructure had an upper deck for railroad tracks and a lower deck for foot and carriage way. The two decks were connected together at each side by posts and diagonal rods to form the deep stiffening truss. The bridge was completed in March, 1855 and stayed in service for forty-one years. Finally, it was replaced with an arch bridge due to the continuous increase in railroad loading.

Fig. 1.10 The Niagara Suspension Bridge, Niagara Falls, 1855

In 1846, the Covington and Cincinnati Bridge Company asked Roebling to prepare the design for the suspension bridge over the Ohio River between Cincinnati and Covington (Fig. 1.11). Nevertheless, it took the bridge company ten years to resolve issues with opposition and allocate a budget for the project. The work started in September 1856 and was completed in another ten years due to the Civil War crisis. The bridge has a main span of 322 m. When opened on New Year's Day, 1867, it was the longest suspension bridge in the world.

The Brooklyn Bridge project was set to go in May, 1867. John Roebling who was appointed the chief engineer for the project completed the bridge drawings and submitted his report in less than three months. In his report he emphasized the role of the stays to guard against vertical and horizontal oscillations. He indicated in his report that the floor in connection with the stays will support itself without any assistance from the cable. Moreover, if the cables were taken out, the bridge would deflect in the center but would not fail. The bridge was designed with a central span of 486.5 m and side spans of 284 m; the width was 26 m; and provided with four cables each containing approximately 5434 parallel galvanized steel oil coated wires. Unfortunately, while conducting surveys for the bridge project, Roebling sustained a crush injury to his foot when a boat squashed it between the timbering. Three weeks later on July 22, 1869, he died as a consequence. Washington Roebling succeeded his father as a chief engineer for the project. Despite challenging problems during the construction of foundations he was able to complete the construction project. The bridge was opened on May 24, 1883 and was considered the most noteworthy structure of its era.

Fig. 1.11 The Cincinnati-Covington Bridge, Ohio, 1867

Fig. 1.12 The Brooklyn Bridge, New York, 1883

1.3 THE ORDISH-LEFEUVRE BRIDGE SYSTEM

The Ordish–Lefeuvre system may be viewed as an early form of the cable-stayed bridge design. It was developed by Rowland Ordish and William Le Feuvre in 1858. This system is illustrated in Fig. 1.13. It has stays (rigid rods) running directly from the top of the pylons to the bridge deck. Each stay consists of a flat wrought iron bar. The catenary cables support the center of the bridge as well as the stays at several points through direct tension members (suspenders).

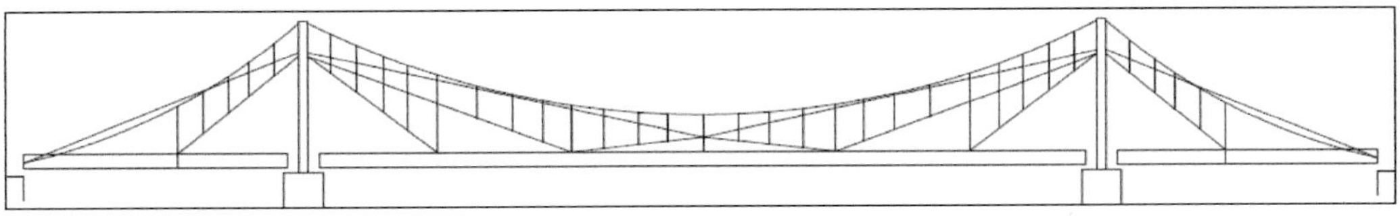

Fig. 1.13 Illustration of the Ordish-Lefeuvre system

Examples of the Ordish-Lefeuvre construction system include the Albert Bridge (Fig. 1.14) over the Thames at Chelsea in England, which has a center span of 122 m and two side spans of 47.25 m each. The bridge was designed by Ordish and Lefeuvre in 1863 and finished in 1873. The London Council, in 1973, added two concrete piers, which converted the central span into a simple beam bridge. Hence, today the bridge consists of three different design styles (Troitsky, 1988).

The Franz Joseph Bridge over the Moldau at Prague, Czech Republic was designed by Ordish. The design, which used the Ordish–Lefeuvre system, featured a combination of stay and suspension cables, which held the diagonal stay rods. The main span was 100 m long and 9.76 m wide, while the entire structure was over 240 m long (Fig. 1.15). The pylons were of cast iron on masonry piers. The cable was made of twelve main chains, measuring 0.10 m by 0.0254 m, with a sag of 18.30 m. Emperor Franz Joseph, after whom the bridge was named, attended the ceremonies for its opening on 13 May 1868. The straight rod bars used for the stays were replaced in 1898 by wire rope when the bridge was strengthened. The bridge was demolished in 1941 and replaced with a modern one.

Fig. 1.14 The Albert Bridge, Chelsea, England, 1873

Fig. 1.15 The Franz Joseph Bridge, Prague, Czech Republic, 1868

1.4 THE RUNYON-WILLIAM CABLE-STAYED BRIDGE SYSTEM

Between December 1888 and March 1893, Edwin Runyon received six U.S patents for his concept of a cable-stayed bridge Historic American Engineering Record (HAER, 1996). Patent No. 394,940 presented his overall concept for a cable-stayed bridge. The main features of the bridge are the tubular floor beams, deck cables, stay cables and their connections, and the pylon details. The deck beams consist of pipes with a “head” or cap on both ends. The patent includes horizontal deck cables that run longitudinally beneath the bridge deck. Apparently, the wooden deck surface rests directly on these deck cables, which in turn transfer the load to the floor beams. The center deck cable rests in saddles attached to the floor beams with no positive connection. The two outer cables sit in castings at the end of each floor beam, but U-bolts secure the cables to the floor beams. The patent description implies that the deck cables were the first elements to span the river during construction, providing an attachment point for the floor beams. The patent also states that the deck cables could replace longitudinal stringers. It will be illustrated later that Runyon did not follow this assumption in the construction of the Bluff Bridge. The patent bridge’s main stay cables are anchored directly to the pylons at one end and to a deck beam at the other end. Runyon indicated the objective of his patent is to make a single span from shore to shore, or at most to have but one pier embedded in the river, because of the difficulty in sinking cofferdams and finding strata of sufficient density to form stable anchors for the piers. The concept is illustrated in Fig. 1.16.

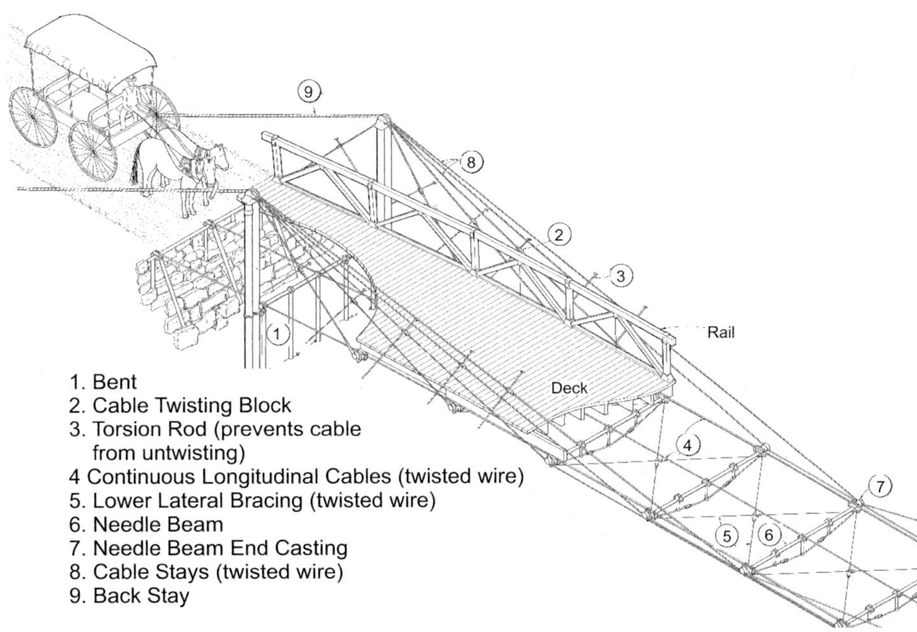

Fig. 1.16 Runyon’s Patent for a Cable-Stayed Bridge, Texas 1888 (Courtesy, HAER TX-72)

The floor beam was further presented in another patent (N0. 400,874) Runyon defined it as “Needle Beam”. The Needle Beam is composed of a horizontal pipe section and a lower chord of about 25 strands of No. 9 wire, separated by three vertical castings. The bowstring action of the beams provides substantial rigidity and bending resistance. The ends of the needle beams are fitted with a complex set of castings, which include attachment points for the cable, the deck cables, lateral X-bracing and the main stay cables (see Fig 1.17).

Runyon patented his cable tensioning device in 1889. The main objective of the patent was to invent a simple mechanical device to tighten cable wires and thus give the required tension to the cables,collectively, so that they could withstand the strain of the bridge weight properly. The patent

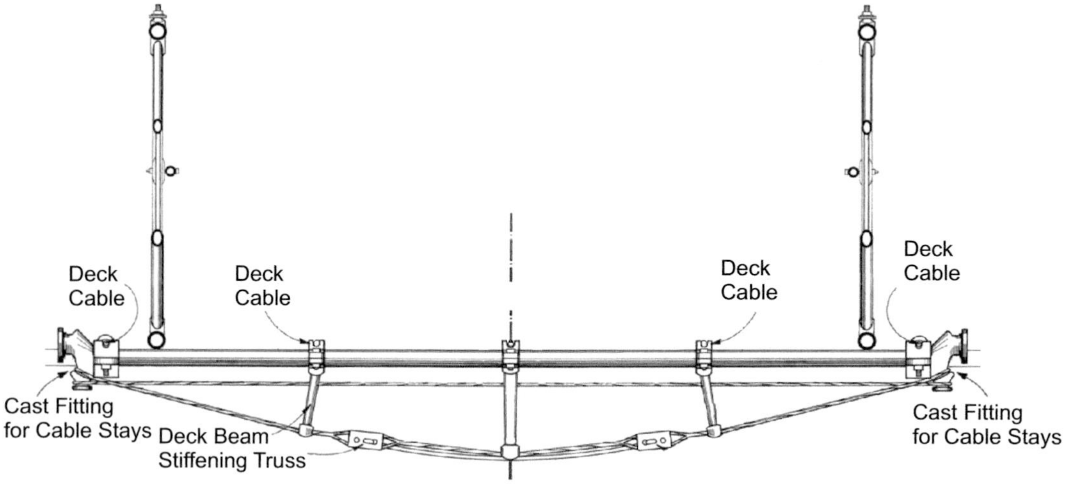

Fig. 1.17 Runyon's Patent for Floor Beam, Texas 1889 (Courtesy of HAER TX-72)

was defined as a device for twisting wire cables of suspension bridges. The device is simply a two-part casting to twist the wires with the help of the rectangular "separator blocks". The two halves of the casting are secured around the separator block, and handles are inserted to twist the assembly. Evidently the two-part casting was intended for repeated use, while the separator blocks remained on the twisted wires. As described in the patent, Runyon intended that one end of a metal tie-rod be bolted to one of the holes in the separator blocks while the other end extends down to and is adapted to be connected with the tower-cables or needle-beams of the bridge. The nature of the device limited its use to bundles of wires that could be twisted by hand. This in turn probably limited the scale of the bridges constructed with the device (HAER, 1996). The concept of the cable twisting process is illustrated in Fig. (1.18).

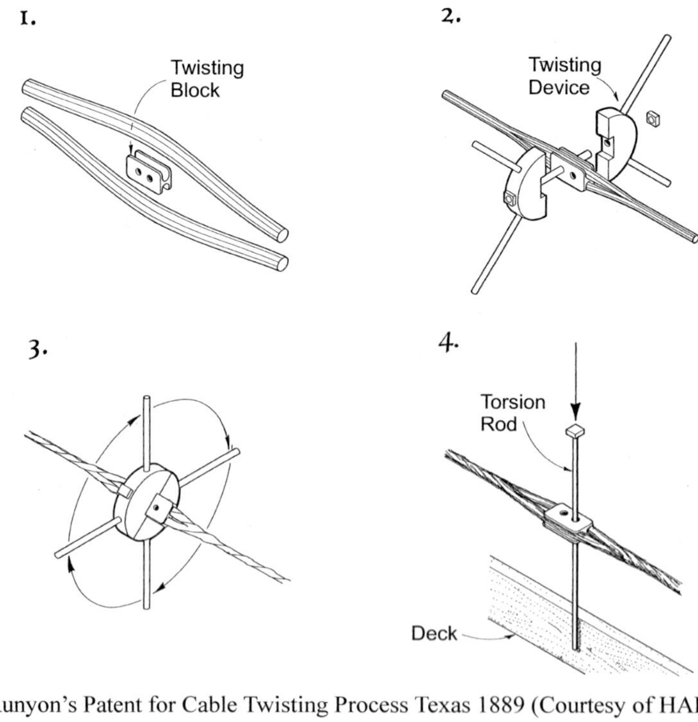

Fig. 1.18 Runyon's Patent for Cable Twisting Process Texas 1889 (Courtesy of HAER TX-72)

During the 1890s, Edwin E. Runyon and William Flinn constructed a group of innovative cable-stayed bridges in north central Texas that are all based on the bridge system patented by Runyon. The Bluff Dale Bridge (Figures 1.19 and 120) was completed in 1890 as a result of the collaboration of these two constructors. Bluff Dale Cable-Stayed Bridge is the second oldest surviving cable-stayed bridge in Texas and perhaps in the United States as well. The Runyon Bridge Co. consisting of Eswin Runyon and William Flinn of Weatherford, Texas, constructed the Barton Bridge, another cable-stayed bridge, several months prior to Bluff Dale Bridge. The pylons, cables and floor beams at Barton Creek survive in their original form and currently remain on private property, while the bridge was abandoned sometime after 1936.

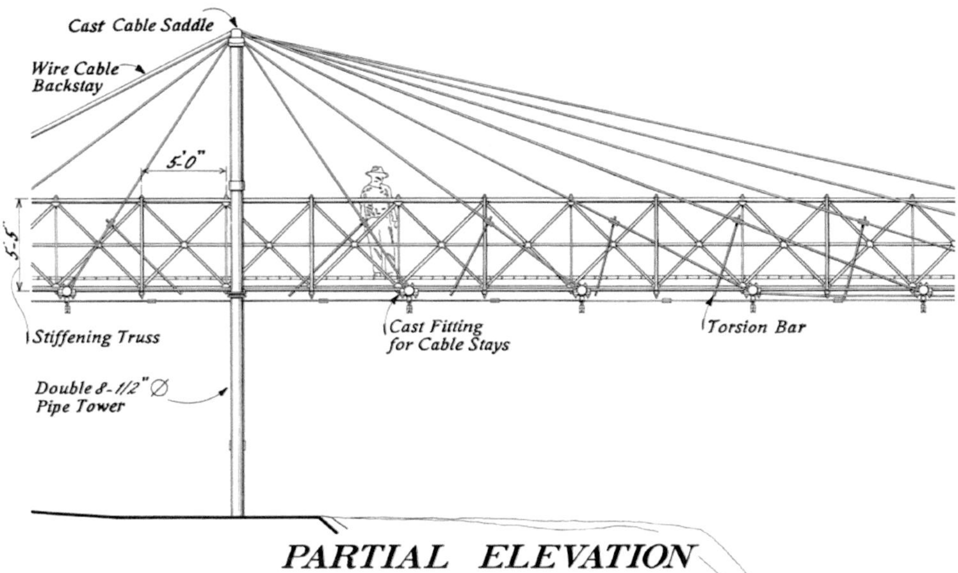

Fig. 1.19 Bluff Dale Bridge with Metal Truss (Courtesy of HAER TX-36)

Fig. 1.20 Bluff Dale Bridge after completion (Courtesy of HAER TX-36)

Bluff Dale Bridge has a main span of 42.67 m and side spans of approximately 9.14 m each. The spans are supported by cable stays of two types fixed and continuous, arranged in a fan pattern as shown in Fig. 1.19. The cables were originally hand-twisted strands of heavy-gauge wire. Rectangular separator blocks were placed between the wires and used to twist the strands. The stay cables were composed of heavy gauge parallel wire strands. The pylons consisted of two 0.20 m diameter wrought-iron pipes arranged along an axis perpendicular to the length of the deck. They rose approximately 4.25 m above the deck surface. The saddles consisted of two stacked castings on top of the pipes as shown in Fig. 1.21. The top casting supported the five continuous stays that were bundled together to form the backstay. The lower casting, referred to in a patent as a cross-piece, was used for the fixed suspension cables, which flanked the pylons on either side. A horizontal pipe was attached as a lateral bracing between the pylons.

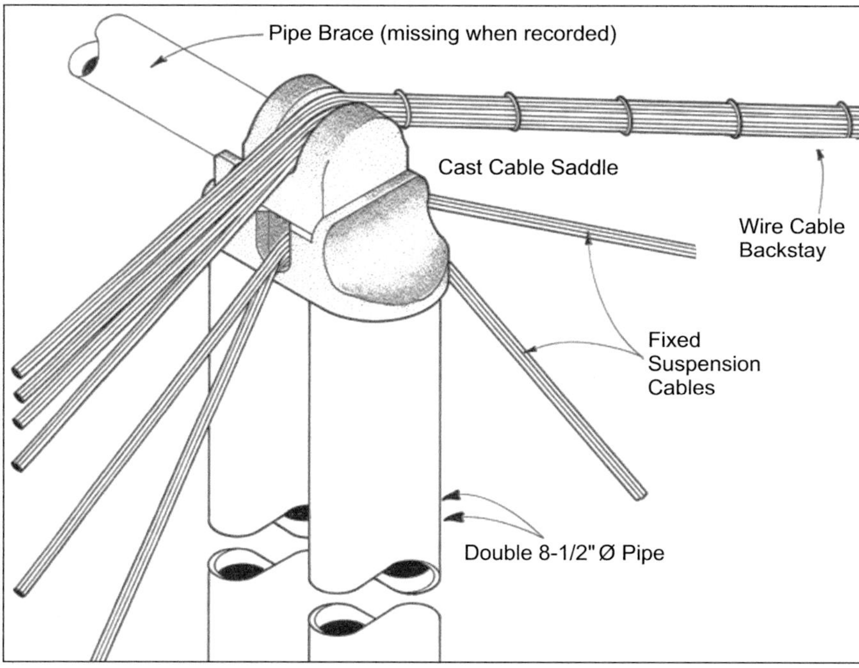

Fig. 1.21 Connection of Bluff Dale Bridge at the Pylons (Courtesy of HAER TX-36)

While the patent states that the deck cables could replace longitudinal stringers, a traditional wooden stringer and decking system was used for the Bluff Dale Bridge. This modification perhaps came as a result of the collaboration of Edwin Runyon and William Flinn. Although there is no evidence of later collaborations between the two constructors. Flinn built many other suspension bridges in Texas, and the Flinn-Moyer Company completed repairs to the Bluff Dale Bridge in 1899, replacing the original wooden truss with the surviving metal truss of pipe and rod sections as shown in the image of Fig. 1.22 (Brown, 1998). The bridge was relocated in 1935. During relocation, all the stays were replaced by modern wire rope.

1.5 THE ARNODIN-GISCLARD BRIDGE SYSTEM

The French famous engineer Ferdinand Arnodin developed a suspension bridge system that combined the suspension curved cable with the stays. In this system, the deck load for about a quarter point of the main span at each side of the pylons is carried directly to the pylons by inclined stays, and

Fig. 1.22 Steel Longitudinal stringers of Bluff Dale Bridge (Courtesy of HAER TX-36)

the main cables support the middle portion of the central span. This loading arrangement, while distorting the cables' curve from the catenary, greatly reduces the load upon them. The first bridge that Arnodin designed for this type of construction is the Saint-Ilpize Suspension Bridge (Fig. 1.23), which was completed in 1879. This bridge, which is still in service, is 4m wide and has a central span of 70.67m and side spans of about 15.25 m each; crossing the river at an elevation of 26 m. Similar bridges include the bridge at Lamoth, France (Fig. 1.24), completed in 1883 with a span of 115 m, a width of 5.5 m; and demolished in 1977. Another bridge is the one over the Saone River at Lyons with a main span of 121 m. In 1904 Arnodin constructed the Bonhomme Bridge over the Blavet River (Fig. 1.25) with a main span of 163 m and side spans of 37 m each. The main span central part was carried by five continuous main cables on each side, and the two end portions were supported by six stays on each side (Troitsky, 1988).

Fig. 1.23 Saint-Ilpize Suspension Bridge, France, 1879 (courtesy of Philip Bourret)

Fig. 1.24 The bridge at Lamoth, France, 1883-1977

Fig. 1.25 The Bonhomme Bridge, France, 1904

Arnodin was also a pioneer in designing and constructing transporter bridges. A transporter bridge carries freight or passengers from one bank of a stream or river towards the other bank by a gondola that was hung from a suspension bridge metal girder stiffening system. Arnodin applied the combined system of the main curved cable with stays in many of these bridges such as the Newport Bridge in Wales, Britain. The proposed transporter design features a main span of 196.56 meters and pylon heights of 73.6 meters (Fig. 1.26). This location was chosen due to the extremely low river banks at the intended crossing point. In order for ships to travel under a conventional bridge, it would be necessary to build an extremely long approach ramp, and during low tide, the location would not be suitable for a ferry. Of the three historic transporter bridges that still stand in Britain, this one is the oldest and biggest. There are eight other bridges of this type left in the world. The bridge was completed in 1906.

For the transporter bridge at Nantes, Arnodin came up with yet another creative design. He took Poyet's proposed system to put a fan-shaped arrangement of cable stays to support the bridge's girder (Fig. 1.27). The bridge was completed in 1903 and had a primary span of 141 meters. In 1958, the bridge was destroyed and a box girder bridge was built in its place.

Fig. 1.26 The New Port Bridge, Wales, UK

Fig. 1.27 The Transporter Bridge at Nantes, France, 1903

Another French engineer named Albert Gisclard created a suitably rigid suspension system in 1899, which was then adjusted for use on railroad bridges. The Ordish principle and this system are comparable in certain ways. This system, as depicted in Fig. 1.28, comprises two main stays that run sloping and then extend horizontally across the entire span. At various points over the span, additional inclined stay cables are fastened in tension with the primary stays. At these locations, either the main stay is fastened directly to the deck, or suspenders supporting the weight of the deck are attached as well.

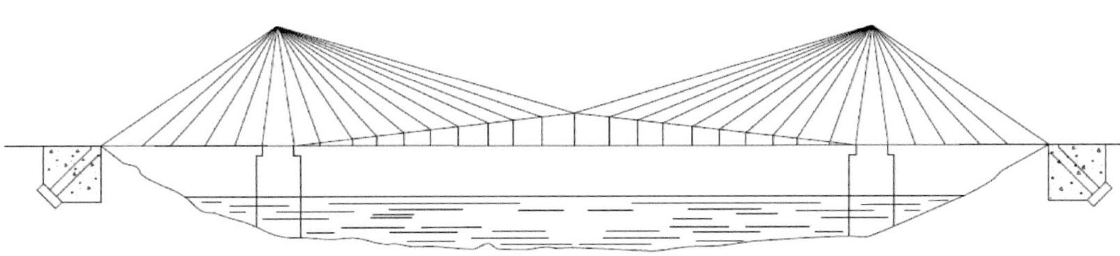

Fig. 1.28 Illustration of the Gisclard suspension system

The Cassagne Bridge, which is still in operation, is an illustration of this kind of architecture (Fig. 1.29). Built between 1905 and 1908 by the Ferdinand Arnodin firm, it has a central span of 156 meters and was designed in 1896 by Albert Gisclard. Regretfully, Gisclard lost his life on October 31, 1909, along with five other individuals, in a rail disaster that occurred during the bridge's weight test.

Fig. 1.29 The Cassagne Bridge, France, 1909

Leinenkugel le Cocq, a fellow French engineer, advanced the Gisclard system in 1925 by designing the Lezardrieux Bridge (Fig. 1.30a), which features stays from both pylons overlapping, as seen in Figure 1.30b. According to Troitsky (1988), this approach proved to be very cost-effective and only produced slight deviations. This bridge served as the model for modern cable-stayed bridges, which use cables that resemble fans.

Fig. 1.30 The Lezardrieux Bridge, France, 1925

References

Billington D.P. and Nazmy A., History and Aesthetics of Cable-Stayed Bridges, *Journal of Structural Engineering 117, No. 10: 3103–3134*, October, 1991.

Brown, M.M. Nineteenth-century cable-stayed Texas bridges. *In The Fifth Historic Bridges Conference*. Columbus, Ohio: Burgess & Nipple, 38–46, 1998.

Curtis, W.S., Inflexible Suspension Bridge, *The Civil Engineer and Architects Journal, 23,* 1837.
Drewry C.S. A Memoir on Suspension Bridges. pp. 25–26, Longman, Rees, Orme, Brown, Green and Longman, London, 1832.
Feige, A., The Evolution of German Cable-Stayed Bridges—An Overall Survey, *Acier-Stahl-Stahl, No. 12*, December 1966.
Historic American Engineering Record (HAER TX-36) Bluff Dale Suspension Bridge. Washington D.C.: Library of Congress, Prints and Photographs Division,1996.
Motley, T., On a Suspension Bridge Over the Avon, Twerton, *Mechanics Magazine*, London, 29 (790), 468, September 1838..
Navier, E.L.M.H. Rapport a Monsieur Beequey et mémoire sur les ponts suspendus. Paris: Imprimérie Royal, 1823.
Ponaldy W. and Scalzi J., Construction and Design of Cable-Stayed Bridges, second edition, John-Wiley, New York, 1986.
Steinmann, D.B. and Watson, S.R., Bridges and their Builders, Dover, New York, 1957.
Troitsky, M.S, Cable-Stayed Bridges, Van Nostrand Reinhold, New York, 1988.

CHAPTER 2

Fundamentals of Cable-Stayed Bridges

2.1 INTRODUCTION

During the last four decades, cable-stayed bridges became a favorable method of construction for spanning natural barriers such as wide rivers, canals, or deep valleys. Hundreds of bridges of this class were built across the United States and worldwide. These bridges as well as their predecessors, the suspension bridges depend on high-strength steel cables as main structural elements. As illustrated in Figure 2.1a, in suspension bridges, the weight of the superstructure is supported at relatively short spaces by vertical steel ropes suspended from the main cables that run between pylons and are typically fixed to large anchorages at each end. Suspension bridges can accommodate spans ranging from 300 m to 2000 m. On the other hand, under the action of vertical loads, the superstructure of the cable-stayed bridge transfers the load to the tower (pylon) through the stay cables, which are always in tension (Figure 2.1b). The longitudinal girders of the superstructure are subjected to bending and axial loading and the pylons transmit the load to foundations under mainly axial action.

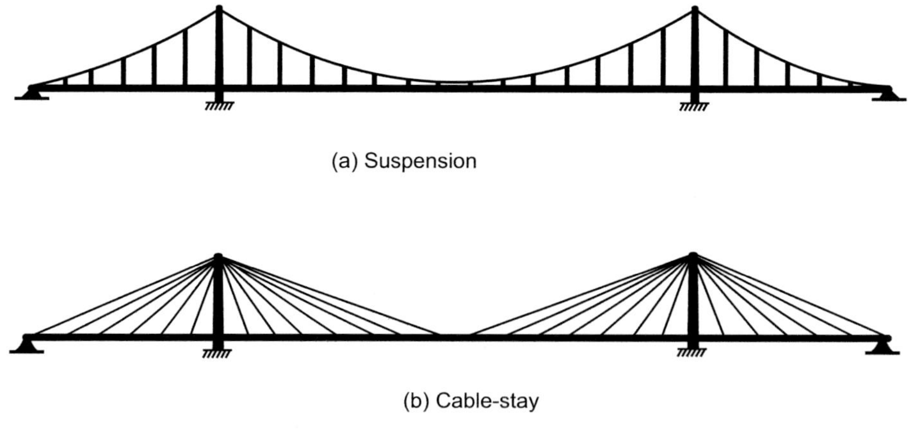

(a) Suspension

(b) Cable-stay

Fig. 2.1 Basic configurations of cable supported bridges

This structural configuration provides relatively inflexible supports at several locations along the span and hence ensures significant reduction in the bending moments of the superstructure longitudinal girders. Moreover, the bending moment can be controlled to be almost uniform along the girder length by optimizing the spacing between the points of cable-to-girder attachments.

Cable-stayed bridges come with main spans ranging between 100 m and 1100 m depending on the material used for the superstructure. This span range fills the gap between continuous girder bridges and suspension bridges with spans up to 200 m. Basic advantages of cable-stayed bridges include but are not limited to:

(i) Their economical utilization of construction materials. Early studies (Troitsky, 1988) indicated that in the span range of 215 m to 245 m, cable-stayed bridges are 5% to 10% more economical than other types.

(ii) Their graceful and aesthetically pleasing appearance. Cable-stayed bridges benefit from their appealing aesthetics. The pylons with their manifested rising and impressive shapes, combined with the attractive arrangements of the cable stays contribute to the striking appearance of these structures and have a significant role in developing their beautiful projected image. Also, cable-stayed bridges proved clearly that drastically unsymmetrical spans, such as the Frank Gatski Memorial Bridge crossing the Ohio River in West Virginia (Fig. 2.2), have an influential visual potential.

(iii) Their increased stiffness over suspension bridges. This increased stiffness results in higher natural frequencies for the cable-stayed bridges, in particular torsional modes of vibrations, which increases their aerodynamic stability over suspension bridges.

(iv) Their fast methods of construction as compared to other types of long span bridges such as suspension bridges, which require very challenging construction steps including but not limited to spinning of the main cables and the need of massive anchorages for their fixities; or arch bridges, which require scaffolding or support during erection as they are not stable until erection of the arch is completed. On the contrary, cable-stayed bridges can be easily constructed by free cantilever construction without any auxiliary supports with minimal equipment.

(v) The inspection and maintenance of cable-stayed bridges is relatively easier and less expensive than other long span bridges, particularly suspension bridges.

Fig. 2.2 Frank Gatski Memorial Bridge

It is due to the above advantages that cable-stayed bridges have gained an increasing popularity among the bridge engineering community and there is currently a rapid evolution in the construction of such bridges around the globe. The number of major cable-stayed bridges worldwide has increased from about a couple of hundred by the end of the nineteen eighties to more than 1500 nowadays. Nevertheless, one must keep in mind that Cable-stayed bridges are highly statically indeterminate structures and evaluation of their three-dimensional non-linear performance for different kinds of loads requires employing computational methods such as the finite element method or matrix methods for structural analysis. These methods would not have been very popular at present without significant progress in the information technology and software industry. The main purpose of this chapter is to present structural principles of cable-stayed bridges and their performance for different structural configurations.

2.2 HOW DOES A CABLE-STAYED BRIDGE FUNCTION?

The basic structural form of a cable-stayed bridge is illustrated in Fig. 2.3. The cables are connected diagonally to the superstructure, henceforth referred to as the deck, providing a series of elastic supports. The pylons form the principal load bearing member. The deck, the cables, and the pylons are under predominant axial forces, with the cables under tension and both the pylon and the deck are under compression. Horizontal components of cable forces are balanced in the form of compression thrusts in the superstructure, while the vertical components are transmitted directly through the pylon to the foundations. The design of the bridge is carried out such that the static horizontal forces resulting from dead load are balanced to minimize the height of the pylons. The self-balancing nature of this kind of bridge contributes mainly to its structural advantage over the suspension bridge, which is characterized by its bulky, expensive main cable anchorage.

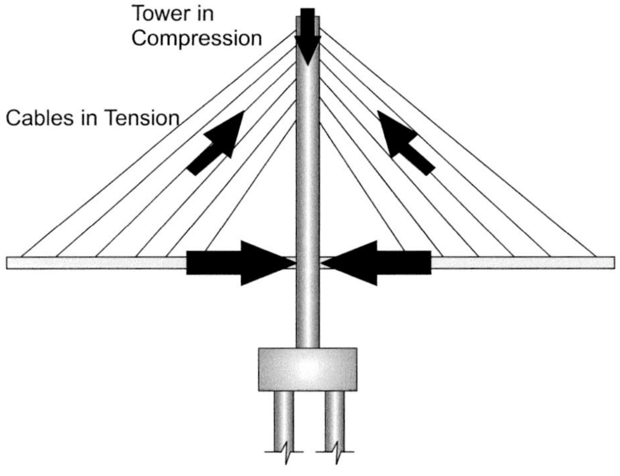

Fig. 2.3 A typical flow of forces for cable-stayed bridges

2.3 SUPPORT CONDITIONS

Variations in the support conditions of the bridge deck can drastically change the performance of the cable-stayed bridge under different loading conditions. Longitudinal support conditions include the floating system illustrated in Figure 2.4a, which allows free longitudinal deformation. This system does not have a connection between the pylon and the main girder at their intersection and comprises movable bearings at the end piers. This system renders a flexible system that reduces the fundamental frequency of the bridge in the longitudinal direction and hence reduces the longitudinal seismic demands. However, braking forces must be transferred directly into cables and pylons. Therefore, the pylons under this configuration will exhibit large bending moments and the superstructure will exhibit large longitudinal movements, particularly due to seismic loads that will require multi cell modular expansion joints at the interface with the approach spans. One way to reduce the longitudinal movements is to have a fixed bearing at one of the end piers as shown in Figure 2.4b or to introduce fixed bearings or at least one fixed bearing at the connection between the pylon and deck (Figure 2.4c), which will provide minimal longitudinal movements but will introduce rigidity to the structural system and will increase the fundamental frequency in the longitudinal direction. Another option is to partially restrain the connection between the pylons and the deck using hydraulic buffers (Figure 2.4d). These are shock absorber devices that allow free movements during normal operations and a temporarily rigid connection during extreme loading from an earthquake, or a heavy wind blow. Therefore, they render a floating system for long-term loads. This system is advantageous compared to a full floating system since through this system the braking forces are transmitted directly to the pylons. Specific bearing types, such as spherical bearings, can also be used to achieve the partial restraint system. Spherical bearings are sliding bearings that can transfer remarkably high forces from the bridge deck into the substructure without causing noticeable resistance because they can accommodate arbitrary rotations in all directions thanks to an internal spherical joint.

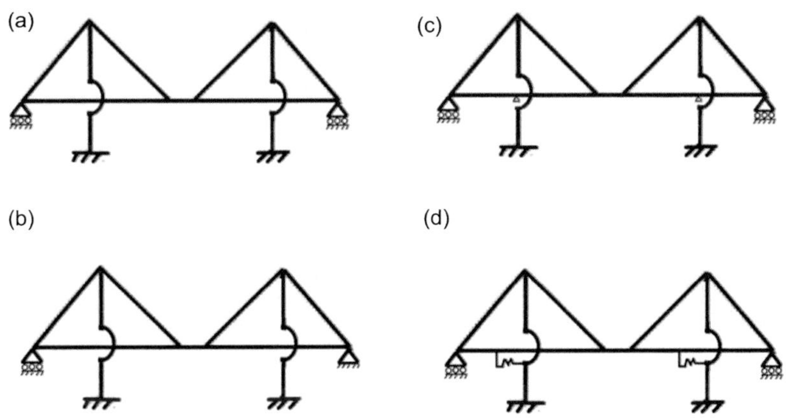

Fig. 2.4 Typical support conditions for cable-stayed bridges

2.4 CABLE STAYS CONFIGURATIONS

2.4.1 Evolution of the Multi-stay Cable System

Since the chosen pattern determines the overall design of the bridge, the cable arrangement in cable stayed bridges is what makes them most distinctive. The number of cables has a crucial role in determining the structure's stiffness, thus it is necessary to consider this while choosing the cable layout. A limited number of cables will cause significant cable forces to be transferred to the pylons, necessitating the use of bulky, intricate anchorages and extra material to transfer the forces throughout the deck's cross-section. Numerous cables will make anchorages easier and can be thought of as a continuous elastic support for the deck, and reduce the resulting bending moments.

Early cable-stayed bridges were designed with one cable at each pylon (Figure 2.5a). To decrease the bending moment of the deck structure and the individual cable forces, engineers started to adopt two to six cables for design stay supports in the main span (Figure 2.5b). The spans between the stay supports were between 30 and 60 m. The stays for this design act as elastic intermediate supports in lieu of piers. This arrangement requires much bending stiffness of the deck and slender pylons, since they are not subjected to large moments. Construction methods with such a number of

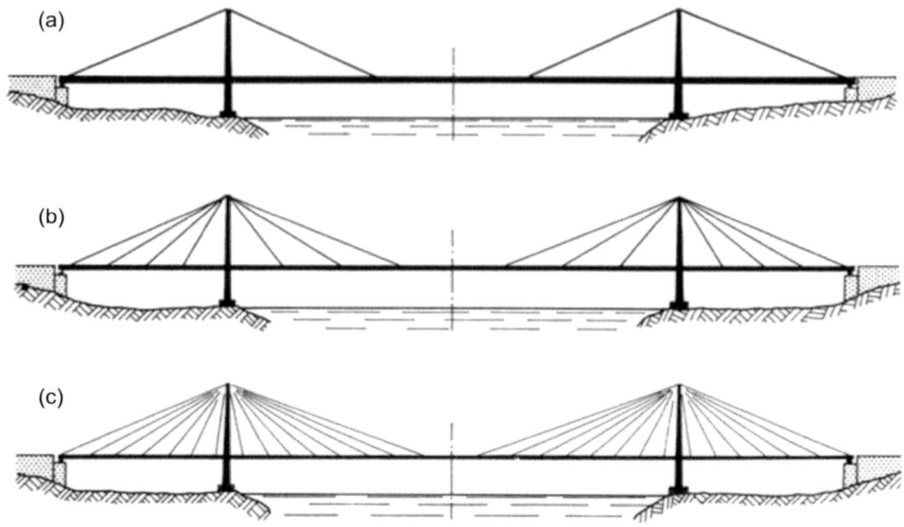

Fig. 2.5 Development of the multi-stay-cable concept

cables were still very challenging as the stay forces were large. This necessitated the use of several ropes to build each cable.

With the advent of computers, more cables were used with spacing in the range of 6 to 12 m. This multi-stay cable system configuration (Figure 2.5c) reduced both the bending moment in the longitudinal beam and the axial forces in the cables tremendously. This development into a multi-stay cable system led to a new triangular structure system with the deck structure acting as the compressive chord member, which does not need much bending stiffness, because the triangle pylon-stay-chord provides sufficient stiffness so that the bridge satisfies performance requirements for live loads. The dimensions of the longitudinal beam are governed in this case by transverse bending due to wind and seismic loads. Moreover, it must be stiff enough to preclude buckling due to the large compressive forces created by the inclined cables. The pylons in contrast should be stiff to take up the bending moments due to live loads. The multi-stay cable system allows for simplified construction methods including free cantilever erection. The multi-stay-cable system is different from traditional systems of beam girders, arches, or Suspension bridges with stiffening girders. It is characterized by less deflection under highway traffic loadings than slender continuous beams or suspension bridges. It also exhibits enhanced dynamic behavior due to its high system damping. These features are contingent on the use of cables with high stiffness and adequate inclination so that vertical deflections are limited to small values.

2.4.2 Longitudinal Arrangements

In determining the configuration of the stays, establishing appropriate stiffness to the main bridge girder is always significant, while maintaining aesthetic considerations of the designed bridge. In general, cable arrangements are mainly dependent on local conditions for the ratios between the main and side spans. Therefore, a harmonic arrangement of the cables is significant not only for the aesthetic quality of such bridges but also for stiffness, robustness, and structural integrity and, therefore, the choice should be made with care and perseverance. Traditionally, the longitudinal arrangement has been classified into two main types, the fan (Figure 2.6a) and the harp (Figure 2.6b). Other systems such as the star pattern (Fig. 2.6c) were also used for early bridges in Germany for aesthetic purposes (Leonhardt and Zellner, 1980).

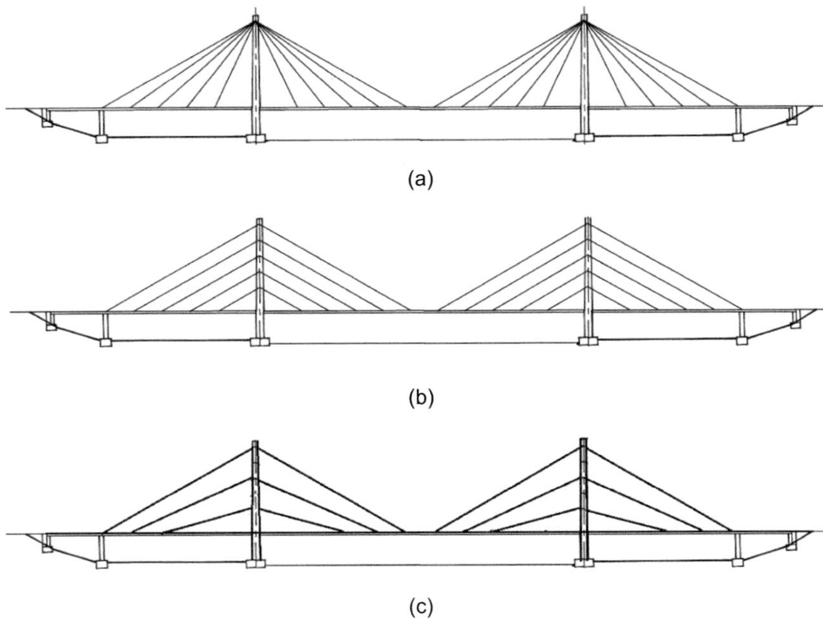

Fig. 2.6 Longitudinal arrangements of cable stays

In the fan-shaped configuration all cables join at the head of the pylon, a concept that was adopted from suspension bridge construction. This arrangement is structurally effective as it transmits the loads from the superstructure to the pylons with the lowest horizontal thrust to the longitudinal girder, which results in smaller cable cross-section and minimal moment applied to the pylon. Hence, it is very efficient in terms of saving material. It is also advantageous in the sense that it requires only one cable anchorage in the pylon. However, the concentration of the anchorages causes construction difficulties when the number of cables is large. Also, there were obvious difficulties with the protection of cables at the pylon head against corrosion. Unfortunately, this arrangement does not provide the best aesthetical solution because the cable lines will intersect when the bridge is viewed from skew angles

The harp arrangement requires that the cables run parallel to each other and connect to the pylons at discrete heights equally distributed over them, which requires taller pylons than in the fan arrangement. This system is advantageous from an aesthetic perspective because all cables look parallel in view under a skew angle. Nevertheless, it is not efficient for long spanned bridges as it results in high compressive normal forces in the deck, which requires more steel for the cables. Also, by anchoring the cables at different levels, large bending moments are produced in the pylons. Moreover, this system relies on the bending stiffness of the pylon and deck for equilibrium under non-symmetrical live loads. Therefore, from a technical and economical perspective the harp system is inferior to the fan shaped arrangement.

Collecting the cables at one point at the pylon head has been viewed as impractical from a construction point of view. For future cable replacements, it is more appropriate to distribute cable anchorages vertically along a certain length of the pylon head. This results in a very popular cable configuration, which is a widely adopted system for the design of cable-stayed bridges, namely the semi-fan arrangement as shown in Figure 2.7, which also improves the appearance of the bridge. The cable anchor points are usually spaced at 1.5-2.5 m vertical intervals to provide enough space for anchoring. This system offers a much-reduced concentrated force at each anchor point. Also, the large number of stays provides continuous elastic support and ensures a uniform distribution of the axial force through the deck, which results in lighter sections and simpler construction methods.

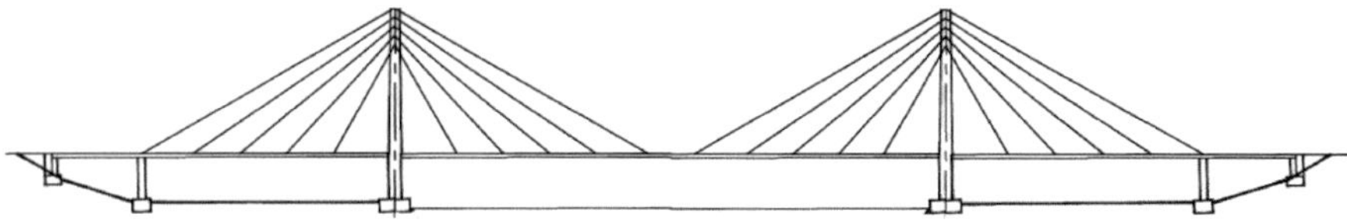

Fig. 2.7 Semi-fan arrangements of cables

2.4.3 Transverse Arrangements

There are two basic configurations for the cables in the transverse direction: the single plane, and the double plane. The single plane system (Fig. 2.8a) is achieved by placing the pylons in the center of the deck and the stays are located in a single vertical strip, which provides additional deck width that is not used by traffic.

In this case, cables must be protected against traffic accidents by guardrails at a sufficient distance from the stays. This arrangement is adequate for divided highways where a central wide meridian strip can be used for locating the pylons in the center. The major disadvantage of this system is that the cables do not supply any rotational restraint to the deck and therefore this system cannot resist torsional loading from eccentric live loads. This necessitates a box girder with high torsional rigidity to accommodate unsymmetrical loading, which may require extra material for the deck structure. The quantities and costs for the box girders make this system uneconomical for relatively long spans. This system, however, is very advantageous due to its aesthetic nature. A single plane of cables offers a great aesthetic unobstructed view and ensures avoidance of cable crossings. It also provides a perception of lightness and gives the structure a stylish look.

The double plane system (Fig. 2.8b) has been used most often because of its high torsional rigidity. This arrangement requires two planes of cables and two pylons, usually located just outside the railing of the bridge deck. It can also use double inclined planes connected from the edge of the deck to either an A-frame or inverted Y-frame pylon.

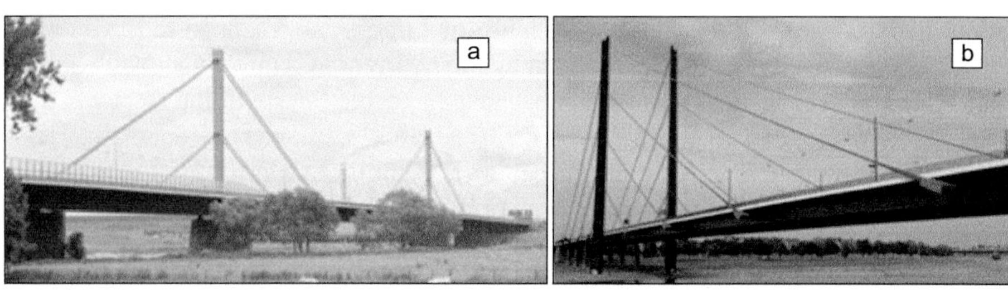

Fig. 2.8 Examples of bridges with single and double plans of cable stays

The cable anchorage may either be located outside the deck structure (Fig. 2.9a) or may be installed on the main girder as shown in Fig. 2.9b. The first location is aesthetically more attractive in the sense that no area of the superstructure is obstructed by the cables. Nevertheless, this layout requires cantilevers to transfer the loads from the superstructure main girder to the cables. Moreover, the width of the pylons piers for this layout needs to be increased as the pylons stand apart and outside the cross-section of the bridge.

2.5 CABLE-STAYED BRIDGE SPECIFICS

Traditional cable-stayed bridges as shown in Figure 2.10(a) are symmetric and consist of one central span, two side spans, and two pylons. Back stays have a major role in the structural behavior

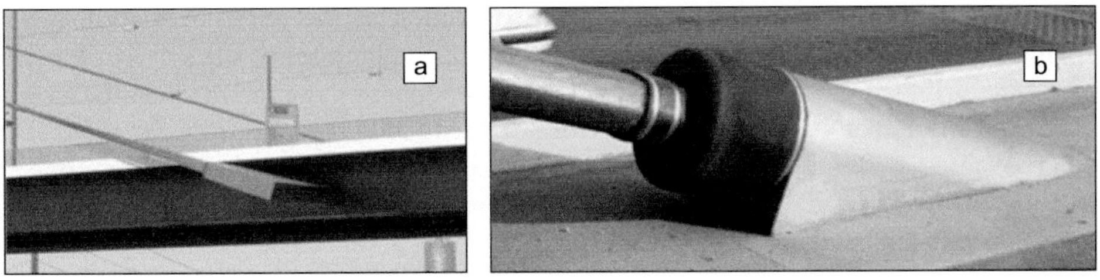

Fig. 2.9 Different layouts of cable anchorage to superstructure

of the bridge and are expected to exhibit very high stress variations due to live loads from which fatigue issues might arise. Two live load scenarios are investigated to understand this behavior. We first discuss loading the main span, which induces a downward deflection in the loaded span and deflection of both pylons towards the main span due to tension increase in the corresponding cable stays. In this case, the back stays that are attached to the anchor piers will be subjected to high tension and will also control the pylon deflection towards the loaded span and induce high bending moments in pylons. Nevertheless, the stay cables supporting the side spans exhibit small tension variations, and side spans are likely to deflect upward as shown in Figure 2.10(b). On the other hand, when a side span is loaded, it is expected to deflect downwards with a tension increase in all cable stays that suspend it. This will produce deflection of the pylon towards the loaded side span. This may result in a reduction of the tension in the main span stay cables and an upward deflection of the main span as shown in Figure 2.10(c). An efficient procedure to improve the structural behavior of a cable-stayed bridge is to provide intermediate supports in the side spans to control moment and deflection of the main span as shown in Figure 2.10(d). For this configuration, when the main span is loaded (Figure 2.10e), all stay cables anchored in the side spans behave as back stays and since the short spans between intermediate piers reduce the upward deflections to low levels, other stay cables in the side spans will also get engaged and the structure thus behaves as if the side spans

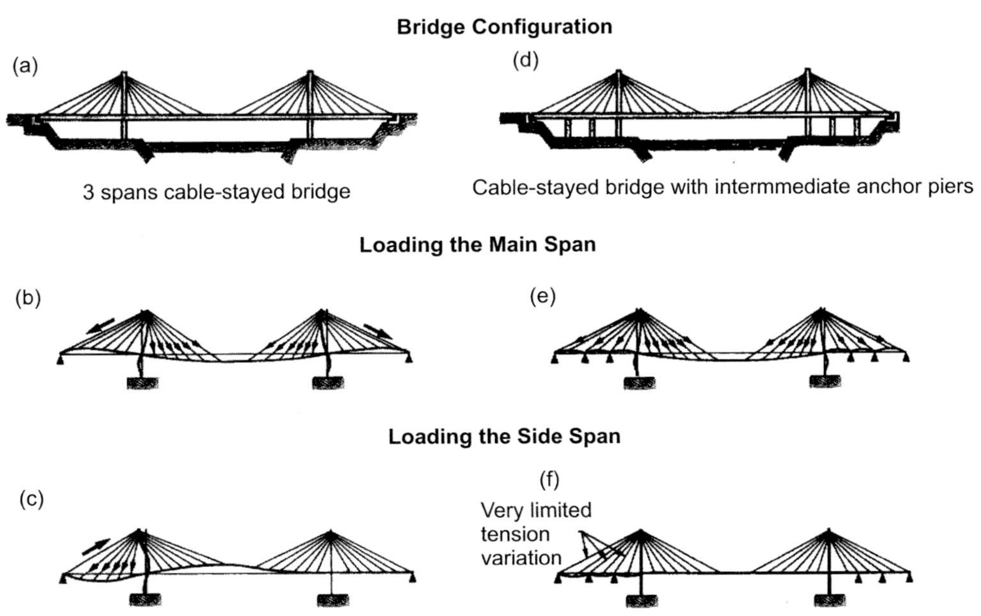

Fig. 2.10 Structural behavior of cable-stayed bridges (Virlogeux, 2001)

are totally rigid. This configuration will result in reduction of the pylon deflection towards the main span and the deflection of the main span itself. The anchorages of the stay cables within the pylon can be distributed over a large distance without producing large moments in the pylon due to the presence of more than one back stays in front of stay cables coming from the main span. This solution is advantageous compared to the traditional 3 span configuration that requires concentrating anchorages in the pylon heads over a small distance to reduce the high bending moments. Moreover, when a side span is loaded there is an inconsequential effect on the cable-staying system as the load is distributed among the intermediate piers (Figure 2.10f). A layout of a cable-stayed Bridge with 3 anchor piers is depicted in Figure 2.11.

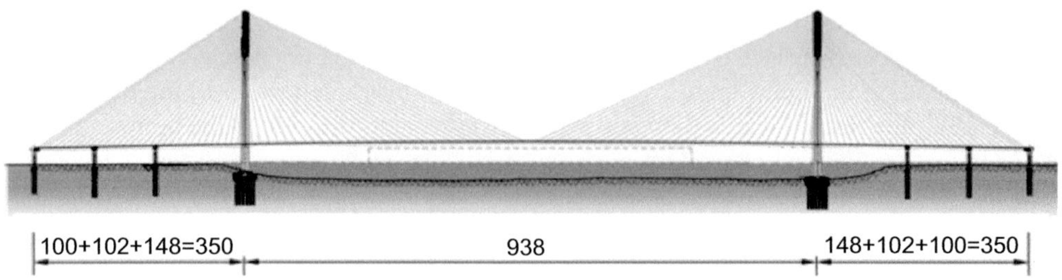

Fig. 2.11 A layout of a cable-stayed Bridge with 3 anchor piers (Gongyi et al., 2022)

Asymmetric cable-stayed bridges as illustrated in Figure 2.2 consist of a single pylon, one main span, and one side span. Very similar results are obtained with a single pylon cable-stayed bridge when intermediate piers are provided in the side span. Multiple-span cable-stayed bridges as shown in Figure 2.12 consist of more than one main span.

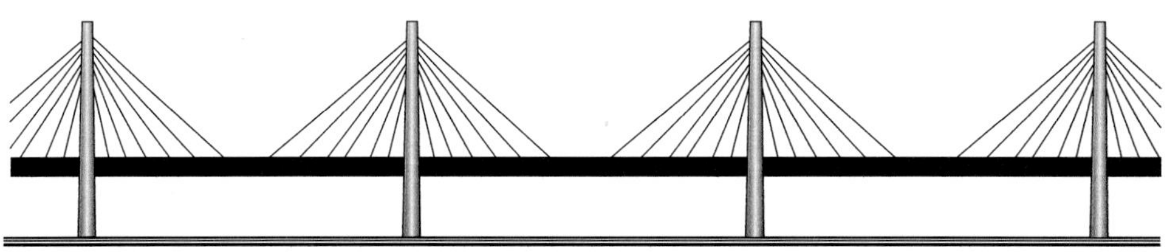

Fig. 2.12 Cable-stayed bridge with multiple spans

It is important to note that the performance of a cable-stayed bridge with multiple spans is completely different from the traditional 3-span cable-stayed bridge. When a span is loaded, tension is increased in the cable stays which suspend it as it deflects downwards, and the adjacent pylons deflect towards the loaded span. The adjacent spans may move upwards due to the deflection of the loaded span and their pylons deflect slightly in the opposite direction (Figure 2.13). This global deflection is controlled by the rigidity of the deck and pylons. There is no back-staying effect to limit deflections and deformations and to improve the efficiency of the cable-staying system. This configuration results in having each structural member subjected to significant bending moments, in one direction and the other, resulting in high stress variations. Therefore, the main challenge for this kind of cable-stayed bridge is to optimize the design to control efficiently, deflections and bending moments produced by live loads and to take advantage of the pier rigidity.

One optimal solution is to design a rigid connection through the deck between the pier and the pylon, which increases the structural efficiency by transferring part of the flexural effects to the pylons. Nevertheless, the structural system must accommodate deck length variations due to thermal loads, and concrete creep and shrinkage.

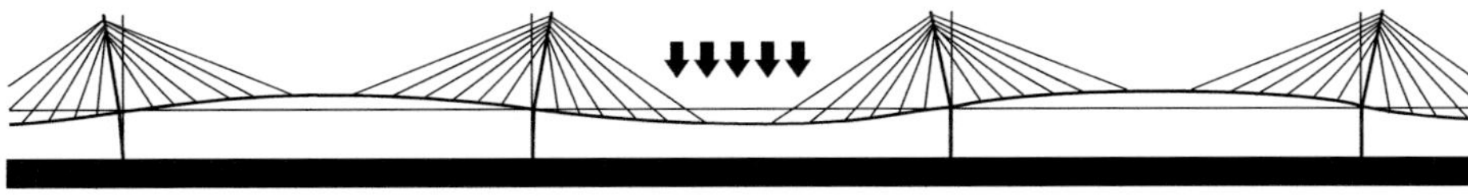

Fig. 2.13 Deflection of cable-stayed bridge with multiple spans

2.6 EFFECTIVE MODULUS OF CABLES

For a preliminary analysis, it is necessary to evaluate the stiffness of the cables. In order to provide a minimal vertical deflection, the stiffness of the cables is influenced by the applied stress, the projected horizontal length of the cable, and the modulus and area of the steel. The effective modulus was created by Ernst (1965) to illustrate how cable stiffness is affected by sag.

$$E_{ff} = \frac{E}{1 + \dfrac{\gamma^2 + L_p^2 E}{12\sigma^3}} \quad ...(2.1)$$

where,

E = Modulus of elasticity of straight cables
E_{eff} = E-modulus of cable with sag
γ = Specific weight of the cable
L_p = projected horizontal length of the cable
σ = tensile stress in the cable

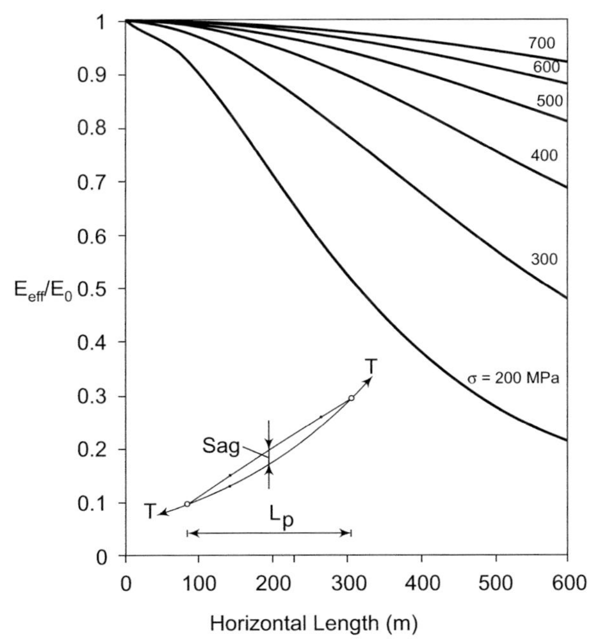

Fig. 2.14 Effective modulus of cables

The equivalent modulus for the cable is suitable for revealing the influences of different parameters on the stiffness of the cables. Figure 2.14 has been constructed using this approach. It can be shown that the stiffness of such cables increases with the third power of the steel stress and decreases with the second power of the horizontal span length due to the sag effect i.e. change of sag by change of stress. Therefore, a high stress state is needed in the cables to achieve a satisfactory effective modulus and consequently a high stiffness. It is important to mention that high strength steel used for cables nowadays can achieve an ultimate strength in the range of 1600 to 1800 MPa. Assuming a factor of safety of 2.5 for service loads, it can be shown that we can only achieve a little reduction in the cable stiffness using current construction practices.

References

Ernst, H.J. The modulus of Elasticity of Ropes taking into Account the Sag (in German), *Bauingenieur* pp. 52–55, 1965.

Gongyi, X., Yanfei, Z., Huiyue, H. and Yang, L. Innovation Design for Qingshan Yangtze River Bridge, *Structural Engineering International*, 32: 2, 243–246, 2022.

Leonhardt, F., and Zellner, W. Cable-Stayed Bridges, IABSE Surveys S-13/80, Zurich, 1980.

Troitsky, M.S. *Cable-Stayed Bridges Theory and Design*, Van Nostrand Reinhold, New York, 1988.

Virlogeux, M. Bridges with Multiple Cable-Stayed Spans *Structural Engineering International*, 11:1, 61–82, 2001.

CHAPTER 3

Stay Cables

3.1 INTRODUCTION

Stay cables are the most significant structural elements of the cable-stayed bridge as they transfer the weight of the deck to the foundations via the pylons. Stay cables are arranged in a nearly straight configuration and are required to act in tension even under seismic loads. This arrangement is reflected into horizontal compressive forces in the deck. The number and arrangement of the stay cables contribute to the redundancy of the cable-stayed bridge system. A cable-stayed bridge is designed such that the loss of an individual stay cable does not result in major damage or failure of the bridge. It is essential that stay cables are durable and easy to maintain. They also must be designed to be restressable and replaceable.

Several types of cables have been used as stays in cable-stayed bridges. Efficiency and usefulness of cables depends mainly on their internal compositions. Nevertheless, the main element of a cable is the steel wire, which is characterized by higher tensile strength than that of ordinary structural steel. Wires are produced from steel rods that are sent in coils to the wire mill where they go to a continuous wire drawing machine. The wire drawing machine reduces the coil to a wire and draws it through a series of successive dies, each of which reduces the wire to a smaller diameter. This drawing process not only reduces the cross-sectional area of the rod by 65 to 75% but also improves the internal structure of the steel and increases the tensile strength.

While the single wire is the basic element of a cable, a strand, which is an assembly of wires formed in a shop or on site, also used for the structure of the cable. Strands can come in different configurations. A helical strand is an assembly of wires formed helically around a center wire in one or more symmetrical layers. A parallel wire strand has all its wires arranged in a parallel arrangement. A locked-coil strand is another type that was produced primarily in Germany and used extensively in Europe. This chapter discusses different cables and strands configurations.

3.2 MAIN ELEMENTS OF THE STAY CABLE

The stay cable as illustrated in Figure 3.1 comprises several elements. The prestressing strand, wire, or bar of a stay cable is defined as the main tension element MTE. It is designed to transmit the tensile load to the superstructure and pylons. A sheathed strand is a prestressing steel element made of an assembly of wires formed around a center wire and coated with a corrosion-inhibiting material and covered by an extruded polyethylene or polypropylene compound.

The free length is the length of cable beyond the cable anchorage and transition zones. The main tension elements are located within this length inside a stay pipe, which is made of either high-density polyethylene (HDPE), high density polypropylene (HDPP), steel or other materials. It provides part of the corrosion protective system, and controls temperature and wind induced

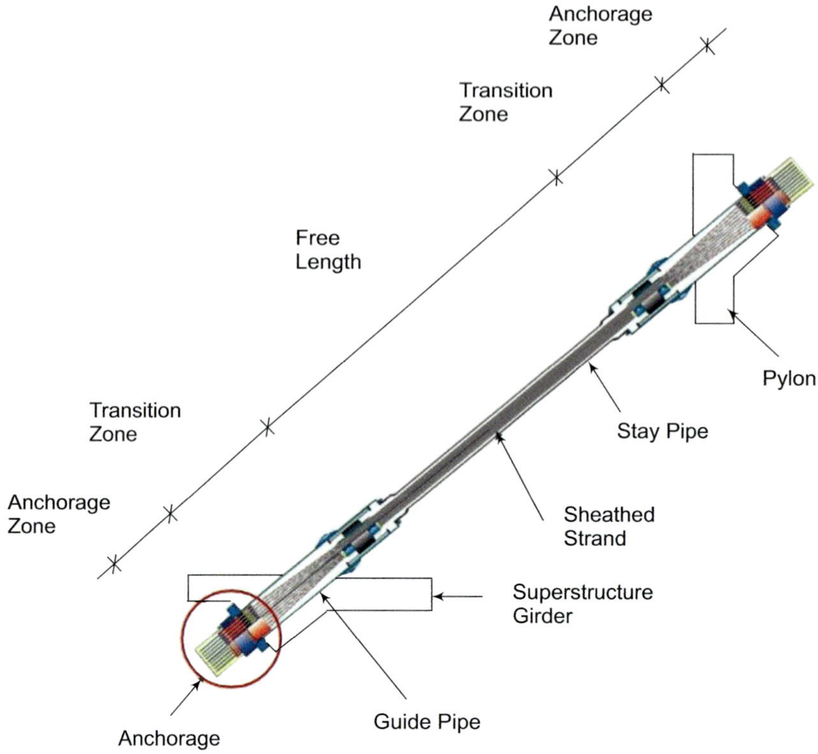

Fig. 3.1 Main elements of the stay cable (Anchorage detail is courtesy of DSI)

vibrations. The main tension elements have their positions fixed relative to the stay pipe by means of non-loading bearing devices called centralizers.

The stay cable includes two end anchorages. Those are devices comprising all components necessary to hold the force in a stressed cable stay and located at the bridge superstructure and pylon. A stressing end anchorage permits stressing of the stay cable while a fixed-end anchorage is located on the other side. Three different types of anchorages are used nowadays in construction. They transfer permanent fatigue loss in the MTE by one of the following mechanisms:

- Wedges for strands.
- Button heads for wires.
- Nuts for bars.

Some other anchorage systems employ resin and cement grouts (bond sockets) or hybrid systems that include one of the three types listed above and bond.

The length of cable within the hardware that serves to anchor the cable tension elements is called the anchorage zone. The transition length is the length of cable where the MTE is deviated from its arrangement in the free length to their arrangement in the stay cable anchorage. Each anchorage contains an elastic bearing namely deviator, which is located at the end of the stay cable free length. The deviator serves to deviate the path of the MTE at the ends of the transition zone to form a compact bundle of parallel elements in the free length. It also laterally guides the stay cable to protect the anchorage from bending and transverse stresses. The strand wedges, ring nut and anchor plate are the main elements of the anchorage that transfer the cable force to either the pylon or the superstructure (see Figure 3.2).

Some designs include saddles at the pylon. Saddles are structural devices that deviate a stay cable continuously from the deck through the pylon and back to the deck without breaking the continuity of the MTE (Figure 3.3).

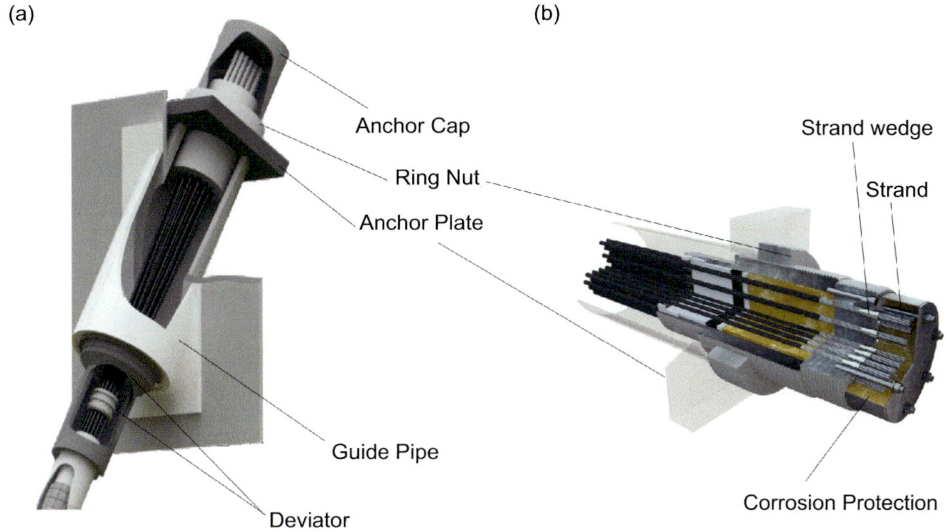

Fig. 3.2 Detail of Anchorage (Courtesy of Freyssinet)

Saddles shall be tested for fatigue and strength in accordance with the criteria of section 4.3.4 of PTI DC45.1-18 (PTI, 2018). Tests are required to qualify the stay and saddle system as a prequalified unit for general use. Changes to the saddle details, materials, exit geometry, MTE type, or covering shall require retesting for prequalification.

3.3 MATERIAL PROPERTIES

The main tension elements for stay cables are wires, strands, or bars. Wires used in stay cables conform to ASTM A421/A421M, "Standard Specification for Stress-Relieved Steel Wire for Prestressed Concrete," Type BA. Figure 3.4 displays a typical stress-strain curve for a galvanized prestressing steel wire. The first portion of the curve is a straight line up to the proportional limit indicated by point B. Beyond the proportional limit, the stress-strain relationship becomes nonlinear until the ultimate strength, or point C, is reached.

It can be observed that there is no yield point as in the relationship for structural carbon steel. Instead, the 0.2 percent offset yield stress defined as the proof stress is established. ASTM A416 and A421 require that the yield strength be measured by a 1.0 per cent extension under the load method. This stress is also defined as the limit of proportionality stress. Therefore, if a straight-line EF is drawn parallel to the initial straight-line portion of the curve, at a horizontal distance to its right, equal to a strain 0.002 mm./mm., the point E, where that line intersects the stress-strain curve, is called the 0.2% proof stress.

Relaxation is a material property of prestressing steel wire and behaves inversely to concrete creeping. The term "relaxation" describes the reduction of the existing stress at a constantly applied material strain. Either a stress-relieved (normal-relaxation) or a low-relaxation wire for prestressed concrete may be used for stay cable application because the creep (or relaxation) is negligible under normal working stress (PTI, 2018).

According to ASTM A421/A421M, the minimum ultimate tensile stress f's = 1655 MPa for both stress-relieved and low-relaxation wire. The minimum yield stress for all wires, measured by the 1.0% extension under load method, shall not be less than 85% and 90% of the specified minimum tensile stress for stress-relieved wire and low relaxation wire respectively. The elastic modulus shall not be less than 200 GPa.

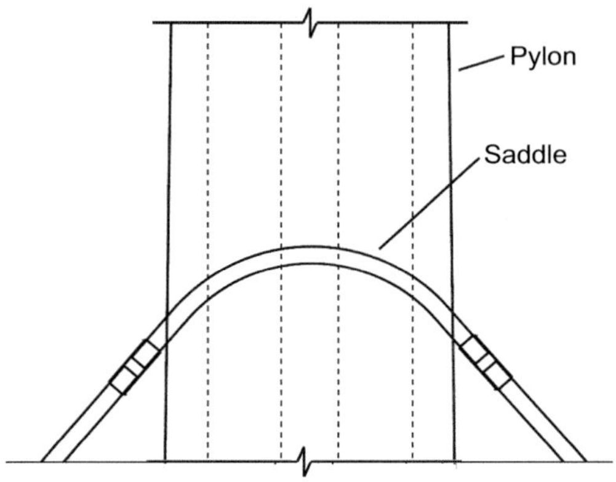

Fig. 3.3 Cable saddle at pylon

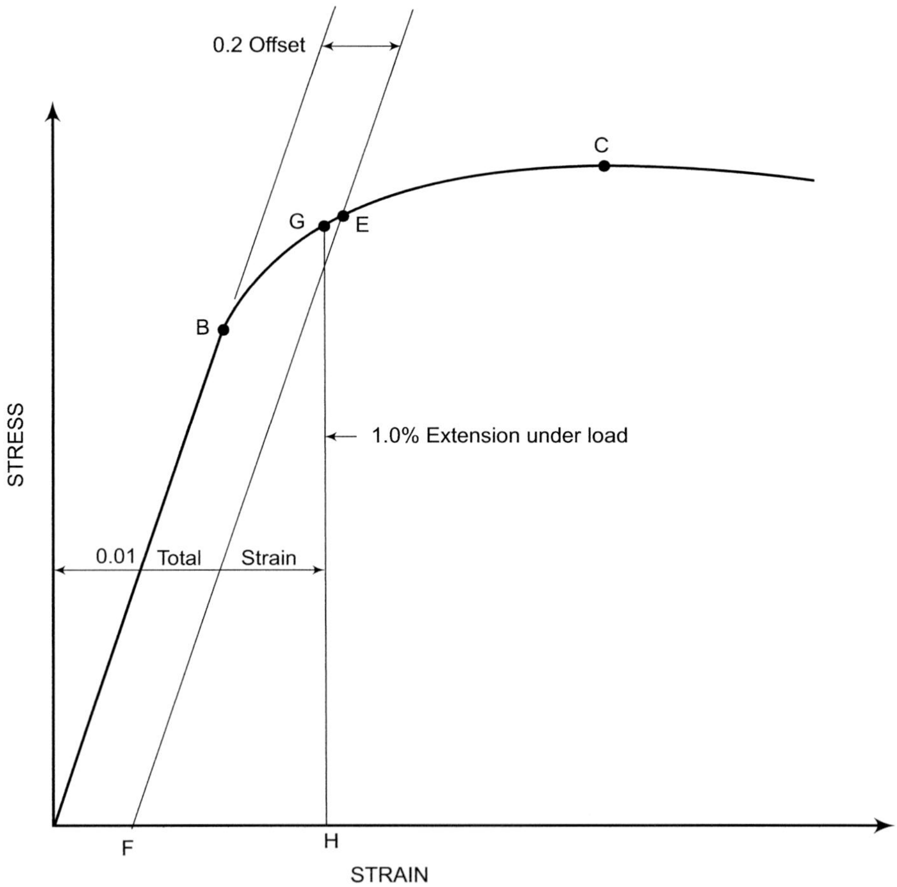

Fig. 3.4 Typical stress-strain relationship for prestressing steel wire used in stay cables

According to PTI (PTI, 2018), the minimum ultimate tensile stress f's = 1860 MPa for the low-relaxation strand. The minimum yield stress shall not be less than 90% of the specified minimum tensile stress. The elastic modulus shall not be less than 197 GPa. An epoxy-coated strand shall conform to ASTM A882/ A882M, "Standard Specification for Epoxy-Coated Seven-Wire Prestressing Steel Strand." It shall be of the type in which the interstices of the strand are filled with epoxy, and the strand shall be of a weldless, low relaxation grade.

3.4 CORROSION PROTECTION

The MTE of the stay cables if not protected adequately, may suffer pitting corrosion and loss of strength and consequently may require replacement. Typical design and fabrication of MTE is to provide a redundant corrosion protection system. Redundancy is satisfied by requiring at least two nested qualified barriers in any corrosion protection system. Internal barriers shall completely encase the MTE for the full free length and the anchorage length. The external barrier shall completely encase the inner barrier for the full free length. The barriers shall be arranged such that if a corrosive environment breaks through the external barrier, the interior barrier will protect the main tension element. Increasing the number of effective barriers increases the redundancy and hence enhances the credibility of the corrosion protective system (PTI, 2018). Internal barriers include:

- A layer of zinc coating applied to the steel wire surface. It provides corrosion protection of the steel wire at the exposed ends in anchorages before the anchorage cap is installed. The zinc coating also provides protection to steel wires if the sheathing is damaged.
- An HDPE or HDPP sheathing on the individual MTE. The HDPE or HDPP sheathing shall be produced by an extrusion process. The minimum thickness of the sheathing shall be 1.25 mm.
- A filler material between the MTE and sheathing to prevent migration of water from the surface into the individual wires of the MTE. The filling material is either a soft material or a hardening material. Soft filling materials include wax, grease, or soft resins

Exterior barriers include a stainless steel, PE or PP stay pipe encapsulating the entire bundle of steel tension elements. This provides corrosion protection and mechanical protection during handling and installation. The pipe may have a blocking agent injected (See Figure 3.5).

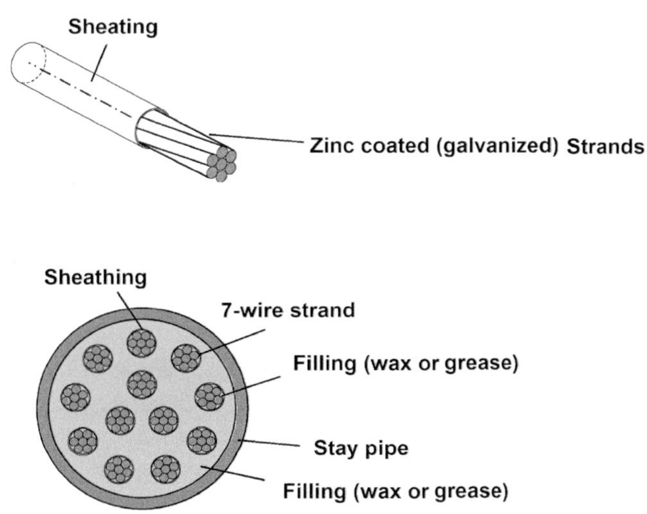

Fig. 3.5 Corrosion protection of stay cable

According to PTI DC45.1-18, internal barriers shall be subjected to a salt fog test. The specimen shall be prepared using samples of the MTE wire, strand, or bar. Tests in the fog cabinet could be conducted for specimens that are individually protected (such as greased and sheathed, epoxy-coated, galvanized) on each barrier element at a time. To qualify as a barrier, each separate barrier must protect the MTE without assistance from other barriers. Therefore, each barrier is tested independently. External barriers shall meet the requirements. If the external barrier is made compatible with the MTE by virtue of the bonding filler, then the external barrier will develop tension strains as the cable is loaded. This strain may affect the performance of the barrier and should be modeled in these tests.

According to PTI DC45.1-18, one fully assembled stay cable anchorage with a transition zone, a minimum of 1 m of free length, and all caps and seals, coatings, and coverings that will be installed in the actual application shall be subjected to a leak test. The specimen shall be placed in a chamber so that the anchorage assembly, transition zone, and all connections between the transition zone and free length of the cable are subjected to not less than a 3 m head of water and dye solution for a continuous period of 96 hours. At the end of this period, the tested anchorage assembly shall be dissected to inspect the MTE for signs of dye. Specimens will be acceptable if visual inspection shows that dye has not reached the MTE.

3.5 TYPES OF CABLES

Stay cables are classified into the following categories:

- Parallel-bar cables
- Parallel-wire cables
- Stranded cables

3.5.1 Parallel-bar Cables

Parallel-bar cables are formed of high strength round rods or bars, parallel to each other and are kept in position inside its duct by means of steel or polyethylene spacers. Standard bar stay cables usually comprise 7-10 round steel bars that are usually delivered with diameters of 26.5 mm, 32 mm or 36 mm and made from steel with a yield stress of 1080 MPa and a tensile strength of 1230 MPa. The bars come in lengths of 15-20 m and are usually joined by threaded couplers. The bar bundle is placed inside a steel tube that is filled with cement grout after erection. This technique was developed by Dywidag and was applied to the construction of a few bridges in Europe and to the Dame Point Bridge in Florida (Loizias and McCabe, 1989). This bridge in particular used cables consisting of coupled ASTM A722, Grade 150 thread bars pressure grouted within steel pipes. All bars have a diameter of 31.75 mm. The number of bars per cable (Figure 3.6) varies from 7 to 9 in accordance with the final cable force. The steel pipe section varies from 168 mm to 219 mm depending on the number of contained bars. Steel spacers were used to maintain the correct position of the bars relative to each other and the pipe. This system is hardly used any more.

3.5.2 Parallel-wire Cables

Parallel-wire cables are generally composed of ASTM A421, Type BA, 7 mm diameter wire with a minimum tensile strength of 1655 MPa. The wires are button headed and are individually anchored in the anchor socket. Each stay comprises a bundle of 7 mm diameter wires. The number of wires in one cable ranges from 50 to 350. The fabrication process is accomplished in hexagonal configuration to achieve uniform stressing in cables (Figure 3.7). The wires are kept in place by twisting a steel rope around the bundle. The duct pipe is made of polyethylene (PE) or stainless steel.

The Fred Hartman Bridge in Texas (Svensson and Lovett, 1995) adopted this stay system. The stay cables for this bridge consist of parallel wires with HiAm (high amplitude) anchorages in polyethylene pipes, injected with cement grout after installation for corrosion protection. The basic design idea of HiAm anchorages (Figure 3.8) is to anchor the individual wires gradually by lateral

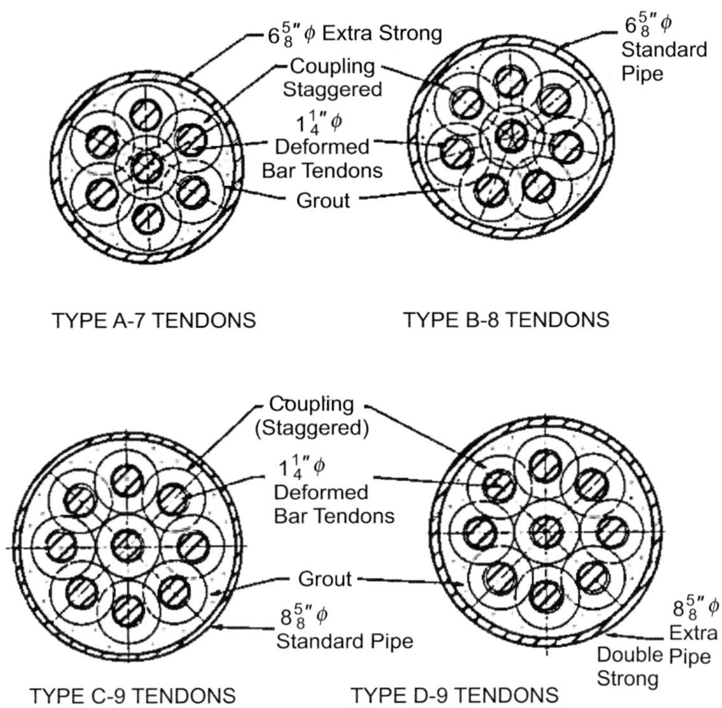

Fig. 3.6 Parallel-bar cables used in Dame Point Bridge (Loizias and McCabe, 1989)

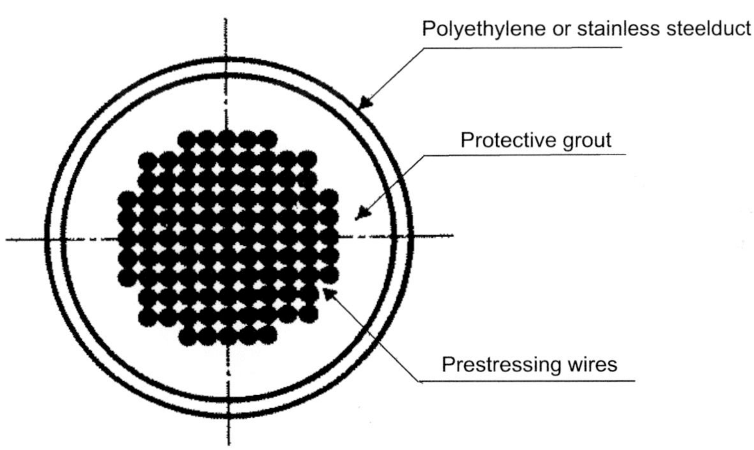

Fig. 3.7 Schematic of a parallel-wire cable

pressure exerted by small steel balls, into which the wires are broomed-out inside a conical anchor head.

In early cable-stayed bridges of this type, the cables used were made of non-galvanized high-strength parallel wires protected by cement grout and a PE-pipe. This led to cable rupture of some of these bridges such as the Zârate-Brazo Largo Bridge, Argentina built in 1971 (Henrik et al., 1999). Evaluation of cables for this bridge in 1996 revealed that a combination of corrosion and fatigue damage caused the failure of one cable, and large damages to a number of other cables. The corrosion was due to insufficient performance of the corrosion protection of the original cables. The cement

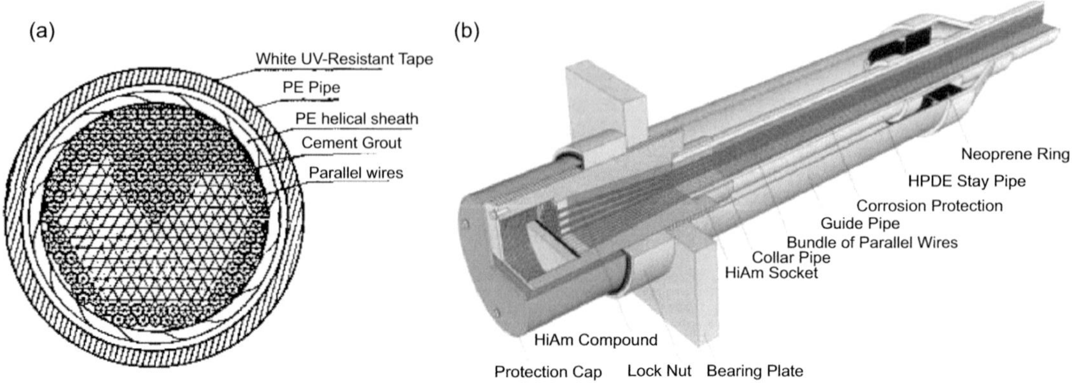

Fig. 3.8 HiAm parallel-wire cable: (a) cross-section; and (b) anchorage

grout, which was supposed to be the main active corrosion protection, was insufficient. Nowadays, it is regarded as mandatory to use galvanized wires and a grout of a non-cracking, ductile material.

3.5.2.1 New Parallel Wire (PW) Stay Cables

The new PW system was developed with a tensile strength up to 1770 MPa. They are also characterized by a protective polyethylene cover extruded directly onto the wire bundle so that no void exists between the wires and the surrounding pipe. Eliminating the spiral rope and the voids for cement grouting rendered the New PW Cable more compact than traditional PW cables. The largest stays can contain up to 400 wires and a coating of high-density polyethylene (HDPE) is applied in the factory using the continuous extrusion process. An example of the utilization of a new PW stay cable is in the cable system developed by OVM. This system has been used in recent projects in China and other parts of the world. Cable protection is achieved through several methods (Figure 3.9). The steel wires are either galvanized or epoxy coated. Sheathing is achieved through two PE layers and the bundle of wires are wrapped inside a high strength polyester belt.

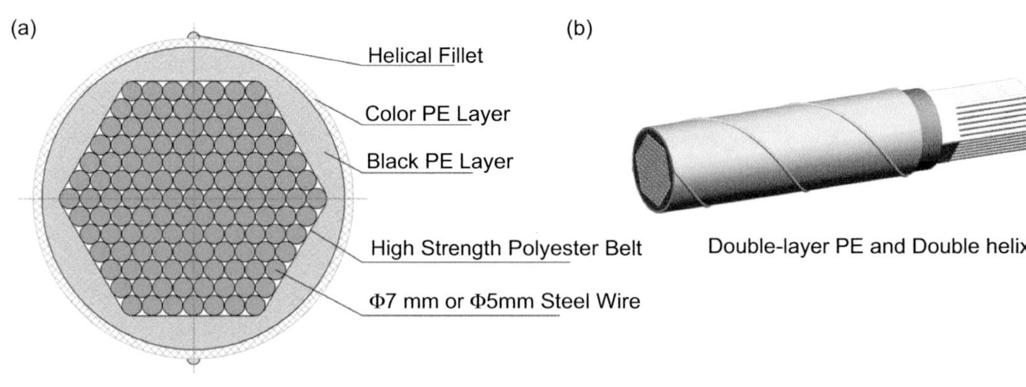

Fig. 3.9 OVM parallel-wire cable: (a) cross-section; and (b) isometric view

This system is characterized by prefabrication on a production line in a factory and packed in reels for transportation to the site. It comes with its cold casting heading anchorage as illustrated in Figure 3.10.

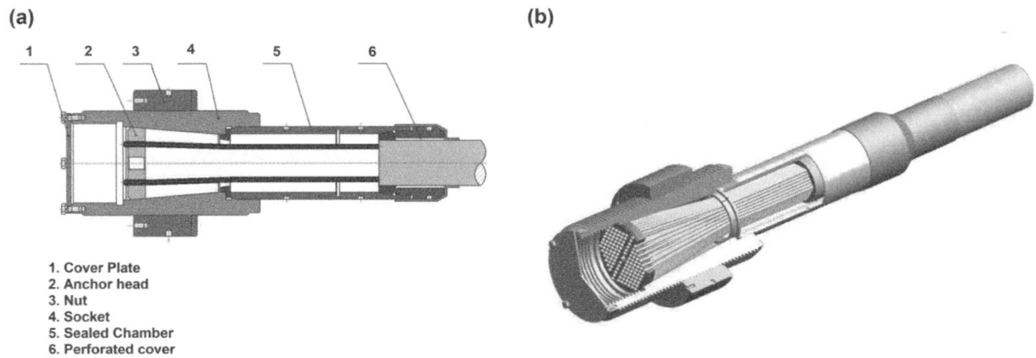

Fig. 3.10 OVM parallel-wire cable anchorage: (a) Elevation; and (b) isometric view

3.5.3 Stranded Cables

Several strand configurations fall under this category. They include:

- Helical or spiral strands
- Locked coil strands
- Parallel wire strands

3.5.3.1 Helical or Spiral Strand Cables

The helical strand is an assembly of wires formed helically around a center wire in one or more symmetrical layers as illustrated in Figure 3.11. Structural helical strands are composed of wires that have been individually coated (galvanized) with a layer of pure zinc. The coating gives the steel wires some sort of protection against various corrosive agents in the environment. The weight requirements of the zinc coatings for various size wires are specified in ASTM A586. Helical strands have been manufactured with 19 to 277 wires depending on the diameter and breaking strength required. They are furnished from 13 mm to 102 mm diameter with a tensile strength of 1570 N/mm2 to 1770 N/mm2. The nominal modulus of elasticity for the multi-wire helical strand is

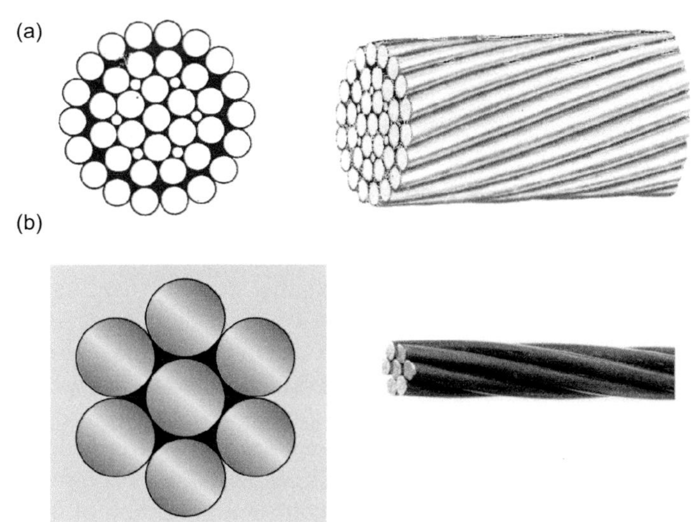

Fig. 3.11 Helical strand cables: (a) typical strand cable; and (b) simple 7 wire strand

usually 15-25% below the value for straight wires. Hence, a typical value for the nominal modulus of elasticity of the helical strand is 170 GPa.

3.5.3.2 Locked-coil Strand Cables

The locked coil strand was developed in Germany and used in all early cable-stayed systems there. Additionally, it is also a helical-type strand. As shown in Figure 3.12, it consists of an outer layer of z-shaped wires that are locked under load with each other and provide protection against water penetration and a superior barrier against external corrosive elements and reduce internal corrosion while giving this product a desirable aesthetic and modern look. At early stages of its application, it had additional z-shaped layers that are not used any more. The inner layers are round wires. The layers are formed such that the lay direction of one layer is opposite to the lay direction of the subsequent layer. The locked-coil strand provides a modulus of elasticity of about 189 GPa. They are furnished from 20 mm to 102 mm diameter with a tensile strength of 1000 N/mm2 to 1300 N/mm2. Density of this strand is approximately 90% compared to approximately 70% for the helical strand with all round wires. The locked-coil strand is manufactured in the United States primarily for tramway applications. Nevertheless, it has never been used for cable-stayed bridges.

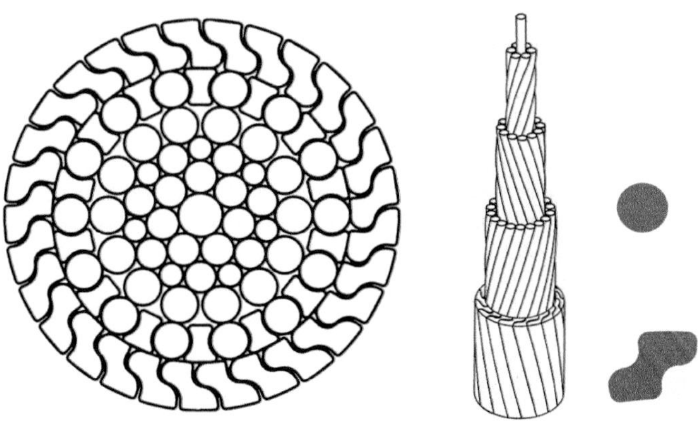

Fig. 3.12 Locked-coil strand cable

3.5.3.3 Parallel Strand Cables

Parallel-strand cables are similar in their configuration to parallel-wire cables except that the individual 7 mm wires are replaced by 15.7 mm diameter seven-wire strands, which are usually galvanized. The individual wires with 5 mm diameter are cold-drawn and thus have an increased tensile strength of 1870 N/mm2 compared with the 7 mm wires with a tensile strength of only 1655 N/mm2. The strand bundle can typically comprise up to 127 strands.

Figure 3.13 illustrates an example of the DSI parallel strand system. As shown the strands run parallel to each other over their free lengths and are encased inside an outer HDPE pipe. They spread out as they approach the anchorages. At their ends, strands are anchored with specially treated 3-part wedges which are characterized by high fatigue resistance as shown in Figure 3.14. Prior to 1997, stay cables in the US had the pipe filled with cement grout or wax as the corrosion barrier. The 7 wire parallel strands used were either black or epoxy coated. Spacers were provided within the sheathing to ensure encapsulation of the strands with the grout material. After 1997, grout inside the stay cable was eliminated in most cable-stayed bridges. The corrosion protection of individual monostrands is achieved through three main elements, as shown in Figure. 3.15. First, galvanization or epoxy

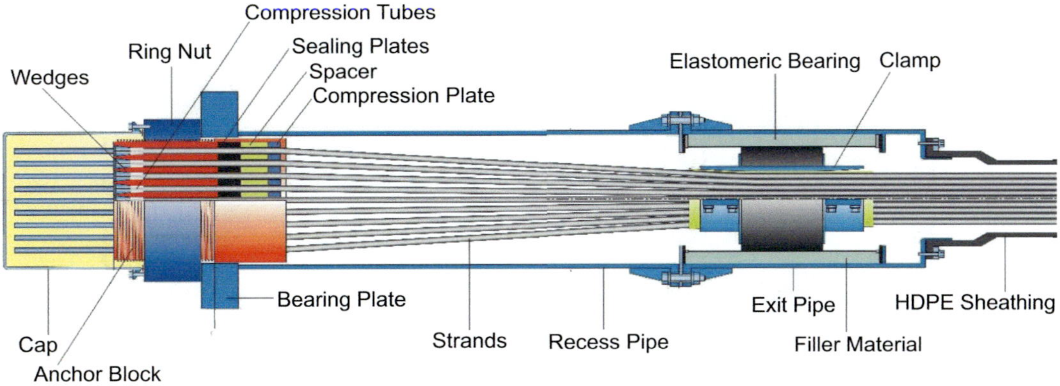

Fig. 3.13 Typical elevation of a parallel strand cable (courtesy of DSI)

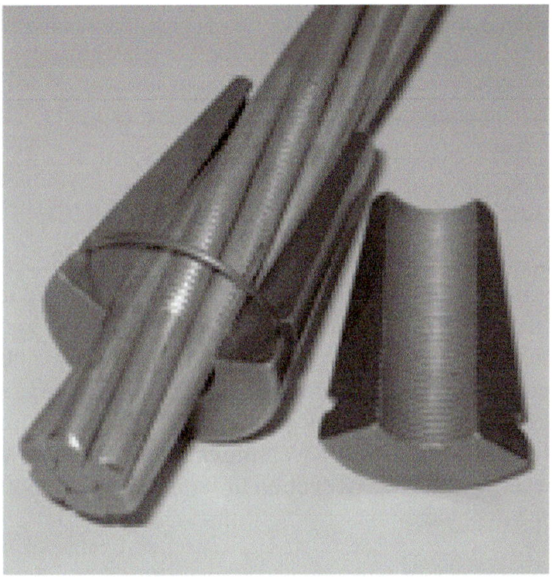

Fig. 3.14 Wedges for galvanized strands (courtesy of DSI)

coating of individual wires (or epoxy coating); next, wax filling of intersections between wires; and utilizing extruded PE-sheath. Furthermore, all monostands are placed into an outer PE pipe.

A parallel-strand cable might be manufactured as a complete unit, as described for a parallel-wire cable, but it is more common to insert and stress the seven-wire strands one by one. Therefore, the fabrication of parallel strand cables takes place on site by installing individual strands one after another. An advantage of completely shop-fabricated cables is the low weight of the individual strands, which will significantly reduce the size of the stressing equipment but increase somewhat the amount of work to be conducted on site.

3.6 CONNECTION TO PYLON

The anchorages at the deck and pylons should accommodate length adjustments during installation and the repair of cables. It is important to prevent bending stresses in the wires or strands at the

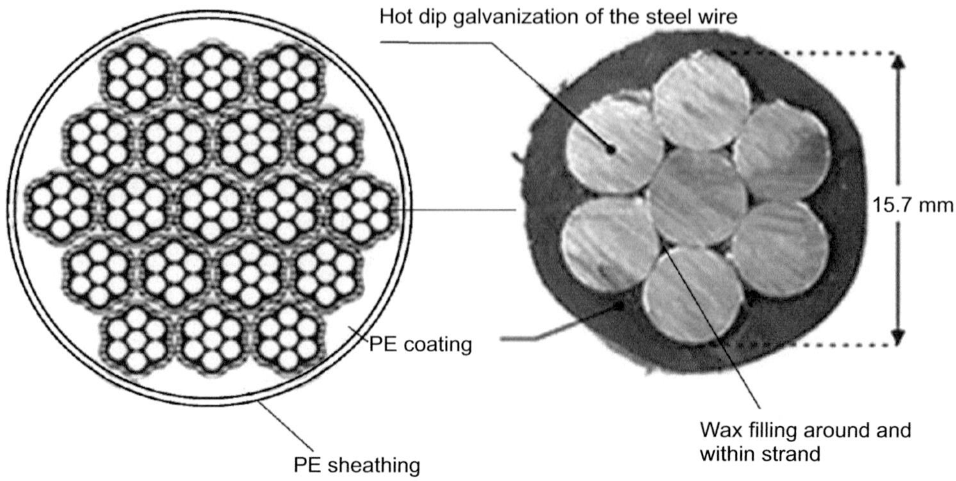

Fig. 3.15 Corrosion Protection of parallel strand cable

socket caused by small oscillations or changes in sag. In order to avoid wind-induced resonance oscillations in the cables, the anchorage should also include dampers. Chapter 6 goes into further detail about cable dynamics.

3.6.1 Direct Cable Anchorage at Concrete Pylon

Concrete pylons are widely used for cable-stayed bridges. Cable anchorage at the pylons can be achieved through different methods. The simplest approach is displayed in Figure 3.16a, where the cables cross each other, in steel pipes. The configuration displayed in Figure 3.16b is more practical for concrete pylons that usually include a box section, where the horizontal component of the cables is taken up in the longitudinal box walls by prestressed bars. All anchorages are easily accessible. In order to change the length of the cable, a jack must be applied. This is much simpler to achieve at the pylon than it is at the deck anchorages. By employing the configuration shown in Figure 3.16c and putting short steel beams inside the box section to support the horizontal component of the cable forces, prestressing can be avoided.

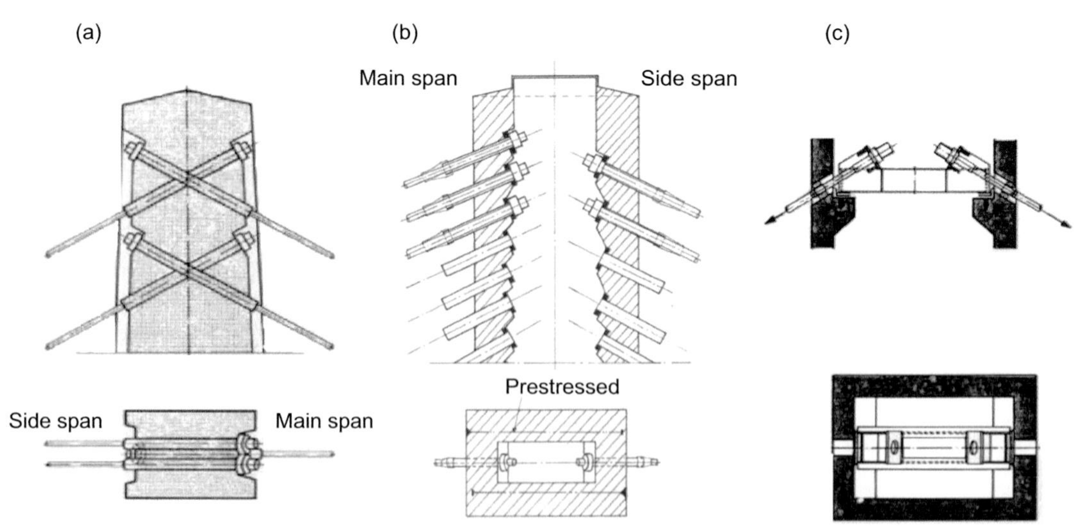

Fig. 3.16 Corrosion Typical stay cable arrangements at concrete pylon heads

The arrangement of Figure 3.16b is investigated further. If cables are individually anchored inside the box cross-section of a concrete pylon the force of each cable is carried by longitudinal and transverse prestressing as illustrated in Figure 3.17. A strut and tie mechanism can be applied. Two virtual struts' forces will be the longitudinal and transverse prestressing.

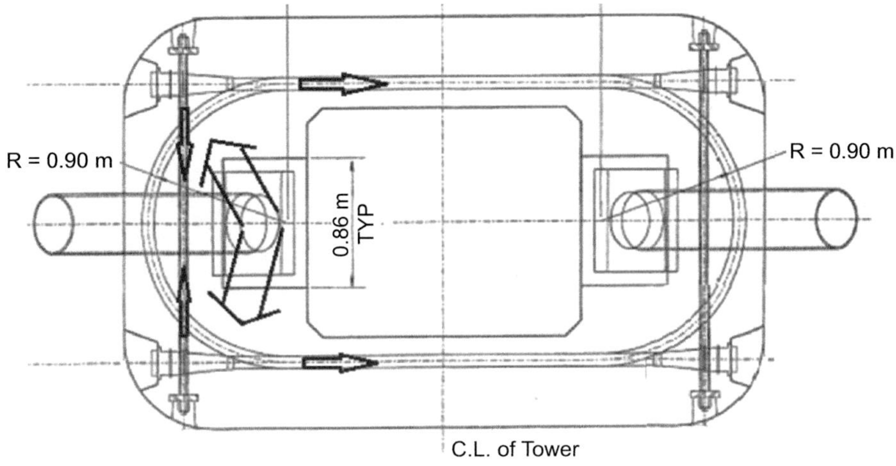

Fig. 3.17 Cable anchorage with prestressing

3.6.2 Composite Cable Anchorage at Concrete Pylon

Composite stay cable anchorages have been preferred for which steel anchor boxes are cast into concrete. They transfer the vertical component of the cable force to the concrete while carrying the horizontal component. They can also transmit uneven cable forces when a cable is lost or replaced. Shear studs hold the steel anchorage units to the inner walls of the cable anchorage chamber. In this way they ascertain the exact placement of each anchor head during construction. Figure 3.18 illustrates a steel frame unit for the William Natcher Bridge (Chandra and Hsu, 1999). Each of the twelve steel frames that make up the pylon heads on this bridge supports two side span cables and two main span cables. Two built-up channels with flanges angled to match the inner pylon walls' slope make up a frame. The inclined flange and cap plates are fastened to the pylon walls using shear

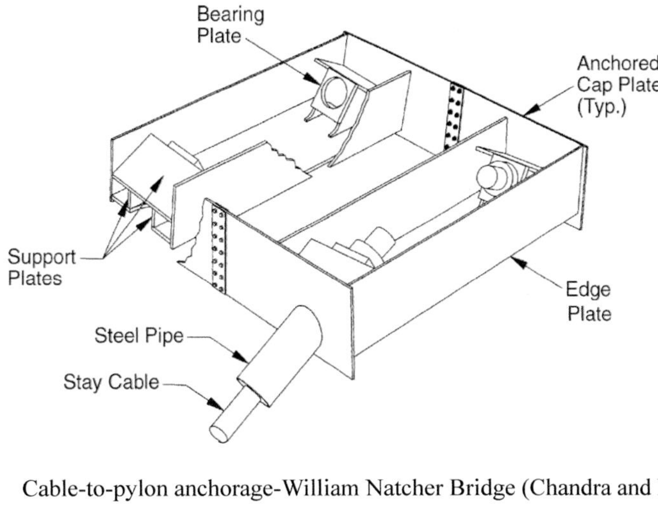

Fig. 3.18 Cable-to-pylon anchorage-William Natcher Bridge (Chandra and Hsu, 1999)

studs, and a steel pipe-equipped cap plate is welded to each end of the channels. The channel flanges are positioned between inclined support plates, which the cable bears. Figure 3.19a displays a photo of the John J. Audubon Bridge steel anchorage frames prior to their erection inside the pylon. For this bridge, the anchor boxes serve as tension ties for opposing cables, and are secured by shear studs located on three sides of the box. Figure 3.19b displays an idealization of the pylon cable anchorage for this bridge (Schemmann et al., 2008).

Fig. 3.19 Composite Cable-to-pylon anchorage

3.6.3 Cable Saddles at Concrete Pylon

Instead of generating large tensile forces, the saddle system transferred the load of the stay cables to the pylon concrete through direct radial compressive stresses, taking advantage of the concrete's inherent ability to efficiently handle large compressive loads. These stresses are more efficient in transferring loads resulting in reinforcement and cross-sectional area reduction of the pylon concrete. Saddles must be designed to ensure safe transfer of stay cable stresses into the pylon structure in the erection and final stage. Transfer of differential forces from the stay cable into the pylon is ensured by friction. The coefficient of friction of the saddle system is ascertained by tests mentioned in sections 4.3.3 and 4.3.4 of PTI DC45.1-18. The system must be designed to allow individual cable strand replacements. This can be achieved through two alternatives. Cable saddles in which the individual MTE elements are isolated through the saddle and are not in contact with each other are referred to as isolated MTE elements. This alternative comprises an external stainless steel curved guide pipe. Inside this outer sheathing, a series of individual stainless-steel tubes carries each of the seven-wire strands of the stay cable's main tensile element (Figure 3.20).

The individual strand sleeves remain parallel to each other throughout the length of the saddle pipe, and their ends are flared to avoid damaging the strands' epoxy coating during their field installation. The space between the individual sleeves is filled with a structural grout with a high compressive strength to provide a load transfer path for the radial compressive stresses generated by each individual strand as it passes through its curved sleeve inside the saddle. This configuration has been employed recently for the I-280 Veterans Glass City Skyway, Ohio, and the Penobscot Cable-stayed Bridge, Maine.

VSL has created a saddle system (Figure 3.21) that consists of a steel box with V-shaped guide voids for each individual strand and is filled with an ultra-high-performance-concrete (UHPC) matrix. Through increased friction, this system enhances the ability to transfer differential forces resulting from uneven loading between bridge spans while preventing damage from fretting under cyclic loading. Individual guiding and encapsulation of the strands allows strand-by-strand installation, inspection, and replacement. This system has passed extensive fatigue testing in accordance with fib Bulletin 30 (fib, 2005) recommendations.

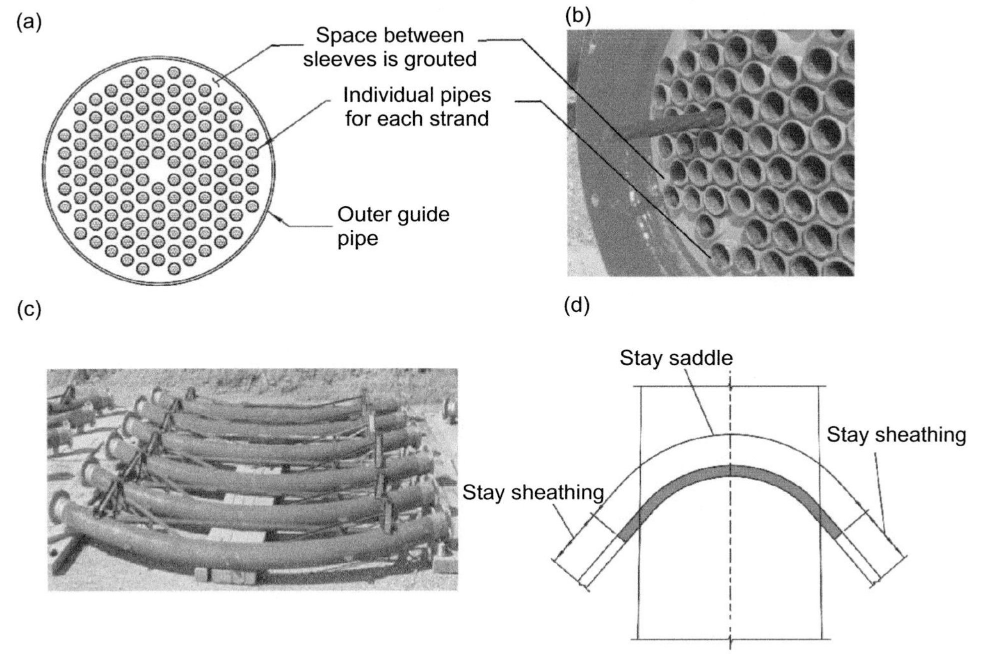

Fig. 3.20 Saddle configurations with individual elements inside individual tubes: (a) cross-section; (b) photo of cross-section; (c) saddles in the field prior to erection; and (d) schematic of stay saddle inside pylon

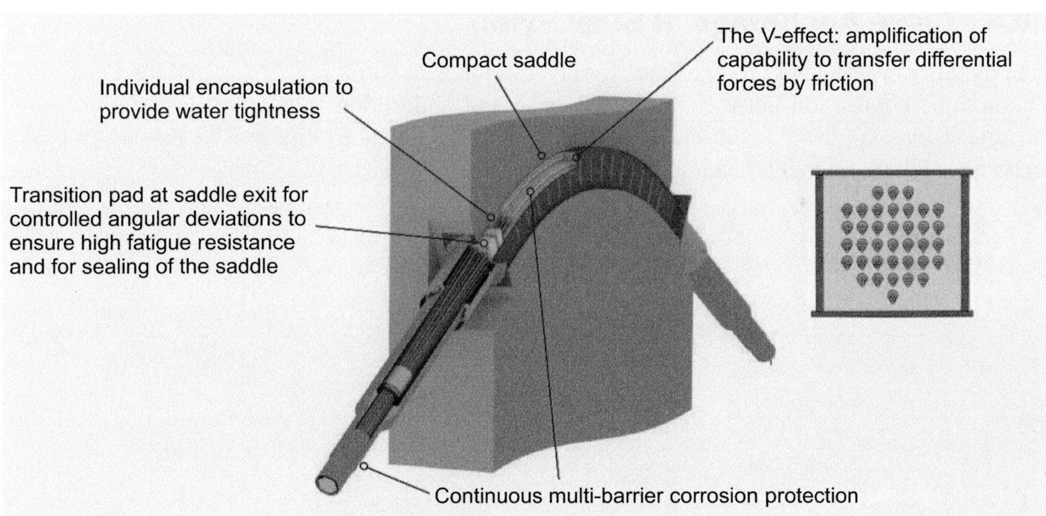

Fig. 3.21 VSL cable saddle

A second alternative comprises double steel pipes, wherein the saddle pipe with a bundle of tensile elements (grouted inside) is placed inside a guide pipe installed into the pylon structure. Transfer of differential forces from the saddle to the guide pipe may be achieved by a shear key or another mechanical connection (see Figure 3.22). Saddles in which a bundle of MTE in contact with each other pass through the saddle are referred to as bundled MTE elements. PTI DC45.1-18 requires that Cable bend radii shall be no less than 2 m for saddles with isolated MTE elements,

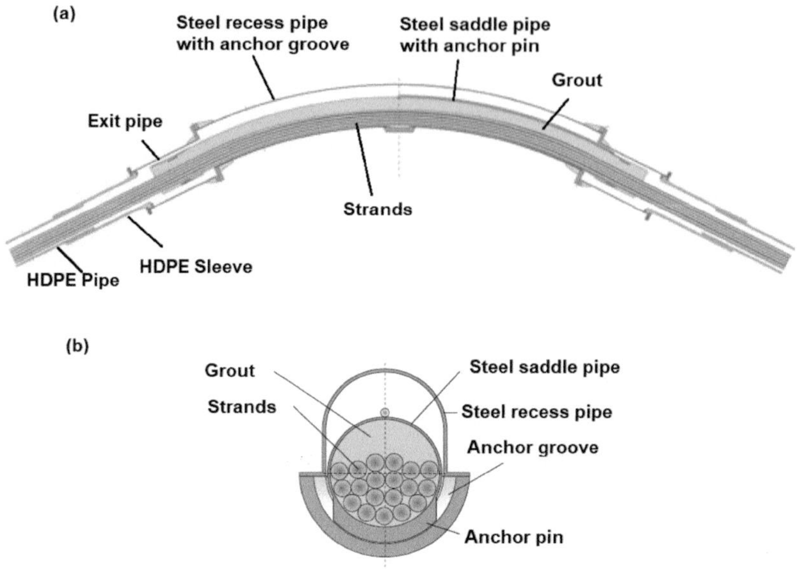

Fig. 3.22 Saddle configuration: bundle of tensile elements grouted inside saddle pipe (courtesy of DSI)

and no less than 4 m for saddles with bundled MTE elements unless fatigue and static testing of the proposed saddle design demonstrates that the specified fatigue and static strength can be achieved in the presence of bending and radial compressive stresses.

3.6.4 Cable Anchorage at Steel Pylon

Steel pylons were designed for early cable-stayed bridges in Germany, Japan, and the USA. Since a monolithic connection between the pylon base and underlying foundation is usually desired, concrete pylons are more favorable nowadays for this application. Figure 3.23 displays a typical anchorage at the steel pylon head.

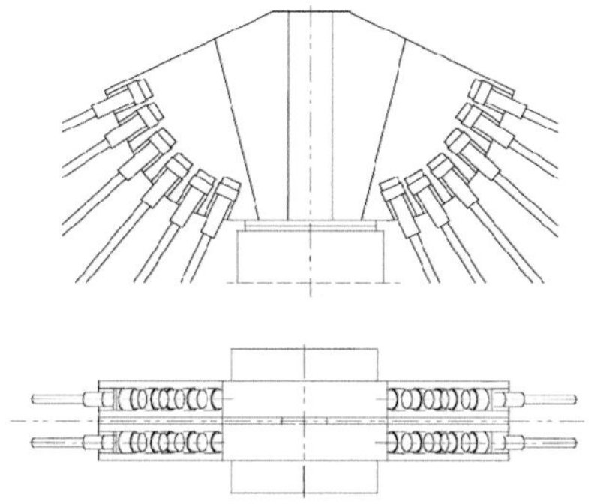

Fig. 3.23 Typical cable fan anchorage at a steel pylon head

3.7 CABLE ANCHORAGE AT THE SUPERSTRUCTURE

A typical anchorage at the concrete superstructure is illustrated in Figure 3.24. A sturdy steel pipe with the proper inclination is embedded into the concrete. The diameter is sufficient to insert the cable's anchor socket through the pipe, insert shims, or tighten the anchor nut to adjust the length. The steel pipe is long enough above the road level to shield the cable from errant vehicles. A thick, soft neoprene pad covers its top, serving as a damper to prevent the cable from flexing. A rubber sleeve is also added to seal the top.

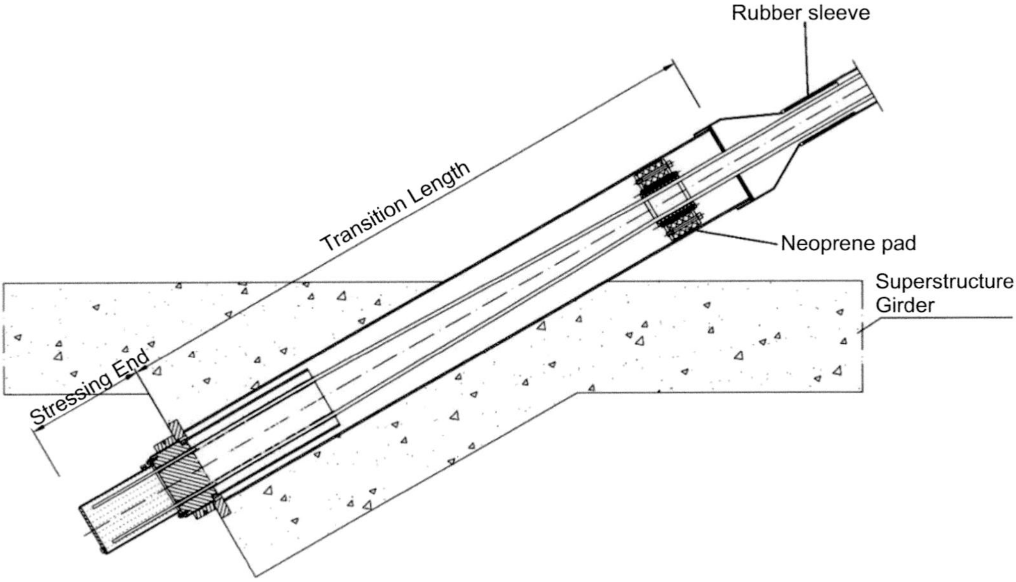

Fig. 3.24 Typical anchorage at concrete superstructure

Figure 3.25 displays anchorages for a composite or steel superstructure. The connection is a simply bolted splice between the connection plate and the girder web.

The connection plate is a flat steel plate that can pass through a slot in the top flange of the edge girder as an extension of the edge girder web. A bolted connection is preferred to prevent stress concentration and cracking in the weldment. At the opposite end of the plate, a thick-walled pipe is welded between two prongs that form a tuning fork shape. After the cable's anchor head is inserted into the steel pipe, it is supported to bear against the pipe's end by a ring nut or shim plate. To stiffen the connection and lower the necessary thickness of the connection plate, additional plates can be welded to the pipe or the connection plate. The concrete deck is poured around the connection plate, and these tapered plates stop above the top of the cast-in-place portion.

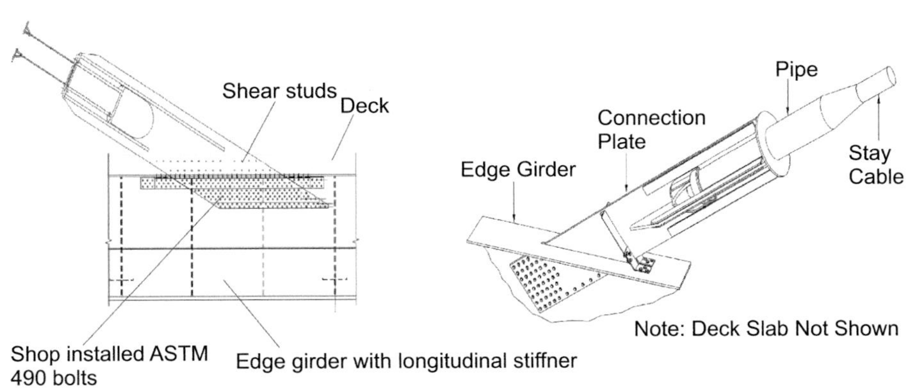

Fig. 3.25 Typical anchorages at the steel superstructure: (a) Olivier-Charbonneau Bridge; and (b) William Natcher Bridge

3.8 DESIGN REQUIREMENTS OF STAY CABLES

Stay cables are designed following the concepts of Load and Resistance Factor Design (LRFD). The load factors and combinations for the limit states contained in Tables 3.4.1-1 and 3.4.1-2 of the AASHTO LRFD and as supplemented by both PTI DC45.1-18 and fib Bulletin 30 must be used for different limit states. Design of cables for wind loads and vibrations is discussed in Chapter 7.

Design of the main tensile element (MTE) for stay cables must satisfy Eq. (3.1) for each of the strength, extreme event, and fatigue limit states (PTI, 2018).

$$\eta \sum \gamma_i Q_i \leq \phi R_n \qquad ...(3.1)$$

where, γ_i denotes the load factor, ϕ is the resistance factor, Q_i is the force effect, axial load, or axial plus bending, as applicable, R_n is the nominal resistance and η is a factor relating to ductility, redundancy, and operational importance as discussed in section 1 of AASHTO LRFD Specifications (AASHTO, 2020). Since PTI DC45.1-18 cover redundant designs of bridges based on the loss of one cable as discussed in this section, it recommends a factor of $\eta = 1$. PTI DC45.1-18 provides resistance factors to be applied for the limit states of Table 3.4.1-1 of AASHTO depending on whether the evaluation is for only axial stresses or combined axial and bending stresses.

3.8.1 Axial and Bending Effects

The cross-section of a stay cable is typically sized such that the maximum axial stress in the stay cable under service conditions (SLS) does not exceed specified limits. Fib Bulletin 30 recommends the maximum axial stress be limited to 50% of the ultimate tensile strength as required by ASTM A421/A421M. It is contingent that the stay cable passes rigorous testing requirements, as specified by both PTI and Fib. Loadings of stay cables during construction or cable replacement should not introduce inelastic deformations in the stay cable system, and a verification of axial stresses against permissible stresses is often sufficient. The permissible axial stresses during construction and cable replacement is typically limited to 60%–70% of the ultimate tensile strength.

The intention of the PTI DC45.1-18 provision is to include local bending stresses in the axial limit state, rather than to establish limit states for axial-flexure interaction. The axial-only resistance factor is based on the level of maximum unfactored stay demand from the live load plus wind on both the structure and live load. The axial only ϕ factor is 0.75 for LL + WS+WL demand less than 2.5% of the minimum ultimate tensile strength (MUTS), and 0.65 for LL + WS+WL demand greater than 7.5% of MUTS, with a linear variation in between these two values. In the absence of project specifications by the owner, unfactored LL + WS+WL stress can be taken from AASHTO Service I. The combined axial plus bending condition should be verified for all unit stay element angular deviations setting $\phi = 0.78$. The maximum resistance factors for extreme and fatigue limit states are 0.95 and 1 respectively. Bending stresses in the free length typically occur near anchorages, saddles, and attachments to the stay cables which impose some lateral displacement. Bending stresses can be calculated using the following engineering mechanics-based equation from the stay cable curvature κ determined for a cable under the design axial force with flexural stiffness:

$$f = E\kappa y = My/I \qquad ...(3.2)$$

where, f is the flexural stress, y is the relevant distance of the strand extreme fiber to the neutral axis for bending, κ is the local curvature in the cable at the local section being examined, E is the modulus of elasticity of the stay cable material as defined in section 3.3, M is the bending moment exerted onto the section and I is the flexural stiffness of the stay cable. PTI DC45.1-18 provides some guidance regarding calculating the flexural stiffness as follows. For stay cables with bare MTE and fully grouted cables, the stiffness may be calculated based on the composite steel-grout section using the nominal cable diameter. Assume that y represents 50% of the nominal cable diameter. For individually protected MTE without grouting, the stiffness may be assumed to be the sum of the stiffness of each individual MTE. If the MTE is a wire or bar, y may be assumed to be 50% of the nominal diameter of the MTE. If the MTE is a strand, y may be assumed to be the center-to-center distance between the king wire and an exterior wire.

Cable saddles and transition details must be designed to preclude slip and fretting of the cable for all fatigue and strength limit states. PTI requires a resistance factor of 0.67 used against the nominal saddle friction coefficient as determined by testing. Saddles must be checked for slip at the extreme limit states using a resistance factor of 0.95.

3.8.2 Design for Lateral Loads

Both PTI DC45.1-18 and fib Bulletin 30 assume a lateral load to be caused by an angular kink of 2.5×10^{-2} radians occurring at the entrance of the stay cable into the anchorage. PTI DC45.1-18 further assumes such angular deviations may occur during construction before the guide deviators are installed or during service conditions if the guide deviators are removed or replaced with special damping devices to control cable vibrations. Therefore, PTI DC45.1-18 requires that stay cable anchorage assemblies and all their components including the deviators be designed for a minimum lateral load of 2.5% of the maximum static cable force at either construction or service conditions.

3.8.3 Fatigue Limit State

According to AASHTO LRFD, the fatigue load shall be a single design truck having a constant spacing of 9 m between the two rear axles while occupying a single lane in each traffic direction. The single lane shall be that which results in the maximum effect on the stay cable(s) under consideration. Since the fatigue provisions of AASHTO LRFD were developed for short spans, therefore, to account for the longer spans of cable-stayed bridges in the absence of rigorous analysis, PTI DC45.1-18 requires that the fatigue design truck calculated design value must be multiplied by a factor of 1.4.

Since η and $\phi = 1.0$ for the fatigue limit state. The equation that governs fatigue is expressed as:

$$\gamma(\Delta F) \le (\Delta F)_n \qquad ...(3.3)$$

where $\gamma = 0.75$ is the load factor; (ΔF) is the stress range due to the passage of the fatigue load multiplied by the factor 1.4; and $(\Delta F)_n$ is the nominal fatigue resistance. To evaluate the fatigue performance of the cable, two values need to be evaluated, the constant amplitude fatigue threshold $(\Delta F)_{TH}$, which is equivalent to the infinite life nominal resistance and the finite life fatigue stress range nominal resistance. $(\Delta F)_{TH}$ is evaluated according to Table 5.1 of PTI DC45.1-18. 50% of the value in that table should be considered to account for the maximum stress range being twice that of $\gamma(\Delta F)$. Therefore, for infinite life,

$$(\Delta F)_n = 0.50\ (\Delta F)_{TH} \qquad ...(3.4)$$

and for finite life,

$$(\Delta F)_n = \left(\frac{A}{N}\right)^{1/3} \qquad ...(3.5)$$

where, A is a constant evaluated according to Table 5.1 of PTI DC45.1-18,

$N = 365$ days × Ny × 1 cycle × $(\text{ADTT})_{SL}$, Ny is the design life of the bridge in years usually taken as 75 unless another value is specified, and $(\text{ADTT})_{SL}$ is the average daily truck traffic in one direction in one lane which may be evaluated in the absence of data according to section C3.6.1.4.2 of AASHTO LRFD.

If $\gamma(\Delta F)$ is less than the value obtained by equation 3.4 then infinite life governs for the nominal fatigue resistance; else if $\gamma(\Delta F)$ is less than the value obtained by equation 3.5 then finite life governs for nominal fatigue resistance. If $(\Delta F)_n$ does not equal either equations 3.4 and 3.5 then $\gamma(\Delta F)$ is too high and measures must be taken to reduce it.

3.8.4 Cable Replacement or Loss of Cable

PTI DC45.1-18 stipulates a cable exchange load case and provides a load combination case for cable replacement that must be satisfied for design as follows:

$$1.2\ DC + 1.4\ DW + 1.5\ (LL + IM) + \text{Cable Exchange Forces} \quad ...(3.6)$$

In (3-6) LL = vehicular live load; IM = vehicular dynamic allowance; DC = dead load of structural components and nonstructural attachments; DW = dead load of wearing surfaces and utilities. PTI DC45.1-18 assumes reduction of the live load in the area of the cable under exchange. Hence, at least one live load lane shifts away from the cable under exchange.

PTI DC45.1-18 also requires that a cable-stayed bridge must be able to survive the loss of one cable without the occurrence of structural instability. The impact dynamic force resulting from the sudden rupture of a cable can be estimated as twice the static force in the cable applied at both the top and bottom anchorage locations. Alternatively, this force can be determined by nonlinear dynamic analysis of a sudden cable rupture, but in no case less than 1.5 times the static force in the cable. PTI DC45.1-18 provides the following load combination for the extreme event of a cable loss from the remaining structure:

$$1.1\ DC + 1.35\ DW + 0.75(LL + IM) + 1.1\ CLDF \quad ...(3.7)$$

Where, CLDF = Cable Loss Dynamic Forces. Full live loads placed in actual striped lanes must be applied in the above equation. The load factor of 1.1 on the Cable Loss Dynamic Force is to account for a variation of the final cable force after construction relative to the force level assumed in the design.

3.9 PROTECTION AGAINST FIRE AND BLAST

Cable-stayed bridges may likely be exposed to more future fire risks as the volume of traffic carried by them increases. The large number of trucks carrying flammable goods crossing bridges every day increases the potential for serious truck fires. Such fires can result in cable failure, which may need long-lasting repairs and will affect the overall behavior of the bridge. Structural stability is therefore not generally a significant issue if the bridge is designed to allow for the loss of one stay cable as described above. Nevertheless, some bridges may be located near fuel depots or oil refineries, hence in this case the bridge will be crossed by many tank trucks carrying flammable materials. Therefore, PTI DC45.1-18 requires that cable-stay systems be qualified by physical testing in a qualified independent testing laboratory to establish a fire rating. Two tests must be conducted. The fire rating (endurance) is a measure of fire resistance for the stay cable protection to a level that will retain its ability to carry the applied design load. For this test, a stay cable specimen must demonstrate fire endurance of 30 minutes or greater, as determined by the time required to reach a temperature of 300°C measured on the outer layer of the cable. Following the fire endurance test, a tensile test must be carried out where the stay-cable specimen of the same materials and fabrication details must be placed in a tension frame and, and uniformly heated to a minimum temperature of 300°C. At this temperature the specimen must be tensioned to 0.45 MUTS. Alternatively, the cable may be tensioned first and then heated to the required temperature and shall last for no less than 30 minutes with no temperature drop below 295°C during this period. The load test can be considered acceptable if the cable is able to resist the applied load for 30 minutes without slip or failure of anchorage components.

Apart from fires, explosions also threaten the security of bridge structures. Over the last several decades there have been an increasing number of terrorist attacks that have resulted in enormous losses. As a result of these incidents, the engineering community has started realizing the significance of designing structures to resist the effects of blast loads. In the US, research was focused on the development of mitigation strategies to improve the performance of a variety of different bridge types to potential terrorist courses of action. Concrete pylons for cable-stayed bridges have been investigated to very severe close-in blasts (Williamson et al., 2010). Nevertheless, there is limited research available on the explosion protection of stay-cables.

Recently the design of the Rod El Farag Axis Bridge crossing the Nile River in Cairo, Egypt, has adopted a newly developed fire and blast protection system for multi-strand stay-cables. The bridge which was opened to traffic in 2019 is a double pylon cable-stayed bridge with a main span of 300 m and a side span of 120 m on each side. Each pylon constitutes three columns, including a plane with two cables in the middle and cable surfaces on both sides of the column with a total of 160 stay cables on the bridge. The fire protection systems were mandated to safeguard cables within the initial 8 m height of each stay above the deck, while also withstanding a 1100°C fire for a minimum of 90 minutes without raising the strand temperature beyond 300°C. These requirements exceed the fire resistance requirements specified in PTI DC45.1-18. Moreover, the cable should be subjected to a load test of high-temperature strength, in which the unprotected cable should be able to resist an applied load of 45% MUTS for 30 minutes without failure of the anchorage components with a minimum 300°C temperature field. Additionally, the blast protection system should be applied to the initial 3 m vertical height of each stay cable above the top of the deck. Furthermore, the number of damaged strands in the cable should not exceed 50% if a cable is exposed to a blast impact of 50 lbs of TNT. (Zou et al., 2022). Figure (3.26) illustrates the protection system, which includes an anti-blast layer of painted steel tube applied for blast protection within the initial 3 m of vertical height above the deck. The strands bundle of the cable is wrapped with a thermal insulation blanket inside the blast protection tube up to the bottom of a high-density polyethylene (HDPE) pipe. It is also wrapped with the same insulation blanket inside the waterproof cap up to 8 m of the vertical height. The HDPE pipe is also wrapped with a thermal insulation blanket and protected by the steel tube. The Rod El Farag Axis Bridge is considered unique in that the stay cables are designed for fire and blast protection systems, which is rare in other existing bridges.

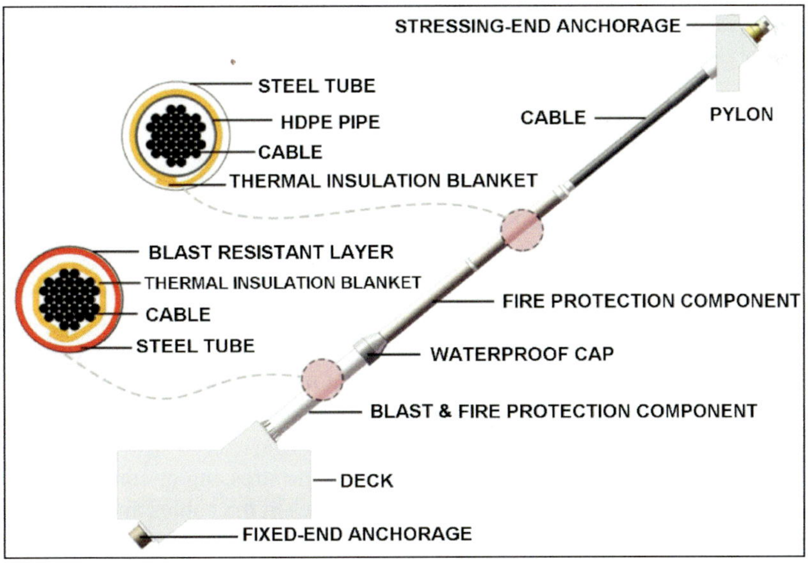

Fig. 3.26 Fire and blast protection systems for Rod El Farag Axis Bridge, Egypt (Zou et al., 2022)

3.10 CABLE INSTALLATION

3.10.1 Parallel Wire Cables

Parallel wire cables were outlined in section 3.2.2. They consist of straight parallel round wires inside a PE-sheath. They are completely shop-fabricated and transported to the site on reels. The cables on reels are raised on the bridge deck and are unreeled on auxiliary carriages. In the next step, the passive cable anchor head is drawn up to the pylon head with a crane. The anchor head is then driven into its position inside the pylon head using a tie rope passing through the anchorage pipe and set into its final position at the pylon head using shims. A crane, spreader beam and a pull-rod are then used to drag the anchor head into its final position at the beam. A hydraulic jack is then used to pull the anchor head to its anchorage position and secured with shims.

3.10.2 Parallel Strand Cables

Parallel strands stay cable systems are usually assembled on site. Installation is carried out strand-by-strand using extremely compact equipment. The steps involved in the installation are as follows:

- The anchor heads are set at the pylon and at their final beam positions
- The stay pipe (sheathing) is assembled on the bridge deck then raised to its inclined position as shown in the photo in Figure 3.27.

Fig. 3.27 Photo of stay pipe being raised to its inclined position (Courtesy, MnDOT)

- Each strand is cut to its proper length from a large coil and pushed up the stay pipe into the anchor box in the pylon head using a light pusher as shown in Figure 3.28
- The crew then installs the end of each strand into the bottom anchor plate (Figure 3.29)
- Wedges are then installed onto each strand to keep it from slipping when it is stressed. All single strands are stressed immediately after being installed in the cable anchorages using a mono-jack, thus optimizing the entire installation process.
- Equalization of the forces in all strands at the end of the stressing operation must be ensured. In addition, the influence of temperature and load changes in the cables and in the structure during stressing must be eliminated. After stressing, the individual strands are closely packed (Figure 3.30).

Fig. 3.28 Photo of strands being pushed through the stay pipe to the pylon anchor box (Courtesy, MnDOT)

Fig. 3.29 Photo of installation of the strand end into the bottom anchorage (Courtesy, MnDOT)

Fig. 3.30 Photo of strands after stressing (Courtesy, MnDOT)

- Another stay pipe goes up and the process is repeated. Once the structures are completed, cable forces can be adjusted by restressing or destressing the complete cable via the ring nut.

References

American Association of State Highway and Transportation Officials, *AASHTO LRFD Bridge Design Specifications*, Washington D.C., 2020.

American Society for Testing and Materials, A416/A416M, *Standard Specification for Low-Relaxation, Seven-Wire Steel Strand for Prestressed Concrete*, West Conshohocken, PA, 2021.

American Society for Testing and Materials, A421/A421M, *Standard Specification for Stress-Relieved Steel Wire for Prestressed Concrete*, West Conshohocken, PA, 2021.

American Society for Testing and Materials, *A586, Standard Specification for Metallic-Coated Parallel and Helical Steel Wire Structural Strand*, West Conshohocken, PA, 2018.

American Society for Testing and Materials, *A882/A882M, Standard Specification for Filled Epoxy-Coated Seven-Wire Steel Prestressing Strand*, West Conshohocken, PA, 2021.

Andersen, H., Hommel, D.I. and Veje, E.M., Emergency Rehabilitation of the Zarate-Brazo Largo Bridges, Argentina, *Cable-Stayed Bridges-Past, Present and Future*, Proceedings of IABSE Conference, Malmo, Sweden, 1999.

Bonzon, W.S., The I-280 Veterans' Glass City Skyway: New Landmark Cable-Stayed Bridge, Ohio, *Structural Engineering International*, Volume 18, Issue1, 43–48, 2008.

CEB-FIP, Acceptance of Stay Cable Systems Using Prestressing Steels, *federation Internationale Du Beton (fib)*, 2005.

Chandra, V. and Hsu, R., The Innovative William Natcher Cable-Stayed Bridge, IABSEConference-Cable-Stayed Bridges-Past, Present and Future, Malmo, Sweden, 1999

DSI, DYWIDAG Multistrand Stay Cable Systems, www.dsiamerica.com, 2022.

Henrik, A., Hommel, D.L. and Veje, E.M., Emergency Rehabilitation of the Zarate-Brazo Largo Bridges, Argentina IABSE Conference-Cable-Stayed Bridges-Past, Present and Future, Malmo, Sweden, 1999

Loziias, M.P. and McCabe, R.J., Design and Construction of the Dame Point Concrete Cable-stayed Bridge in Jacksonville, Florida, *Proceedings of the International Bridge Conference*, IBC-90-61, Pittsburgh, 1990.

OVM, Parallel Wire Stay Cable System-Design, Manufacture, Installation, Maintenance, *Liuzhou OVN Machinery CO. Ltd,* China, 2016.

Post-Tensioning Institute (PTI*). Recommendations for Stay-Cable Design, Testing, and Installation.* Farmington Hills: PTI 2018. Standard No. DC45. 1–18.

Schemmann, A.G., Bergman, D.W. and Shafer, G., John James Audubon Bridge-Design-Build Delivery of the Longest Span Cable-stayed Bridge in the United States, *Proceedings of the International Bridge Conference*, IBC-90-61, Pittsburgh, 2008.

Svensson, H.S. and Lovett, T.G., The Twin Cable-Stayed Composite Bridge at Baytown, Texas, *International Association for Bridge and Structural Engineering*, IABSE 60–90, 1990.

Williamson EB. e, Blast-Resistant Highway Bridges: Design and Detailing Guidelines. *Transportation Research Board*. 2010.

Zou, Y., Qin, Lei, Sun, L., Pang, R. and Chen, L et al., Fire and Blast Protection for Multi-Strand Stay Cables of Rod El Farag Axis Bridge, *Structural Engineering International*, Volume 33, Issue 1, 179–182, 2023.

CHAPTER **4**

Proportions and Sizing of Cable-Stayed Bridges

4.1 INTRODUCTION

Cable-stayed bridges have undergone fast development during the past forty years. The cable-stayed system is advantageous compared to other types of long span complex bridges. This is because it is characterized by lighter decks. The tensile cable forces in it are balanced with compression within the deck and pylon thus eliminating traditional large anchorages as in the case of suspension bridges. The evolution in the production of high-strength steel cables has positioned the cable-stayed system as an alternative to the suspension system in spans that exceed 1000 meters. The design, structural detailing and construction methods for cable-stayed bridges are simplified, thus resulting in a bridge that is aesthetically and economically favorable. Proportioning and sizing of different elements of a cable-stayed bridge is a significant process that should be considered in the early phases of the design process. In general, the stiffness of the pylon, the size of the back-stay cables, and the inclination of the stay cables are the principal factors that affect the stiffness of the bridge superstructure. Bending moments associated with dead loads can be manipulated in the longitudinal girders by cable adjustments. Generally, the deck can be treated as a beam on elastic foundations. The main parameters that govern the selection and sizing of different components of the bridge are discussed in this chapter. Considerations in the selection of types and sizes of main components such as the superstructure, the pylons, the foundations, and lengths of spans including their ratios are outlined.

4.2 SUPERSTRUCTURE CONFIGURATIONS

The superstructure of a bridge can be categorized into three main types: steel, concrete, or composite. The selection of a certain type is dependent on factors such as the main span and other economic considerations such as material availability and preferences.

Additionally, the superstructure needs to meet certain specifications, including ease of cable-stay anchorage, aerodynamic stability, and lightness. Depending on the deck suspension system, the superstructure is designed with either a single central cable plane or two outer cable planes. To withstand asymmetrical loads in a single cable plane configuration, the superstructure needs to be highly rigid torsionally. In this arrangement, the deck needs to be large enough to support the median strip's stay cables and shield them from passing vehicles. Because the cables provide a stiff support along each edge and the deflection is minimal in a configuration with two outer cable planes, torsional rigidity is not required. As a result, unsymmetrical loading only results in a very slight transverse inclination of the deck Furthermore, no torsional rigidity is required for aerodynamic safety. Generally, a large concrete slab with or without ribs around the edges is adequate if the bridge's width is almost 17 meters (Figure 4.1a). If present, the edge rib secures the buckling safety and permits the cables to be anchored at any point. Such a design is illustrated in the Evripos Bridge in Greece. Cross girders are required for bridges with wider decks. They should be spaced 3 to 5 meters apart to allow for an easy longitudinal span of the orthotropic steel plate or concrete slab, as well as the longitudinal movement of most of the steel stiffening ribs or reinforcing bars, which aid in supporting compressive normal forces (Figure 4.1b and c).

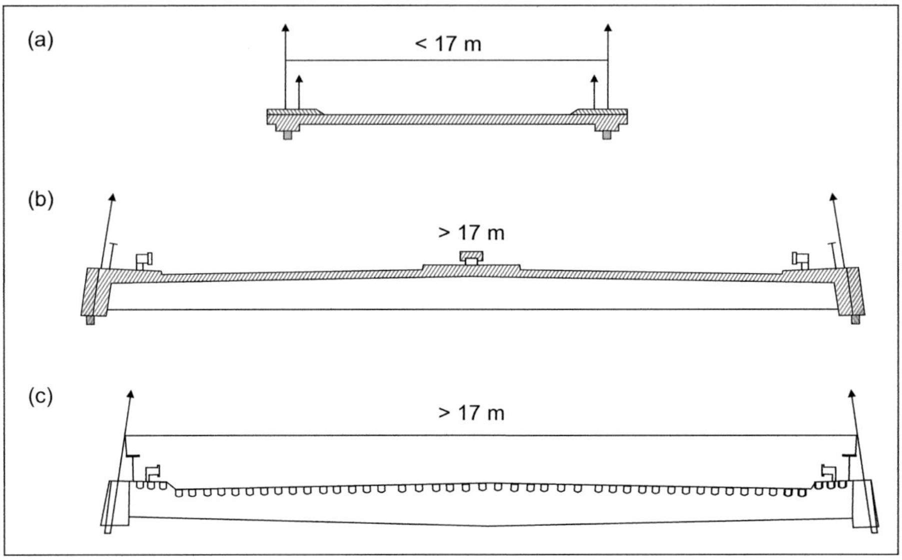

Fig. 4.1 General configurations of cross-sections for cable-stayed bridges superstructures

The concrete cross section deck is applicable for main spans close to 550 m. The all-steel bridge with an orthotropic plate deck becomes mandatory for large spans to reduce the dead loads. The composite deck falls in between the ranges of the other two types as depicted in Figure 4.2 which summarizes the range of application of each type.

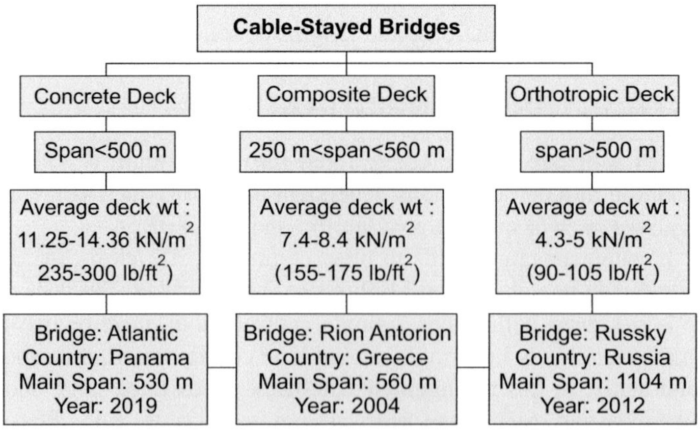

Fig. 4.2 Ranges of applications of different cable stayed bridges decks

A steel cross-section can be opened as illustrated in Figure 4.3a. It is convenient for two outer cable planes configuration. It consists of two main edge girders that support an orthotropic deck. The two outer cable plane configurations can also come with a closed steel cross-section that consists of a box girder as shown in Figure 4.3b. This section is characterized by its high torsional stiffness and is hence used for two and one cable planes. Recent designs for cable-stayed bridges in China and Japan employed two separated steel box girders attached with a transverse steel beam with a box section (Figure 4.3c). The Stonecutters Bridge in Hong Kong is one example of such a design.

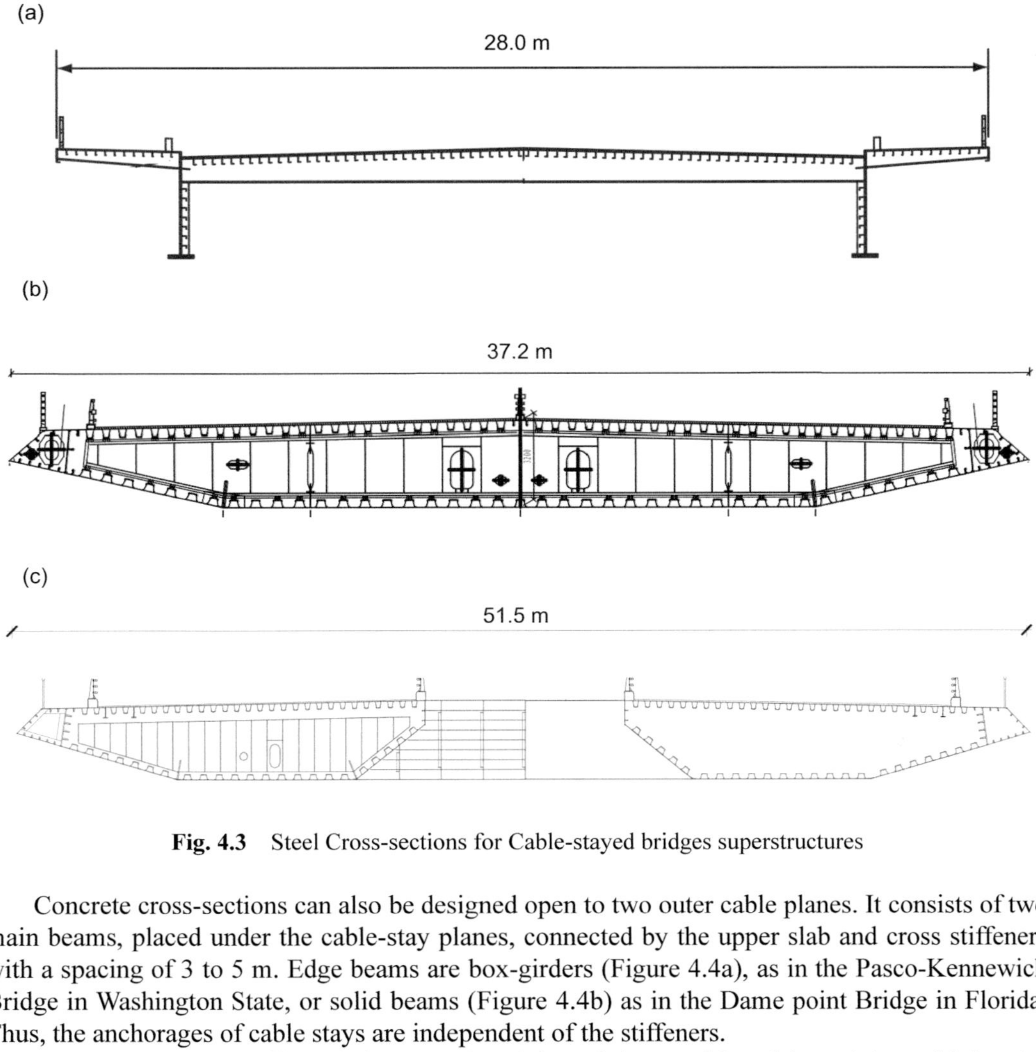

Fig. 4.3 Steel Cross-sections for Cable-stayed bridges superstructures

Concrete cross-sections can also be designed open to two outer cable planes. It consists of two main beams, placed under the cable-stay planes, connected by the upper slab and cross stiffeners with a spacing of 3 to 5 m. Edge beams are box-girders (Figure 4.4a), as in the Pasco-Kennewick Bridge in Washington State, or solid beams (Figure 4.4b) as in the Dame point Bridge in Florida. Thus, the anchorages of cable stays are independent of the stiffeners.

There are two types of box girders: 3 or 4-web box girders; and 2-web box girders with interior triangulation. Three-web box girders have two main disadvantages: difficult access to cable stays anchorages, usually placed under the center web (Figure 4.5a), and cross-sectional deformability, due to the transfer of suspension forces from the cable-stays to the lateral webs. The first disadvantage can be remedied by a double plane cable-stay configuration, or by splitting the central web, thus obtaining a four-web box-girder (Figure 4.5b). The Aomori Bay Bridge in Japan is an example of such a configuration. Two-web triangulated box-girders are generally lighter than multicell systems, and cable-stay tensioning is created systematically from within the deck, which makes the process

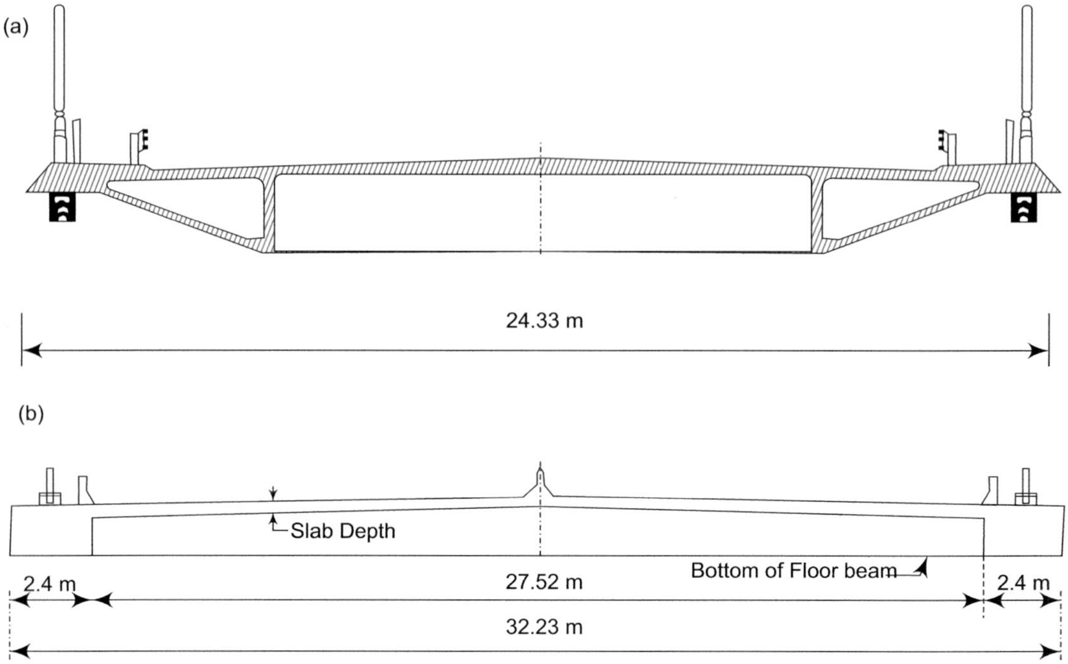

Fig. 4.4 Opened concrete cross-sections for cable-stayed bridges

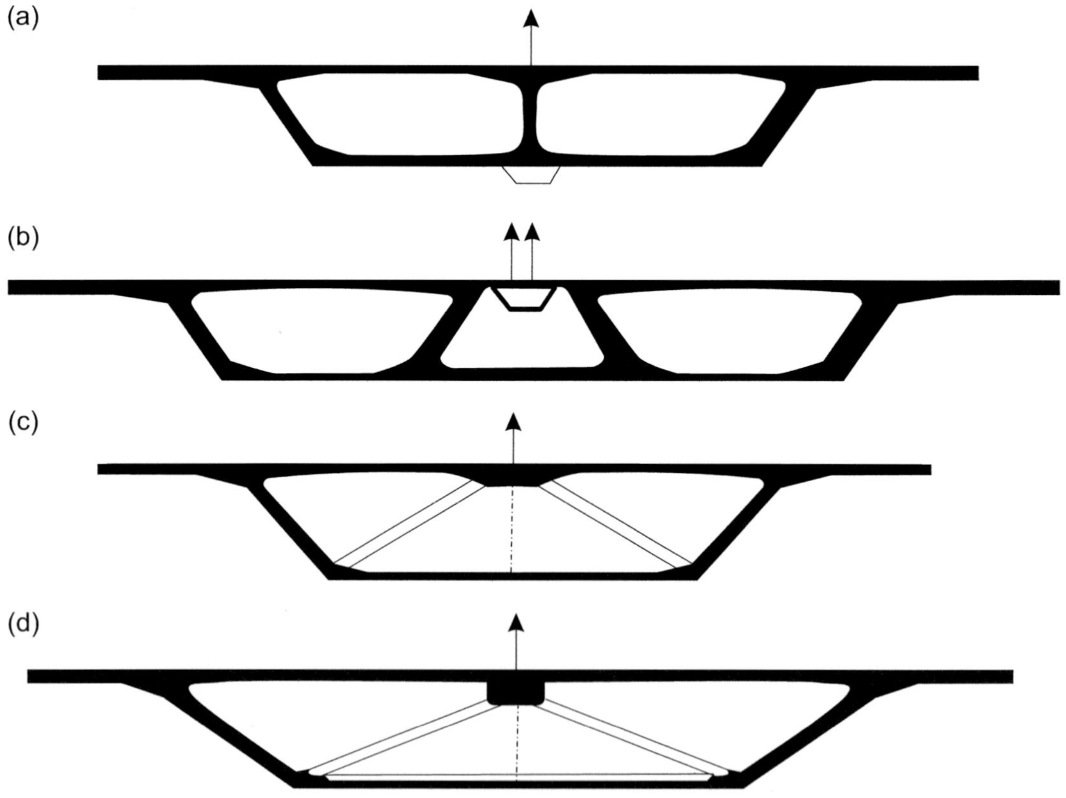

Fig. 4.5 Box girder cross-sections for cable-stayed bridges

much easier. A configuration of a triangulated box-girder with two inclined webs, with inclined struts as interior stiffening is shown in Figure 4.5c The Brotonne Bridge in France is an example of such a configuration. Triangulated box-girder can be designed with two inclined webs, with inclined struts as interior stiffening, and cross stiffeners at bottom slabs. The Sunshine Skyway Bridge in Florida and the Penobscot Bridge in Maine are examples of such configurations.

It is evident from a comparison of the orthotropic deck to an equivalent concrete slab road deck that the former is more expensive due to its high fabrication costs. The additional costs associated with the greater area of cable stays, as well as the pylon and foundation requirements to support heavier dead loads, might offset the cost savings achieved by using a composite concrete slab instead of a steel deck (see Figure 4.2). Composite superstructure cross-sections are only used for the two outer cable planes. They consist of two edge steel girders that are attached transversely by a steel beam as shown in Figure 4.6. This steel grid system supports a concrete roadway slab. It is very prudent for this kind of construction to keep away tensile stresses from the concrete slab. This is achieved in the longitudinal direction by the compressive forces from the inclined cables in the transverse direction; the floor beam is designed as a simply supported one with a bending moment that forces the concrete at the top flange to be always in a state of compression.

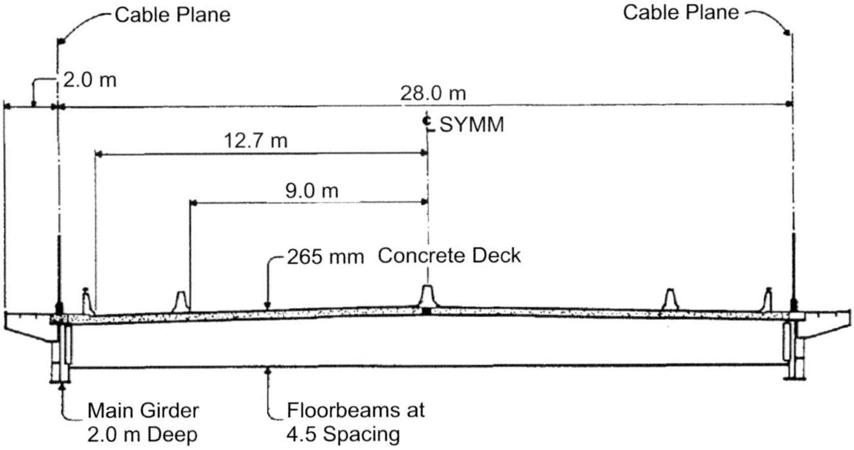

Fig. 4.6 Composite cross-section for cable-stayed bridges superstructure

A hybrid system is sometimes used. It comprises a steel main span and concrete side spans that act as a counterweight to the lighter steel main span. Normandy Bridge in France, Stonecutters Bridge in Hong Kong, and Shinminato Bridge in Japan are all examples of bridges employed in the design of a hybrid system with a steel main span and concrete side spans

4.3 SUPERSTRUCTURE DEPTH AND STIFFNESS

According to Post-Tensioning Institute (PTI DC45.1-18) cable-stayed bridges are required to be designed for cable loss and cable replacement load cases. Therefore, spacing between individual cables along the deck is sometimes reduced in order to allow the girder to span the missing cable. For bridges with two planes of cables, a 1.8 m deep girder is adequate when cables are spaced 12 m to 14 m apart. These requirements account for cable loss and cable replacement load cases. Box girders supported by a single cable plane would be about 3 m to 3.5 m deep with a cable spacing of about 6 m. The depth of the box girder is probably governed by the torsional strength requirement rather than the cable replacement or cable loss cases.

4.4 PYLON GEOMETRY

Various structural solutions are used for the pylons: single pylons, double-leg portals (vertical, slightly angled, free-standing, or interconnected as a portal frame, with "A", "H", "Y", or inverted "Y" shape. A pylon is typically a vertical element that is sustained in the longitudinal direction by the cable stays, which also provide resistance against forces induced by horizontal loads. The type of suspension, central one-plane cables or two outer planes, closely influence the transverse design of pylons. In general, pylon selection is dependent on several factors such as type of superstructure and suspension system; span length; aerodynamic stability; ease of construction; overall cost; and accommodation for future inspection and maintenance. Figure 4.7 displays different pylon geometries. The pylons can be continuous above and below the deck supporting both the deck and

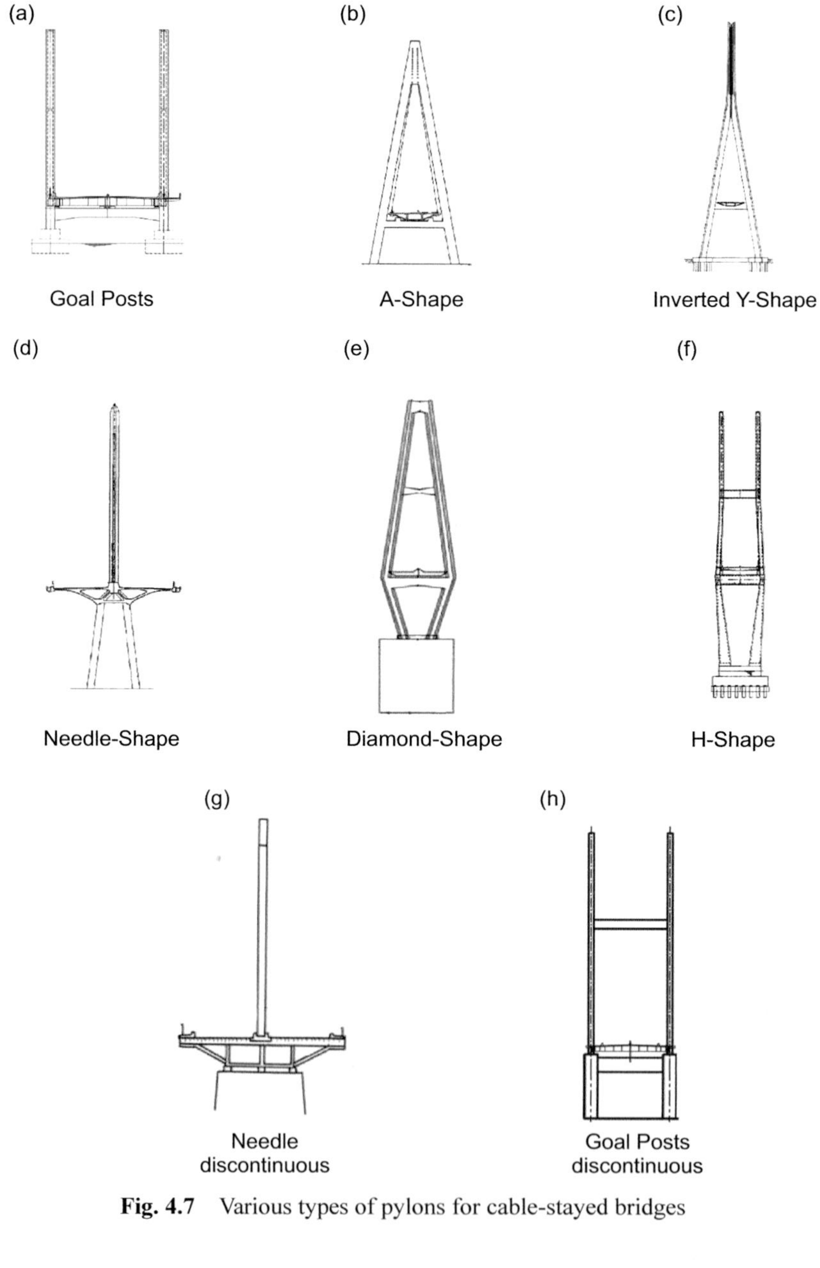

Fig. 4.7 Various types of pylons for cable-stayed bridges

the cables as shown in Figure 4.7a through f, or the upper part can support only the cables while the deck-girder is supported directly by piers as shown in Figure 4.7g and h. This arrangement is very convenient from both aesthetic and economic perspectives, whenever the pylon strength and elastic stability can be achieved transversally. Nevertheless, it is not recommended in seismic areas. A single plane of stays with a needle shaped pylon is suitable for concrete box girders. The Goal Posts (Figure 4.7a) and the H-shape are generally easier for construction. A horizontal beam connecting the pylon legs at the top may be helpful if the height of the pylons is significantly greater than the width of the bridge. This will allow the cable plane to have a slight inclination. Because all cables join at a single pylon top, A-shaped pylons for long spans improve the aesthetics of these bridges (Figure 4.7b). The 404-meter Pont de St. Nazaire Bridge in France is a notable example. The A-frame pylon is suitable for inclined stay arrangements and very efficient aerodynamically because of the inclined planes of stays. The inverted Y-frame where the vertical leg, containing the stay anchors, extends above the bifurcation point is a variation of the A-frame. Examples of the inverted Y-frame include but are not limited to the pylons for the Normandy Bridge, France and the Rama VIII Bridge, Bangkok, Thailand. Both the A-Shaped and inverted Y pylons require a wide footprint, particularly when a high navigation clearance to the deck is required. The diamond-shape pylon (in Figure 4.7e) was developed to overcome this deficiency by breaking the pylon legs at or just below the deck to produce inward-leaning legs to the foundation. However, this modified arrangement results in reduction in the pylon's stiffness in the transverse direction which may result in significant deflections under wind or seismic loads. This deflection can be mitigated only with a considerable increase in the stiffness of the lower section of the pylon leg below the deck. Nevertheless, this arrangement was applied for the pylons of several recently built cable-stayed bridges such as the Tatara and Meiko Gand Bridges in Japan, and the Industrial Ring Bridge, Thailand.

4.5 SIZING OF PYLONS

The following factors need to be taken into considerations when sizing pylons (FHWA, 2012):

- Larger cross-sections are selected when the design calls for a greater laterally unsupported length. Likewise, a larger cross-section is generally suitable when the pylon's axial and bending loads are high.
- The H-shaped pylon is the least favorable in terms of providing aerodynamic stability but tends to be more economical for construction.
- The vertical cable spacing in the pylon should be selected to accommodate the cable anchors and cable installation requirements. The number of cable planes is an additional factor that would control the pylon's lateral dimensions.
- Adequate space should be provided to allow access for construction, inspection, and maintenance. The space should also allow for material and equipment to be delivered from the deck or ground level to the cable anchors for repair and replacement work.

4.6 INCLINATION OF PYLONS

Some cable-stayed bridges have been built with the pylon inclined backwards. The Bratislava Bridge in Slovakia and the Alamillo Bridge in Spain are examples of such designs. While backward inclination gives the bridge more thrill, it can, however, be proved that there is no economic advantage. The back stays become shorter and steeper. Construction of the pylon is more challenging than if it was built vertically and then tilted. An inclination in the other direction towards the main span, also makes no sense and also brings no aesthetical advantages.

4.7 PYLON HEIGHT ABOVE DECK

The height of the pylons has an influence on the longitudinal compressive forces in the bridge deck and the amount of steel required for the cables. The higher the pylon, the smaller will be the quantum of cable steel required. This will also reduce the compressive forces. The height of the pylon above the deck is determined by the location of the highest stay cable connection. The higher the cable connection point, the lower the compression in the deck and the smaller the quantities of cable steel. The theoretical quantities of cable steel in the stay cables of the fan system can be obtained using the following equation (Gimsing, 1980):

$$Q = PL\left(h + \frac{L^2}{12h}\right)\frac{\gamma}{\sigma} \qquad ...(4.1)$$

And for harp system:

$$Q = PL\left(\frac{h}{2} + \frac{L^2}{8h}\right)\frac{\gamma}{\sigma} \qquad ...(4.2)$$

In (4.2) and (4.3):

Q = theoretical quantity of cable steel
P = vertical components of the cable forces
L = main span length
γ = density of cable material
σ = allowable cable stress

These correlations are depicted in Figure 4.8, which indicates that, considering the quantities required for the pylon, its ideal height is between 20% and 25% of the main span length. To ensure

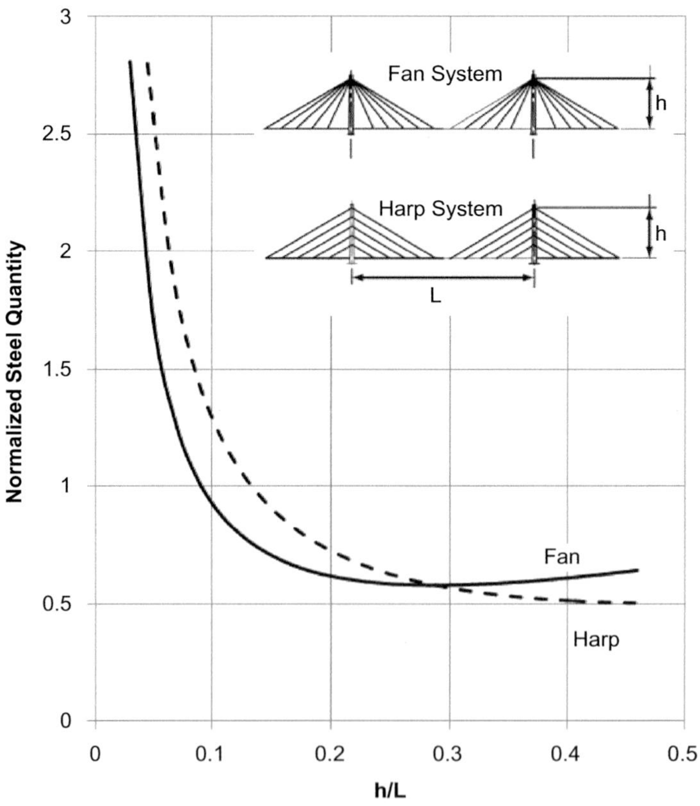

Fig. 4.8 Relation between Pylon Height and Cable Steel Quantity

that live loads in the main span do not result in significant bending moments in the pylon and instead cause the back stay cables to act, the pylons should be slender and have a small bending stiffness in the longitudinal direction.

4.8 STAY SPACING AND INCLINATION

To accommodate the longitudinal girders' capacity, the stay anchor spacing along the deck must be restricted to a maximum of 5 to 15 meters. If a stay is abruptly lost, this range will maintain the longitudinal girders' capacity below critical levels. This spacing range allows the deck to be installed using the free cantilevering method without the need for additional supports because it is small enough. While 10 to 15 meters is better suited for steel or steel composite construction, 5 to 10 meters will be needed for heavier concrete structures.

The bridge's overall stiffness will be determined by the pylon's height. The required stay size will decrease and the pylon height will increase as the stay angle (α) increases. However, as each stay lengthens, the deck's deflection will increase. When the expression $1/(\sin \alpha \times \cos \alpha)$ is likewise a minimum, then so is the weight of the stay and the deflection of the deck (Farquhar, 2008). Thus, in theory, staying at 45 degrees is the most effective. A stay inclination between 25 and 65 is quite reasonable in practice and won't materially reduce the design's efficiency. The outer stay that connects the deck panel next to the main span's center to the top of the pylon will be inclined at an angle of 25 degrees. The stay closest to the pylon will be inclined at 65 degrees. This is consistent with the ideal ratios of 0.25 and 0.2 for pylon height above deck to main span length, as shown in Figure 4.8.

4.9 BACK SPAN TO MAIN SPAN RATIO

The anchor (back stay) cables are always exposed to stress changes depending on the position of the live load on the bridge. This is illustrated in (Figure 4.9), which depicts the deflection of three-span cable-stayed bridges under the effect of live loads. Initially under permanent loads all the stays are in tension. Live loads in the central span are transferred by the forestays to the pylon heads and then anchored by the back stay cable in the hold-down abutment or anchor pier. This scenario will increase the tensile stresses in the back stay. If the live load exists in the side span, it will be

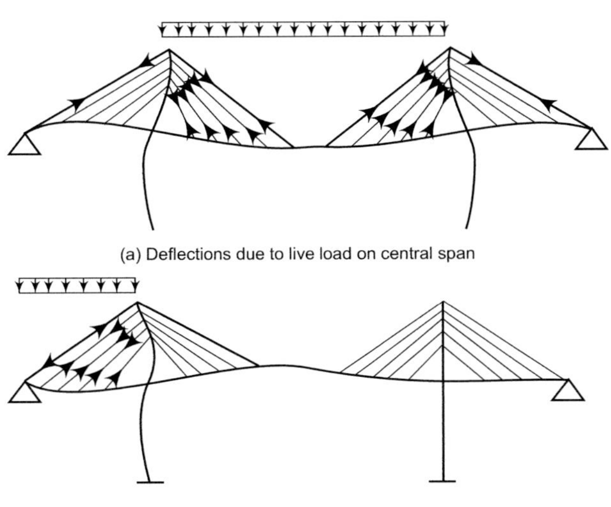

Fig. 4.9 Deflections of cable-stayed bridges due to live load

transferred through tension in the side span cables to the pylon head and from there via compression in the back stay to the anchor beam, i.e., reduction of the initial tensile stress due to permanent loads.

Consequently, it is evident that the back stay cables, which support the tower head and hold it in place at the anchor pier, are primarily affected by changes in stress caused by the side span to main span ratio. It is evident that while a live load in the side span reduces back stay stresses, a live load in the main span will cause those stresses to increase. Of all the cables, the back stay cables experience the highest stress amplitudes, which need to be safely maintained below the cables' fatigue strength. Both the maximum tensile force due to permanent loads and the fatigue stress change are governed by the ratio of the back span to main span (l/L). Long main spans will result in high tensile forces in the back stays due to loads. On the other hand, long side spans will result in high stress changes for fatigue. The intent of the design is to find the l/L ratio that will result in a cable cross-section that can accommodate both the maximum tensile force and the fatigue stress change without reducing the effective modulus in the back stay under the effect of live loads in the side span. The ratio l/L further

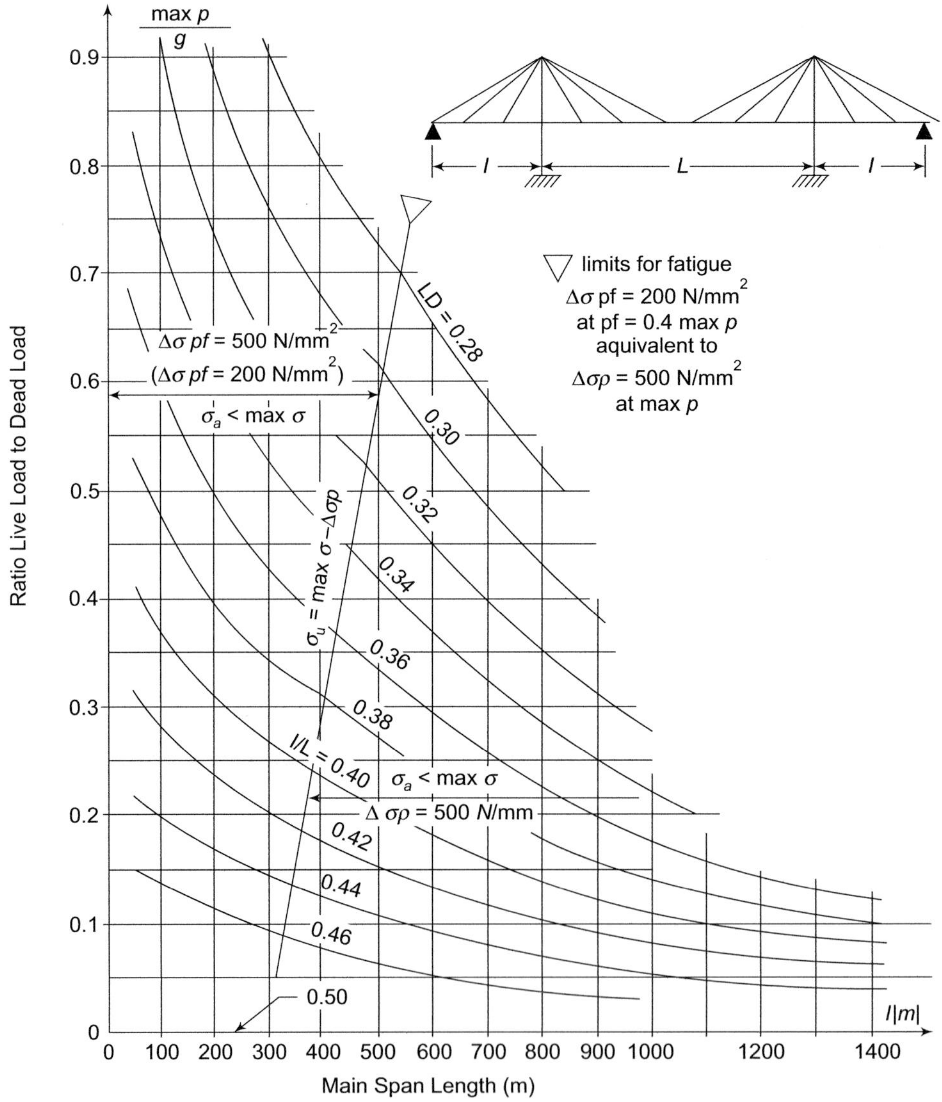

Fig. 4.10 Relationship between span ratios and LL/DL ratios (Leonhardt and Zellner, 1980)

influences the amount of vertical anchoring forces at the anchor pier. This anchor force decreases with an increasing l/L ratio.

Leonhardt and Zellner (1980) developed charts for the relationship between the main span length, in meters (Figure 4.10), and the ratio of live load to dead load (LL/DL) for different values of the ratio of the back span to main span (l/L). The straight line intersects the curves at the points where a fatigue stress change of the back stay cable of 200 Mpa is reached under 40% live load. For concrete road bridges the ratio of live load to dead load is 0.2. This value corresponds to l/L = 0.42 for a typical main span of about 400 m. For railway bridges LL/DL is about 0.6 corresponding to a l/L value about 0.30. Currently most of the road cable-stayed bridges have their back spans laid out to be approximately 45% of the length of the main span. It is worth noting that this span ratio will likely cause uplifts under permanent loads at the anchor pier that needs to be resisted by a tie-down or counterweight.

To summarize the optimum main span to side span ratio is sensitive to the proportion of live load to dead load. Of course, this will vary depending on whether a concrete, steel or composite deck is being proposed. For a bridge with a steel superstructure the length of the back span will be smaller than for a similar span bridge with a concrete superstructure. When establishing the conceptual arrangement of the bridge it is important that the ratio between the back span and the main span be less than 0.5 to give a clear visual emphasis to the main span. This ratio is equally as important structurally as it influences the uplift forces at the anchor pier and the load range within the back stay cables supporting the top of the pylon. Most cable-stayed bridges have their back spans laid out to be approximately 45% of the length of the main span. The benefit of staying close to this span ratio is that the anchor (or back stay) cables can be kept taut all the time. This span ratio will likely cause uplift at the anchor pier that must be resisted by a tie-down or counterweight.

References

Farquhar, D.J., Cable Stayed Bridges, ICE Manual of Bridge Engineering, Institution of Civil Engineers, 2008.

Federal Highway Administration, *Design Criteria for Arch and Cable Stayed Signature Bridges,* FHWA-NHI-023, Washington DC, 2012.

Gimsing, N.J., History of Cable-Stayed Bridges, *Proceedings IABSE Conference Cable-Stayed Bridges-Past, Present and Future*, Malmö, Sweden, 1999.

Leonhardt, F. and Zellner, W. Cable-Stayed Bridges, *IABSE Surveys* S-13/80, Zurich, 1980.

CHAPTER 5

Evolution of Cable-stayed Bridges in Europe

5.1 DISCHINGER'S FIRST MODERN CABLE-STAYED BRIDGE

The advancement of cable-stayed bridge technology owes much to Franz Dischinger, a pioneering German structural engineer, who collaborated with the firm of Dyckerhoff and Widmann A.G. since 1913, primarily focusing on concrete structures. In the 1930s, Dischinger had become intrigued with the prestressing ideas pioneered by Eugene Freyssinet (1879-1962). This led Dischinger to patent the technique of external prestressing (where the prestressing bars or tendons are not encased in concrete). He then started to study long span suspension bridges with an aim to design very-long-span bridges for heavy live loadings. He observed that previous stay and cable bridges were technically flawed and visually disturbing. On the basis of his studies, he published an article in 1949 wherein he proposed a series of ideas all based upon the Arnodin-Gisclard system (see section 1.5). For his design, Dischinger suggested a curved cable and vertical suspenders to support the central part (between 40% and 50% of the span) of the main span while the cable stays carry the remaining parts near the pylons. He proposed in a paper (Dischinger, 1949) two alternatives for the Koln-Mulheim suspension bridge to be built over the Elbe River in Hamburg. Although never adopted for actual construction, the proposals by Dischinger in his paper definitely had a significant impact on the subsequent introduction of the pure cable-stayed bridge.

Around 1953, Dischinger worked on a proposal with the German Construction Company DEMAG for the design of the Stromsund Bridge in Sweden. The design standards by the Swedish government called for a suspension bridge with a central span of 160 m and side spans of 80 m. Dischinger submitted two alternatives, one of them for a cable-stayed bridge with two pairs of cables radiating from the tops of each side of two pairs of pylons. This design proved very economical and was selected by the Swedish authorities. The construction of the bridge was concluded in 1955 and it was opened in 1956. The bridge as shown in Figure 5.1 has a main span of 182.6 m and two side spans of 74.7 m. The stays are in the form of converging bundles, consisting of two top and two bottom cables anchored together at the top of the pylon. The pylons are in the form of trapezoidal frames hinged at the base to allow rocking in the longitudinal direction of the bridge. The deck as shown in Figure 5.2 consists of a reinforced concrete slab supported on longitudinal steel plate girders and cross girders and has a depth of about 3.20 m, i.e., 1/58 of the span, which is considered by no means a very economical design, in comparison with continuous girder bridges that may require more material for such a span (Feige, 1966).

Fig. 5.1 The Stromsund Bridge, Sweden, 1956

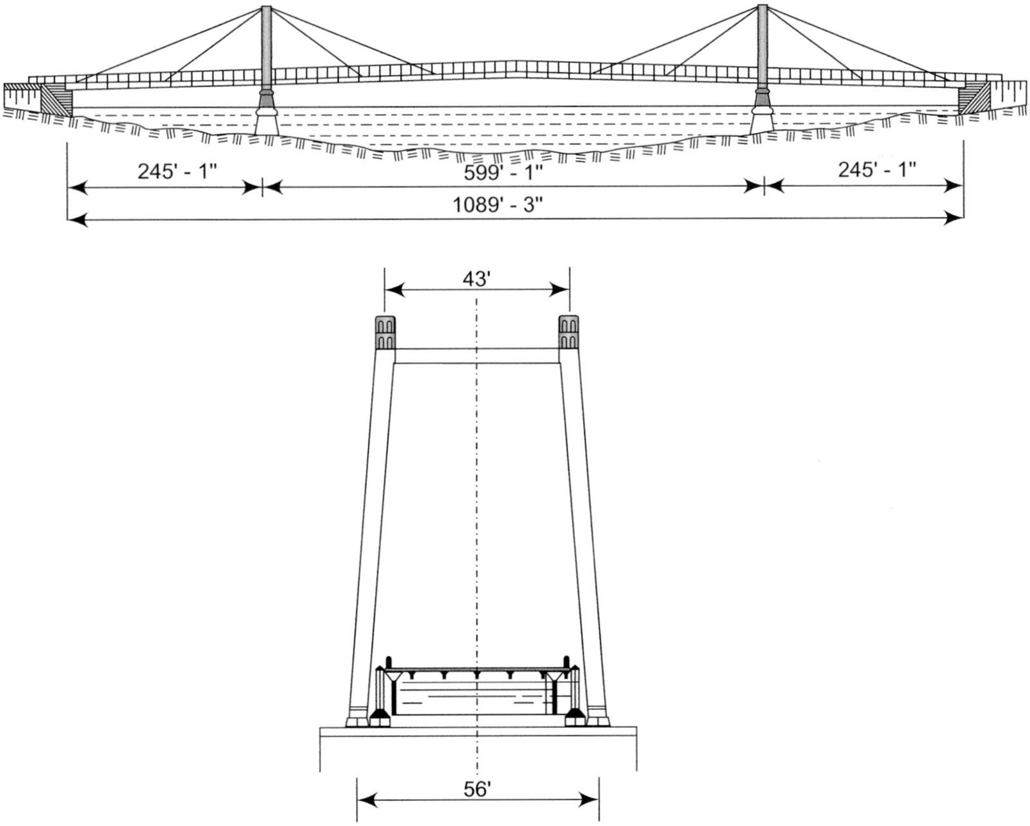

Fig. 5.2 Stromsund Bridge General Arrangement

5.2 EVOLUTION OF MODERN CABLE-STAYED BRIDGES IN GERMANY

The design of the Stromsund Bridge came at the right time. After the Second World War, 1500 bridges in Germany had been destroyed. Dischinger's cable-stayed bridge design along with efficient methods of construction by DEMAG opened a new era for cable-stayed bridge design.

Dischinger's ideas had been adopted by other German engineers for proposing several cable-stayed bridges in postwar Germany in the early 1950s. Hence, in the first two decades after the construction of Stromsund Bridge, the evolution of cable-stayed bridges was dominated by steel bridge designs (This differentiation between steel and reinforced concrete cable-stayed bridges is in the context of the deck since it is well understood that although the Stromsund Bridge had steel pylons, subsequently most of the bridges built later used reinforced concrete pylons).

5.2.1 The Theodor Heuss Bridge

The Theodor Heuss Bridge (North Bridge) across the Rhine at Dusseldorf (Figure 5.3) was the second modern cable-stayed bridge and the first one to be built in Germany. This bridge which opened to traffic in 1957 has a main span of 260 m and side spans of 108 m. The bridge is 26.6 m wide and its cross-section consists of two main steel box girders with an orthotropic deck. Each 40 m high pylon is fixed to the girders (see Figure 5.4).

Fig. 5.3 The Theodor Heuss Bridge (North Bridge) across the Rhine at Dusseldorf, Germany, 1957

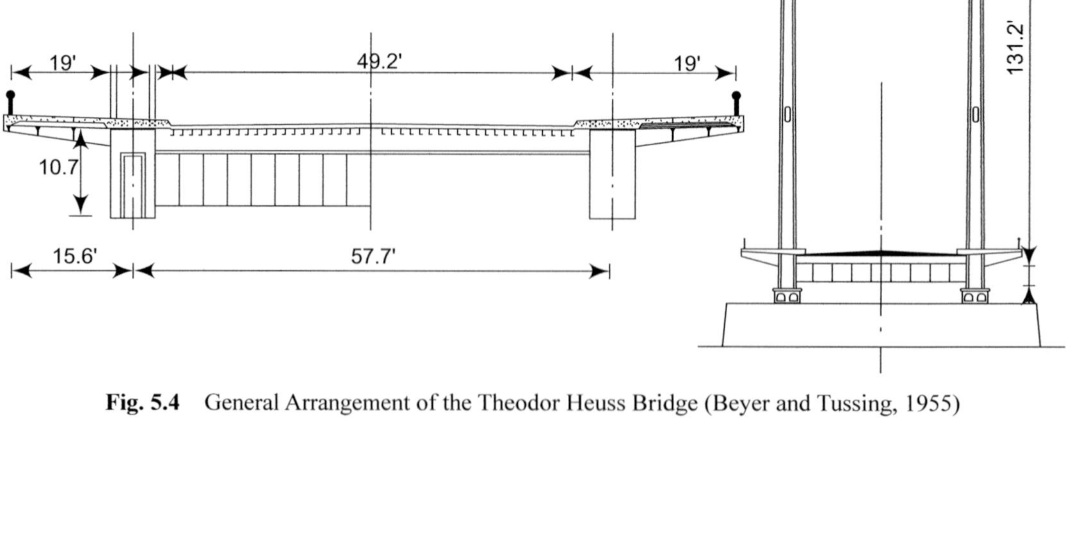

Fig. 5.4 General Arrangement of the Theodor Heuss Bridge (Beyer and Tussing, 1955)

5.2.2 The Severin Bridge

The Severin Bridge across the Rhine at Cologne (Figure 5.5) is the second cable-stayed bridge erected in Germany. This bridge has a main span of 302 m, including a length of 121.7 m without intermediate support. The bridge is characterized by an A-shaped pylon about 62 m high, and all the cables are fixed directly to the top. The pylon is rigidly attached to the pier. This bridge is 29.5 m wide; its cross-section is like that of the Theodor Heuss Bridge in that the two main girders are also steel box-section members with a depth of 4.60 m (Figure 5.6), which is equivalent to 1/66 of the length of the span. The bridge was completed in 1959.

Fig. 5.5 The Severin Bridge across the Rhine at Cologne, Germany, 1959

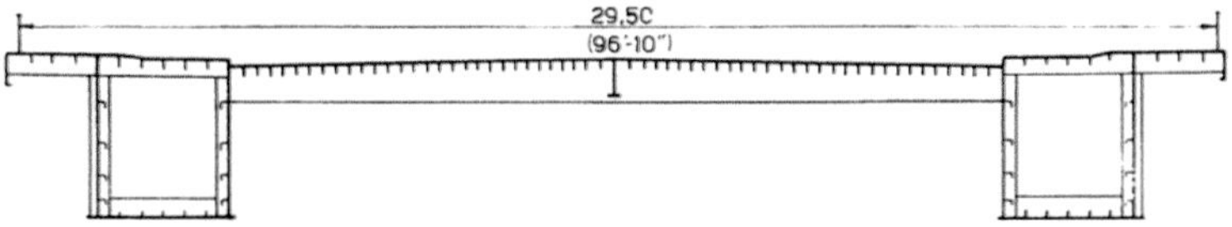

Fig. 5.6 Severin Bridge Cross-section

5.2.3 The Norderelbe Bridge

The Norderelbe Bridge followed in 1962 (Figure 5.7). It was characterized by its upward extended pylons and the star-pattern of the stay system. The main span is 172 m long and the girder is 3m deep,

Fig. 5.7 The Norderelbe Bridge, Germany, 1962

i.e., 1/57 of the main span. The bridge is 30.70 m wide. The deck consists of a grillage comprising an 8.20 m wide box girder and two single web lateral girders interconnected at intervals of about 21.5 m by solid web diaphragm-type cross-girders (see Figure 5.8). The 53 m high pylons are fixed to the girders and the top cables are fixed to the pylon, whereas the bottom cables have movable bearings at their points of attachment to the pylons.

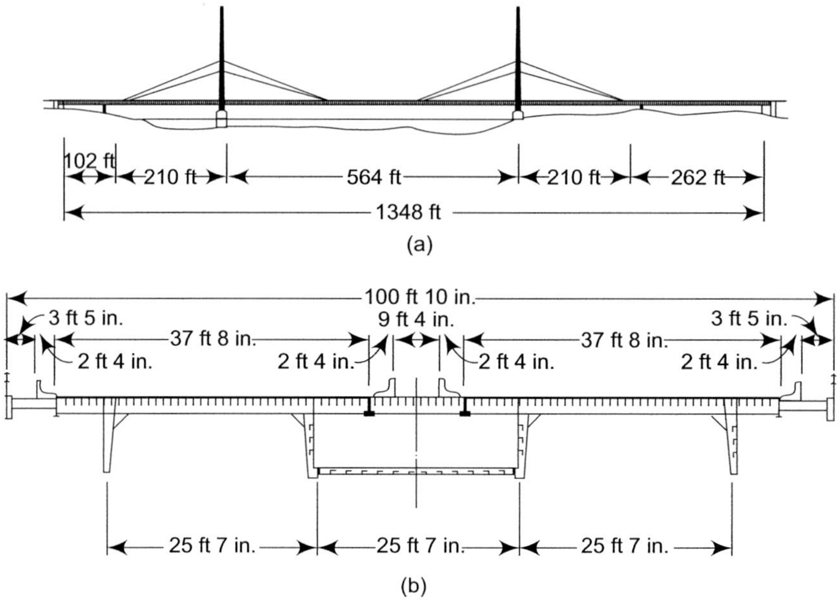

Fig. 5.8 The Norderelbe Bridge: (a) Elevation; and (b) Cross-section

5.2.4 The Julicherstrasse Bridge

The Julicherstrasses Bridge at Dusseldorf (Figure 5.9) was completed in 1963 to link eastern Dusseldorf to its western counterpart with a main span of 98.70 m. The deck is an orthotropic steel box girder, as shown in Figure 5.10, with a depth varying between 1.25 m and 1.65 m, and a

Fig. 5.9 The Julicherstrasse Bridge at Dusseldorf, Germany, 1963

width of 26.40 m. It has upper orthotropic slab cantilevers on both sides of the box girder stiffened transversely by brackets. The vertical steel pylons are 16.5 m high and attached rigidly into the deck and supported to the reinforced concrete piers by fixed and moveable bearings in the east and west sides respectively. The stays are formed of six locked-coil cables 74 mm in diameter, anchored to the deck, and run on a saddle at the pylon to allow relative movements over the pylon head.

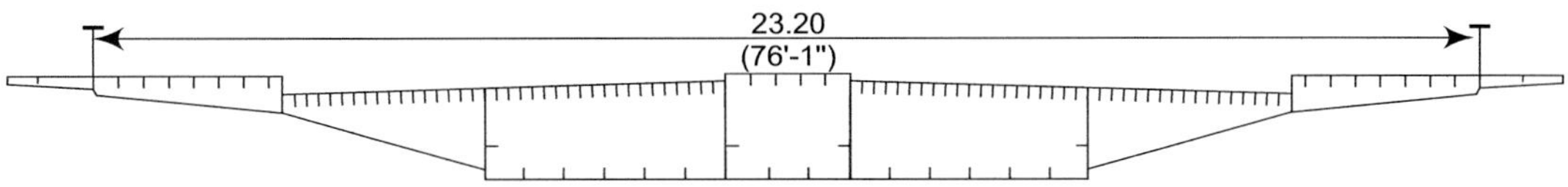

Fig. 5.10 Orthotropic Box Girder for the Julicherstrasse Bridge

From 1964 to 1966, 3 more bridges were built including the bridge over the Rhine near Leverkusen having a main span of 280 m.This bridge was built in 1965. The bridge over the Rhine near Maxau in 1966 had a main span of 175.20 m. It was characterized by an orthotropic deck on a torsional rigid box girder with two cantilevers as shown in Figure 5.11. The bridge at Ludwigshafen has a main span 138.0 m. In 1967 two more bridges over the Rhine River were built.

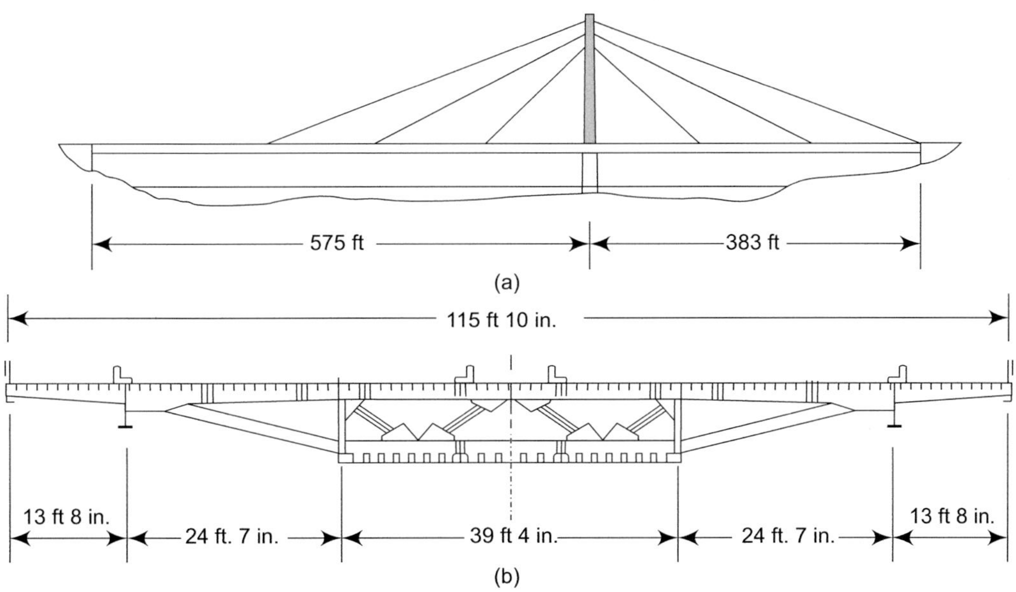

Fig. 5.11 General Arrangement of the Bridge near Maxau: (a) Elevation; and (b) Cross-section Germany, 1966 (Thul, 1966)

5.2.5 The Rees Kalkar Bridge

The Rees Kalkar Bridge is characterized by a harp stay system (Figure 5.12), which reflects the beginning of the computer-aided design era in the late 1960s. The bridge has ten stays, one above the other, on each side. This arrangement ensures that the girders have a semi-continuous elastic support from the stay system, rather than being suspended at a few distinct points such as in earlier bridges. This bridge has two individual pylons on each side. The main span has a length of 255 m, and the girder is about 3.5 m deep, corresponding to 1/73 of the main span. The bridge is 18.8 m wide. The bridge deck is of the orthotropic type, which is resting on the top of two main girders of the single-web type (see Figure 5.13).

Fig. 5.12 The Rees Kalkar Bridge, Germany, 1967 (Courtesy, Herrard Elizabeth Taubenheim, 2009)

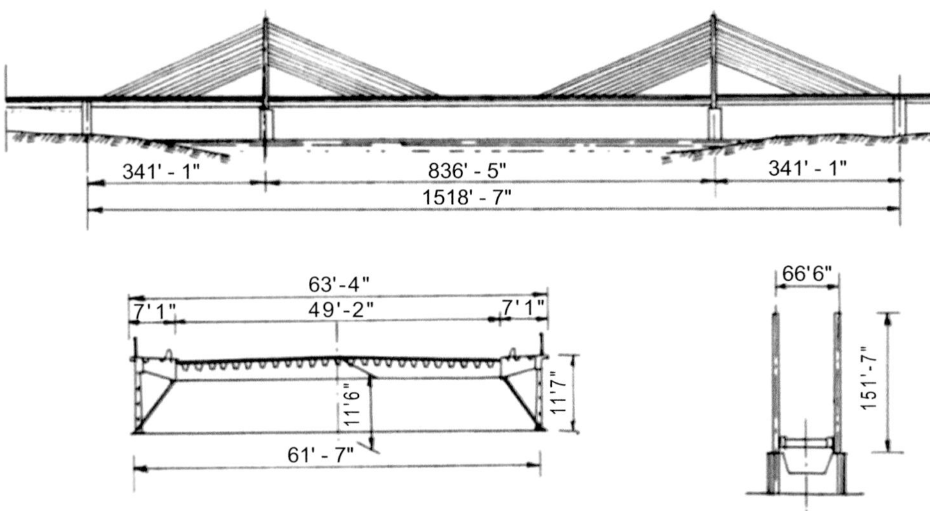

Fig. 5.13 General Arrangement of the Rees Kalkar Bridge

5.2.6 The Friedrich Ebert Bridge

The above-mentioned principle of suspending the girder at many points has been further extended in the Friedrich Ebert Bridge over the Rhine at Bonn (Figure 5.14). Friedrich Ebert Bridge has a single stay plane comprising twenty cables. The points of attachments are about 4.5 m apart along the girder, while the connections between the cable stays and the pylon are spaced at intervals of 1 m. The steel pylons are about 54 m high and are fixed to the piers. Since this bridge used a single plane of stays, the deck has been designed to have considerable torsional rigidity and consists of a 12.6 m wide box section with an 11.85 m wide cantilever on each side. The main span is 255 m long; the depth of the girder is 4.20 m, corresponding to 1/61 of the span (see Figure 5.15). It is important to note that this bridge contains a total number of 80 stays, which reflects a highly statically indeterminate structure that cannot be resolved by manual calculations. The design wouldn't be achieved without the aid of electronic computers.

Fig. 5.14 Friedrich Ebert Bridge over the Rhine at Bonn, Germany, 1967

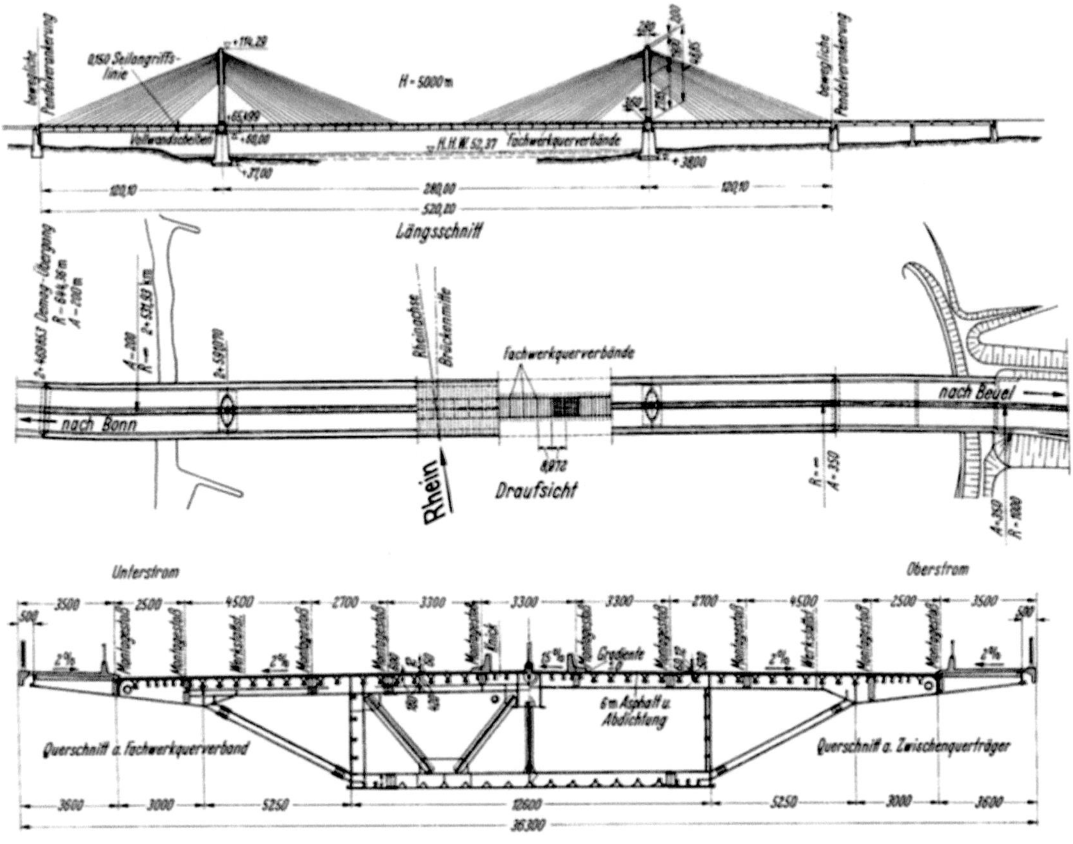

Fig. 5.15 General Arrangement of the Friedrich Ebert Bridge

5.2.7 The Knie Bridge

The Knie Bridge is the second bridge over the Rhine River at Dusseldorf. Similar to its predecessor, this bridge adopted the harp-type shape, which was mandated by the city authorities. The bridge has an asymmetrical layout (Figure 5.16) with a span of 320 m. When opened in 1969 it was considered the world's longest-spanning cable-stayed bridge. The bridge is provided with intermediate supports under each cable anchor point in the side span to obviate any lifting that might occur, which rendered a slender deck with a main girder depth of around 3.2 m, i.e., 1/100 of the span. The bridge deck consists of two plate girders connected by a floor beam with an orthotropic plate, which provides for six lanes of traffic, and a sidewalk. The pylon is 114 m high and consists of two free-standing posts that were constructed without any cross beams, which was mandated by the authorities to distinguish them from the conventional pylons for suspension bridges (see Figure 5.17).

The Knie Bridge continued to be a world record for only one year as it was superseded by the Duisburg-Neuenkamp Bridge across the Rhine River.

Fig. 5.16 The Knie Bridge, Dusseldorf, Germany, 1969

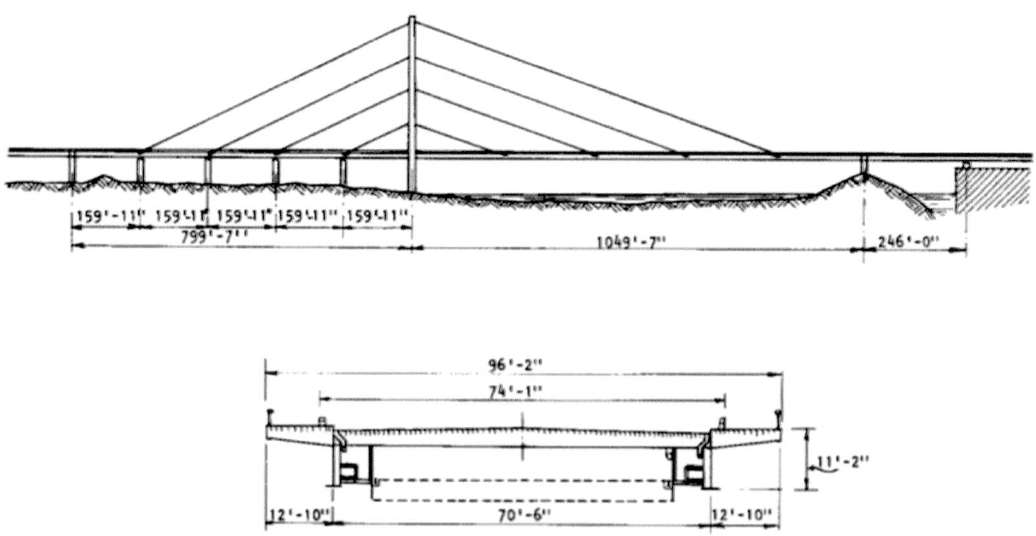

Fig. 5.17 General Arrangement of the Knie Bridge

Fig. 5.18 The Duisburg-Neuenkamp Bridge, Germany, 1970 (Courtesy, Nicolas Janberg, 2004)

5.2.8 The Duisburg-Neuenkamp Bridge

The Duisburg-Neuenkamp Bridge (Fig. 5.18) was opened to traffic on 16th October 1970 with a main span of 350 m. The main girder, as shown in Figure 5.19, is a two-cell box type, 12.70 m wide; its outer web plates have a constant depth of 3.75 m. The roadway and footway deck plates, which cantilever out from the girder, have an overall width of 36.30 m accommodating three traffic lanes for each direction.

The two steel pylons are 48 m high and are rigidly attached to the main girder and connected to the supporting piers by two moveable bearings on the east side and two fixed bearings on the west side. All the supporting cables are each composed of nine locked-coil wire ropes that are clamped at intervals of about 17 m. At each end the individual ropes are anchored to the main girder of the bridge and pass through the pylons at different heights to achieve a harp shaped arrangement with a divergent pattern. There are two expansion joints installed at the two ends of the bridge permit expansion movements of 120 mm and 350 mm at the western and eastern end respectively. The bridge is considered a vital link at Europe Highway No. 3, which runs from Stockholm via Hamburg, Hanover and crosses the Rhine at Duisburg and continues via Antwerp, Lille, Paris, and San Sebastian to Lisbon.

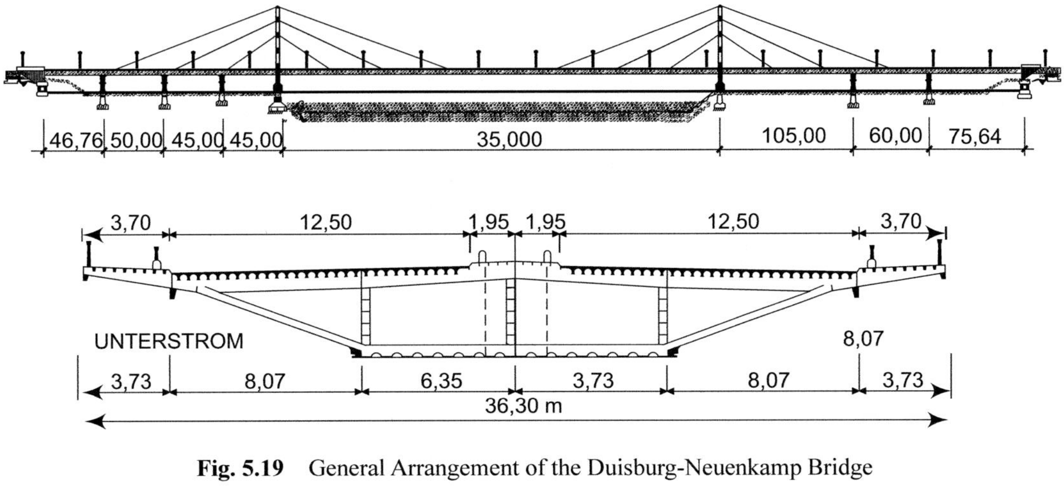

Fig. 5.19 General Arrangement of the Duisburg-Neuenkamp Bridge

5.2.9 The Kurt-Schumacher Bridge

The Kurt-Schumacher Bridge between Mannheim and Ludwigshafen across the Rhine is the first bridge to use a parallel wire strand (Figure 5.20). This bridge, which was opened in 1972, has a total length of 433.45 m. The main steel span, which is 287 m long, is balanced by two post-tensioned reinforced concrete side spans spanning 60.16 m and 65 m respectively. It is an asymmetrical stayed girder bridge with a single pylon supporting five suspension cables. The steel pylon is A-shaped with a height of 77.93 m. The three cables on the side of the river support the continuous parts while the others are used to restrain the Anchorage devices. The cables are disposed of in parallel bundles.

Fig. 5.20 The Kurt-Schumacher Bridge between Mannheim and Ludwigshafen, Germany, 1972

5.2.10 The Kohlbrand Bridge

The Kohlbrand Bridge (Figure 5.21) in the port of Hamburg has a span of 325 m. It is characterized by its diamond shaped pylons and multi-cable system with double cable plans.

The superstructure (Figure 5.22) is a trapezoidal orthotropic box girder 3.45 m high, having a top chord 17.6 m wide and a bottom chord 5.98 m wide. The orthotropic deck comprises a steel

Fig. 5.21 The Kohlbrand Bridge, Germany, 1974

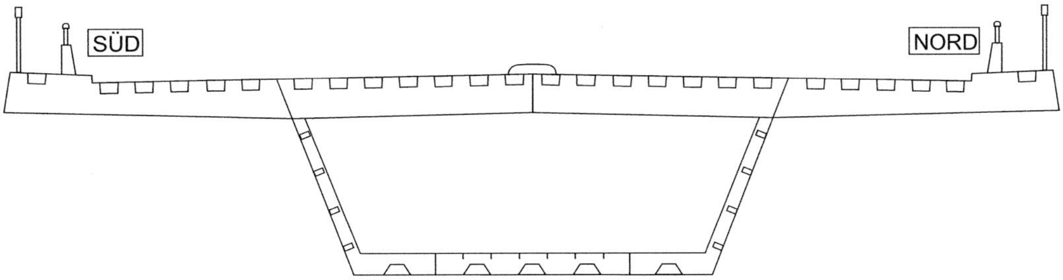

Fig. 5.22 Superstructure of the Kohlbrand Bridge

plate, having a thickness of 12 mm at the roadway and 9.5 mm at the sidewalks. The stiffening ribs are 143 mm high with a wall thickness of 5.6 mm. The floor beams are spaced at 1.95 m and 2.13 m. They are used to carry the orthotropic deck. The two pylons are A-shaped, possessing two transverse beams. The cables range from 54 mm to 110 mm in diameter.

5.2.11 The Oberkassel Bridge

The Oberkassel Bridge shown in Figure 5.23 is the third bridge to be built over the Rhine River at Dusseldorf in 1976. This bridge was designed to replace an old truss bridge. It had to be built beside the existing bridge and then moved 47 m in the direction of the river to reach its final position. To facilitate the construction, one pylon was chosen in the middle of the 520 m length. The bridge has four cables symmetrical to the pylon. Using one plane of cables made it necessary to design a box girder with high rigidity as shown in Figure 5.24. The overall width of the orthotropic box girder is 35 m and depth is 3.15 m resulting in a depth-to-span ratio of 1/82. The total length of the bridge is 590.5 m with a major span over the Rhine River of 257.75 m and five anchor spans of 51.5 m each. The back stays are attached to the piers by tie-down linkages. They are continuous through the pylon on the saddles. The pylon height is 100 m above the deck.

Fig. 5.23 The Oberkassel Bridge, Germany, 1976

5.2.12 The Neuwied Bridge

The Neuwied Bridge built in 1978 was characterized by two nearly equal spans and an A-shaped longitudinal pylon (Fig. 5.25). It spans two wide branches of the river Rhine; the island in between

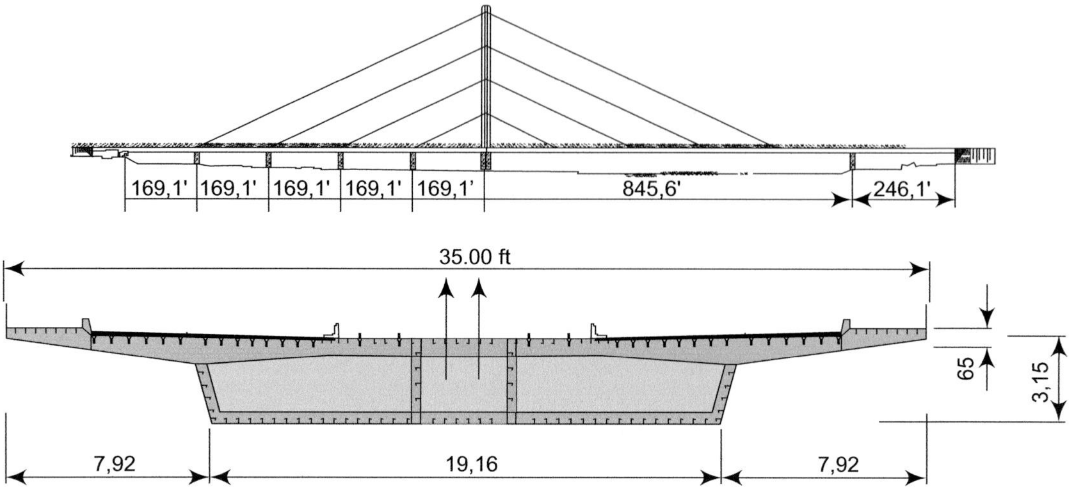

Fig. 5.24 Elevation and cross-section of the Oberkassel Bridge

Fig. 5.25 The Neuwied Bridge, Germany, 1978 (courtesy of Structurae)

was used to erect a longitudinally oriented A-shaped pylon. The bridge, which has one central plan of cables is 485.60 m long and divided into three spans: one 235.20 m long light on the main waterway directed towards Andernach; another secondary with a span of 212 m directed towards Neuwied; and another of 38.40 in between located on Weissenthurm Island (see Figure 5.26).

Each leg of the pylon stands on a pier with sufficient transverse length to support the box girder. The box girder comprises a 16.3 m wide orthotropic box section provided with 9.1 m and 10.1 m wide cantilevers on the two sides. The girder depth is 2.45 m, corresponding to 1/96 of the span. The back stays are attached to the piers by tie-down linkages as shown in Figure 5.49.

5.2.13 The Flehe Bridge

The Flehe Bridge opened in 1979 (Figure 5.27). This bridge has a 364 m steel main span and a 780 m concrete approach span. It is a single pylon cable-stayed bridge that has in each direction three vehicle lanes and a hard shoulder. This was the first German bridge to employ a reinforced concrete pylon, of an inverted Y configuration and 160 m height. The backstays are anchored in the first 240 m

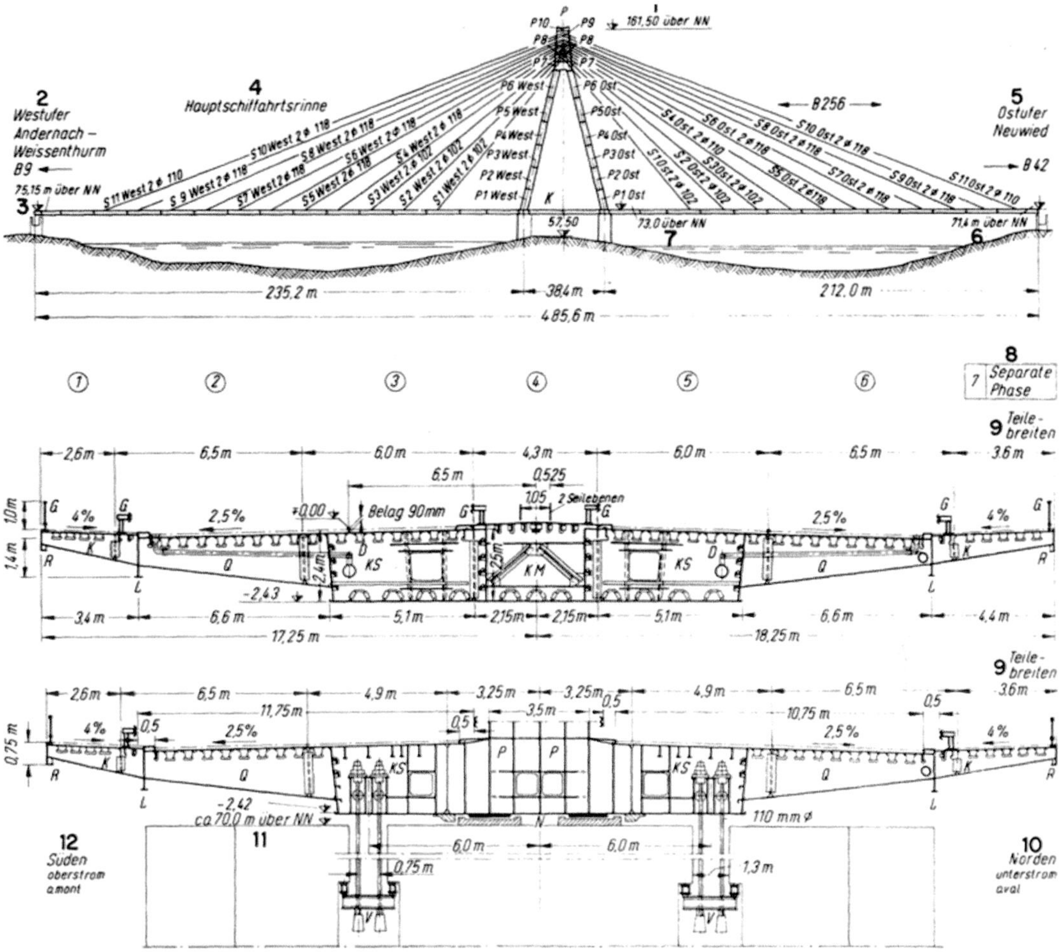

Fig. 5.26 General Arrangement of the Neuwied Bridge (Idelberger, 1979)

Fig. 5.27 The Flehe Bridge, Germany, 1979

of the approach spans, which serve as a counterweight for the main span. The main girder is a two-cell box type with a depth of 3.8 m corresponding to 1/100 of the main span (see Figure 5.28). The stays each comprise 19 locked coil ropes with a diameter of 60 to 90 mm. Some of the stay cables were exchanged under a rehabilitation plan that was completed in 2009 (Gurtmann et al. 2010).

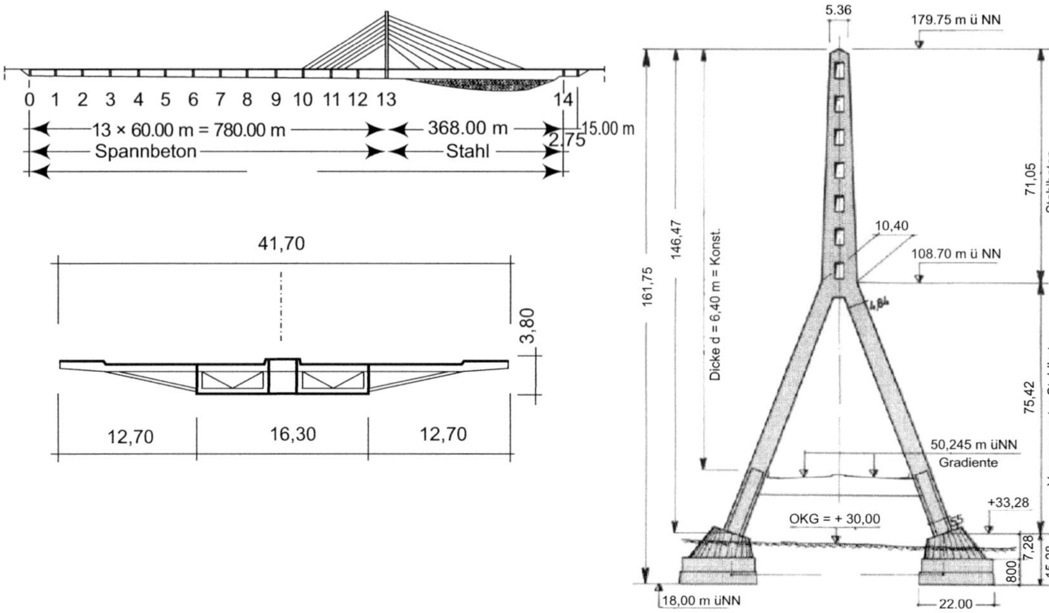

Fig. 5.28 General Arrangement of the Flehe Bridge (Gurtmann et al., 2010)

5.2.14 The Metten Danube Bridge

The Metten Danube Bridge, (Figure 5.29) was opened to traffic in 1981. This bridge is part of the A3 Motorway in Germany. This bridge was unique at its time in Germany in the sense that the individual fore and backstays were sized as prestressed concrete members as opposed to a stay cable. The 600 m long and 30 m wide bridge has 7 typical spans of 68 m, and a river span of about 145 m. The 4.2 m prestressed concrete box girder cross-section is continuous over the full bridge length. The fore and backstays supported by the single concrete pylon substituted for an additional pier in the river. The same concept was used for the design and construction of the Ludwig Erhard Bridge that was opened to traffic in 1987. This bridge used a pylon with two planes of cables. The total length of the bridge is about 295 m. The bridge was built to cross several railway intersecting lines (Figure 5.30) at that location. A pylon with a prestressed concrete backstay at the center was an optimal solution to replace two piers whose construction would interfere with the rail lines.

5.2.15 The Obere Argen Valley Bridge

An inventive concept was used in the design of the Obere Argen Valley Bridge. This six-lane bridge spans a valley that is 730 meters wide and only 45 meters deep at its deepest point above the Obere Argen River. For a distance of roughly 440 meters, the soil conditions between the eastern abutment and the western bank of the river are suitable for supporting a continuous girder on level foundations. Over a distance of 290 meters, the slope between the western bank of the

Fig. 5.29 The Metten Danube Bridge, Germany, 1981

Fig. 5.30 The Ludwig Erhard Bridge, Germany, 1987

river and the western abutment is constantly eroding at a rate of 10 to 20 centimeters per year. Thus constructing foundations is impossible. Therefore, the steel superstructure is stayed from above and from underneath, (Figures 5.31 and 5.32), to overcome the creeping slope. The superstructure is supported by stays and by struts at six distances of about 43 m each as shown in Figure 5.33. The cross-section is a 3.7 m deep single-cell box cantilevered at both sides with supporting inclined elements. The three V-shaped struts supporting the superstructure are designed as pendulum piers in the longitudinal direction and thus transfer only vertical loads. They transfer their reaction through six locked coil cables with a 126 mm diameter. The bridge was completed in 1990.

Fig. 5.31 Obere Argen Valley Bridge, Germany,1990

Fig. 5.32 Illustrating under-deck stays for Obere Argen

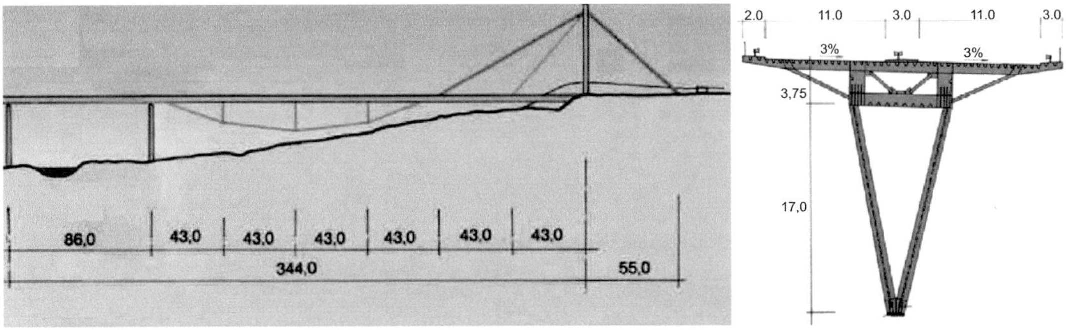

Fig. 5.33 Elevation of Obere Argen Bridge

5.2.16 The Brücke der Deutschen Einheit Bridge

The Brücke der Deutschen Einheit Bridge in Bavaria was completed in 1992 (see Figure 5.34). This bridge has two planes of cables that are supported by a two-leg single pylon, which is located at the center between the piers. The total length is 239 m, and the two spans are 74 m and 75 m respectively. The superstructure comprises a reinforced concrete deck supported by two edge reinforced concrete girders and transverse reinforced concrete floor beams that are spaced 5 m apart.

Fig. 5.34 The Brücke der Deutschen Einheit bridge, Germany, 1992

5.2.17 Rhine River Bridge at Ilverich

The location of the Rhine River Bridge at Ilverich, also namely Flughafenbrücke Bridge, obliged the designer to limit the pylon height. A normal height in the range of 110m was required. Nevertheless, only 81 m was permitted because of air traffic safety requirements. Therefore, an innovative design for this bridge subdivided each pylon into two V-shaped legs; the backstay and forestay cable forces are carried by a tie (Figure 5.35). The bridge is running from abutment to abutment with a total length of 1286.5 m and is composed of approach spans on the left bank of the Rhine spanning 7 × 63 = 441 m; the cable-stayed central span measures 287.5 m (Figure 5.36); the approach spans on the right bank of the Rhine are 8 × 63 m + 54 m = 558 m each. The superstructure is prestressed concrete for the approach spans and steel for the central span (Saul et al., 2002). The pylons are also made of steel, and the stays consist of locked coil cables.

Fig. 5.35 The Flughafenbrücke bridge, Germany, 2002

Three bearings are arranged on the pylon piers: a middle bearing under the pylon with a maximum of 114 MN surcharge, fixed on the right bank of the Rhine and transversely fixed on the left bank, as well as two multirotational bearings near the outer webs and the main girder with a maximum load of 34 MN. The superstructure, which is 38.5 m wide in total, comprises the median 4.00 m, the lanes $2 \times 13.5 = 27.00$ m, and the pedestrian and cycle paths $2 \times 3.75 = 7.50$ m. The superstructure is a two-cell orthotropic steel box type, 16.5 m wide; its outer web plates have a constant depth of 3.75 m. The orthotropic deck plate cantilever extends 11.25 m outward from the girder on both sides. The superstructure depth is approx. 4.1 m or 1/71.5 of the main span. The plate thicknesses of the orthotropic deck and boxes ranged from 14 mm to 10 mm. The box girders are reinforced with a K-bracing in the cable entry area. The diagonal struts from the cantilever to the bottom of the steel box as well as the K bracing ensure distribution of force to the inner and outer webs and flanges of the box girder.

5.2.18 The Berliner Brucke Bridge

The Berliner Brucke Bridge was built in 2006 in the middle of the Händel-city of Halle (Saale) to replace an old truss bridge. The bridge crosses partly electrified tracks of the facilities of the Deutsche Bahn passenger transport company north of the main train station as shown in Figure 5.37. The bridge is characterized by a central pylon and a composite stiffening truss. It is the first road bridge of this type in Germany. The bridge has a curved alignment, and the pylon is skewed to the bridge center line (Brixner et al., 2007).

The abutments had to be located at the end of track to permit right of passage on both sides, which resulted in two spans of 86.85 m and 84.15 m, and a bridge length of 171.0 m. The A shaped pylon is in the middle and has a height of 73.25 m above the top of the foundation. There are six pairs of cables that transmit the load of the superstructure to the pylon designed in a fan-shaped asymmetric configuration (see Figure 5.38c and d). The superstructure is made of three trapezoidal, welded hollow boxes and a reinforced composite concrete deck with boxes passing through the shear studs. This bridge received the German Bridge Building Award in the Road and Rail Bridges category in 2008.

5.2.19 The Strelasund Bridge

The Strelasund Bridge (Second Strelasund Crossing), which was opened to traffic in 2007 serves as one of the elements of the Strelasund roadway network. The crossing consists of six individual

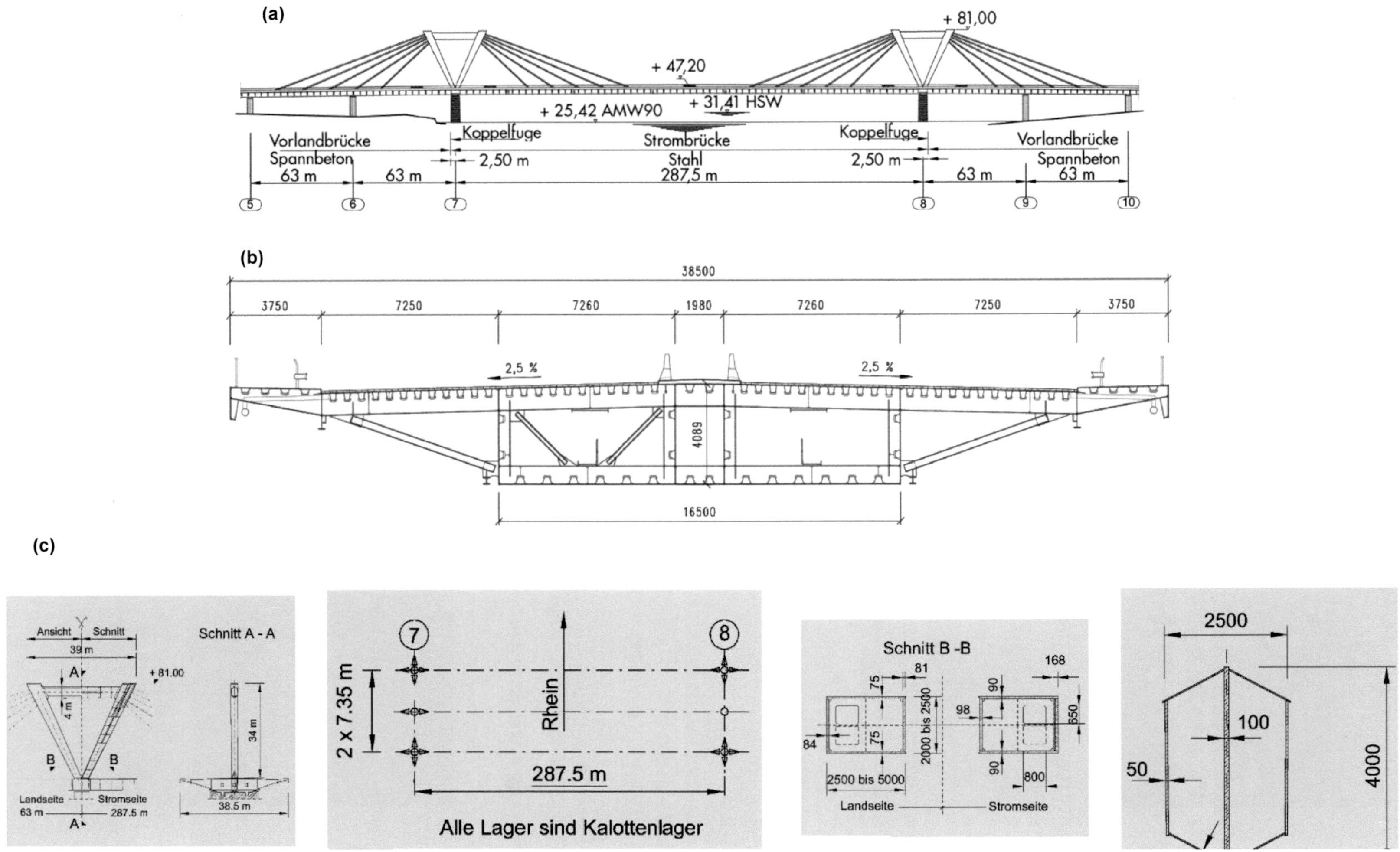

Fig. 5.36 General Arrangement of the Flughafenbrücke Bridge: (a) Elevation; (b) cross-section; and (c) pylon detail, bearings, and pylon tie beam (Saul et al., 2002)

Fig. 5.37 The Berliner Brucke Bridge, Germany, 2006 (Brixner et al., 2007)

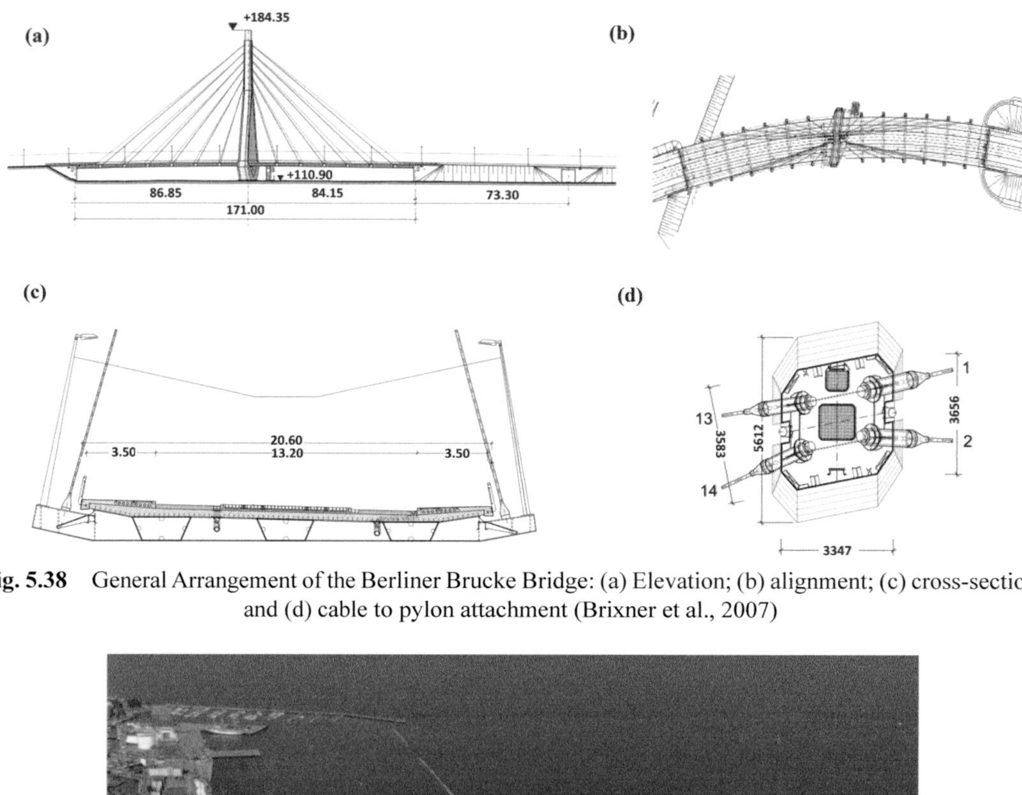

Fig. 5.38 General Arrangement of the Berliner Brucke Bridge: (a) Elevation; (b) alignment; (c) cross-section, and (d) cable to pylon attachment (Brixner et al., 2007)

Fig. 5.39 The Strelasund Bridge, Germany, 2007 (Courtesy of Structurae)

bridges (Figure 5.39), four prestressed concrete bridges, one steel composite bridge and a cable-stayed bridge with a total length of 2830 m. The crossing runs across the eastern part of the city of Stralsund, the Ziegelgraben waterway and the Dänholm and Strelasund island.

The cable-stayed bridge has a main span of 198 m and a side span of 126 m as shown in Figure 5.40a. The main girder is a 3.15 m deep three-cell steel box with an orthotropic deck (Figure 5.40b)

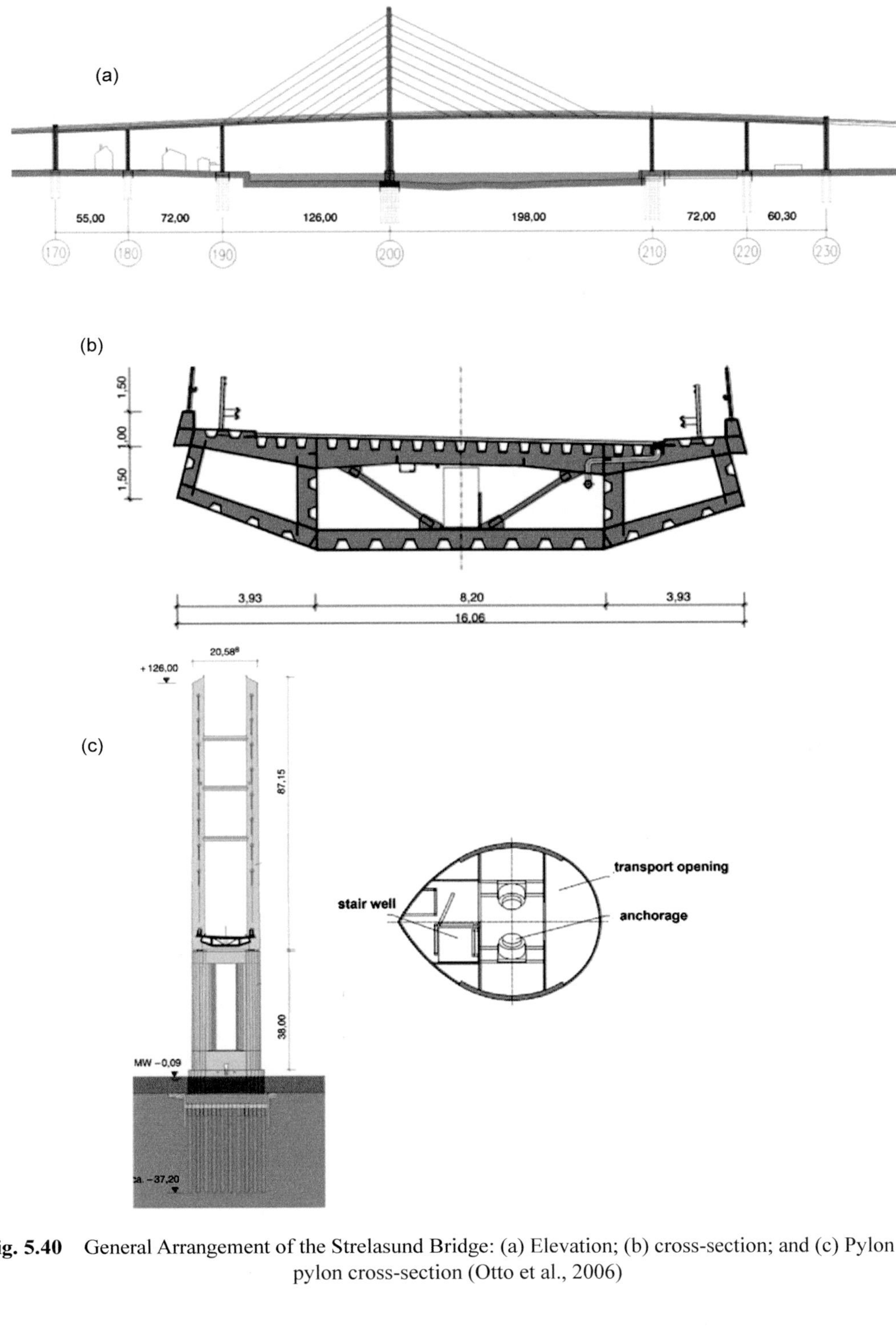

Fig. 5.40 General Arrangement of the Strelasund Bridge: (a) Elevation; (b) cross-section; and (c) Pylon and pylon cross-section (Otto et al., 2006)

covered by an 8 cm asphaltic wearing surface. Cross frames were placed at 4.0 to 4.4 m from the main girder to provide a means for stiffening the bridge cross-section. Each alternate cross frame is provided with diagonals made of round tube sections. The steel pylon is 128 high and comprises two parallel vertical legs connected by three struts to ensure stiffening and stability converting it to a robust frame structure. The superstructure girder is fixed to the pylon legs, and both are supported on hinged pot bearings. It rests on a reinforced concrete pier. The stay cables have a harp arrangement made of parallel-strand cables for the first time in Germany (Otto et al., 2006).

5.2.20 The bridge at Niederwartha, Dresden

The bridge at Niederwartha, Dresden (Figure 5.41) was completed in 2008. This bridge which crosses the river Elbe at Niederwartha is the first cable-stayed bridge in Saxonia. The total length of the crossing is 366 m including a 274.5 m cable-stayed bridge. The main span of the bridge is 192 m and the back span is 82.5 m as shown in Figure 5.42.

Fig. 5.41 The Bridge on River Elbe at Niederwartha, Germany, 2008 (Courtesy, Nicolas Janberg)

The height of the A-shaped pylon is 77 m above the terrain. The upper part includes a steel chassis for cable anchorage. The pylon legs are made of hollow sections 3.7 m wide and wall thickness of 50 cm. The transition of the pylon-to-pylon head where the steel box starts is at 50 m over terrain. The cables are made of locked coil spiral strands. Anchorage at the pylon head is achieved through hammer cable caps as illustrated in Figure 5.42. The damping elements at the pylon and at connection to the superstructure are designed to be moveable to facilitate the magneto-

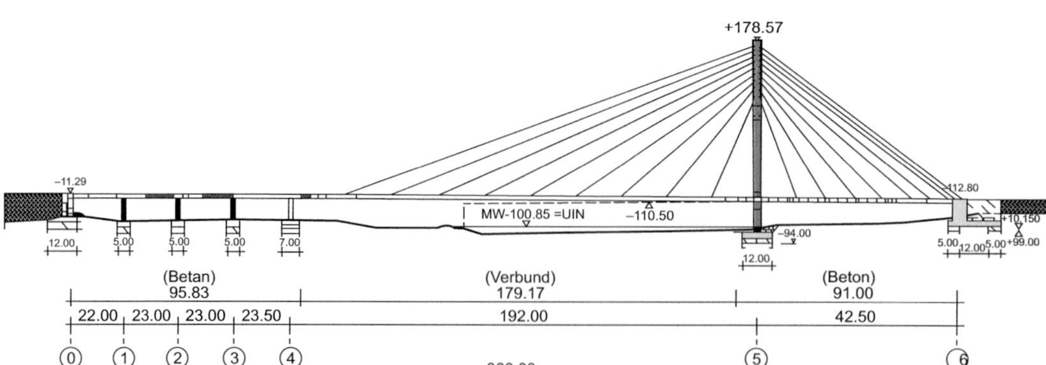

Fig. 5.42 Elevation of the Bridge on River Elbe at Niederwartha (Eilzer et al., 2006)

inductive inspection of the cables. The overall width of the superstructure cross-section is 12.50 m between the railings, the construction height is 2.00 m. The composite superstructure cross-section comprises 1.70 m high steel girders and a 30 cm thick composite deck (Figure 5.43). The 1.4 m deep floor beams are spaced at 12.0 m. At the cable anchorage locations, the floor beam converts to a truss-type cross girder as illustrated in Figure 5.44 (Eilzer et al., 2006)

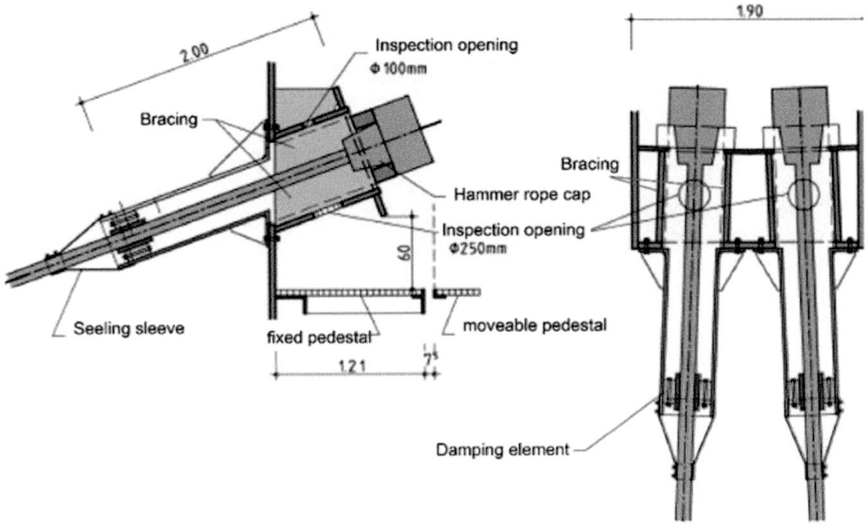

Fig. 5.43 Cable anchorage detail for on River Elbe at Niederwartha (Eilzer et al., 2006)

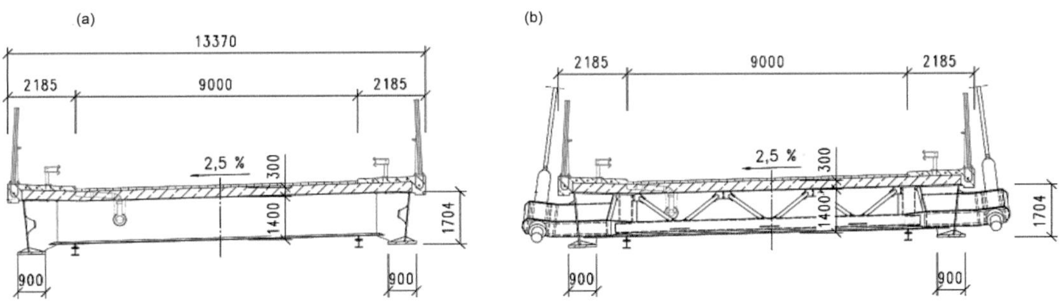

Fig. 5.44 Cross-section for on River Elbe at Niederwartha: (a) Standard; and (b) truss-type at cable anchorage (Wagner, 2013)

5.2.21 The Bridge at Wesel

The completion of the Rheinbrücke Wesel bridge in Wesel in 2009 replaced an earlier design that had become overcrowded due to the volume of traffic passing through on a daily basis. To avoid blocking the navigation channel with two piers, a single-pylon cable-stayed bridge was chosen (Figure 5.45). The approach spans had to be constructed out of relatively heavy prestressed concrete in order to provide the necessary counterweight for anchoring the long backward cables, while the main span was constructed out of steel with an orthotropic bridge deck in order to reduce its weight. With the main suspended span measuring 376 meters, the crossing has a total span of 773 meters (refer to Figure 5.46).

Fig. 5.45 The Wesel Bridge, Germany, 2009 (Lockmann and Marzahn, 2013)

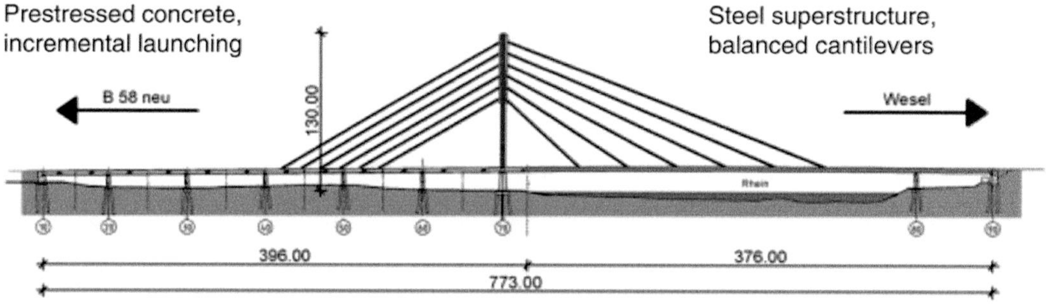

Fig. 5.46 Elevation of the Wesel Bridge (Lockmann and Marzahn, 2013)

The cross section used an orthotropic bridge deck on top of a 13.8 wide three-cell box-type girder. On both sides of the box girder, the bridge deck has 7.71 m long cantilevers 7.71; it is supported by 0.34 meters by 0.40 meters struts spaced four meters apart equally. The middle cell, which is 2 m wide, is intended to anchor the cable forces, and the cable plane is positioned in the center of the central reserve between the two traffic directions (Figure 5.47).

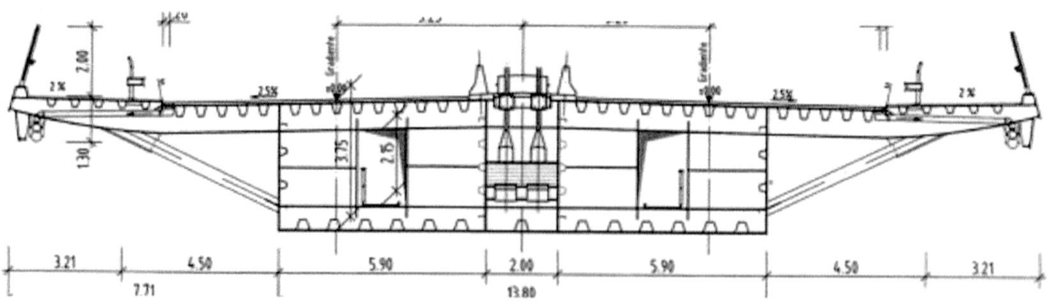

Fig. 5.47 Cross-section of the Wesel Bridge (Lockmann and Marzahn, 2013)

Due to the small median, the pylon had to be placed off the cross section; as a result, an inverted Y shape was selected. To make construction at the pylon's legs easier, the link between the concrete and the steel section of the deck was designed as a stiff, uniformly prestressed structure that was moved 12 meters off the pylon section toward the main span. This bridge is the second in Germany to use parallel strand cables. A total of six sets of cables were selected to support the bridge. Each of the six stays is made up of three cables arranged in two planes. The groups were arranged in the shape of a fan across the main span, while the remaining material forms a harp on the back side of the pylon. (See Figure 5.48)

Fig. 5.48 Wesel Bridge Cable configuration

5.2.22 The Neckar Bridge at Zwingenberg

The Neckar Bridge at Zwingenberg was opened to traffic in 2011. The bridge replaced the ferry system transferring passengers and tourists to its left. This bridge is considered a landmark of the town of Zwingenberg (Figure 5.49). The total bridge length of 216.5 m is divided into single spans of 37.0 m – 121.5 m – 56.5 m as shown in Figure 5.50.

Fig. 5.49 The Neckar Bridge at Zwingenberg, Germany, 2011 (Courtesy, Municipal Administration of Zwingenberg)

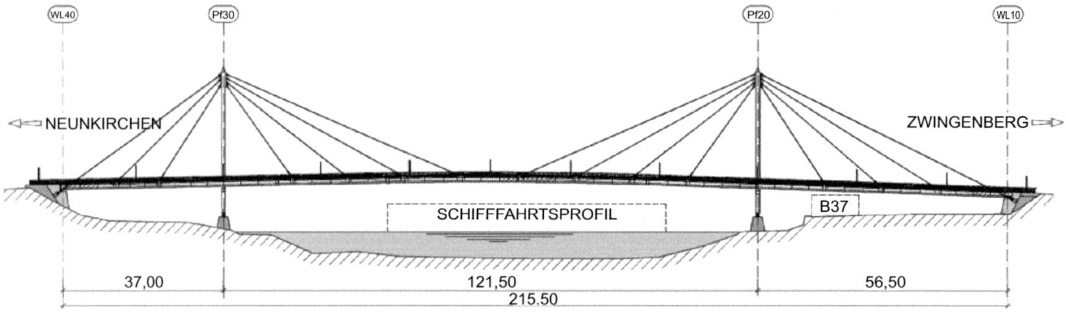

Fig. 5.50 Elevation of the Neckar Bridge at Zwingenberg (Stelzer and Dorrer, 2011)

The superstructure consists of two external welded hollow boxes and a floor beam that has its upper flange integrated with an orthotropic deck The hollow boxes are attached through cross beams at regular distances of 4.25 m. The orthotropic deck is integrated with the upper flange of the floor beam. The welded hollow boxes have dimensions of 1.24 m × 1.16 m (downstream) and 2.26 × 1.19 m (upstream). The bottom plates of the box girders are curved with a radius $R = 1.93$ m (downstream) or $R = 6.20$ m (upstream). The web plates are inclined at 7.3° (see Figure 5.51). The bridge has two A-shaped pylons that are made of 0.914 m tubular sections with wall thicknesses, 30 mm and 36 mm east and west respectively. The pylon legs are attached to the upper surface of the reinforced concrete piers via articulated bolted connections. The height of the pylon is 33.95 m from the pier's upper surface. There are 30 stay cables transferring the superstructure load to the pylon. They are made of locked coil strands with diameters of 70 mm and 95 mm respectively. The stay lengths are between 23.50 m and 60.40 m. The cables are attached via gusset plates fixed to the superstructure and the pylon.

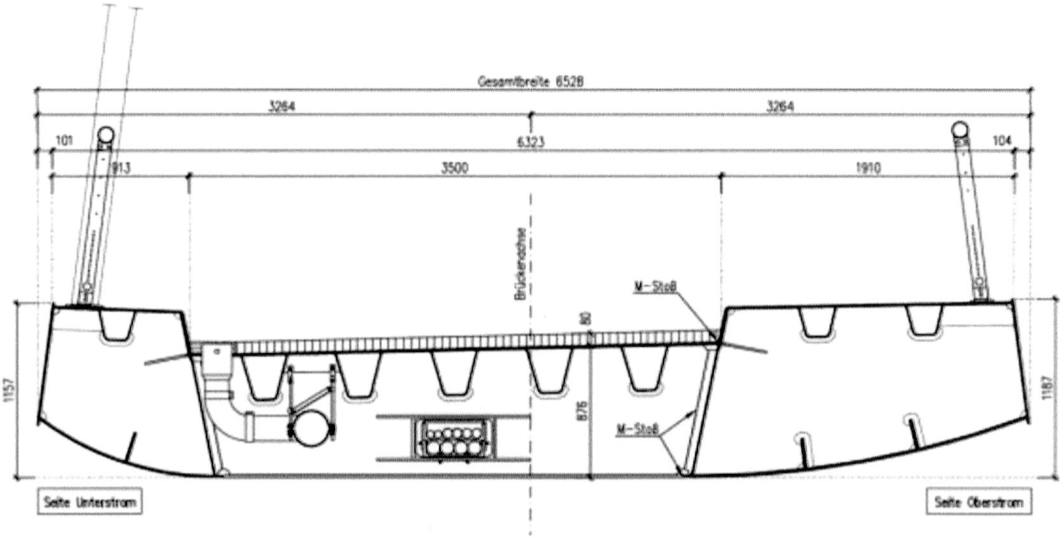

Fig. 5.51 Cross-section of the Neckar Bridge at Zwingenberg (Stelzer and Dorrer, 2011)

5.2.23 The Schönebeck Elbe Bridge

The Schönebeck Elbe Bridge (Figure 5.52) is part of a 1128.5 m long bridge consisting of a 330.5 m long southern foreshore bridge, a 305 m long river crossing and the 439 m long northern foreshore bridge.

Fig. 5.52 The Schönebeck Elbe Bridge, Germany, 2013 (Courtesy, Leonhardt, Andra and Partners)

The river crossing is a single-sided cable-stayed bridge with a single pylon, a span of 185 m and three side spans with a span of 40 m each. The stay cables are arranged in two cable levels fan-shaped configuration. The optimal cable spacing was set at 18.5 m. The main span's superstructure is designed as a welded, single cell characterized by high torsional stiffness and the resulting lower susceptibility to vibration. The superstructure at the side span is a composite section consisting of two longitudinal steel beams attached transversely with a floor beam (Figure 5.53). The superstructure girder is fixed to the pylon transverse beam, on hinged pot bearings. The lower part of the A-pylon, up to the point of combination of its shafts, is a reinforced concrete construction. The upper portion of the pylon is a steel inner box girder and double-sided concrete flanks forming a composite cross-section.

The total height of the pylon is 73 m above ground, whereby the maximum spread of the shafts above the foundation is 23 m. The transverse beam supporting the superstructure is located at a height of 8 m above the ground. Above the transverse beam, the pylon has a hollow cross-section of 3.5 m in the longitudinal direction and 2.5 m in the transverse direction and a wall thickness of 50 cm. The section, however, is solid below the transverse beam (Eilzer et al., 2010).

5.2.24 The Rudolf-Ihm Bridge at Raunheim

The Rudolf-Ihm Bridge at Raunheim (Figure 5.54) was completed in 2015. The total length of the bridge is 130 m. The superstructure is a reinforced concrete box girder. The bridge has a single A-shaped reinforced concrete pylon, which is located so that the main span and side span are 95 m and 35 m respectively. There are 32 cable stays supporting the superstructure to the pylon. The cables are locked coil ropes utilizing dynamically loadable fittings that were used for a cable-stayed bridge for the first time, characterized by their high fatigue resistance to cyclic loads.

5.3 DEVELOPMENT OF CABLE-STAYED BRIDGES IN EUROPE

This section reviews the evolution of cable-stayed bridges in European countries other than Germany. Bridges are listed chronologically and only those that are distinguished by an exceptional feature are outlined. It was mentioned earlier in section 5.1 that the first modern cable-stayed bridge was opened to traffic in Sweden in 1956, followed by many bridges in Germany and worldwide.

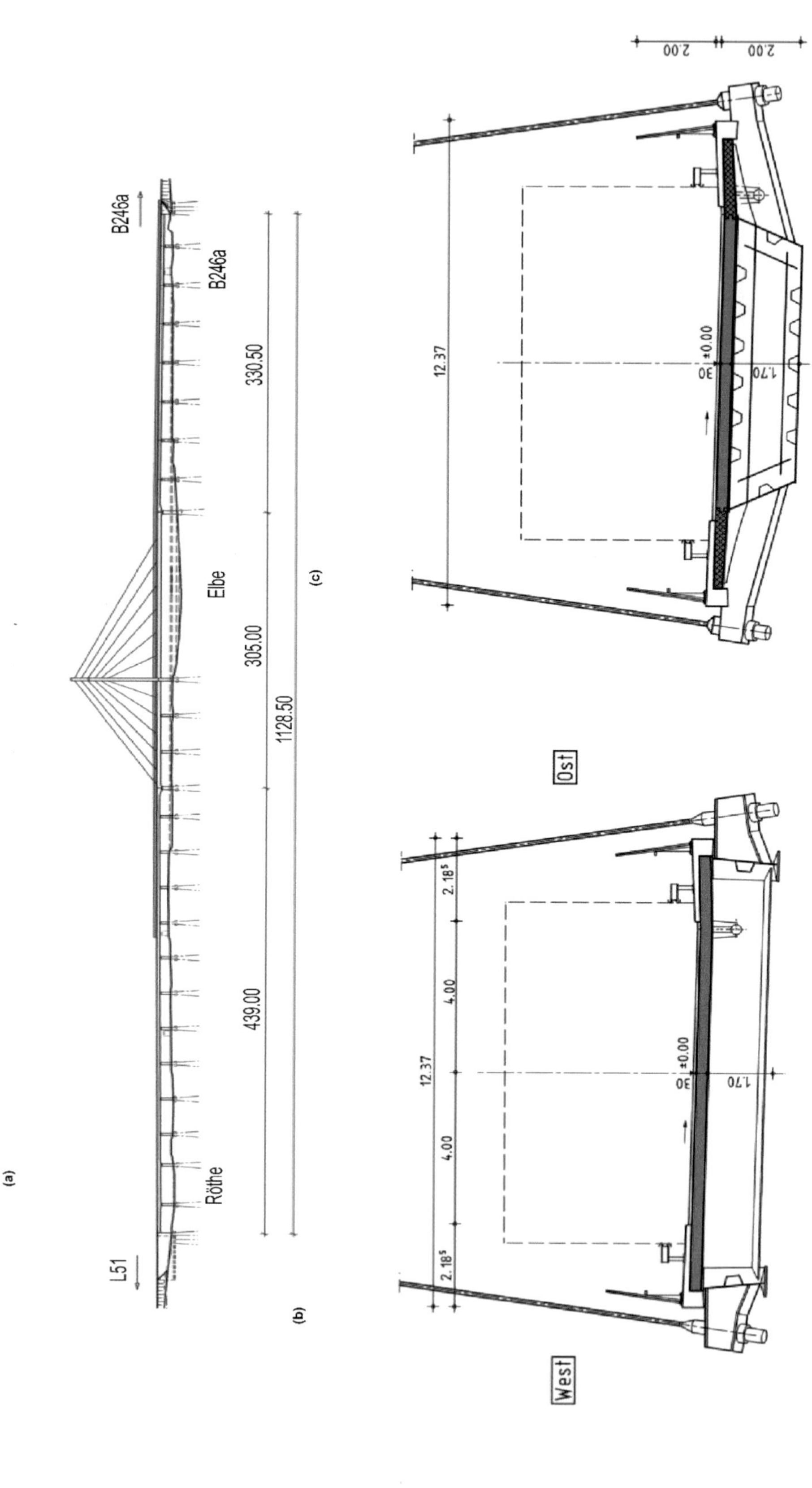

Fig. 5.53 General Arrangement of the Schönebeck Elbe Bridge: (a) Elevation; (b) cross-section at the side span; and (c) cross-section at the main span (Eilzer et al., 2010)

Fig. 5.54 The Rudolf-Ihm Bridge at Raunheim, Germany, 2015 (Courtesy, PFEIFER)

5.3.1 The George Street Bridge

The George Street Bridge at Newport, Wales, opened to traffic in April 1964 over the Usk River, Figure 5.55, is considered Britain's first example of a cable-stayed cantilever bridge construction. The bridge consists of a steel main suspended span of 152.4 m over the river and concrete approach viaducts on both sides with lengths of 189 m and 196 m on the east and west sides respectively.

Fig. 5.55 The George Street Bridge, Wales, UK, 1964

The principal tension members are the steel wire ropes through the concrete support pylons. Their inner ends are attached to the suspended cantilever span and their outer ends to the anchorage spans, which are the first three spans from the riverbank on each side. The deck is 20.1 m wide, which includes a 14.6 m roadway with two lanes in each direction and two 2.7 m footways for pedestrians and cyclists. The reinforced concrete pylons rise 56.4 m above sea level. They are rectangular in section, tapering from 4.1 m × 3 m at the base to 3 m × 2.1 m at the top, with wall thickness varying from 457 mm to 305 mm. Each of the pylons legs has three openings, at 11.7 m centers vertically, through which passes a pair of 76 mm diameter locked coil steel cables. The deck comprises steel edge boxes 11.68 m long by 51.32 m wide by 1.17 m deep with a transverse box in between. The top and bottom flange plates of the deck are stiffened by welded closed type ribs running longitudinally (Fig. 5.56).

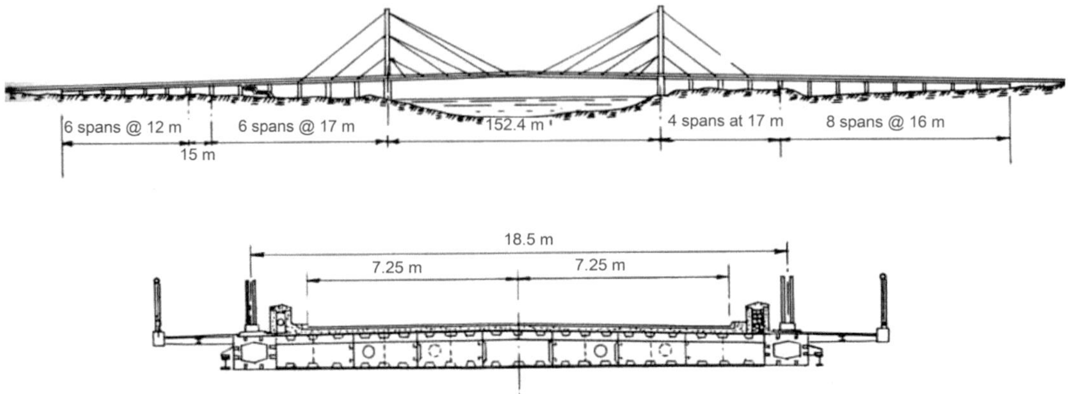

Fig. 5.56 General arrangement of the George Street Bridge

5.3.2 The Saint Florent Bridge

A year after the opening of the George Street Bridge in UK, another cable-stayed bridge was completed in France. The Saint Florent Bridge, built in 1965, crosses the Loire River between Saint-Florent and Batailleuse Island (Figure. 5.57). This bridge has two equal spans, each of 104 m and a single pylon supporting the stay cables. The pylon is a 28 m portal reinforced concrete frame. The composite deck is 9.1 m wide and consists of two longitudinal plate girders that are interconnected by cross-girders and supports a reinforced concrete slab. The bridge has two planes of stays running in a radial pattern on each side.

Fig. 5.57 The Saint Florent Bridge, France, 1965

5.3.3 The Wye River Bridge

A second bridge was built in Wales, UK in 1966. The Wye River Bridge, Figure 5.58, is a major cable-stayed bridge on the Wye-Severn estuary. It consists of a main span of 235 m and two side spans of 87 m each. The bridge has a single plane of stays connecting the deck to the pylons, which are made of a steel box-section. The rigid deck has a steel box girder (Figure 5.59) that consists of stiffened plates with transverse diaphragms spaced at 4.27 m intervals. The deck carries a 4m median; four lanes of traffic with a total width of 14.62 m; 3.66 m bike track; and 3.66 m footpath. The cable stays pass into the box girder through the median and is anchored into the diaphragms of the box girder at one side and run over a saddle on top of the pylons at the other end. The cable comprises 20 galvanized wire spiral strands each approximately 63.5 mm in diameter.

Fig. 5.58 Wye River Bridge, Wales, UK, 1965

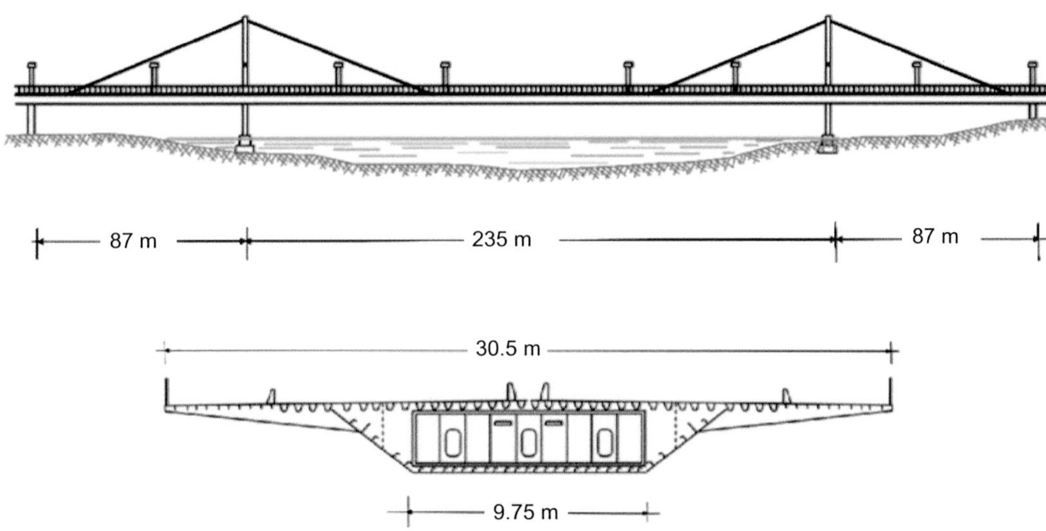

Fig. 5.59 General arrangement of the Wye River Bridge

Two bridges in Italy were designed by Riccardo Morandi who used concrete stays. Concrete stays of cable-stayed bridges are, in essence, post-tensioned tendons cast inside a beam. This concept was applied to the Ansa della Magliana Bridge that was completed in Rome with a main span of 145 m in 1967 and the Polcevera Viaduct (Figure 5.60) that opened in Genoa in 1968 with a main span of 207.9 m. The same concept was also used for the Metten Danube Bridge in Germany in 1981 (See section 5.2.14).

Fig. 5.60 The Polcevera Bridge Italy, 1967 (Courtesy, Nicolas Janberg)

5.3.4 The Massena Bridge

Another significant French cable-stayed bridge was completed in 1971. The Massena Bridge crossing the railway lines coming from the Austerlitz Station in Paris, (Figure. 5.61a), is a one plane cable-stayed bridge that has a total length of 493 m and consists of three spans of lengths 81.1 m, 161.6 m and 80.8 m. Four stays are located on the longitudinal centerline of the bridge to carry the deck load and transfer it to the pylons, which extend to a height of 32.9 m above the deck level. The stiffening girder consists of two steel longitudinal box girders that are attached transversely by solid-web cross girders (See Figure 5.61b).

5.3.5 The Erskine Bridge

The Erskine Bridge, Figure 5.62, is a major multi-span cable-stayed box girder bridge that was completed in 1971. It is spanning the river Clyde in west central Scotland downstream from Glasgow. It is the only bridge in Scotland with single cables over central main supports.

It has a 305 m main span and two 110 m approach spans. The stiffening girder is a single steel box girder with a trapezoidal cross section made from 15 sections of high-yield steel welded continuous box girders. It carries a road deck of 31.25 m, which consists of a two-lane dual roadway, cycle tracks and footpaths on each side. The single cable passing over the saddle of each pylon consists of 4,272 galvanized steel wires of 5mm diameter, arranged in 24 strands. The cables are anchored in the median area between the carriageways. The pylons are 38.1m high and rest on slender reinforced concrete piers. The piers flex with the movement of the deck caused by temperature changes. The design is notable for its economy of materials (Figure 5.63).

5.3.6 The Bratislava Bridge

August 1972 saw the opening of the Danube cable-stayed bridge in Bratislava, Czechoslovakia, a steel cable-stayed bridge with a single cable plane and a single steel pylon angled backward. One distinctive feature of this bridge is the asymmetrical placement of the inclined 84.6 m high A-shaped pylon, which supports a 32 m diameter circular restaurant (Figure 5.64). The bridge comprises four traffic lanes and two pedestrian walkways of 3 m on both sides

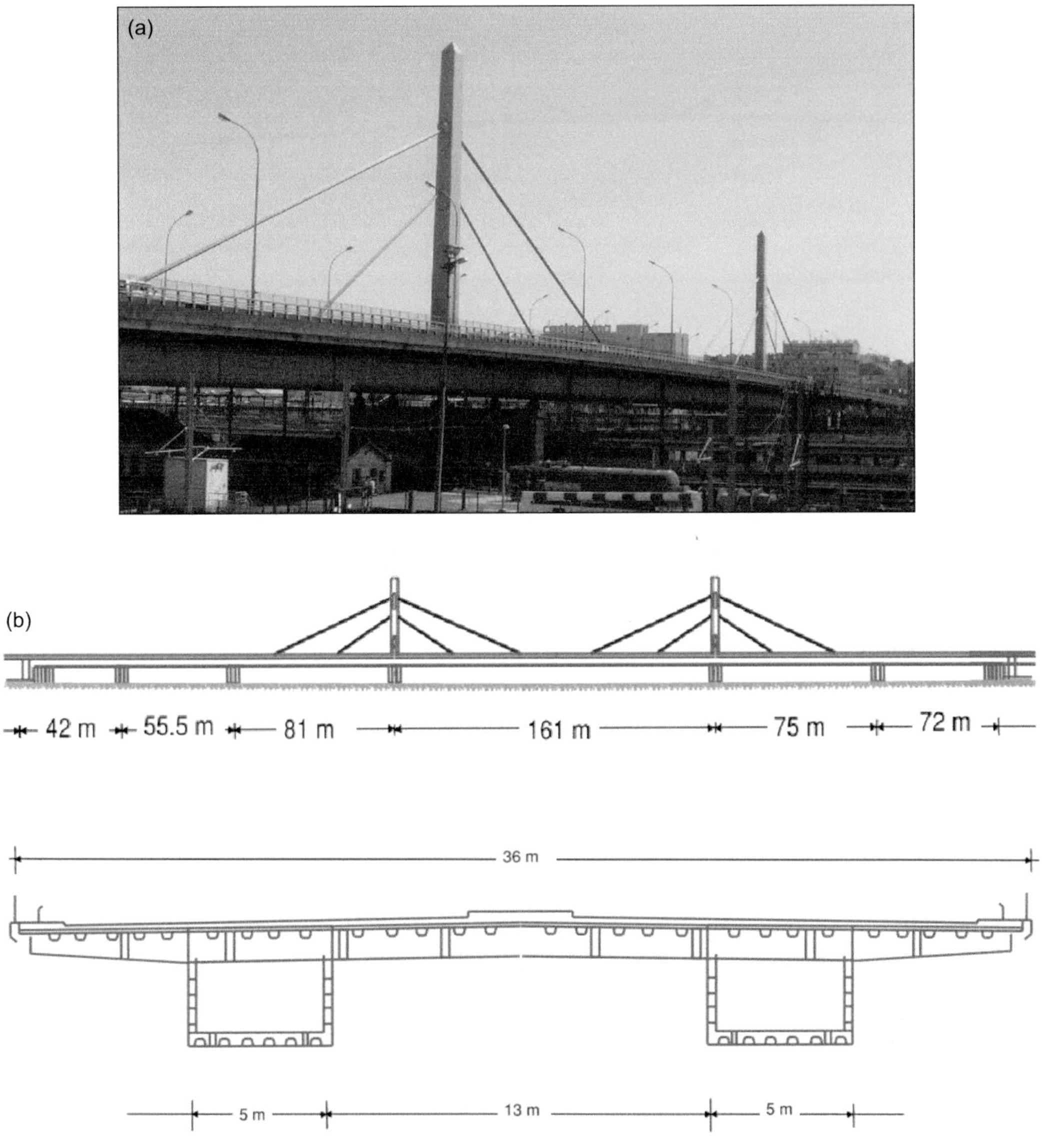

Fig. 5.61 The Massena Bridge France, 1971: (a) General view; and (b) General arrangement

Fig. 5.62 The Erskine Bridge, Scotland UK, 1971

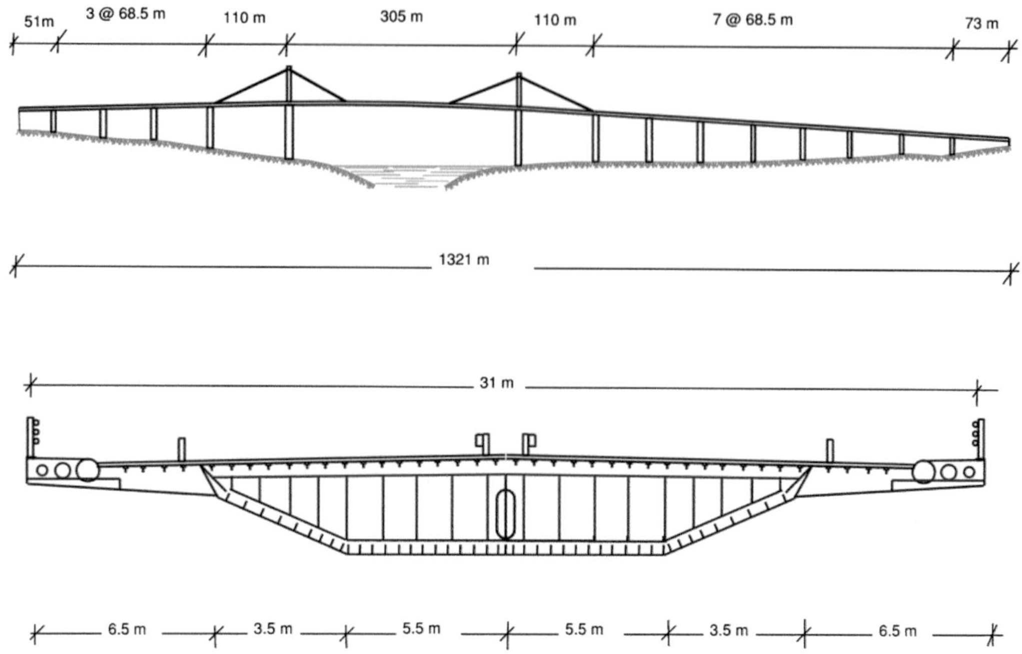

Fig. 5.63 General arrangement of the Erskine Bridge

Fig. 5.64 The Bratislava Bridge, Slovak, 1972 (Courtesy, Nicolas Janberg)

The bridge has three spans of 74.8 m + 303 m + 54 m with a total length of 431.8 m. Three cables support the 4.6 m high steel orthotropic box girder, dividing the main span into four segments of 51.5 m + 70.2 m + 82.6 m + 98.7 m = 303 m. The bridge's single back stay, which is fixed to the abutment, accounts for the major span's unusual length. It was recommended that the navigation opening measured 10 m × 180 m.

A single plane fan cable system supports an orthotropic two-cell steel box girder, which makes up the superstructure of the bridge. The width of the steel box girder is 21 meters (Figure 5.65). The thickness of the orthotropic deck plate is primarily 12 mm, but it was increased to 16 mm, 20 mm, 25 mm, and 30 mm at locations with cable anchorages. The vertical web plates have a thickness of 12 mm, while the bottom flange plate varies in thickness from 12 to 22 mm. The cross frames are spaced at 3 m. The bridge deck is strengthened by longitudinal closed trapezoidal ribs spaced 600 mm apart from 6 mm and 8 mm thick plates. The three vertical web plates, spaced 6.3 m apart and 4.6 m high, as well as the box girder's bottom flange, are stiffened by L-profiles. The tapering rectangular box-sections on the legs of the pylon, one housing a lift and the other a staircase, are rectangular in shape. The cables are made up of locked-coil wire ropes with a 70.3 mm diameter and an outer coating of galvanized wires.

5.3.7 The Voest Bridge

The Voest Bridge (Figure 5.66) is an unsymmetrical cable-stayed bridge that was opened to traffic with a main span of 215 m in December 1972 at Linz, Austria. The steel superstructure has four main girders spaced at 8.40 m and an orthotropic deck (Figure 5.67). The steel single pylon is about 70 m high and has a rectangular box section of 3.50 by 2.60 m, and the cable stays are in the plane of the bridge centerline configured as a harp. The cables consist of 22 locked-coil ropes of 69 mm diameter with a galvanized outermost layer of Z-section wires.

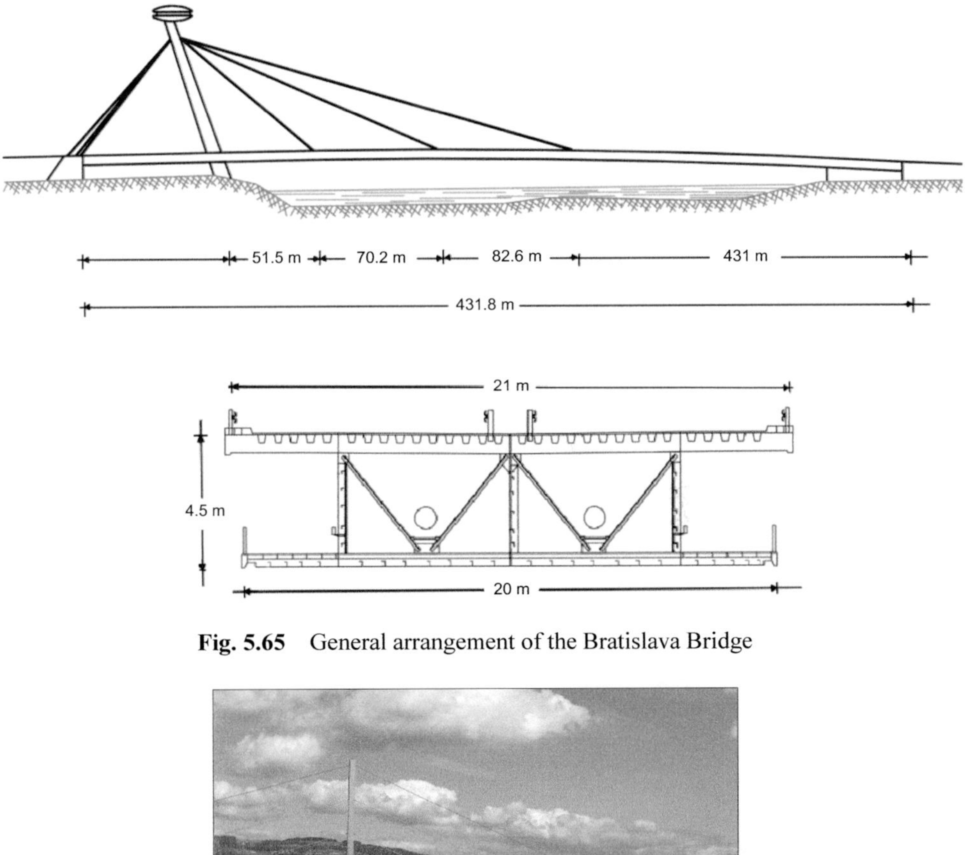

Fig. 5.65 General arrangement of the Bratislava Bridge

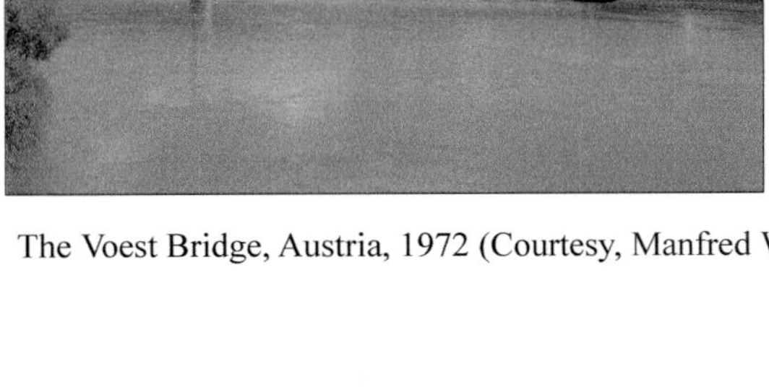

Fig. 5.66 The Voest Bridge, Austria, 1972 (Courtesy, Manfred Weghuber)

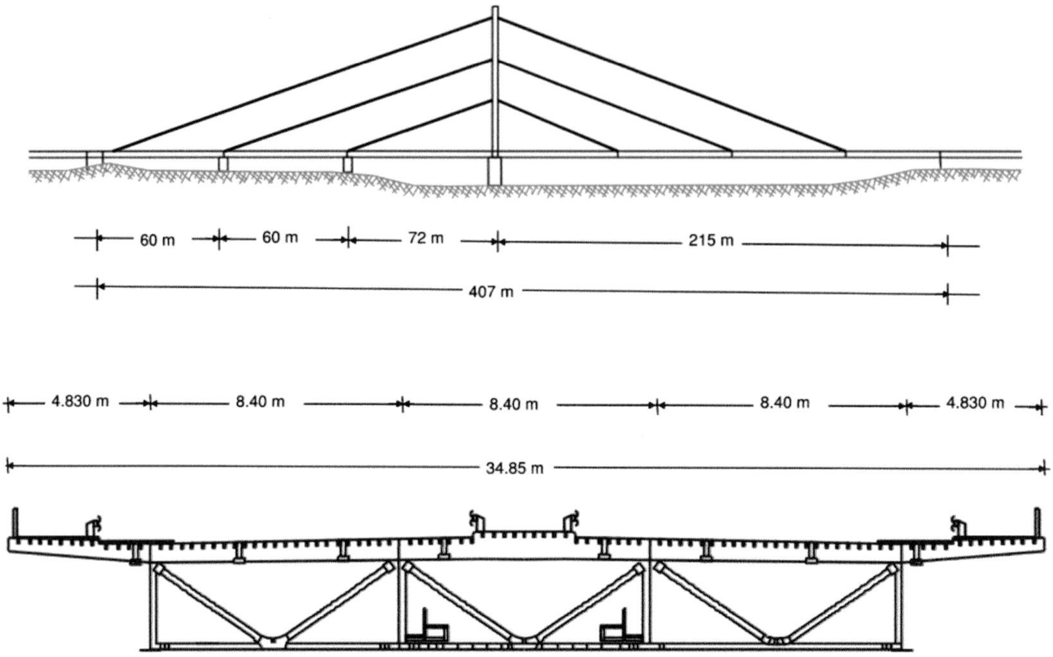

Fig. 5.67 General arrangement of the Voest Bridge

5.3.8 The Heer-Agimont Bridge

The Heer-Agimont Bridge over the Meuse, Figure 5.68, is the first cable-stayed bridge in Belgium. The bridge, which was completed in 1973, came to replace an old Bailey-type bridge that was constructed in 1948 (Warolus, 1975). The design of this structure reflected a deck of composite steel-concrete construction, steel pylons and steel cased cables, arranged in two parallel systems. The

Fig. 5.68 The Heer-Agimont, Bridge,Belgium, 1973 (Courtesy, Roland Nizet)

bridge has a 124 m main span and two side spans of 39 m. The superstructure consists of two main girders that are solid-webbed, fitted with vertical and horizontal stiffeners, 1.70 m high throughout their length and spaced at 11.90 m. The two main girders are interconnected transversely by secondary trusses (floor beams) that have their upper chords aligned with the upper flanges of the main girders and both are reinforced with shear studs to key them to the concrete deck, which has a uniform thickness of 190 mm (see Figure 5.69). The two-leg steel pylons are 40.70 m high and are interconnected at the deck level by a transverse floor beam. Two planes of cable stays arranged in harp configurations were attached to the opposite faces of the webs of the main girders. The pylons are a work of art, however, attached by means of coupling Ferrules to ease any potential future replacement.

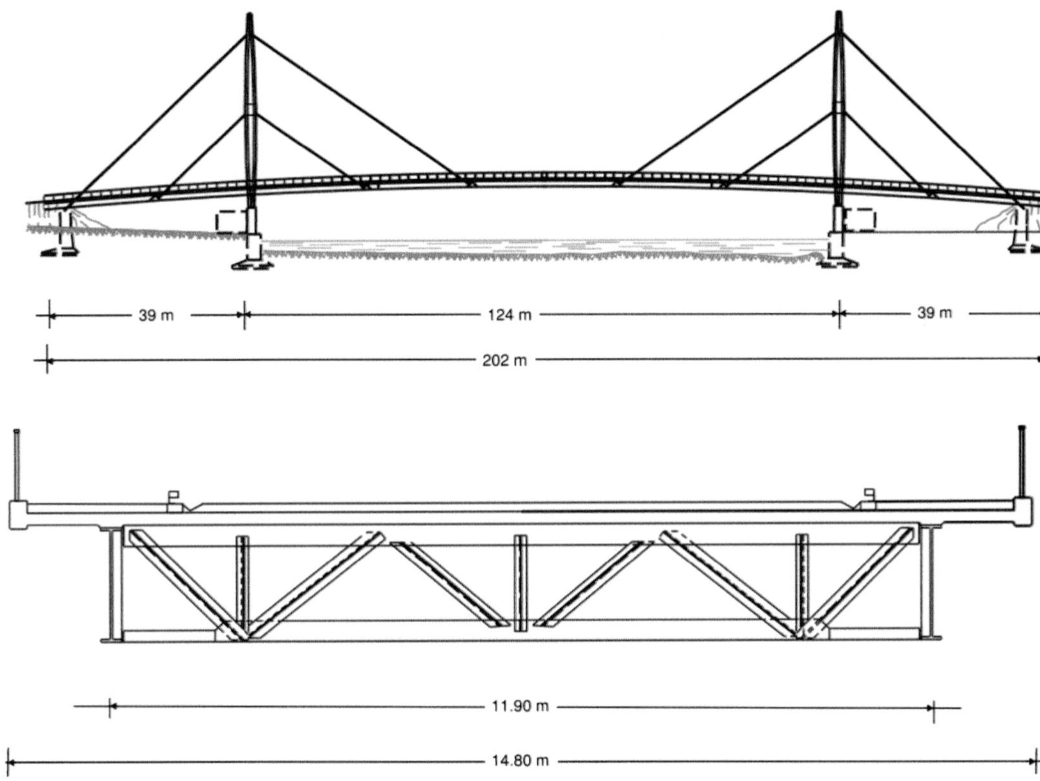

Fig. 5.69 General arrangement of the Heer-Agimont Bridge

5.3.9 The Danube Canal Bridge

The Danube Canal Bridge in Vienna, Austria, Figure 5.70 was completed in 1975. This bridge crosses the Danube Canal at an angle of 45°. The requirements by the authorities were to maintain a navigation channel of at least 25 m during construction. Due to the skewed arrangement and the navigational clearance, it was decided to construct the bridge initially parallel to the Danube Canal followed by rotation to the final position. The bridge has three bays with spans of 55.7 m, 119 m and 55.7 m. It has two planes of eight parallel cables anchored at the deck at 42 m from the pylons. The superstructure consists of a cast in situ prestressed concrete box section with a depth of 2.80 m and a width of 5 m, which cantilevers at the road level 5.4 m on both sides. Precast struts were installed as a means of stiffening the cantilever. The stays consist of eight locked-coil cables of a nominal diameter of 72 mm. The U-shaped pylons, which built into the deck, bear on hexagonal piers, each founded on pile foundations (see Figure 5.71).

Fig. 5.70 The Danube Canal Bridge, Vienna, Austria,1975 (Courtesy, Walter H. Mickerts)

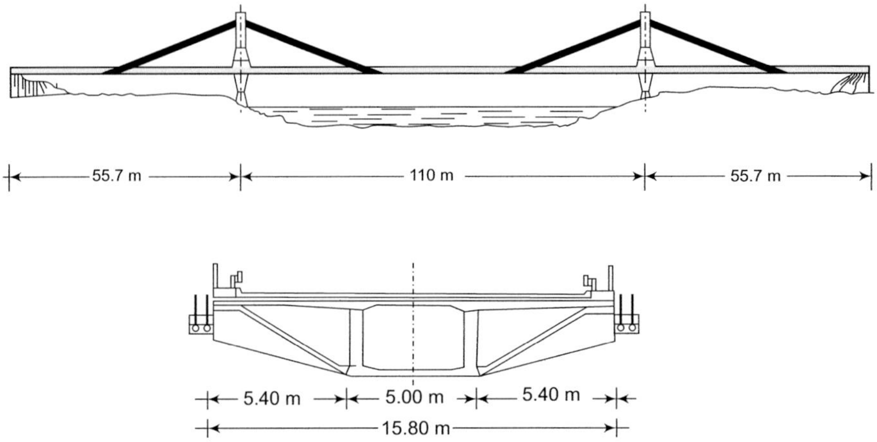

Fig. 5.71 General arrangement of the Danube Canal Bridge

5.3.10 The Saint Nazaire Bridge

When opened in 1975 in France, the Saint Nazaire Bridge, Figure 5.72, became the world's longest-spanning cable-stayed bridge with a main span 404 m (Sanson, 1976). The total length of the structure is 3356 m, which is subdivided into a northern approach viaduct consisting of 22 spans, 50.7 m each, with a total length 1115 m; suspended structure consisting of three spans with a total length of 720 m; and a southern approach viaduct consisting of 30 spans, 50.7 m each, with a total length of 1521 m. The central suspending structure consists of a main span of 404 m and two side spans each of 158 m, supported by stay-cables arranged in sloping planes. The superstructure is a steel box-section with an orthotropic deck stiffened by steel diaphragms that are spaced every 4m. The cable system is of the multi-cable fan type, wherein the stay cables are attached at the top to thick steel gusset plates, fixed to either side of the top length of the pylons. Each stay consists of a single locked-coil strand. The cables vary in diameter from 72 mm to 105 mm according to their location in the superstructure. The pylons consist of an upper V-shaped steel portion having a height of 68 m above the piers (see Figure 5.73).

Fig. 5.72 The Saint Nazaire Bridge, France, 1975 (Courtesy, Elizabeth Taubenheim)

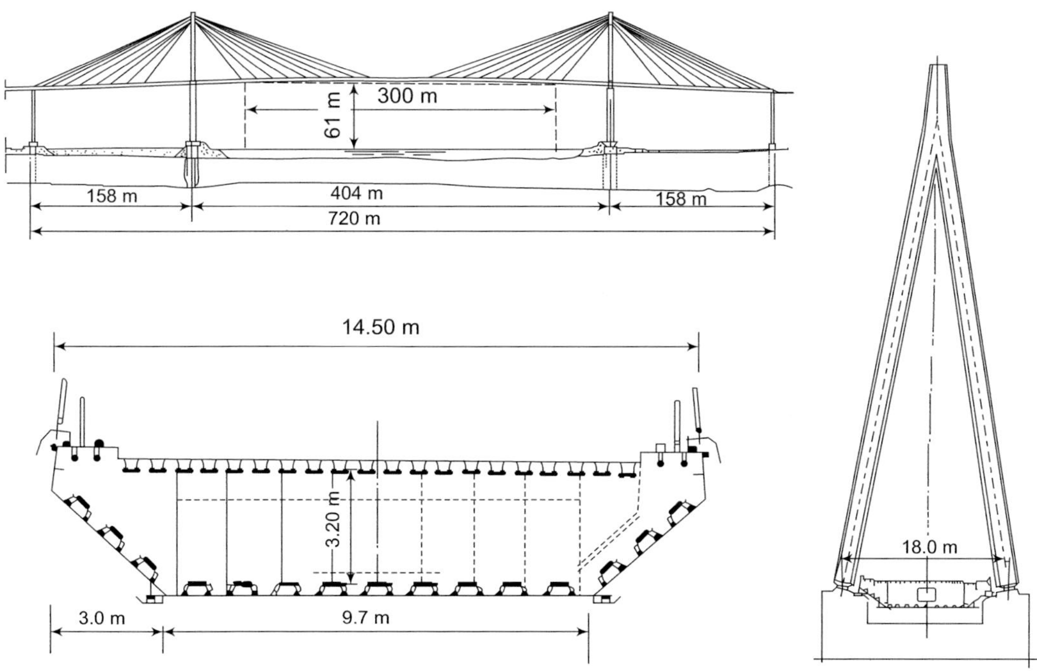

Fig. 5.73 General arrangement of the Saint Nazaire Bridge

5.3.11 The Tacitus Bridge

The Tacitus Bridge in Netherlands was opened to traffic in June 1976. This bridge crosses the Waal River, one of the most significant waterways of Holland, near Ewijk. The bridge has an overall length of 1055 m divided into 10 spans (Figure 5.74). The middle three spans form the cable-stayed structure with a main span of 270 m and two side spans, each of 105 m. The main girder is a box section, 25 m wide and 3.5 m high, having two lateral brackets so that the deck has an overall width of 36.4 m (Figure 5.75). The main girder is stayed in the main span by cables, placed in the bridge centerline. The stays pass through hinged saddles in the 50 m high pylons, which rise from the bridge deck at the piers on both sides of the main span. The steel pylons are located on the bridge centerline over the main span piers. The pylons are made of rectangular box-sections, tapering 2.50 m (measured parallel to the bridge) × 1.50 m at the base to 3.50 m × 2.50 m at the top. The pylons are hinged to a special diaphragm at the base longitudinally. The diaphragm transfers the pylons' load directly to the bearings. Hinging the pylons at the base eliminated the bending moments at the pylon base and hence saved on material. The cables are 100 mm wire-ropes of the closed interlocking type (Schaaf and Spoelstra, 1976). In 2013, following the construction of a second structure beside it to double the capacity, the bridge was renovated, with the deck raised to increase the clearance for river traffic. It is now one of the country's most important bridges, carrying more than 110,000 vehicles per day while allowing ships to navigate the important waterway which connects the port of Rotterdam to Germany.

Fig. 5.74 The Tacitus Bridge, Netherlands, 1976: Illustrating also the new bridge constructed in 2013

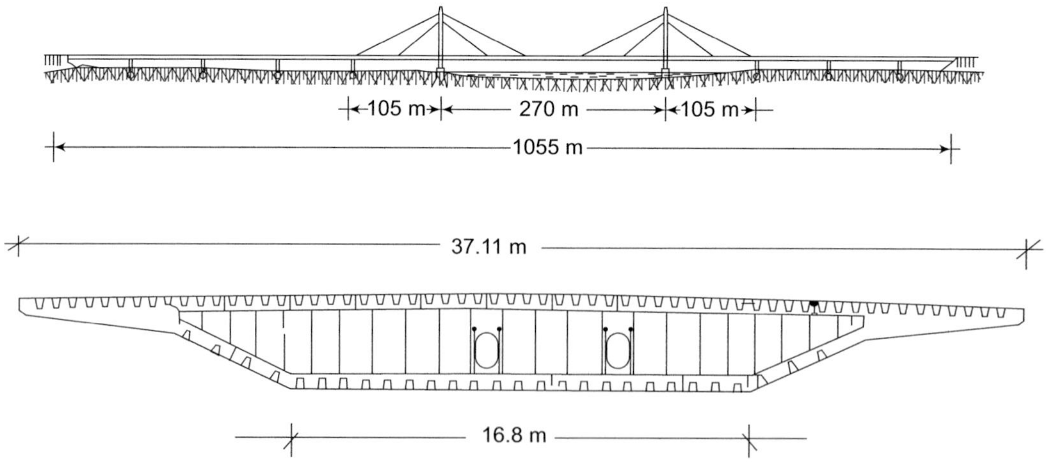

Fig. 5.75 General arrangement of the Tacitus Bridge, 1976

5.3.12 The Brotonne Bridge

The Brotonne Bridge was the first to use a multi-cable semi fan system (Figure 5.76). This bridge is in Normandy, France and spans the Seine River since 1977. Its concrete central span of 320 m was the longest in the world when it was opened to traffic in 1977.

Fig. 5.76 The Brotonne Bridge, France, 1977

A central cable plane design required a rigid single-cell trapezoidal box girder with interior stiffening struts. The section was post-tensioned in three directions, i.e., the webs are vertical, the top and bottom flanges are transverse, and the inclined stiffeners are longitudinal. The inclined stiffeners were necessary to transfer the vertical component of the cable force to the webs. The concrete pylons are fixed to the deck and rise 70 m above it (Figure 5.77). The superstructure is

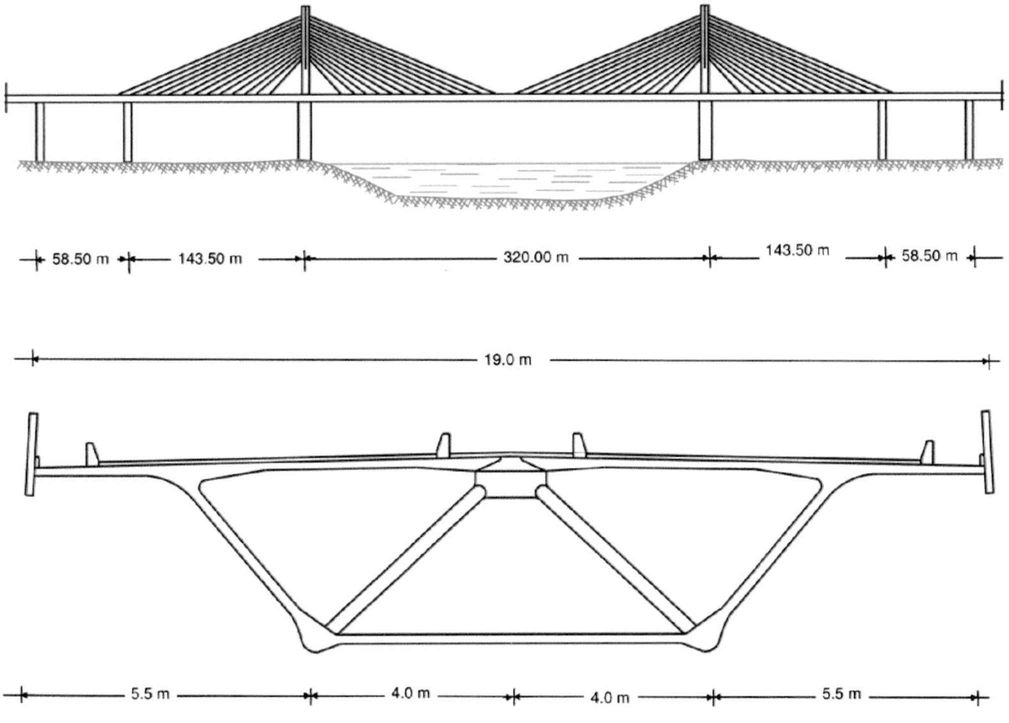

Fig. 5.77 General arrangement of the Brotonne Bridge

supported by 21 stay cables that are fixed on saddles inside each pylon. The cables are anchored into the stiffening girder at intervals of 6 m. They are made of stranded prestressed strands, threaded in steel ducts, and grouted with cement. The number of strands ranges from 39 to 60, from the shortest to the longest cables. The stay anchorages arc spaced every 6 m at the deck. The bridge used the free cantilevering method for construction. The box girder was made partially of prefabricated elements, i.e., the precast webs with projecting reinforcements were placed in the forms and concrete was poured around them.

5.3.13 The Indiano Bridge

The Indiano Bridge is characterized by its earth-anchored back stays (Figure 5.78). It runs across the Arno River in Florence, Italy. The bridge was built between 1972 and 1978 and was designed by Fabrizio de Miranda, a remarkable Italian structural engineer and university professor.

Fig. 5.78 The Indiano Bridge, Italy, 1978

The bridge has a 206 m long main span and two side spans of 70.5 m each. It is characterized by a single plane of stays and its steel pylons were designed to lean slightly backwards. The cables are arranged in a fan configuration. The stays are anchored to the 45 m high pylons. The back stays are attached to the pylons and fixed to anchor blocks that transmit their forces directly to the soil.

5.3.14 The Rande Strait Bridge

Another significant Cable-stayed bridge that was designed by Fabrizio de Miranda is the Rande Strait Bridge in Spain (Figure 5.79). The design for this bridge was finished in 1970 and the bridge was completed in 1977 and opened to traffic in 1981. It is a composite cable-stayed bridge that links Redondela and Moaña through the Strait of Rande. It has a 401 m main span and two side spans each of 147 m. The superstructure consists of two longitudinal steel plate girders attached by transverse trusses at about 3.5 m intervals.

When average daily traffic reached 55,000 vehicles in 2006, it was almost at its effective capacity and frequently caused traffic congestion. Studies of the bridge's viability were conducted in order to boost its traffic capacity. A widening project was determined to be the most advantageous option following comparative analyses of technical, aesthetic, and financial factors. Two new external composite decks that are positioned along the outer edges of the current deck, adjacent to the main piers, make up the widening solution. This system enabled bridge-widening operations without interfering with current bridge traffic or having an impact on the Vigo estuary, an extremely valuable area in terms of the natural environment and landscape.

Fig. 5.79 The Rande Strait Bridge, Spain, 1977

5.3.15 The Tjorn Bridge

Four cable-stayed bridges were completed in Scandinavia between 1981 and 1991. The Tjorn Bridge across the Askerofjord Narrows in Sweden (Figure 5.80) was built to connect Stenungsund on the mainland with the Tjorn island on the western coast of the country. It replaced the Almo Bridge, which collapsed due to a collision of a cargo ship in January 1980. The new cable-stayed bridge was opened to traffic in November 1981, 15 months after the beginning of the work in August 1980. The bridge has a cable-stayed main steel span, concrete pylons and concrete approach viaducts. The steel and concrete parts are connected near the pylons. The overall length is 665 m; the main span is 366 m long; the clear width is 45 m; and the height above water is 45 m. The bridge carries a three laned roadway and a combined footway and cycle track (Brodin, 1982).

Fig. 5.80 The Tjorn Bridge, Sweden, 1981 (Courtesy of Herrad Elizabeth Taubenheim)

The orthotropic deck has a clear width of 15.0 m and is supported by a welded rectangular box girder of width 8.5, and depth of 3.0 m. The box girder is provided with floor beams that are spaced every 4 m. The box girder is cantilevered 3.25 m on both sides and is stiffened by trapezoidal ribs at its top and bottom plates as well as the extensions, while the webs are stiffened by angles (Figure 5.81). The side spans are designed as continuous concrete girders with intermediate column supports at each cable anchor point. The coupling between the viaduct superstructure and the main span is located 10 m from the pylon. It consists of a 3 m long steel segment of the same shape as the main span section that is fastened to the viaduct by dowels and prestressing tendons. The concrete pylons have two parallel constant section legs, 4.0 × 4.5 m, connected by cross beams at two levels. The stay cables consist of two Strands of locked coil, diameter 77-108 mm. The Strands are anchored at the superstructure individually to facilitate their replacement through steel anchorages at four levels of each pylon leg.

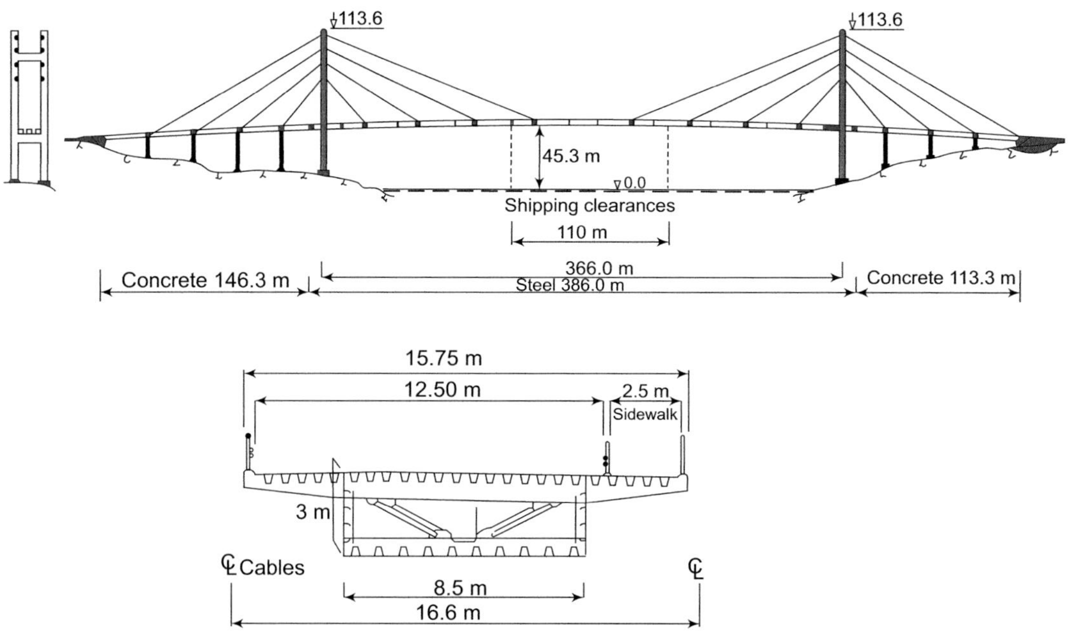

Fig. 5.81 General arrangement of the Tjorn Bridge

5.3.16 The Farö Bridge

The Danish Parliament in 1976 approved building a highway section between Ronnede and Sakskøbing passing east of the city of Vordingborg. The highway required two large bridges: the north bridge connecting the island of Zealand to the island of Farö and the south bridge (Fig. 5.82) linking Farö to the island of Falster carrying the southern motorway, which connects Copenhagen with Germany and the rest of Europe.

The north bridge was executed as a 1.5-km long continuous girder bridge and the south bridge as a 1.7-km-viaduct with a cable-stayed main span. The southern cable-stayed bridge has a central span of 289 m supported by a central cable plane, and two side spans each of 120 m. Construction of the bridge commenced in 1980 and it was opened to traffic in 1985. The bridge is characterized by symmetrical multi-cable fans and displaced end piers. The end pier is located under the anchor point of the second cable, 16 m from the end of the fan. Thus, the three stays at the end of the side span fan are activated as backstays and a special heavy single backstay is avoided. The steel deck

Fig. 5.82 The Farö Bridge, Denmark, 1985 (Courtesy, Svenja Kramarczik)

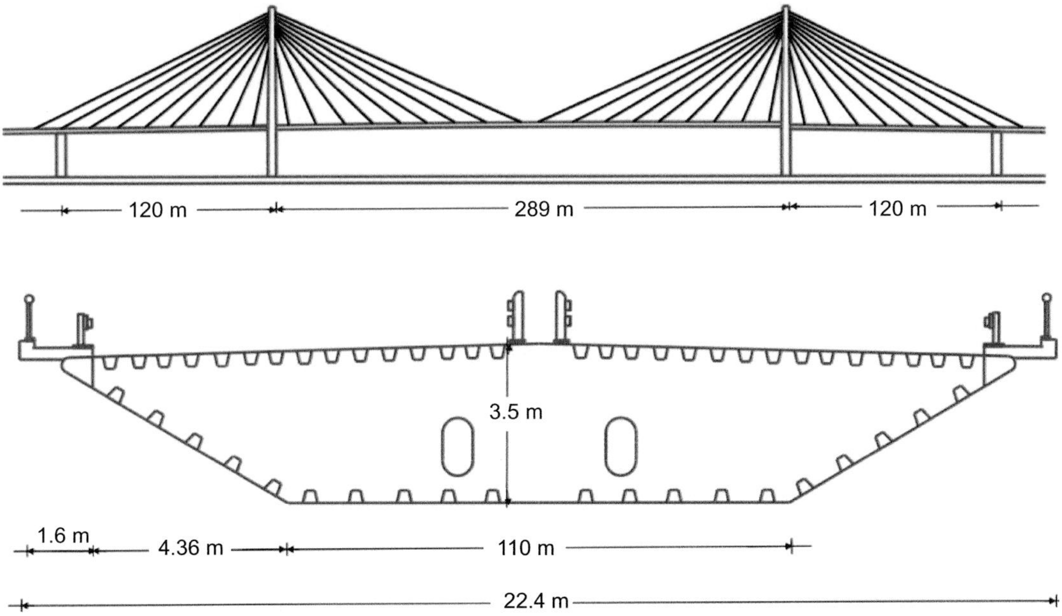

Fig. 5.83 General arrangement of the Farö Bridge

has a trapezoidal box-type cross-section 22.4 m wide at the top of height 3.5 m (Figure 5.83). The diamond-shaped pylons are made of reinforced concrete hollow sections, with the lower portion extended directly to rest on top of the caisson foundations. This is considered a further development of the Kohlbrand pylon, which is supported by intermediate piers that transfer the load to the foundation.

5.3.17 The Skarnsund Bridge

The Skarnsund Bridge built in Norway in 1991 across the Skarnsundet sound, in Inderoy (Figure 5.84) is one of the world's longest concrete cable-stayed bridges, with a length of 1,010m. On its completion, it was the longest cable-stayed bridge in the world with a main span of 530 m and kept the record for two years.

Fig. 5.84 The Skarnsund Bridge, Norway, 1991

It is characterized by very small width-to-span ratios as it carries two lanes for vehicles and one for pedestrians and bicycles. The superstructure is a one box girder 2.15 m deep, i.e., vertical slenderness of 1:247, and 13 m wide, i.e., 1:41 horizontal slenderness Figure (5.85). It has an aerodynamically advantageous triangular cross-section which is solid in the rather short side spans (ratio 0.36) to provide sufficient counterweight. The cables have diameters varying between 52 and 85 mm. The concrete pylons are A-shaped and rise 152 m above sea level and are founded on rock below the seabed. The bridge was built to withstand winds up to 48.5 m/s.

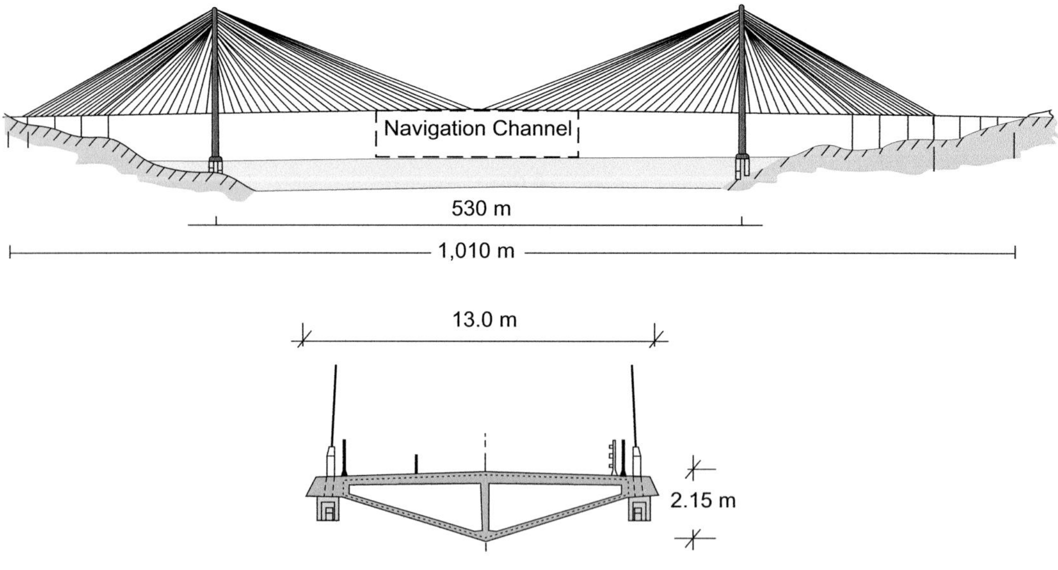

Fig. 5.85 General arrangement of the Skarnsund Bridge

The Helgeland Bridge is another bridge that was opened to traffic in the same year and in the same country, but with a 425 m main span. This bridge has the same features as those of the Skarnsund Bridge but with a 1.2 m deep concrete deck and a width of 12 m.

5.3.18 The Guadiana International Bridge

The Guadiana International Bridge in Portugal (Figure 5.86) is another concrete superstructure bridge. It was completed in 1991 to link Castro Marim in Portugal to Ayamonte in Spain to establish a road connection between the southern sections of the two countries.

Fig. 5.86 The Guadiana International Bridge, Portugal, 1991

The prestressed concrete bridge deck as shown in Figure 5.87 consists of a continuous girder over 666 m with a main span of 324 m. The deck is a prestressed concrete box section, 18 m wide and 2.5 m deep resulting in vertical slenderness of 1:130, and horizontal slenderness of 1:18. Inside the box there are transverse concrete crosses bracing 4.5 m apart to stiffen the cross section. The main span is suspended by two planes of cables at every 9 m of spacing. The two main inverted

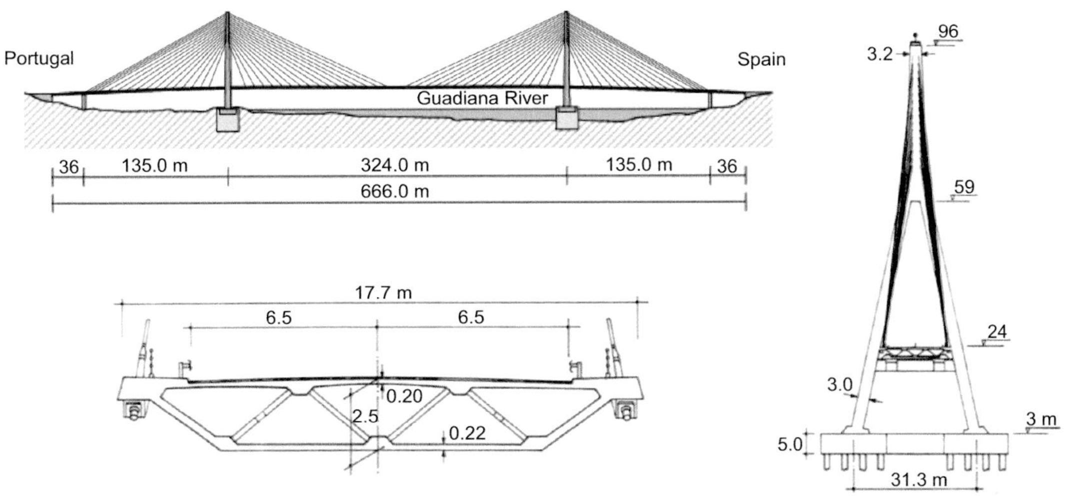

Fig. 5.87 Guadiana International Bridge: Elevation, cross-section, and pylon configurations

Y-shaped concrete pylons are 99 m high and are supported by pile groups, with 2.0 m diameter piles. The cables are composed of parallel wire 15 mm diameter strands (varying from 22 to 55 strands per cable). The cables are attached to the pylons' crests at the center of the bridge.

5.3.19 The Evripos Bridge

The Evripos Bridge crossing the Euboean Channel, 80 km north of Athens (Figure 5.88) was opened one year later in Greece and broke the record for cable-stayed bridges with slender concrete decks. It may have the slenderest concrete deck in the world (Stathopoulos, 2014). This multi-cable-stayed bridge was constructed with a concrete deck slab thickness of only 45 cm in the central span of 215 m resulting in a vertical slenderness of 1:478. The bridge also is characterized by a monolithic connection between the deck and the two concrete pylons.

Fig. 5.88 The Evripos Bridge, Greece, 1992

It comprises three spans, 90 m + 215 m + 90 m, supported on two piers and suspended from the two pylons. The bridge has two approaches, 143.5 m and 156 m long making the total length of the structure 694.5 m. The solid deck slab has a constant depth 0.45 m along the whole length, except for the pylon area where the depth increases linearly to 0.75 m. It is prestressed transversally and longitudinally at local locations only in the central area where the deck axial force is minimal, and in the region of the approach piers characterized by high local bending moments. The stay cables are arranged into two vertical planes following a semi-fan shape and anchored in the deck slab every 5.9 m. The cables consist of seven 21 individual parallel wire galvanized strands with 15.25 mm nominal diameter. The twin pylons are made of hollow reinforced concrete sections with dimensions that vary from 4 m × 4 m × 0.50 m, at the base, to 2.5 m × 2.5 m × 0.40 m, at the top. The columns of each pylon are connected transversally through two double concrete girders, one below the deck and the other under the pylon top. The cables are anchored to an internal steel chassis, which was designed to transfer the cable load to the external layer of concrete. Supporting of the side spans on the piers was achieved by using hinged steel holding down devices. Due to the tensile reaction force applied by the bridge to the piers, they are prestressed vertically and anchored into the soil by micro-piles (See Figure 5.89).

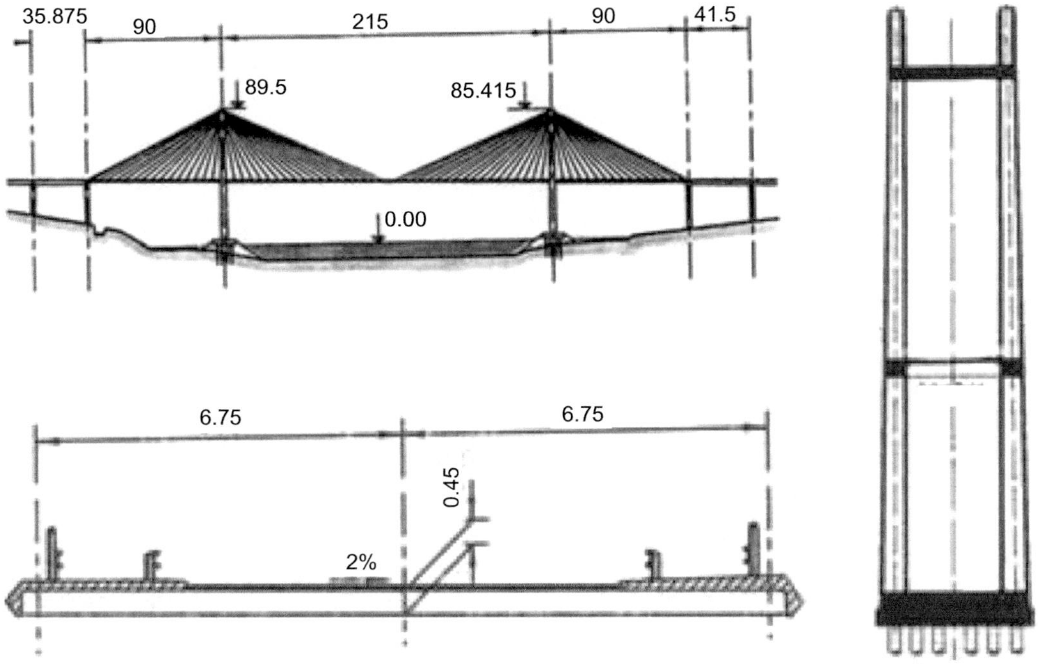

Fig. 5.89 Evripos Bridge: Elevation, cross section, and pylon configuration

5.3.20 The Alamillo Bridge

The Alamillo Bridge was completed in 1992 for the Expo' 92 Seville in Spain (Fig. 5.90). The bridge, which was designed by Santiago Calatrava, an outstanding architect and structural engineer, is distinguished by the lack of cable backstays. The bridge's total length is 250 m with a main span of 200 m. The width of the bridge is 32 m accommodating six lanes, one central walkway located on a higher level than the traffic lane. The bridge stiffening girder is a complex beam consisting of a central box girder, which is a hexagonal steel box beam 4.4 m. deep, 5.6 m. wide; transverse steel ribs 12 m long, positioned at every 4 m along the bridge. The deck is a 230 mm deep concrete slab on top of the ribs attached to them by shear connectors. The leaning pylon is an irregular roughly hexagonal cross section with average dimensions 12 m × 8 m comprising a stiffened steel plate and a concrete fill that provides weight and strength to the pylon (see Figure 5.91). The pylon rises 134.25 m above the roadway inclined at 32 degrees to the vertical. The main span is supported by two vertical cable planes (13 pairs of stays every 12 m) located at the sides of the walkway effectively a single cable plane. The structural system relies on the weight of the pylon to achieve equilibrium. The rotation induced by the cable stays is counteracted by the rotation about the base of the pylon due to the force action of the pylon mass. The use of the inclined pylon is obvious: to mobilize enough mass to cause a counter-acting rotation to that induced by the cable system. When the bridge is over or under loaded relative to the equilibrium case, then the foundation must mobilize a resistance to the overturning moment that these loadings induce. The Alamillo Bridge with its appealing appearance, designed by an architect, mobilized other architects in other parts of Europe to get involved in shaping and designing cable-stayed bridges.

5.3.21 The Erasmus Bridge

Ben Van Berkel, an architect from Holland, designed the Erasmus Bridge, which was opened in 1996. This bridge, which is considered Rotterdam's landmark, is the first in the Netherlands which for the most part has been shaped by architects.

Fig. 5.90 The Alamillo Bridge, Spain, 1992

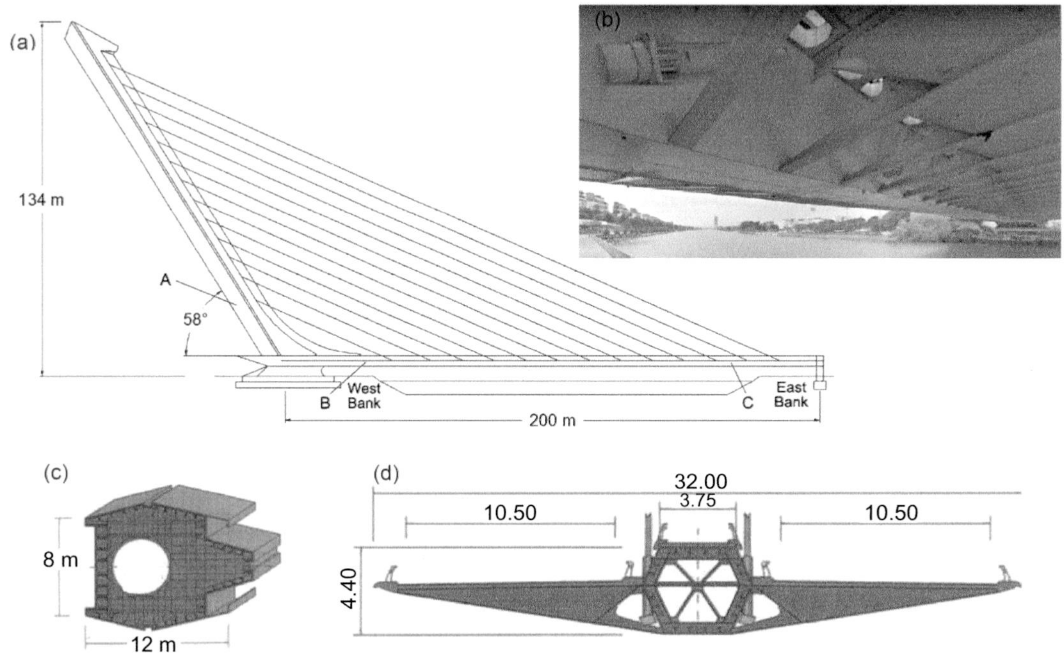

Fig. 5.91 General arrangement of the Alamillo Bridge: (a) elevation; (b) photo of longitudinal girder; (c) pylon cross section; and (d) cross-section

The bridge spans the river Maas (Figure 5.92). It has two abutments and four piers. In the south-north direction, it consists of a composite approach span, a 50 m shipping clearance bascule, a cable-stayed bridge and a concrete side span. The asymmetric cable-stayed structure consists of three continuous spans: The rear span of 73.6 m. the main span of 284 m and a front span of 51.7 m (Figure 5.93). The main supporting structure of the bridge deck consists of two box girders 2.25 m high (span/height ratio of 130) and 1.25 m wide, with floor beams located at 4.9 m intervals and cantilevering at 6.7 m. The pylon, which is asymmetric in shape, is 139 m high. The pylon's height, keeping up with Rotterdam's high-rise buildings, is closely related to the free span and results in a span-to-height ratio of approximately 2 : 1. By applying a backward lean to the pylon and offsetting its top part to act as a counterweight, it was initially designed without back stays. This basic idea echoed the concept that Santiago Calatrava had used for the Alamillo Bridge. However, during a three-month feasibility study, it was recommended that the backstays get added.

Fig. 5.92 The Erasmus Bridge, Holland, 1996

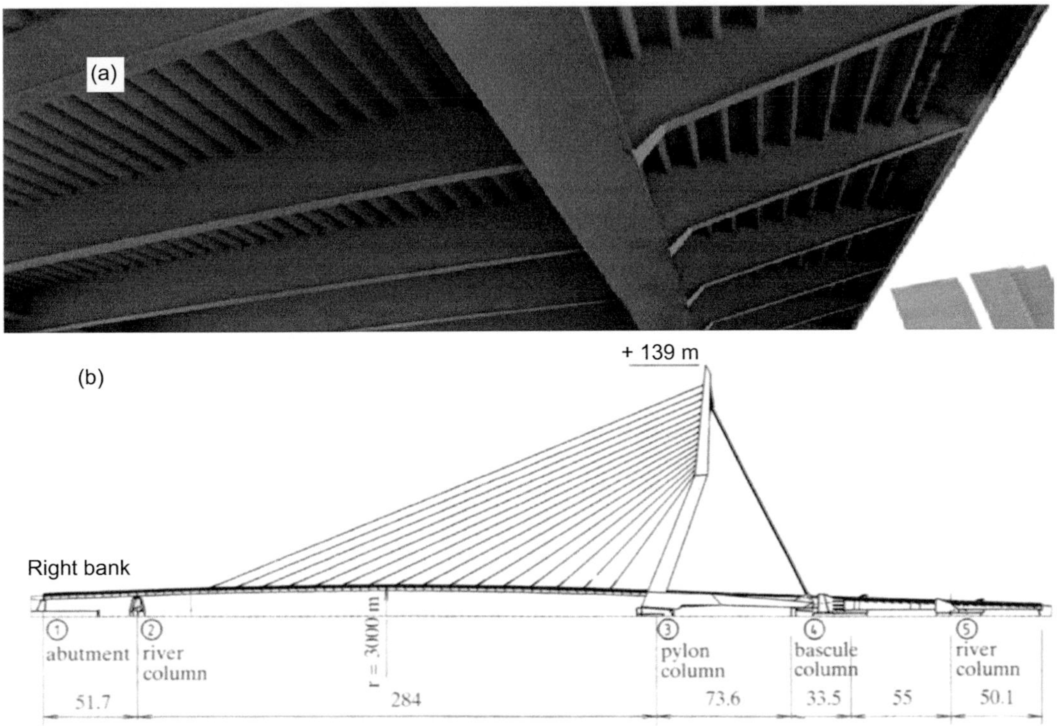

Fig. 5.93 General arrangement of the Erasmus Bridge: (a) photo of the deck; and (b) Elevation (Reusink and Kuijpers, 1998)

5.3.22 Mariansky Bridge

The Mariansky Bridge in Usti nad Labem, Czech Republic (Figure 5.94) is another example of cable-stayed bridges with an unusual appearance. The bridge, which was completed in 1998, has a main span of 123.5 m and a side span of 55.5 m. The deck width is 26.1 m and its depth is 26.1 m. The shape and inclination of the pylon were designed to work without backstay (Crhan et al, 1999). The main span was designed in steel to minimize its weight. The span length and the pylon's shape made it possible to transfer the forces of the suspended span weight without the need of back stays. Like the Alamillo Bridge, this bridge used one steel girder beam with outriggers from both sides as shown in Figure 5.95. This bridge also used an orthotropic deck that was built integrated with the outriggers.

Fig. 5.94 The Mariansky Bridge, Czech Republic, 1998

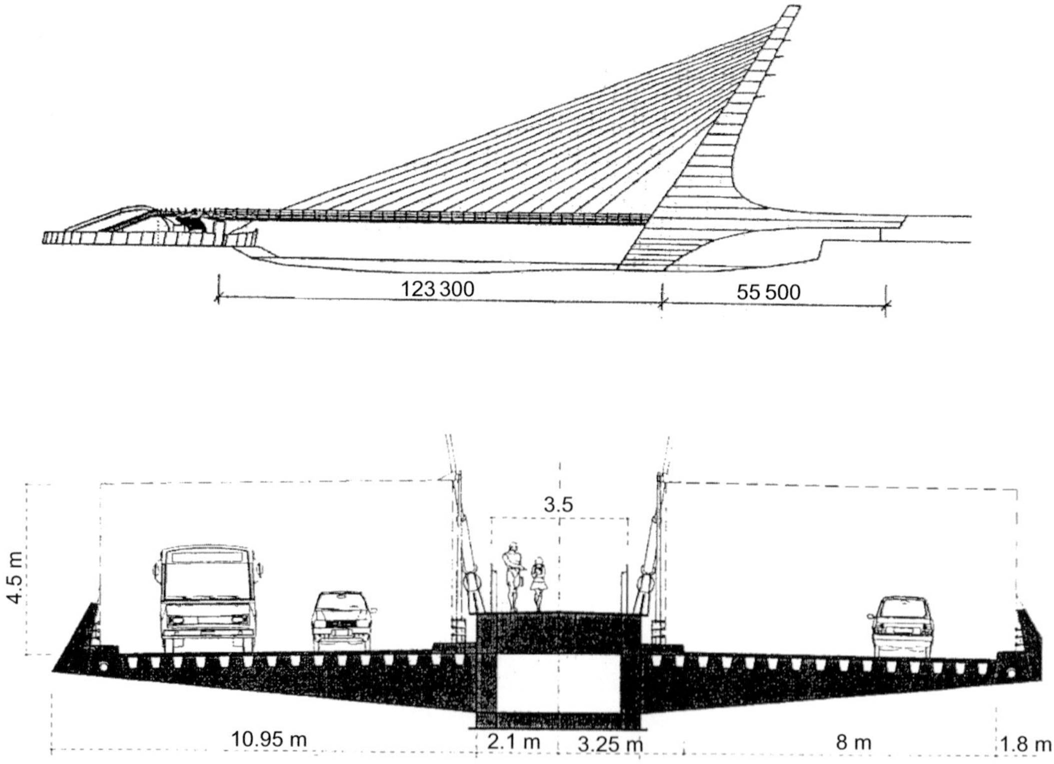

Fig. 5.95 Elevation and cross-section of the Mariansky Bridge

5.3.23 Normandy Bridge

When it was built in 1995, the Normandy Bridge was the largest cable-stayed bridge in the world. This bridge links the industrial center of Le Havre to the tourist area of Honfleur crossing the Seine river in one go over a distance of 2.14 km. The choice of a cable-stayed bridge was an economical and aesthetic solution that met the limitations imposed by the site since a conventional bridge would have required the presence of supports that would have been resistant to boat impact. Also a suspension bridge would have generated significant additional costs related to its maintenance. The bridge is currently the sixth longest cable-stayed span in the world (Virlogeux, 1993). The bridge is 2141 m long from the south abutment to the north abutment with a main span, 856 m long (Figure 5.96).

Fig. 5.96 The Normandy Bridge, France, 1995

The bridge design efficiently combined concrete and steel for the design of the deck as it has a concrete deck in the approach spans, and a steel deck in the central part of the main span only. On the north side, the approach spans are arranged such as the span length of the first one towards the north pylon that is 96 m followed by 14 spans each of 43.50 m length and one span 32.5 m long. The same arrangement is followed in the south approach spans except that nine spans of 43.5 m are arranged. The concrete deck extends 116 m inside the main span from both sides, i.e., the steel part of the main span is only 624 m (Figure 5.97).

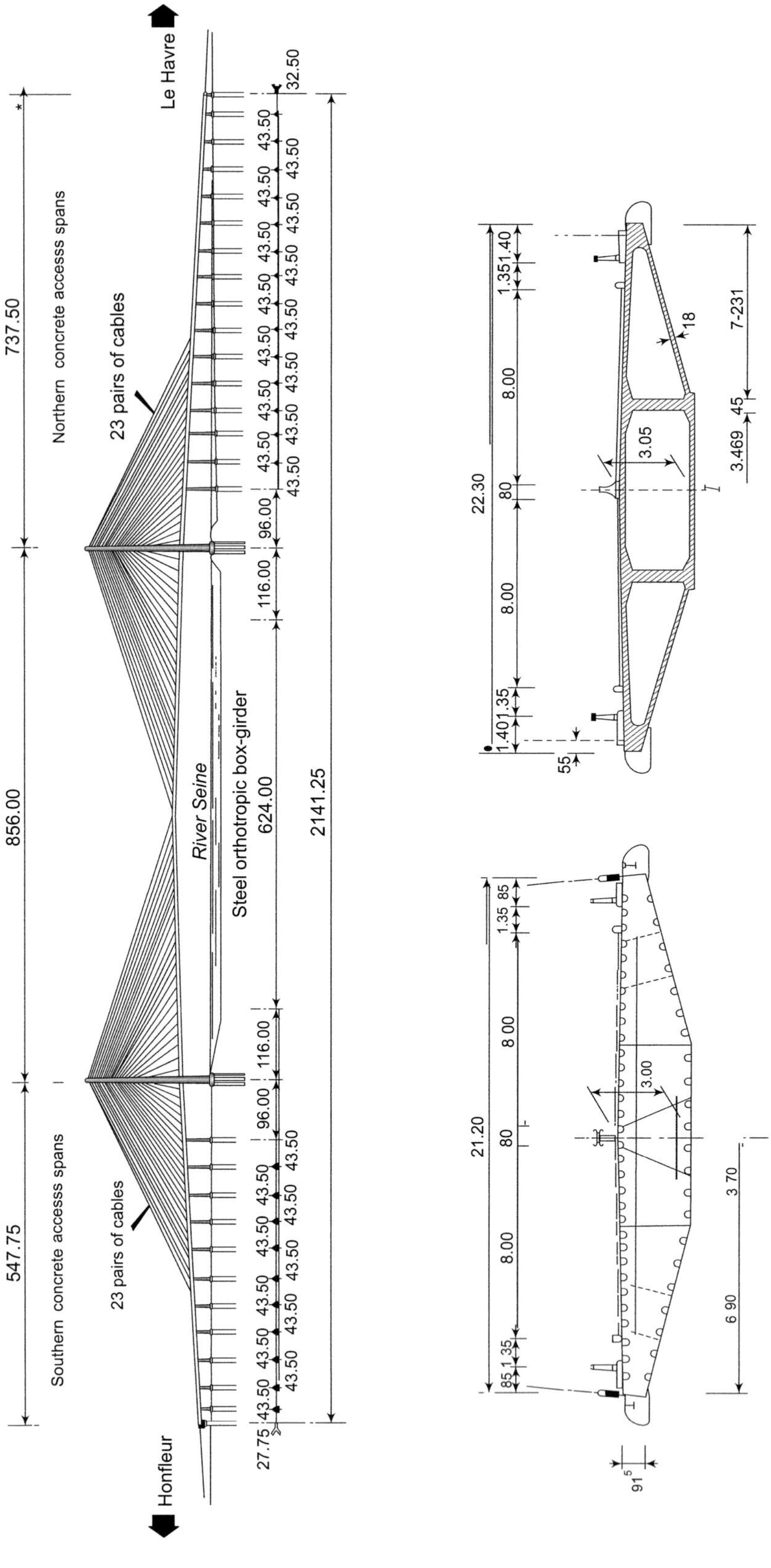

Fig. 5.97 Elevation and cross-sections of the Normandy Bridge

The deck is an orthotropic box-girder that has been designed to reduce wind forces, and to give a high torsional rigidity. The box-girder width is 21.20 m and 22.0 m for the steel and concrete sections respectively. The depth has been limited to 3.00 m to reduce vortex shedding. The steel deck is stiffened by diaphragms 3.93 m apart and trapezoidal ribs. The thickness of the top plate varies from 12 mm to 14 mm and the rib thickness varies from 7 mm to 8 mm. In the bottom and inclined plates, which are 12 mm thick, the ribs are 240 mm deep and 8 mm thick. They are 1.00 m apart.

The pylons have the shape of an inverted Y. This shape limited the pylon height to 214.7 m (155.70 m above deck); and increased the structural capacity of the pylon to wind loads. The deck is monolithically connected to the pylons, i.e., the main span and the pylons therefore act as a frame. This rigid connection limits lateral deflections produced in the main span by wind effects (see Figure 5.98). The inverted-Y shape limits the pylon crest height, and thus transversal wind effects: transversal bending moments are limited at the crest base and with them the second-order effects. As the pylon's lower part has the shape of a triangular frame, with an intermediate bracing where the deck is rigidly connected, it easily resists transverse wind forces, coming either from the cables or through the deck. The cables suspending the main span are interconnected in the pylon crest on the bridge center. This allows significant torsional deformations in comparison with an H-shaped pylon, where the two vertical legs act almost independently. The cables are made of parallel strands, 15 mm in diameter. Each cable comprises 30, 44 or 51 strands according to its position in the bridge.

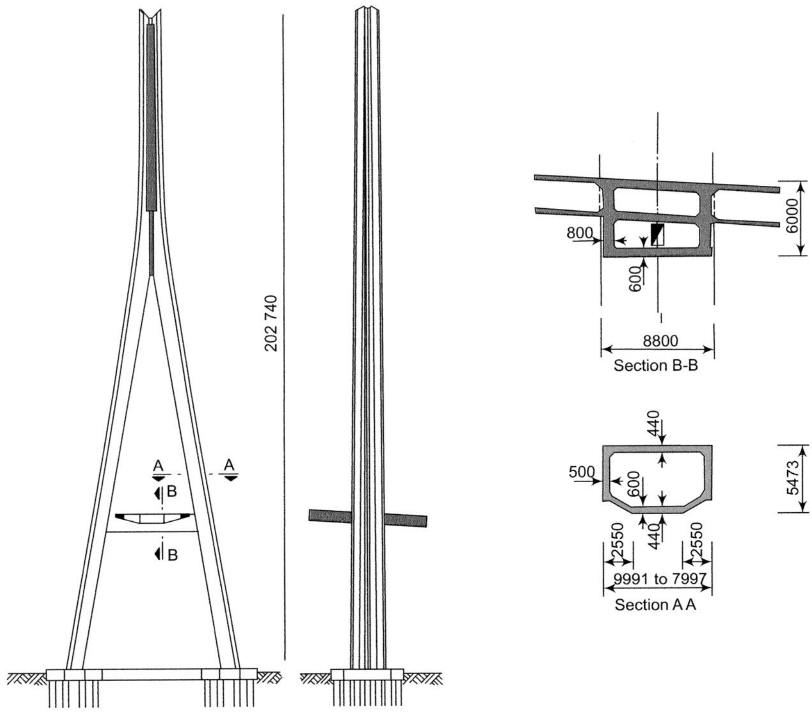

Fig. 5.98 Monolithic rigid connection between Pylon and superstructure-Normandy Bridge

5.3.24 The Vasco da Gama Bridge

The Vasco da Gama Bridge in Lisbon, Portugal, which was completed in 1998 is considered the longest bridge in Europe (Fig. 5.99) with a total length of 17.3 km, including three interchanges. The cable-stayed structure is 824 m long with a main span of 420 m (Figure 5.100a). Since Portugal is a country with high earthquake risk, seismic considerations significantly influenced the design of the bridge. The superstructure is 30.9 m wide and comprises a 250 mm concrete slab supported by two reinforced concrete beams: 2.6 m deep. The longitudinal girders are connected transversely by steel beams at 4.42 m intervals except over the piers, the transverse beams are made of concrete (Figure 5.100 b). A total of 192 cable stays, arranged into two vertical planes, are used in a semi-fan configuration. To counteract possible uplift under service loads, anchoring of the stay cables extends through piers P2, P3, P4 and P5. Earthquake resistance was a major concern for this bridge. A strategy of designing a deck completely independent of the pylon was implemented. However, under high earthquake excitation, seismic energy is dissipated through longitudinal and transverse dampers installed between the deck to the pylons (Figure 5.100 d).

Following this perspective, sliding bearings are provided on the pier. Thus, the deck is stabilized under service loads only by the 96 stay cables linking it to the pylons. The pylons and piers have been designed to be as flexible as possible to reduce the inertial forces during a seismic event. The 150 m high H-shaped pylons are founded on 84 m × 20 m ×14 m caissons designed to resist ship collisions.

Fig. 5.99 The Vasco da Gama Bridge, Portugal, 1998 (Courtesy, Robert Cortright)

5.3.25 The Oresund Bridge

The Oresund Bridge is distinguished as the longest combined road and rail bridge in Europe. It connects Copenhagen, the Danish capital city, and the Swedish city of Malmo. The bridge has a length of 7.8 km from the Swedish coast to the artificial island Peberholm in the middle of the strait, which made it the longest bridge in the world for both road and rail traffic at the time. The crossing is completed by the 4 km tunnel from Peberholm to the Danish coast (see Figure 5.101).

The eastern segment comprises three main parts: the eastern approach bridge with a length 3729 m; the cable-stayed structure at the navigation channel with a length of 1092 m from expansion joint to expansion joint; and the western approach bridge with a length of 3014 m (see Figure 5.102).

The superstructure is designed as a double deck truss with two railway tracks on the lower deck and a four-lane highway on the upper deck (Gimsing, 2011). At completion in 2000 the main span of 490 m was also the longest in the world for a cable-stayed bridge carrying both road and rail traffic. The continuous truss is supported by harp-shaped cable systems with an inclination of approximately 30°, except for the outer regions close to the expansion joints. The cables are supported by 203.5 m high H-shaped pylons. A tie-down system is arranged at the piers closest to the pylons to balance live loads in the main span. Expansion joints are provided at the transition to the approach bridges.

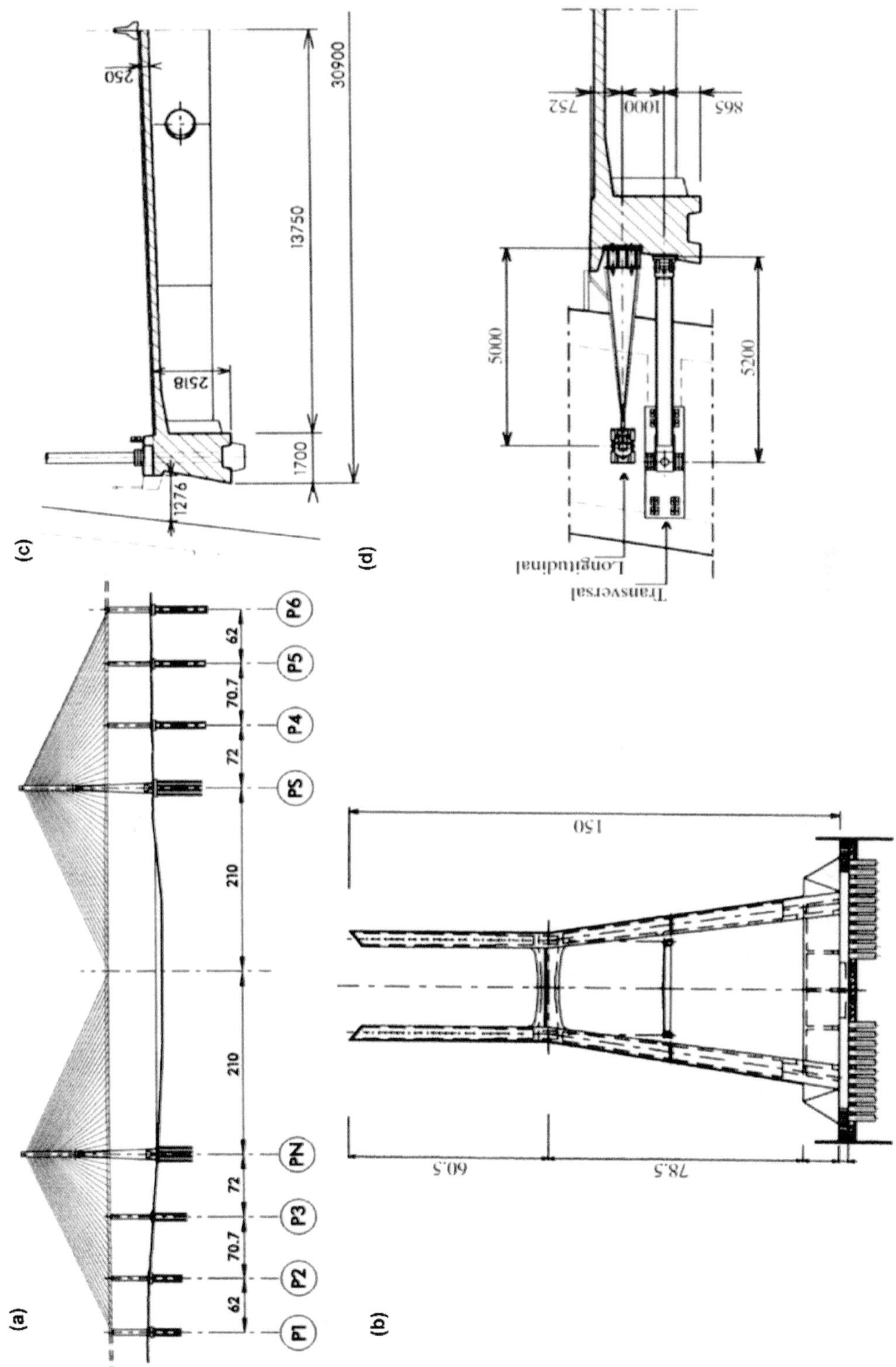

Fig. 5.100 General arrangement of the Vasco da Gama Bridge: (a) elevation; (b) pier; cross-section; (c) deck cross-section; and (d) dampers connections (Capra and Leiville., 1998)

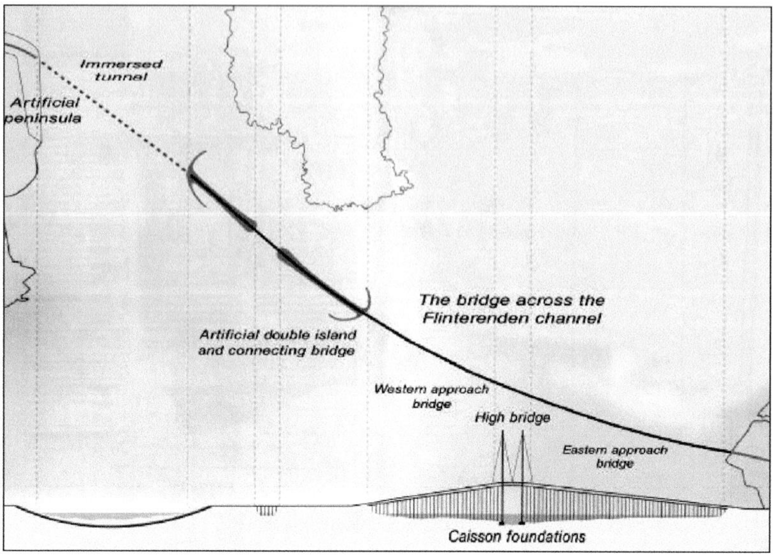

Fig. 5.101 The Oresund Crossing, Sweden-Denmark, 2000

Fig. 5.102 The Oresund Bridge

The superstructure consists of two continuous trusses, spaced at 13.7 m. With the overall design leading to vertical cable planes outside the deck, it was possible to let the pylons consist of vertical free-standing posts above the deck. By ensuring that the centroid of the pylon is positioned in the vertical cable planes the cross section of the pylon is subjected to pure uniform compression from dead load and vertical traffic load acting on the deck. The stay consists of two cables (one on top of the other) each comprising approximately 70 seven-wire strands. The anchor points of the harp cable system extend beyond the side span of 160 m to the approach span of 141 m. Thus, the six stays at the end of the side span are activated as back stays, which has a pronounced influence on the stiffness of the system (Figure 5.103). The main span deck consists of a transversally prestressed concrete slab acting compositely with the top chords of two vertical steel trusses. The lower railway floor is made entirely in steel in the form of a shallow multi-cell box [Figure 5.103 C]. The main trusses are connected transversally by an upper steel floor beam that cantilevers 8.4 m from both sides by triangular outriggers. The floor beam carries a prestressed concrete slab that accommodates a 4-lane roadway, a median and shoulders. At the bottom chord level, the longitudinal trusses are connected by a steel closed box carrying the railway track. The inclination of the truss members is arranged at approximately 30° and 60°, respectively, allowing the flat diagonals to match the inclination of the cable stays. The cables are anchored to the girder on the outriggers with the same inclination as for the flat diagonals. The pylons are designed as clean Hs without an upper cross beam and with the legs disappearing directly into the sea.

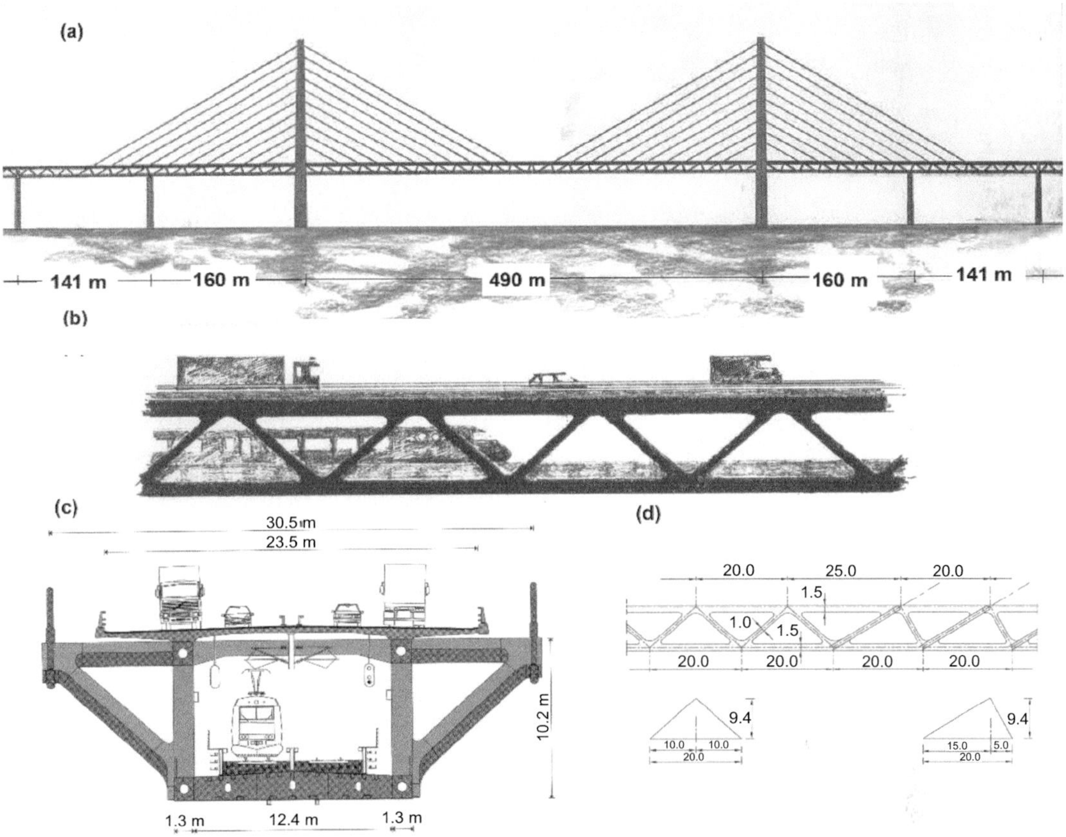

Fig. 5.103 General arrangement of the Oresund Bridge: (a) elevation; (b) realization of truss superstructure; (c) cross-section; and (d) geometry of longitudinal truss at the transfer from approach to suspended span

5.3.26 The Surgut Bridge

About less than two decades ago, Russia entered the list of countries having cable-stayed bridges. More than twenty bridges of this type were completed between 2000 and 2013. The list includes the Surgut Bridge across the Ob River at Surgut (Figure 5.104), which is one of the longest in Siberia. The bridge is 2,070 meters long and the suspended structure is 556 m. It has only one pylon located such that the central span is 408 m long, and the side span is 148 m long. Its central span makes it the longest single-pylon cable-stayed bridge in the world. The 146 m high pylon comprises two legs attached firmly to each other by cross bracings and tie beams. The pylon is made up of steel. The superstructure is a trapezoidal steel box girder. The bridge was opened to traffic in September 2000.

Fig. 5.104 The Surgut Bridge, Russia, 2000

5.3.27 The Rion–Antirion Bridge

The Rion–Antirion Bridge, opened in 2004, is one of the world's longest multi-span cable-stayed bridges and longest of the fully suspended type (Figure 5.105). It crosses the Gulf of Corinth linking the town of Rio on the Peloponnese peninsula to Antirion in mainland Greece. The structure consists of a main bridge, 2252 m long and 27.20 m wide, and two approaches, 392 m and 239 m long, one on each side of the Corinth Strait. The bridge's three central spans, each measuring 560 meters, and its two side spans, each measuring 286 meters, were intended to minimize the number of supports in the strait. The deck is a 27.20 m wide composite steel-concrete structure made up of two longitudinal steel I girders that are 2.20 m deep and braced every 4 m by transverse cross beams. The concrete slab is 250 mm thick (Figure 5.106). The deck is entirely suspended by eight sets of twenty-three pairs of cables, and it is continuous throughout with expansion joints at both ends.

Fig. 5.105 The Rion–Antirion Bridge, Greece, 2004

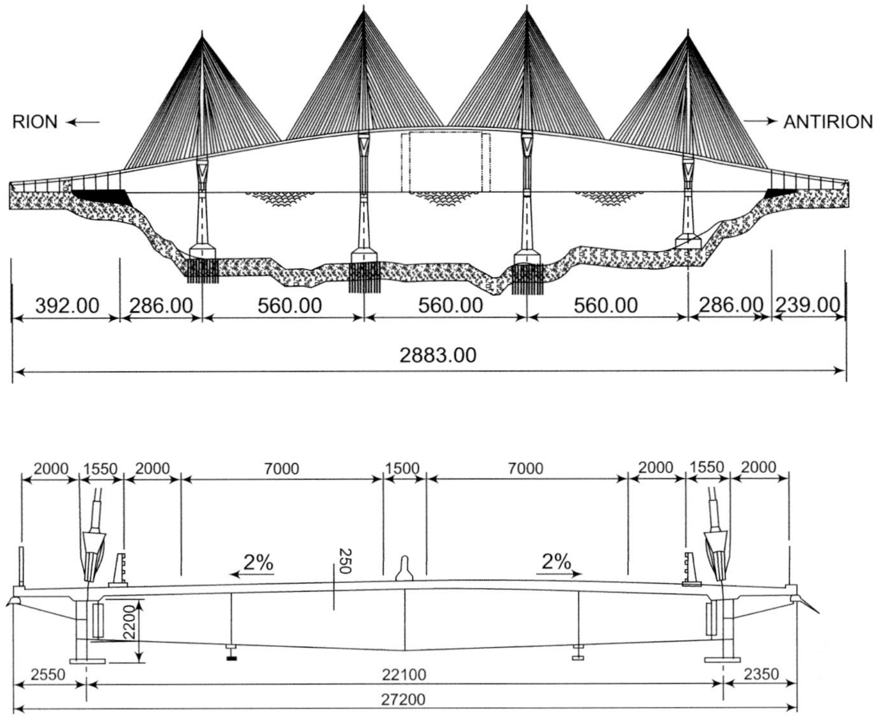

Fig. 5.106 Elevation and cross-section of Rion–Antirion Bridge (Combault and Teyssandier, 2005)

Fig. 5.107 Hydraulic dampers at pylons for Rion–Antirion Bridge (Combault and Teyssandier, 2005).

The deck is free to accommodate longitudinal movements due to thermal and seismic actions and is connected to each pylon in the transverse direction with 4 hydraulic dampers to accommodate seismic excitations as shown in Figure 5.107. The stay cables are arranged in two planes in a semi-fan configuration. They are made of 43 to 73 parallel galvanized strands.

Large concrete substructure foundations measuring 90 m in diameter and 65 m in height support the pylons and disperse all stresses into the soil. The bases of the pylons are made up of a 1 m thick bottom slab and 32 peripheral cells enclosed in a 9m high perimeter wall and covered by a top slab slightly sloping up to a conical shaft. This cone, measuring 38 meters at the bottom and 27 meters at the top, is the deepest pier. It rises 65 meters over the bed and reaches a height of 3 meters above sea level. These bases support 15.8 m high pyramidal capitals that spread to form the 40.5 m wide square base of four concrete legs, via vertical octagonal pylon shafts that are 24 m wide and almost 29 m high (Figure 5.108). In order to provide the top 20 meters of soil with enough shear strength to withstand potential seismic forces, the seabed's upper soil layer is reinforced with hollow steel pipes (Combault and Teyssandier, 2005).

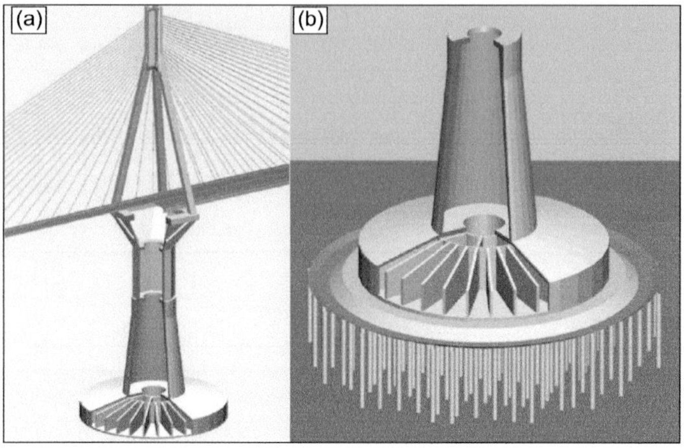

Fig. 5.108 Pylon details of the Rion–Antirion Bridge: (a) pyramidal capital, shaft, and legs; and (b) soil reinforcement at the base

5.3.28 The Millau Viaduct

The Millau Viaduct, opened in France, is another multi-span cable-stayed bridge that was completed in 2004 with a total length of 2460 m (Figure 5.109). The Millau Viaduct is the major bridge on the A75 highway between Clermont-Ferrand and Béziers, which is considered a new link between Northern Europe and Eastern Spain. The bridge is distinguished not only because it is the highest in the world, crossing the Tarn River at a height of 270 m, but also the construction period of only 38 months at very reasonable costs is outstanding. The bridge consists of six intermediate spans 342 m long and two side spans 204 m long. It carries two lanes of traffic in each direction with 3m wide shoulders on each side.

The superstructure comprises a trapezoidal profiled steel box girder with a maximum height of 4.20 m at the axis. The upper orthotropic plates are made up of steel sheets 12-14 mm thick on the major part of the main spans to ensure resistance to fatigue. The longitudinal stiffening of the upper orthotropic decking is provided by trapezoidal stiffeners 7 mm thick and in general 600 mm apart which go through the transverse diaphragms. The sloping base plates of the bottoms of the side box girders consist of 12mm steel sheets and 14-16 mm sheets around the pylons. 6 mm thick trapezoidal stiffeners are fitted at variable centers. The bottom of the box girder consists of steel sheets that have a thickness range from 25 to 80 mm. Two vertical webs 4 m apart, consisting of metal sheets between 20- and 40-mm thick cover the entire length of the structure. They are required for construction by incremental launching to spread out the localized forces on temporary piers, for construction, during deck launching. These webs are stiffened on their lower part by two longitudinal trapezoidal

Fig. 5.109 The Millau Viaduct, France, 2004

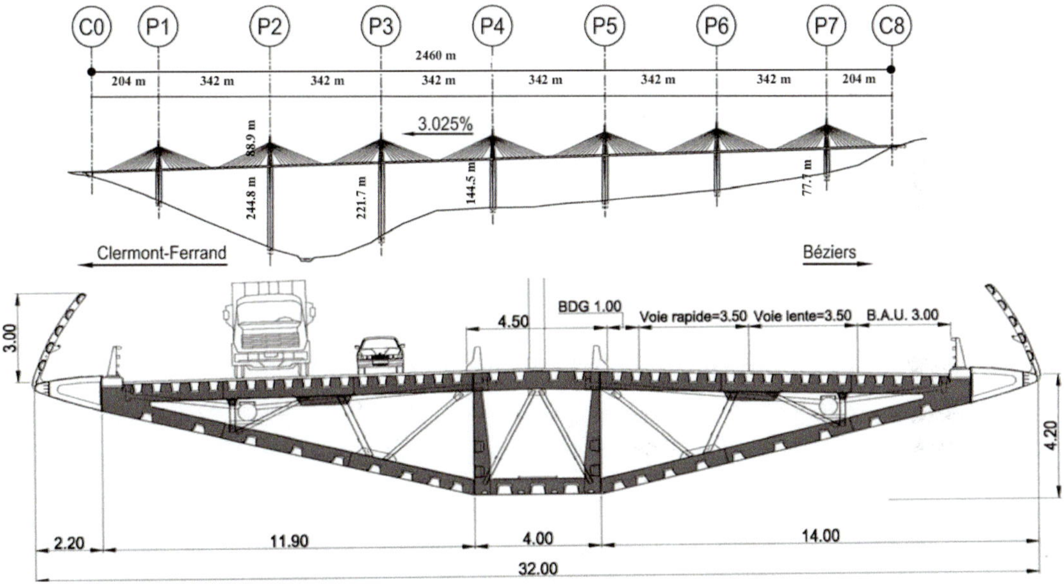

Fig. 5.110 The Millau Viaduct: elevation and cross-section (Virlogeux et al., 2005)

stiffeners. Triangulated cross beams, spaced at 4.17 m intervals in the longitudinal direction are installed to stiffen the orthotropic box (Figure 5.110). The box-girder carries two lanes of traffic in each direction with 3 m wide shoulders to increase the distance between the traffic and the bridge edge, to reduce the risk of vertigo (Virlogeux et al, 2005). It is also equipped, in addition to classical barriers, with wind screens designed to limit the wind velocity on the viaduct to the value at the approach ground level, to avoid wind shocks to vehicles arriving on the bridge; and of fairings intended to improve both the aerodynamic streamlining and aesthetic quality.

The pylons (Figure 5.111) are 90 m high above the deck with an A-shape in the longitudinal direction to achieve high stiffness. They are made of steel based on modification by Eiffel Construction Metallique, the contractor who selected this alternative against a concrete alternative to expedite construction. The pylons are fixed to the deck longitudinally to ensure continuity between the metal

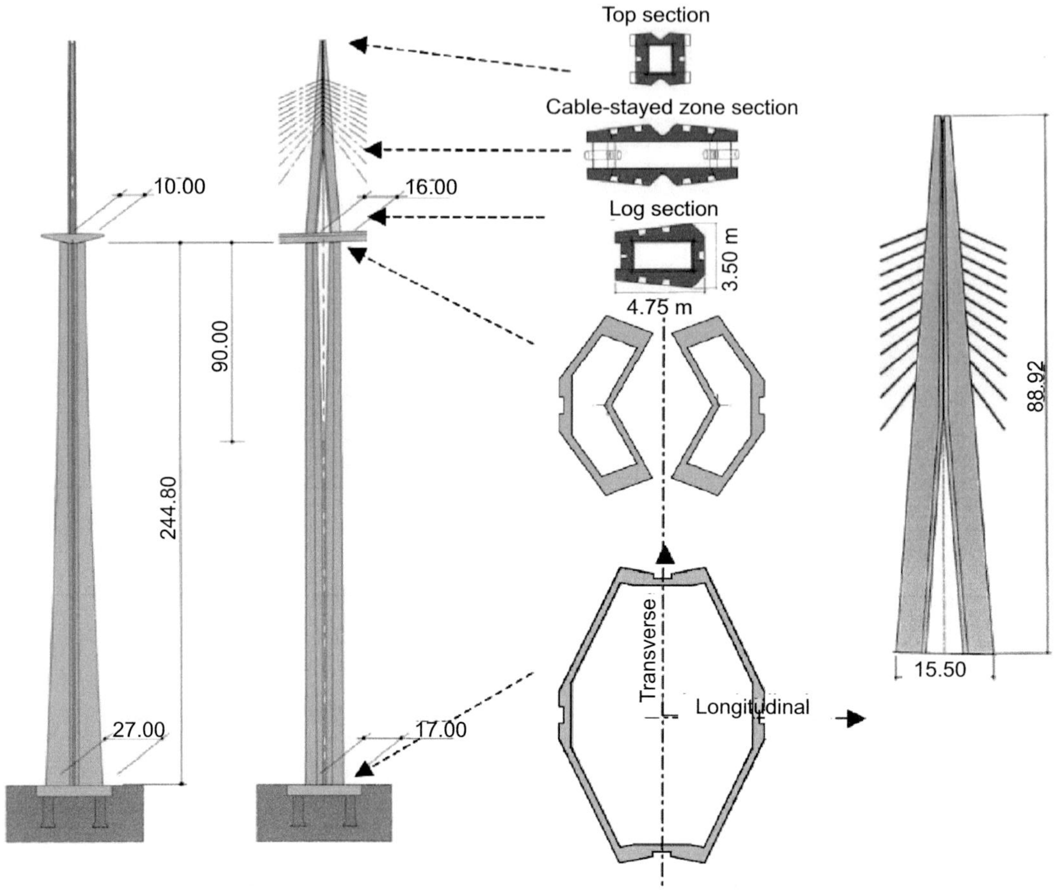

Fig. 5.111 The Millau Viaduct: piers and pylons (Virlogeux et al., 2005)

web sheets of the central box girder and those of the walls of the pylons legs. Transversely, rigidity is provided by a frame which is attached to the bearings on each pier shaft. The legs of the pylons, which are 38 m high, are composed of two stiffened steel boxes that are attached to a 49 m high mast onto which the cables are anchored. The top 17 m of each pylon, whose overall height is 87 m is not structural, but purely aesthetic. The box-girder deck is tied down to the pier by vertical prestressing tendons in line with two fixed bearings on each shaft. Guided spherical bearings, which can resist stresses up to 180 MPA under ultimate loads were used for this purpose. The piers were designed as wide strong box-sections that split into two flexible shafts in the last upper 90 m.

The cables consist of 15 strands of class 1,860 MPa which are super-galvanized, sheathed and waxed. Each cable is protected by a white, overall aerodynamic sheath made of non-injected polyethylene high-density (PEHD). This acts as a barrier to ultraviolet (UV) light and has discontinuous spirals on its surface to combat vibrations resulting from the combined effects of wind and rain. The number of strands making up each cable varies between 45 near the pylons and 91 towards the middle of each span. The cable anchors are adjustable at the deck end and fixed on the pylons (Virlogeux et al., 2005).

5.3.29 The Big Obukhovsky Bridge

Another cable-stayed bridge was opened in Saint Petersburg in December 2004. The Big Obukhovsky Bridge (Figure 5.112) gained its name based on a referendum among residents of Saint Petersburg.

Given that there is an existing three span arch bridge in the city having the name (Obukhovsky Bridge), it was decided to add the word *"Big"* to the new bridge. The Big Obukhovsky is a pair of twin cable-stayed bridges that span the Neva River, each carrying 4 lanes. The total length of the crossing, including exits, is 2884 m. The cable-stayed bridge is 994 m long and comprises 7 spans, 2 × 66 m approach spans at each side; 2 × 174 m side span at each side; and a 382 m navigable span. The height of the spans above the water surface is 30 m, which provides a passage for all ships. The pylons are diamond shaped and made of steel. They are 126 m high. The width of each span is 25 m, and the height is 2.5 m. The superstructure consists of two longitudinal steel girders connected by transverse cross beams spaced every 4.5 m and carrying an orthotropic steel deck that is reinforced with longitudinal ribs. The cables are made of parallel strands each based on seven high-strength wires with a diameter of 5 mm.

Fig. 5.112 The Big Obukhovsky Bridge, Russia, 2004

5.3.30 The Zhivopisny Bridge

The Zhivopisny Bridge, opened in December 2007, is one of the most outstanding new structures in the Russian capital. The bridge has a total length of 800 m, and one central pylon located such as the main span and the side spans are 309.5 m and 390.5 m respectively. What is unique about this bridge is its steel arch pylon (Figure 5.113), which has a peak height of 102 m and weighs 5000 tons. The superstructure is a 45 m wide orthotropic steel box and with a depth of 3.2 m. There are 72 stay cables transferring the superstructure load to the arch pylon. They were installed in only eight weeks using a laser guiding system developed by Freyssinet. IRD type dampers were installed on the longest cables to absorb vibrations.

Fig. 5.113 The Zhivopisny Bridge, Russia, 2007

5.3.31 The Solidarity Bridge

Another significant bridge was completed in Eastern Europe in 2007. The Solidarity Bridge (Figure 5.114) was built as part of the eastern bypass around Płock, along National Road No. 60, Poland. The total length of the bridge is 1200 m. It consists of two different parts, the main bridge over the stream of the Vistula River (615 m) and the access bridge over the flood plains of the left bank of the river (585 m). The total length of the bridge including the right-bank flyover is over 1700 meters and its main span is 375 meters long. The main span is one of the longest in the world among cable-stayed bridges with cables located in a single plane. At the time it was built it was considered the biggest and longest bridge in Poland and in Central Europe. The bridge carries 4 traffic lanes.

Fig. 5.114 The solidarity Bridge, Poland, 2007

The bridge comprises two pylons with a single plane harp cable stay system. The height of the pylons above the level of the deck is 64 m. The deck is 27.5 m wide. The superstructure is designed as a 3-cell steel box girder with an orthotropic deck plate and 3.5 m construction height (i.e. approximately L /100) see Figure 5.115. The central cell was designed to transmit the forces to the cables. The pylons are also designed as steel restraints in the pier. Under the bridge in the pylon axis, a support in the form of three bearings is used: middle ones for transmitting vertical force and two lateral ones. The bearing capacity of the central bearing under the pylon is 110 MN. There are 28 double cable stays transmitting the suspended span to the pylons.

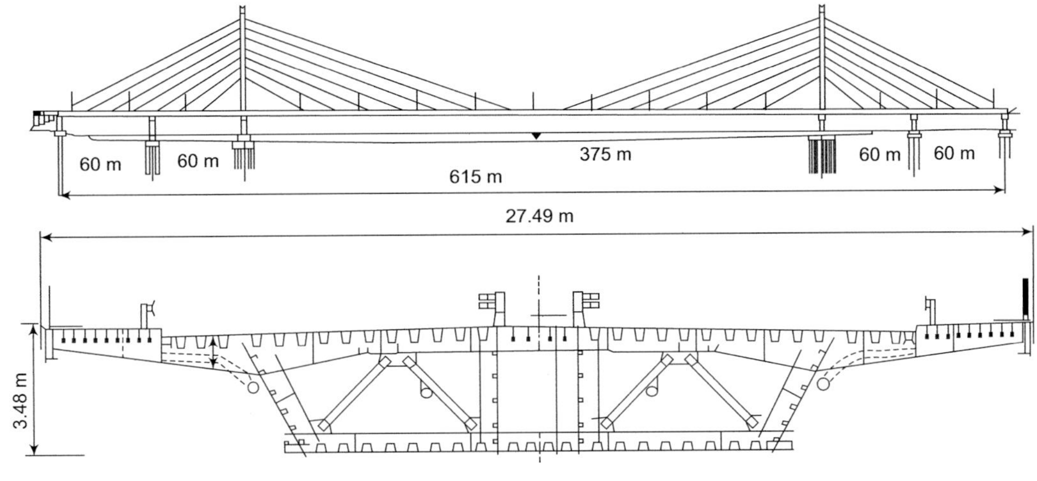

Fig. 5.115 Elevation and cross section of the solidarity Bridge (Zoltowski and Wask, 2006)

5.3.32 The Megyer Bridge

Hungary inaugurated its first cable-stayed bridge in September 2008. The Megyer Bridge (Figure 5.116) was built over the River Danube at the northern most point of Budapest. It is part of the M0 Highway around Budapest and considered the longest bridge of Budapest. The three-span fan-shaped cable-stayed bridge has a symmetric arrangement with a 300 m long middle span and 145

Fig. 5.116 The Megyer Bridge, Hungary, 2009

m long side spans, that results in a total length of 590 m (Figure 5.117). The deck is suspended by two inclined cable planes, each having 44 stay cables, onto two typical, "A"-shaped pylons. The bridge has three continuous steel spans, The superstructure is a composite cross-section consisting of two steel girder I beams that are connected by floor beams every 5 m and an orthotropic deck integral with the superstructure. All side spans and approaches are made up of restressed concrete box girders.

The two pylons of the Megyer Bridge are "A"-shaped frame structures consisting of partially prestressed, reinforced concrete pylon legs having rectangular, box-shaped cross sections. Their height is 100 m above the substructures while the outer horizontal distance between the pylon legs at the bottom is 51.0 m. The outer cross-sectional sizes of the pylon legs parabolically decrease from 5.0 × 4.0 m to 3.5 × 4.0 m parallel to the wall thickness decrease from 1.0 m to 0.5 m. The corner edges of the pylon legs are circularly curved along a 300 mm radius to reduce the wind turbulence effects.

A reinforced concrete, box-shaped beam ties the pylon legs at 55.0 m above the substructure for each pylon. The steel units as the upper anchorages for the stay cables are arranged in pylon leg sections above these tie beams. These anchorage units were positioned and fixed simultaneously with the concreting of the pylon at different elevations. The vertical components of the anchorage forces are transmitted directly to the 0.6 m thick walls of the pylon legs while the horizontal components in the longitudinal direction coming from the two sides are mostly balanced in these steel anchorage units.

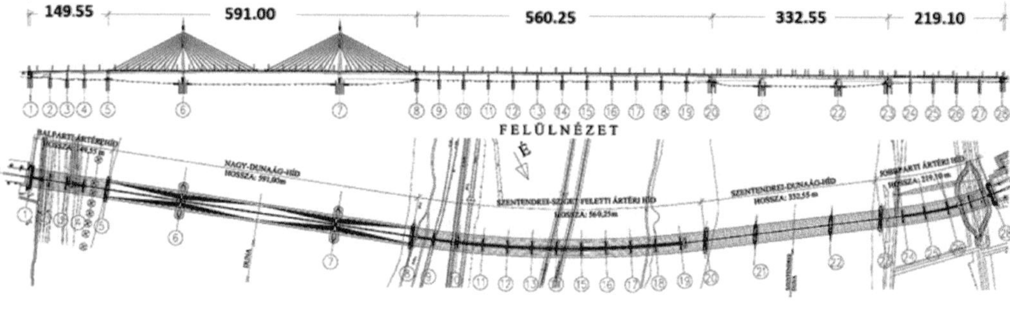

Fig. 5.117 Elevation and plan of the Megyer Bridge (Hunyadi, 2009)

5.3.33 River Suir Bridge

The Republic of Ireland opened a significant cable-stayed bridge link over the river Suir as part of the N25 Waterford Bypass that opened to traffic on 19 October 2009. The bypass forms part of the North/South Strategic Corridor which runs from Belfast via Dublin and Rosslare to Cork, linking the three most populated cities in Ireland and provides access to the key commercial seaports in the east and south. The cable-stayed bridge (Figure 5.118) is 465 m long with a 230 m long main span and a single pylon. It has the longest span and the highest pylon in Ireland. The reinforced concrete pylon is an inverted Y without any intermediate cross beam with an overall height of 112 m above the foundations. There is a total number of 76 cables transferring the load of the suspended span to the pylons. The diameter of cables including the outer sheath is ranging from 355 mm to 455 mm comprising 26 to 55 strands.

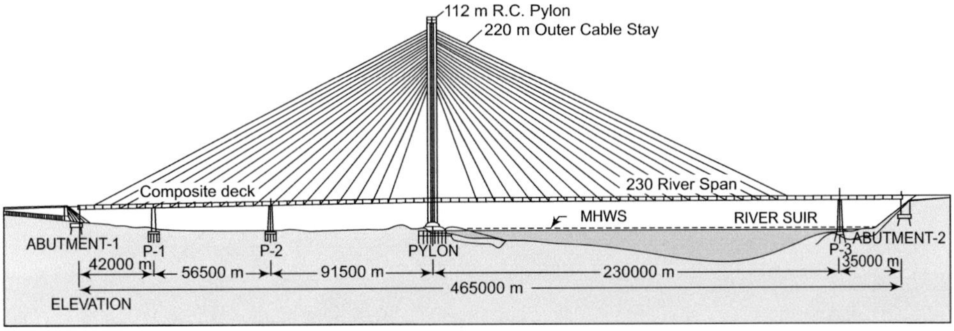

Fig. 5.118 Elevation of the River Suir Bridge, Ireland, 2009

5.3.34 The Murom Oka Bridge

Two bridges were opened in Russia in 2009 the Murom Oka Bridge and the Lazarevsky Bridge. The Murom Oka Bridge is located on The Oka River in Murom, central Russia. The cable-stayed bridge has three pylons, the length is 1391.6 m and the width is 15 m. It consists of 63.0 + 108.5 + 2 × 321.0 + 108.5 + 63.0 m span, designed for 2-lane traffic and pedestrians. The construction length of the bypass, including 3 road junctions and 5 overpasses, is 21.9 km. The bridge consists of 6-spans comprising two 231 m riverbed spans (Figure 5.119). The superstructure is a composite continuous deck, which is supported by fan shaped cables. The cables consist of secured parallel strands. The number of strands varies from 11 to 32.

Fig. 5.119 The Murom Oka Bridge, Russia, 2009

5.3.35 The Lazarevsky Bridge

The Lazarevsky Bridge is another cable-stayed bridge located in St. Petersburg and was opened to traffic five years later after the Big Obukhovsky Bridge. It crosses the Little Nevka River, connecting Krestovsky Island and Petrogradsky Island (Figure 5.120). The bridge replaced an old timber bridge that was built for trams but eventually it was used by pedestrians. One of the main designing challenges was the strict limitation on the superstructure depth. It was limited by the need to maintain the navigation clearance, while on the other hand the deck level was governed by the height of the embankments, which could not be changed since they were considered as historical monuments. Therefore, the vertical deck alignment was shaped as a vertical curve of 1000 m radius. A cable-stayed option was selected over a continuous span option to eliminate intermediate piers.

Another challenge during the designing process was to provide the required rigidity to the deck while simultaneously minimizing its weight to reduce the moments in the pylon elements and balancing the system. Therefore, a single-span cable-stayed bridge with a steel superstructure and steel pylon was selected. The deck is supported by two rows of stays, with five stays in each row. The cables pass through the pylon and are anchored in the reinforced concrete slab of the counterweight, which is located beyond the abutment on Krestovsky Island.

Fig. 5.120 The Lazarevsky Bridge, Russia, 2009

The front arch of the pylon, which is inclined towards the river, carries the dead anchorages. The arch shaped pylon was significant for distributing the forces in an optimal way. The deck consists of two main h-edge girders connected with transverse beams that work integrally with an upper orthotropic deck. The cable stays are made of standard parallel strands and each one comprises from 50 to 73 strands.

5.3.36 The Samuel Beckett Bridge

A cable-stayed-swing bridge was opened in Dublin, Ireland in 2010. This landmark structure was designed by Calatrava and follows his design for the 1992 Alamillo Bridge in Spain. The Samuel Beckett Bridge is spanning the maritime gateway to the city and located east of the city's center and within the 'heart' of the newly developed docklands area, facilitating an important urban transport link for private car use, public transport, cyclists, and pedestrians (Figure 5.121). The bridge includes two pedestrian and cycle tracks and four lanes for vehicular traffic.

Fig. 5.121 The Samuel Beckett Bridge, Ireland, 2010

The Samuel Beckett Bridge has a span of 123 m across the Liffey river. The bridge, which rotates through 90° to maintain shipping movements upstream, has an asymmetric shape, with the base to the cable-stayed steel pylon set outside of the river's central navigation channel. The pylon curves northwards to a point 46 m above the water level with 25 forestay cables set in a 'harp' formation (See elevation in Figure 5.122). The cross-section of the deck consists of a multi-cell box girder, made up of relatively thin steel plates stiffened internally using a combination of longitudinal bulb flats, angle sections and trapezoidal stiffeners. Outriggers are cantilevered from the box section to carry pedestrians and bicycles. The top of the box consists of a 14 mm thick plate with 12 mm trapezoidal stiffeners (See cross section in Figure 5.122). The cable-stays are all locked coil cables with twenty-five 60 mm diameter cables supporting the front span and a total of six 145 mm diameter cables towards the back. The main support in the river consists of bored concrete piles, with a concrete pile cap supporting a circular concrete pier of varying diameter. This houses the hydraulic turning and lifting equipment and the horizontal and vertical bearings, which support the entire bridge while turning. At each end of the bridge, locking pins are moved by hydraulic cylinders and locked into the abutments to enable the bridge to carry traffic (Cutter et al., 2011)

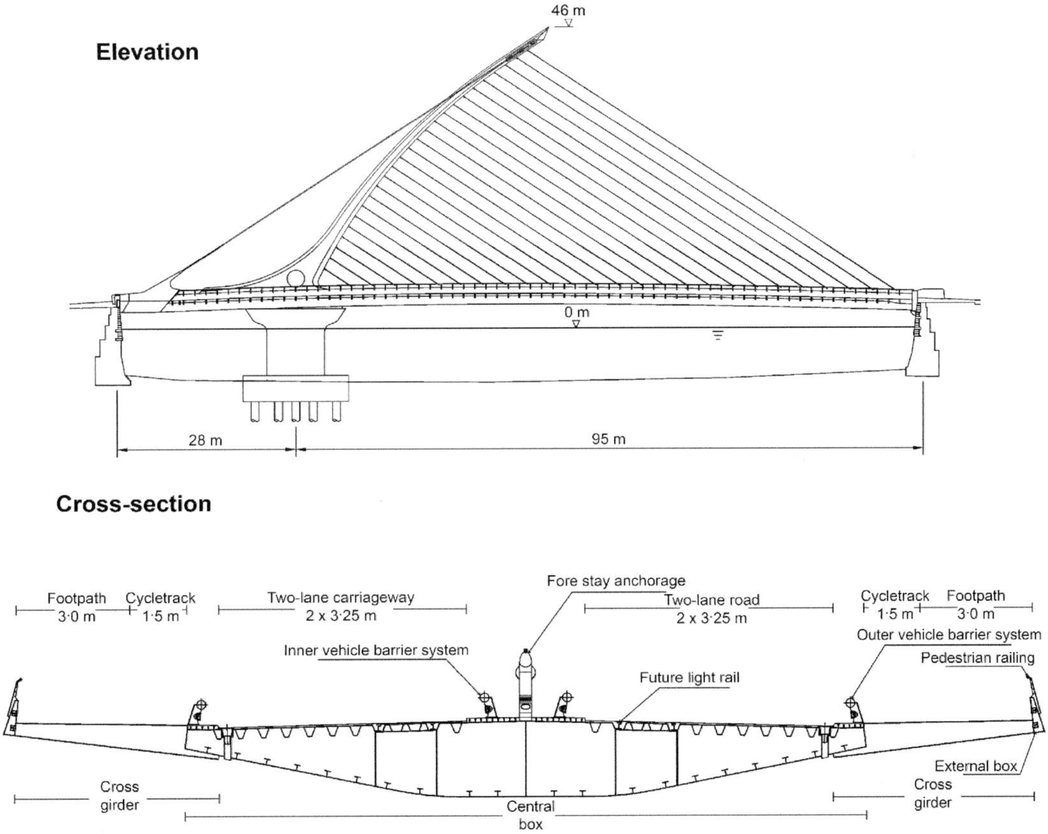

Fig. 5.122 Elevation and cross-section of the Samuel Beckett Bridge, 2010 (Cutter et al., 2011)

5.3.37 The Talavera Bridge

The Talavera Cable-stayed Bridge (Figure 5.123) opened in Spain in 2010 is considered a world record in the type of cable-stayed bridges with a single mast in high-strength prestressed concrete. This 726 m long bridge has been designed to cross the Tagus River by the South Ring Road of

Talavera de la Reina, a middle size city located in central Spain on the route connecting Madrid (Spain) and Lisbon (Portugal).

Fig. 5.123 The Talavera Bridge, Spain, 2010 (Courtesy, estudio A.I.A.)

The bridge is made up of two structures with different configurations. The main one is a cable-stayed bridge 318 m long, which crosses with a single span the widest channel of the Tagus River. The other one is an access viaduct with a length of 408 m, with two extreme spans of 36 m and seven central spans of 48 m. The cable-stayed bridge has a 36 m wide deck, with a double cable-stayed plan with cables that support the deck longitudinally every 7 m and come together in a single pylon. The two planes of cables connect both sides of the deck; at its back zone there are two more planes of cables which balance the pylon by transmitting the forces to the foundation. This is illustrated in the plan view of the bridge in Figure 5.124. The deck is a multi-cell box girder of concrete with

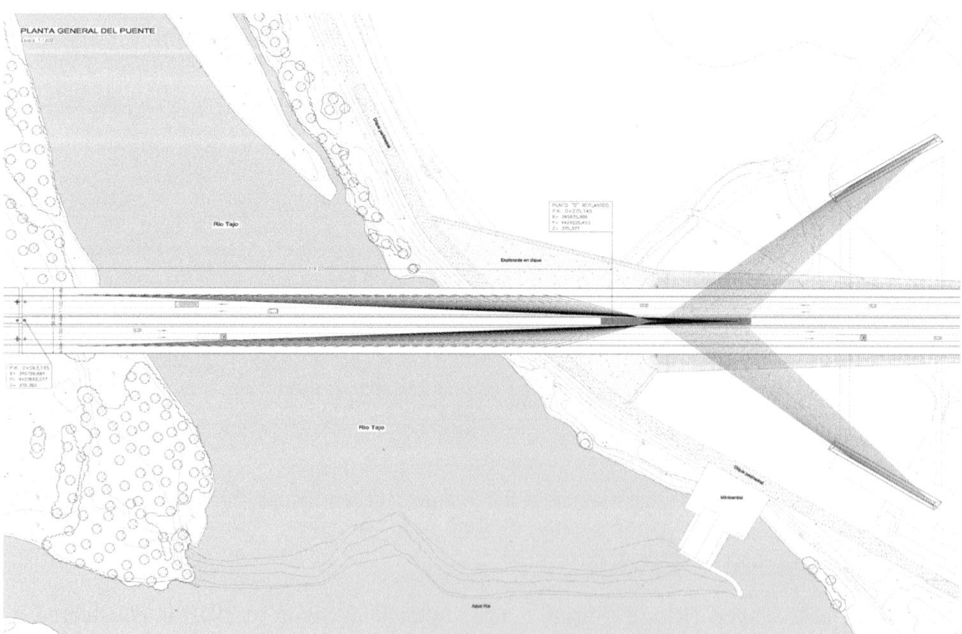

Fig. 5.124 Plan illustrating cable stays configurations of the Talavera Bridge (Courtesy, estudio A.I.A.)

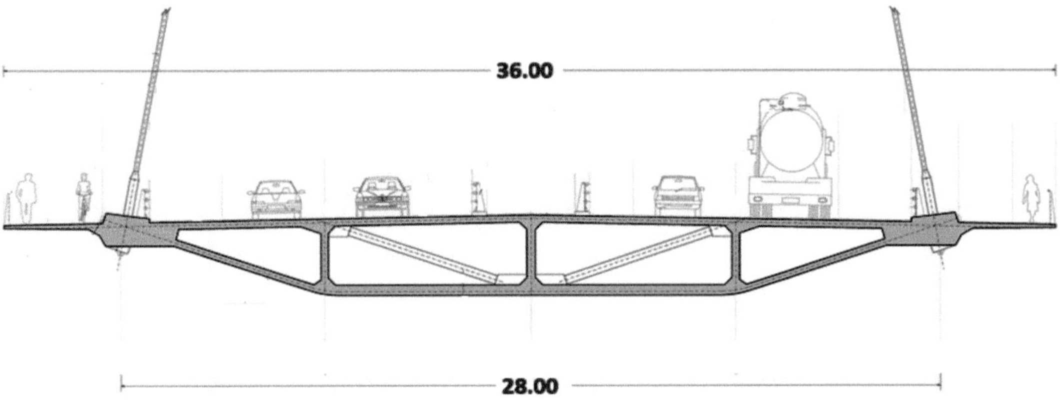

Fig. 5.125 Cross-section of the Talavera Bridge (Courtesy, estudio A.I.A.)

two lateral cantilevers. Its width is 36 m and its depth is 2.77 m. The upper slab of the deck resists vehicular traffic, and the lateral cantilevers are for the pedestrians (Figure 5.125). The pylon is an inclined reinforced concrete mast with a height of 174 meters above the top face of the deck. It is braced by a double plane of back-stay cables that balance the span through anchorages which act as true counterweights. The deck and pylon are made of high-resistance concrete between 70 and 80 MPa, with very low thicknesses to minimize the structure's weight.

5.3.38 The Terenez Bridge

The transport link across the Aulne River in Brittany, northwest France, has had a checkered history. But with the opening of the new Terenez Cable-stayed Bridge at the site in 2011, the location is finally getting a fitting structure. The first Terenez suspended bridge was built, and it opened in 1925. However, the bridge was blown up in 1944 during World War II and had to be rebuilt in 1952 using the original concrete pylons. Unfortunately, the second bridge suffered damage because of alkali-silica reaction in the concrete, which caused numerous cracks in the pylons, and it was deemed necessary to build the new cable-stayed bridge (Figure 5.126)

Fig. 5.126 The Terenez Bridge, France, 2011

The cable-stayed bridge, which is 515 m long, is curved in plan, having a radius of 800 m for the central span, and radii of 200 m in the side spans. The central span is 285 m long and each side has two back spans of 81.25 m and 33.7 m length. The 7.5 m-wide two-lane roadway is flanked on both sides by a 2.4 m-wide path intended for pedestrians, cyclists. The cross section is made of two longitudinal, almost rectangular prestressed beams, connected by the upper slab and multiple steel floor beams; the roadways are supported by the upper intermediate slab, and the sidewalks are outside the stay cables, on cantilevered slabs at the lower level (see Figure 5.127). The two pylons supporting the main span have a distinct lambda-shape. There are 18 pairs of cables that transmit the load of the superstructure to the pylon designed in a semi fan-shaped asymmetric arrangement.

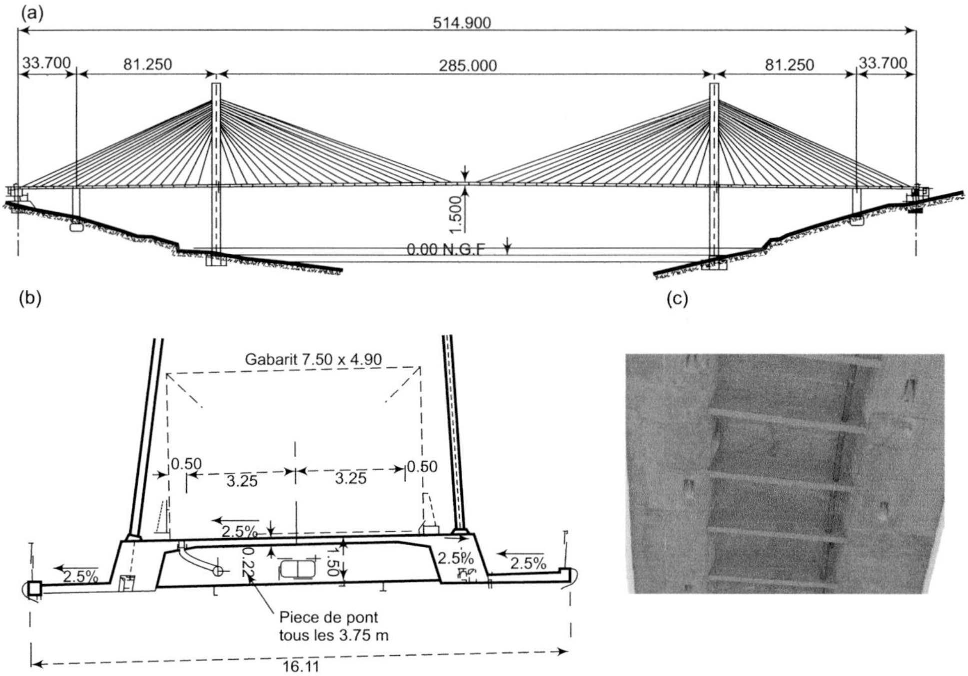

Fig. 5.127 The Terenez Bridge: (a) elevation; (b) cross-section; and (c) photo illustrating the steel floor beams (Virlogeux et al., 2004)

Fig. 5.128 The Ada Bridge, Serbia, 2012

5.3.39 The Ada Bridge

The Ada Bridge across the Sava River in Belgrade, Serbia (Figure 5.128) connects the municipalities of New Belgrade and Cukarica. When completed in 2012, it was considered one of the world's largest single-pylon bridges. The bridge is 929 m long and 45 m wide with six traffic lanes, two pedestrian pathways and a single rail line. It comprises seven spans with a 376 m long main span. The main span stiffening girder (Figure 5.129) is made of steel and supported by 80 stay cables arranged in a single plane semi-fan configuration and fixed to the central inverted Y-shaped pylon, which stretches to 200 m. The main span is counterbalanced by a post-tensioned, reinforced concrete side span of 200 m that forms the northern section of the bridge. The radius of the concrete pylon decreases from 16 m at the base to 4 m at a height of 175 m and a 1.5 m diameter at the apex. The lower part of the pylon is split into two legs situated at a height of 98 m from the base, through which the two central railway tracks pass, while the roadway lanes are located on each side. The pylon is founded 37 m deep on a circular diaphragm wall supported by bored piles.

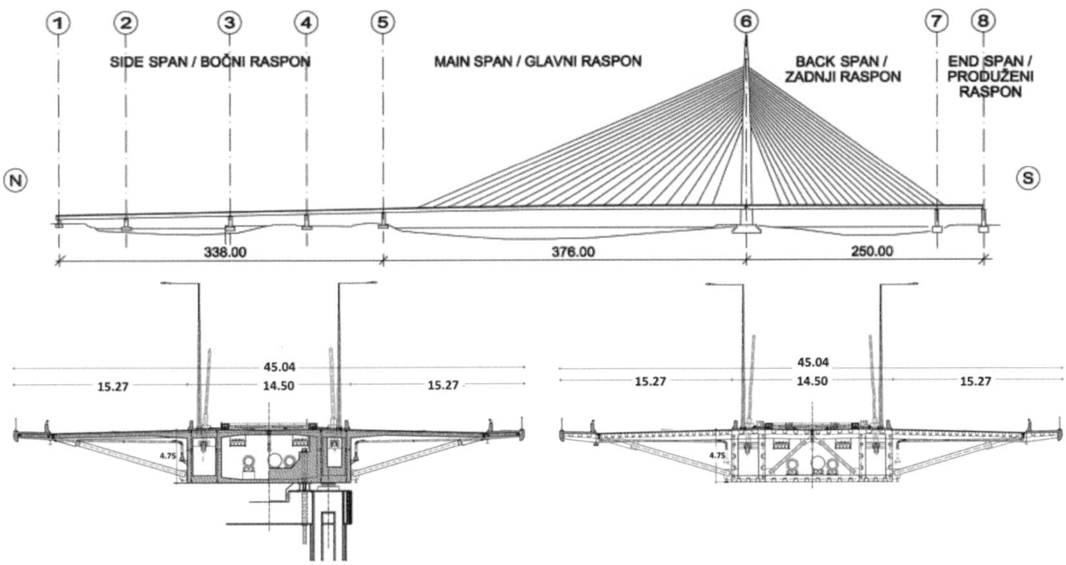

Fig. 5.129 The Ada Bridge: elevation and cross-sections (Steinkuhler and Minas, 2011)

5.3.40 The Russky Bridge

The Russky Bridge opened to traffic in July 2012 m with a main span of 1104 m to become the longest cable-stayed bridge in the world. The bridge, which crosses the Eastern Bosphorus Strait, links Russky Island with the city of Vladivostok in the Asia-Pacific region of Russia (Fig. 5.130).

Construction of the bridge was undertaken in preparation for the 2012 Asia-Pacific Economic Community Summit on Russky Island. The bridge has five end spans on each side. The bridge layout is arranged as follows: one at 60 m, one at 72 m, three at 84 m, 1104 m main span, three at 84 m, one at 72 m, and one at 60 m (Figure 5.131). The bridge carries two driving lanes in each direction and the road width is 21 m. The bridge was designed with a 70 m navigation clearance and storm waves up to 6 m high, and a temperature change from −37° in winter to + 37° in summer. The central span is supported by A-shaped reinforced concrete pylons, which are 324 m high and of varying cross-sections tapering upwards. They are transversely connected by rigid struts at about one third and two third of the height from the deck. The pylons rest on two 13 m reinforced concrete footings in

Fig. 5.130 The Russky Bridge, Russia, 2012

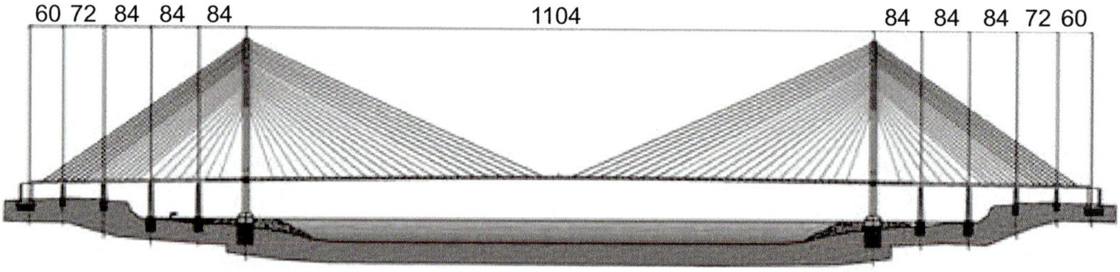

Fig. 5.131 The Russky Bridge: elevation (Lebon and Maillet, 2013)

height, connected to each other by a link beam and supported by 240 drilled shafts of 2 m in diameter and length varying from 20 m to 65 m embedded in bedrock (see Figure 5.132) (Lebon and Mallet, 2013).

The deck has a total width of 25.96 m and a height of 3.20 m and was designed with an aerodynamic profile to resist wind loads with speeds up to 36 m/s and a scaled bridge model was tested in a wind tunnel. The stiffening girder was designed as an orthotropic box girder with a system of cross beams and transverse diaphragms which extends the central span 70 m in the side bays beyond the pylons. The end spans are continuous and made of reinforced concrete to counterbalance the main span load. The side spans rest on reinforced concrete piers and have variable ranges between 60 m and 84 m. In the concrete part of the side spans, the deck is a multicellular prestressed concrete box with 2 lateral vertical webs of 0.6 meters and a central core 0.4 m thick. The upper slabs and lower slabs have a common thickness of 0.3 m (see Figure 5.133).

168 stay cables were employed for the bridge. Each stay cable consists of several parallel strands, each 16 mm in diameter. Each strand is composed of 7 galvanized, 5 mm steel wires, which are made of high-tensile steel. Stay cables consist of 13 to 79 strands depending on their position on the bridge and the design load assigned to each of them. The cables are protected by a jacket made of high-density polyethylene.

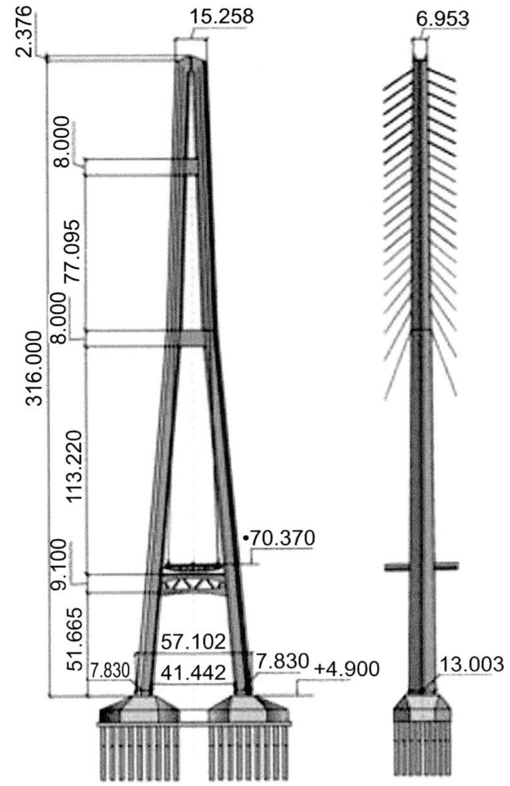

Fig. 5.132 The Russky Bridge: pylons and foundations (Lebon and Maillet, 2013)

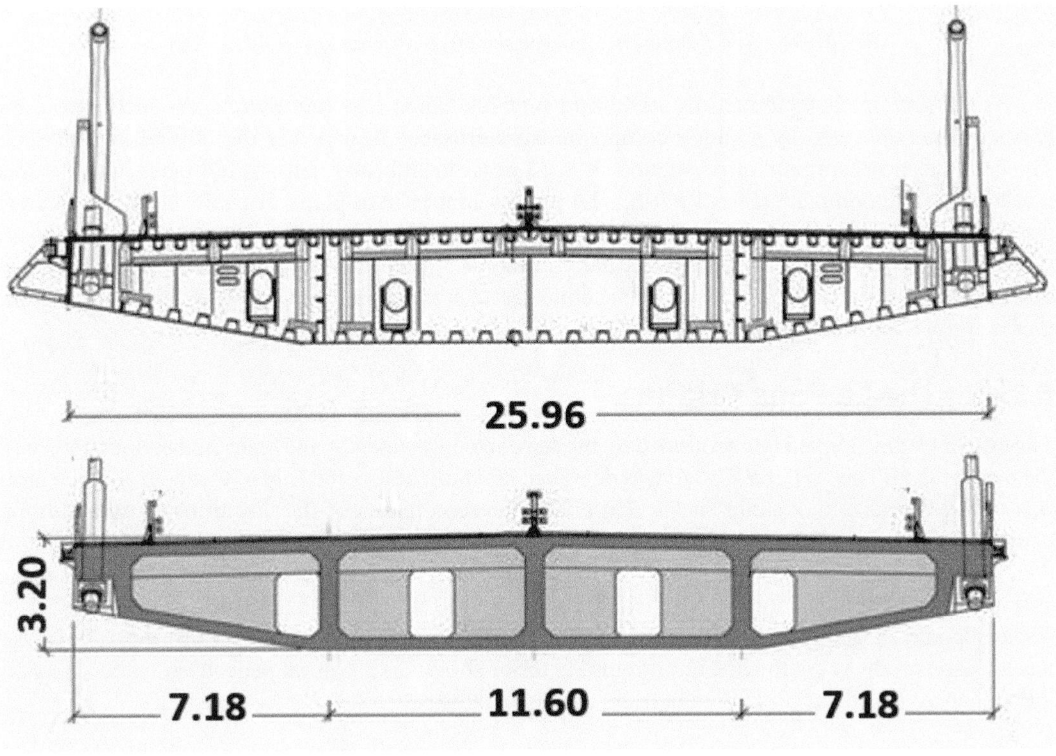

Fig. 5.133 Cross-section of the Russky Bridge: upper steel orthotropic section; and lower prestressed concrete section (Lebon and Maillet, 2013)

5.3.41 The Zolotoy Bridge

Another cable-stayed bridge was also completed in August 2012 in the city of Vladivostok, Russia. The Zolotoy is a cable-stayed bridge across the Golden Horn Bay in Vladivostok (Fig. 5.134). It is the second bridge that was built along with the Russky Island Bridge in preparation for the 2012 APEC summit. With a main span of 737 m it is considered as the world's 17th longest cable-stayed bridge. The bridge carries three lanes in each direction with a width of 28.5 m. It is designed for wind loads with speeds up to 47 m/s and earthquakes with a magnitude of 8. Hence, it is equipped with shock-transmitter devices which allow improved bridge performance under these environmental conditions. In addition to the 737 m long main span, the bridge has two 100 m side spans at each end, one span of 90 m at each end, and a 50 m span at one end with a 42 m span at the other.

Fig. 5.134 The Zolotoy Bridge, Russia, 2012 (Courtesy, MAURER SE)

The bridge is characterized by its 224 m high V-shaped concrete pylons. At each pylon, the two legs are connected by a single composite truss crossbeam, which is directly below the deck. The deck has a streamlined cross section. It is 33 m wide and carries six traffic lanes and provides a vertical navigation clearance of 67 m. The pylons support two plans of cable stays which carry the steel orthotropic box girder deck. While the main span deck is made of steel, the side spans are made of a post-tensioned concrete like Russky bridge. Both the abutment and the first pier have free bearings in a longitudinal direction. The second pier has fixed pin bearings. The third long slender pier is rigidly connected to the deck.

5.3.42 The La Pepa Bridge

The city of Cadiz in Spain is surrounded by the sea and constrained by the restricted connections with the nearest land. The La Pepa Bridge was designed and built across the Bay of Cadiz to add one more link to the Carranza moveable bridge and ended the remoteness of the city through an additional connection to Puerto Real in mainland Spain. The La Pepa crossing has a total length of 1,180 m and covers the main span of 540 m between the pylons, two back spans, each 200 m long, and spans 120 m long for cable anchorage. The whole crossing is 3,092 m long as illustrated in Figure 5.135. The cable-stayed bridge provides a vertical clearance of 69 m for navigational purposes. The bridge has a 6-lane roadway (with hard interior and exterior shoulders). Lateral pedestrian walks (only for maintenance) and wind-protective barriers are also included.

Fig. 5.135 The La Pepa Bridge, Spain, 2015: (a) general view; (b) elevation of the crossing; and (c) plan (A. Martínez Cutillas et al., 2016)

The superstructure comprises a trapezoidal orthotropic steel deck, 34.3 m wide and 3 m deep (see Figure 5.136), which is stiffened by transverse diaphragms every 5 m. The bridge has two diamond shaped pylons with a section in the shape of two combined trapezoids 14.15 × 8.19 at the base as shown in Figure 5.137. The transverse dimension reduces up to the bifurcation of the two base trapezoids into two branches that open for the superstructure to pass through and close at the top where the cable stays are attached. The upper portion houses a steel chassis for cable anchorage. The superstructure is supported by 176 cable stays whose number of strands varies between 75 0.6″ in the 4 first stays, 31 0.6″ on subsequent vertical stays and 78 0.6″ in the inclined stays. They have double protection; each cord is self-protected by galvanized steel and a single pod. The overall sheath has a helical cord to control the effects of aeroelastic instability caused by rain and wind. All tie rods have contact shock absorbers with the deck, axial for the shortest and triaxial for others.

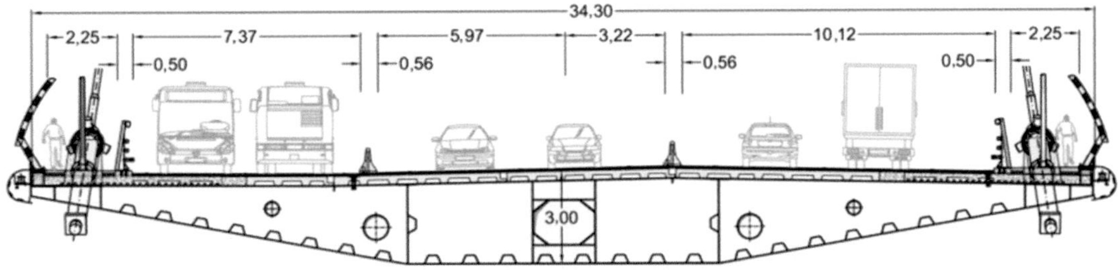

Fig. 5.136 Cross-section of the La Pepa Bridge

5.3.43 The New Forth Bridge

The Firth of Forth separates Edinburgh, the Scottish capital, from England. The downstream crossings of the Forth at Queensferry are a pair of historic bridges: a cantilever rail bridge constructed in the 1880's and the Forth Road Bridge, UK's first long span suspension bridge, which was opened in

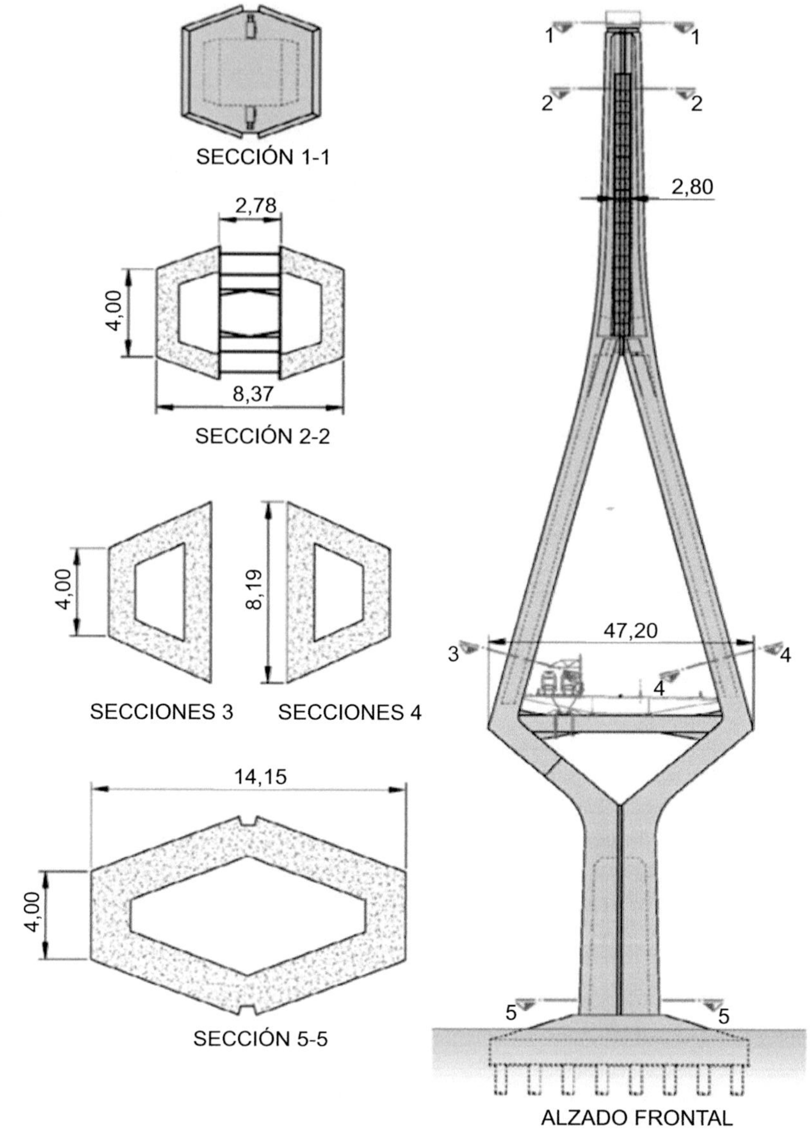

Fig. 5.137 Pylon details of the La Pepa Bridge

Fig. 5.138 Existing and new Forth Bridges: 2017 Queensferry Crossing (left); 1964 Forth Road Bridge (middle); 1890 Forth Bridge (right)

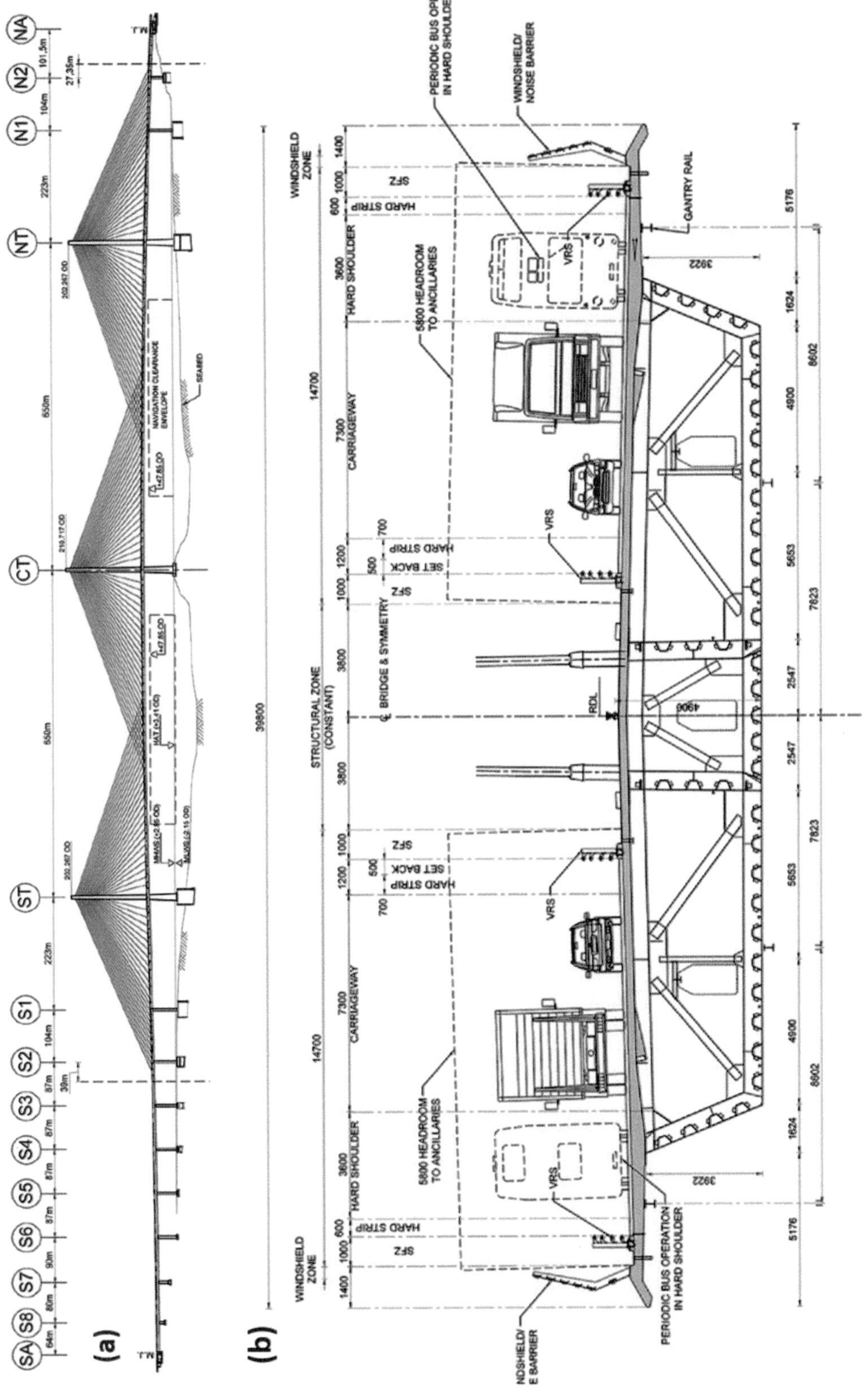

Fig. 5.139 The New Forth Bridge: (a) elevation; and (b) cross-section (Romberg, 2017)

1964. The Forth Road Bridge has successfully carried road traffic across the Forth during that time. Nevertheless, because of increased traffic and the influence of weather in 2005, inspection of the main cables revealed significant corrosion, which if not repaired could lead to the bridge being closed to heavy trucks as early as 2014, and to all traffic by 2019. The great difficulty in rehabilitation without significant disruption to traffic, resulted in a decision by authorities in December 2007 of a replacement crossing to secure the future of cross-Forth travel. The new cable-stayed bridge was formally opened to traffic, to the west of the existing bridges, on Monday, September 4, 2017 (Figure 5.138) as the replacement crossing to the Forth Road Suspension Bridge, exactly 53 years to the day from its opening it in 1964 (Shacman et al., 2019).

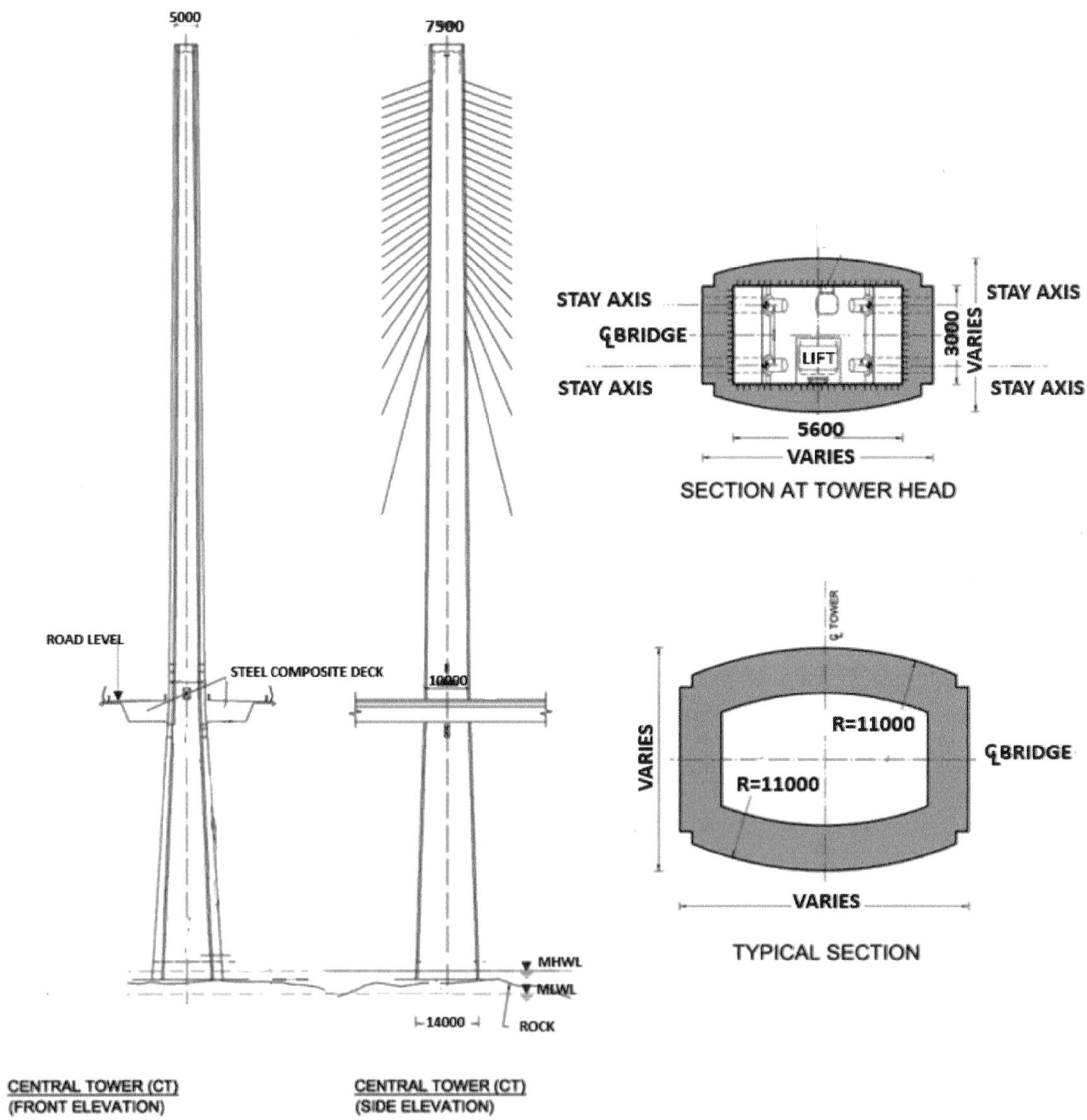

Fig. 5.140 The New Forth Bridges: Pylon (Romberg, 2017)

The cable-stayed bridge has two main spans of 650 m, and the approach twin bridges have a maximum span of 90 m. The entire bridge elevation is shown in Figure 5.139a and the cross sections of the cable-stayed bridge are shown in Figure 5.139b. The superstructure is a trapezoid orthotropic steel section consisting of three boxes (see Figure 5.139b) the central box is about 5.1 m wide and the two edge boxes have identical widths of 10.56 m each. Transverse diaphragm plates are provided for stiffening the superstructure and spaced every 5.0 m. Transverse truss cross-frames are located every 4.05 m comprising an upper plate and two diagonal plates for each box.

The steel box supports a concrete slab that has a varying thickness, transverse pre-stressing, and cantilevered lengths of approximately 5.2 m transversely on each side. The box girder height varies from 4.5 m at the center to 4.0 m at the inclined webs. All plates are stiffened by longitudinal trapezoidal steel ribs. The orthotropic box is dehumidified, and the steel structure is painted to prevent corrosion.

The cable-stayed bridge has three concrete pylons (Figure 5.140), with a height up to 210 m, positioned in the center line of the bridge. All the pylons resist the transverse force, but the central pylon, which is placed on a natural island in the center of the estuary, is the single point of longitudinal restraint (CT in Figure 5.139a). At piers S1 and N1 vertical tie down cables are provided to act as backstay connections at the North and South pylons ST and NT. All longitudinal forces are taken at the central pylon and the torsional moments are taken about the central pylons, piers and abutments of the approach bridges. The longitudinal movements are taken in moveable joints at the South and North Abutments SA and NA. Transversely the bridge deck is supported at all pylons, piers, and abutments. Each span is supported by a fanned central stay cable plane. Each fan consists of twenty-four pairs of stays anchored to the deck and the pylons. The cable pairs are spaced at an approximate transverse distance of 5 m and anchored at a longitudinal spacing of 16.2 m from the central webs of the box girder. At the mid-span, the cable stays overlap to provide longitudinal stability to the central pylon.

5.3.44 The Mersey Gateway Bridge

The Mersey Gateway Bridge is another giant cable-stayed bridge that was opened in the UK two months after Queens Ferry Bridge. The Mersey Gateway is a six-lane road toll bridge across the River Mersey and the Manchester Ship Canal in north-west England (see Figure 5.141). It connects the Central Expressway in Runcorn with the Eastern Bypass and Speke Road in Widnes. The bridge is the largest infrastructure project in the Northwest of England. It was designed to be a landmark structure that is recognizable throughout the Northwest and beyond. It covers around 9 km of road improvements and a series of major new junctions running throughout Runcorn and Widnes,

Fig. 5.141 The Mersey Gateway Bridge, UK, 2017 (Courtesy, Mersey Gateway Authority)

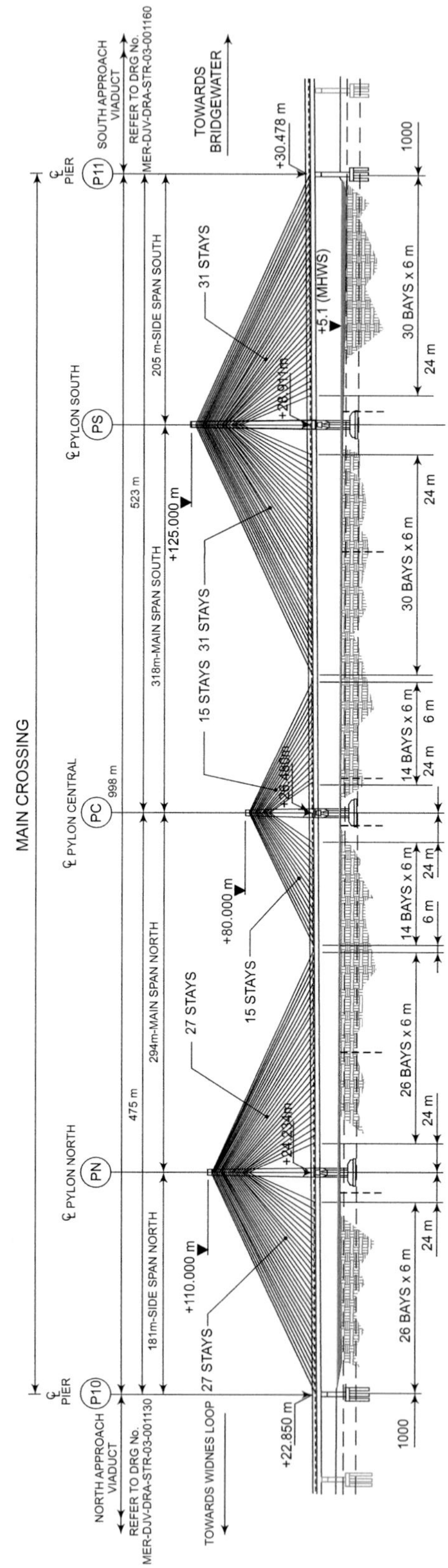

Fig. 5.142 Elevation of the Mersey Gateway Bridge (Courtesy, Mersey Gateway Authority)

England. It will improve journey times and reliability for millions of people and attract massive inward investment and regeneration in the region. The Mersey Gateway Crossing is 2.25 km long and includes a 1 km long three-pylon cable stayed bridge across the river Mersey estuary, together with approach viaducts to the north and south. The main spans are 294 m and 318 m long with approach viaduct spans at around 70 m each. The three-pylon configuration provided a balanced deck, featuring a shorter central pylon as illustrated in Figure 5.142. The main bridge deck is made from reinforced and post-tensioned concrete. The deck features a single central plane of cables and a continuous single cell concrete box girder 4.6 m deep with transverse post-tensioned ribs at around 6 m centers (see Figure 5.143).

The north and south pylons support the deck on bearings whereas the central pylon is monolithic with the bridge deck. The bridge deck, with a combined load-bearing weight of more than 53,000 tons, is supported by 146 stay cables. There are 62 stay cables attached to the south pylon (31 on each side), 54 attached to the north pylon (27 on each side) and 30 attached to the central pylon (15 on each side). The cable sheaths are helix-shaped for wind and rain protection and tested in wind tunnels. They vary in length; with the shortest measuring approximately 41 m and the longest measuring 226 m. Each stay cable consists of 41 to 91 individual steel strands that sit inside a stay pipe. Each strand contains seven wires, which are galvanized, waxed, and coated (Sanders et al., 2019)

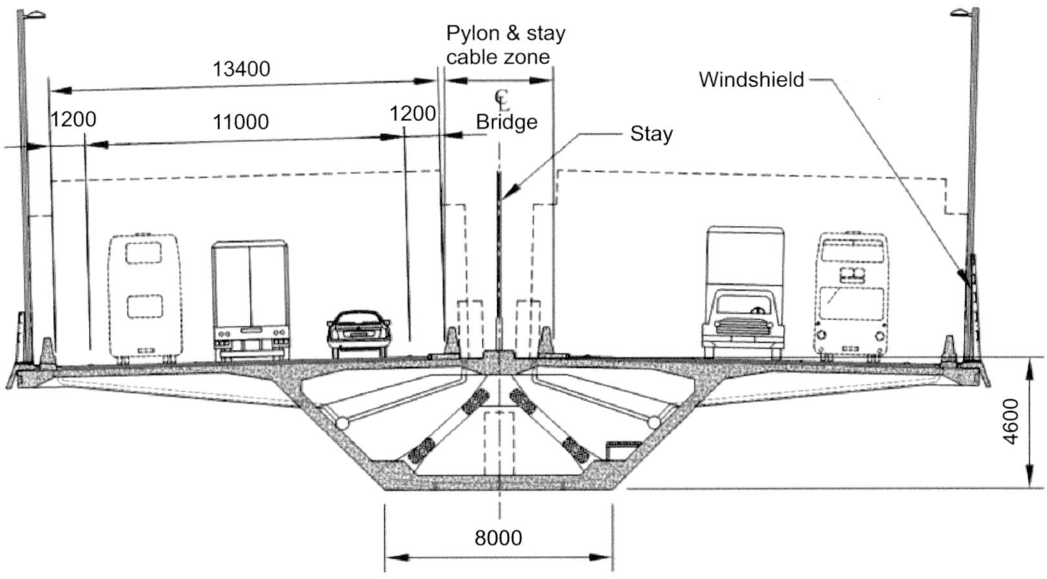

Fig. 5.143 Cross-section of the Mersey Gateway Bridge (Courtesy, Mersey Gateway Authority)

References

Beyer, E. and F. Tussing, Nordbriicke Dusseldorf. *DerStahlbau*, Berlin, Germany, 24(2), 25–33 (in German), 1955.

Brixner, S., Mundecke, M. and Gunkel, F., The new Berlin bridge in Halle – the first German one Cable-stayed bridge in composite construction (in German), *Stahlbau* 76, 2, pp 79–86, 2007.

Brodin, S., New Tjorn Bridge across the Ackerofjord (Sweden), *IABSE* Structures, C-21/82, 40–41, 1982.

Capra, A. and Leveille, A., Vasco da Gama Bridge, Portugal *Structural Engineering International* 4/98 pp 261–262.

Combault, J. and Teyssandier, J.P., The Rion-Antirion Bridge Concept, Design and Construction, *Metropolis and Beyond, Structures Congress*, New York, 2005.

Crhan, J., Geryk, J., Kominek, M., Koucky, R., Rotter, T. et al., New cable-stayed bridge over the Elbe in Aussig, (in German), *Stahlbau* 68, 1, pp 56–61, 1999.

Cutillas, A.M., González-Valerio, J.A.N, Garcia, S.F, Gutierrez, G.O., Design of the Constitución de 1812 cable-stayed Bridge over the Cadiz Bay, *Hormigón y Acero* 2016; 67(278–279): 71–85.

Cutter, J., Flanagan, J., Brown, P., Rando, M., Mo, G., Samuel Beckett Bridge, Dublin, Ireland *Proceedings of the Institution of Civil Engineers, Bridge Engineering*, Volume 164 Issue BE3, pp 133–144.

Dischinger, F.: Suspension bridges for the heaviest traffic loads, *Civil Engineer (Bauingenieur),* pp. 65–75, 107–13 (In German), 1949.

Eilzer, W., Richter, F., Wille, T., Heymel, U. and Anistoroaiei, C. et al., The Niederwartha Elbe Bridge –the first cable-stayed bridge in Saxony, (in German), *Stahlbau* 75, 2, pp 93–104, 2006.

Eilzer, W., Portius, M., Morawietz, M., Stockmann, R., and Heymel, U. et al., New construction of the Schönebeck Elbe Bridge–Design, tender and award *Stahlbau* 79, 2, pp 77–90, 2010.

Feige, A., The Evolution of German Cable-Stayed Bridges: an Overall Survey, *Acier-Stahl-Steel*, 12–66, 1966

Gimsing, N., J., Cable Supported Bridges across Straits in Denmark, *Docomomo* 45–2, 2011.

Gurtmann, S., Hamme, M., Marzahn, G, and Sieberth, S., Rope replacement under traffic at the Rhine Bridge Flehe (in German), *Stahlbau* 79, 9, pp 682–688, 2010.

Hunyadi, M., The northern Danube bridge of the M0 ring road. Pre-planning, licensing plan, commissioning (in Hungarian), *Journal of the Hungarian Group of fib*, 08/3 pp 70–72, 2008.

Idelberger, K., Cable-stayed bridge over the Rhine with A-shaped battery Neuwied – Germany (in Spanish), *Informes de la Construcción* Vol. 32, nº 314, pp 51–59, 1979.

Lebon, J-D, Maillet, V., The Russki Cable-Stayed Bridge to Vladivostok, *TRAVAUX* Number 896, pp 98–106, 2013.

Lockmann, H., and Marzahn, G.A., Spanning the River Rhine with a new cable-stayed bridge, *Steel Construction* 2, 1, pp 9–14, 2009.

Otto, S., Thiele, K., Casper, H-J, Karpa, M., and Sachse, F. et al., The second Strelasund crossing Statics, construction and assembly of the composite and cable-stayed bridge (in German), *Stahlbau* 75, 2, pp 69–81, 2006.

Reusink, J. and Kuijpers, M., Damper against rain-wind-induced vibrations of the stay cables at the Erasmus Bridge, (in German), *Stahlbau* 67, 10, pp 768–775, 1998.

Romberg, M., New Queensferry Bridge in Scotland (in German), *Construction technology (Bautechnik)*, 88, issue 7, pp 470–478, 2011.

Sanders, P., Brennan, G., Wood, H., Banks, J., Martin, J.R. et al., Design and Construction of Mersey Gateway Bridge, U.K., Structures Congress, Orlando, Florida, pp 186–197, 2019.

Sanson, R., Saint-Nazaire-Saint Brevin Bridge over the Loire Estuary (France), *Acier-Stahl-Steel*, No. 5–76, pp 161–167, 1976.

Saul, R., Hopf, S., Weyer, U. and Dieckmann, C., The Rhine crossing A 44–Construction and static Calculation of the current bridge (in German), *Stahlbau* 71, 6, pp 393–401, 2002.

Schaaf, T. and Spoelstra, J.S., Cable-Stayed Bridge over the Waal near Ewijk (Netherlands), *Acier-Stahl-Steel*, No. 1–76, pp 10–21, 1976.

Stelzer, B. and Dorrer G., New construction of a single-lane road bridge over the Neckar in the municipality of Zwingenberg, (in German), *Stahlbau* 80, 5, pp 331–336, 2011.

Steinkühler, M. and Minas, F., Construction of a cable-stayed bridge over the Sava in Belgrade (in German) *Construction technology (Bautechnik)*, 94, Issue 2, pp 93–103, 2017.

Thul, H. Steel Road Bridges in the Federal Republic, *Civil Engineer (Bauingenieur),* No. 5, 169–189, 1966.

Virlogeux, M., Normandie Bridge Design and Construction *Proceedings of the Institution of Civil Engineers-Structures and Buildings* Volume 99 Issue 3, AUGUST 1993, pp. 281–302.

Virlogeux, M., Motard, M., Bouchon, E., Imberty, F., Petitjean, J. et al., Design of the Térénez curved cable-stayed bridge, *fib Symposium Avignon, France*, April 26–28, 2004.

Virlogeux, M., Servant, C., Cremer, J-M., Martin, J-P, and Buonomo, M. et al., France, *Structural Engineering International*, Volume 15, Issue 1, pp 4–7, 2005.

Wagner, P., Development potential in road bridge construction through the use of hybrid support structures – Part 2: Cable-stayed bridges, (in German), *Stahlbau* 82, 11, pp 851–867, 2013.

Warolus, L., Cable-Stayed Bridge over the Meuse at Heer-Agimont (Belgium), *Acier-Stahl-Steel*, No. 5–75, pp 161–166, 1966.

Zolotowski, K. and Wask, T., Cable Stayed Bridge Over Vistula River in Plock. Dynamic Analysis and Site Test *Proceedings of the International Conference on Bridges, Dubrovnik, Croatia,* 431–438, 2006.

CHAPTER **6**

Advancement of Cable-Stayed Bridges in the US and Canada

6.1 CABLE–STAYED BRIDGES IN THE US

The first cable-stayed bridge in the US was built in 1971, exactly fifteen years after the Stromsund, the first world cable-stayed bridge in Sweden. Today, more than sixty cable-stayed bridges are part of the roads and highway network in the US. Construction technology and advanced methods of research in material science have been significant factors for developing cable-stayed bridges in the US. New advanced materials that were recently implemented in bridge technology include Ultra-High-Performance Concrete (UHPC), high strength low alloy steel, stainless steel for reinforcement, and carbon-fiber-reinforced polymers. Advanced methods in structural health monitoring were very important in inspiring the widespread adoption of this type of construction because they facilitated reliable methods of inspection and maintenance of these structures. The new technologies incorporated into hardware over the past 40 years have improved methods of analysis and design of cable stayed bridges. Evolution in methods of analysis such as Finite element methods and their incorporation in state of art the software contributed to the enhancement of this practice in the US and around the world. Spans of highway cable-stayed bridges in the US range from 100 m to 506 m with the exception of the Gordie Howe International Bridge in Michigan expected to be operational in 2024. This range is significantly less than those of bridges in some other parts of the world. The main reason is that the US has been ahead of the rest of the world in building long span bridges. Most of the spans beyond 506 m are covered by suspension bridges since the late thirties of the last century. Suspension bridges have been existing and taken care of for several decades. Examples include but not limited to the Bronx-Whitestone Bridge in New York City, with a main span of 700 m; the Walt Whitman Bridge connecting Philadelphia to Camden New Jersey, with a main span of 610 m; the multi-span San Francisco-Oakland Bay Bridge, with a main span 704 m and more. Currently retiring suspension bridges in the US are being replaced with cable-stayed bridges such as the Waldo-Hancock in Maine (main span 244 m), which has been replaced with the Penobscot Narrows Bridge (main span 348 m). Only selected bridges from different States are discussed in detail herein.

6.1.1 The John O'Connell Memorial Bridge

The Site of the first cable-stayed vehicular bridge in the United States is Sitka, on Baranof Island, in the south-east panhandle of Alaska. The John O'Connell Memorial Bridge (Figure 6.1) was opened to traffic in 1971. The bridge has four 38.1 m approach spans, two 45.7 m side spans, and a 137 m main span (see Figure 6.2). The side and main spans are continuous with expansion joints between the approach and side spans.

Fig. 6.1 The John O'Connell Memorial Bridge, Alaska US, 1971, (Courtesy, SolDuc Photography)

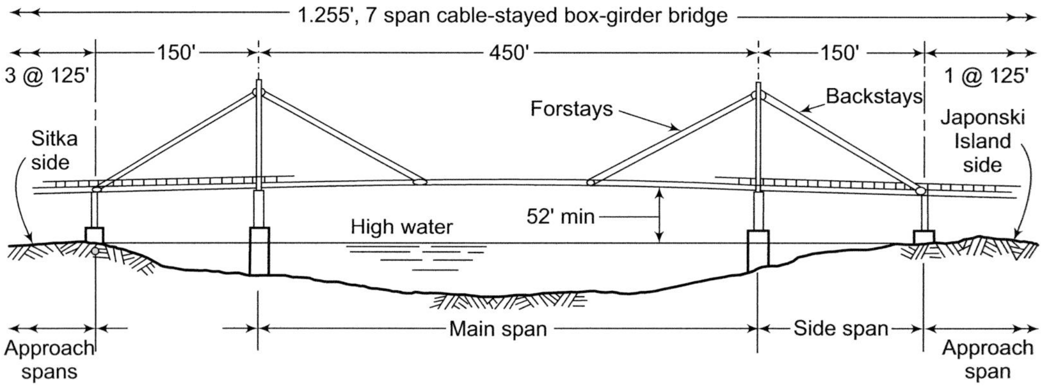

Fig. 6.2 Elevation of the John O'Connell Memorial Bridge (Gute, 1973)

The superstructure consists of two edge box girders with 1.8 m i.e., 1/74 of the main span (Gute, 1973). The width of the box girders is 0.76 m. A conventional reinforced concrete deck slab, composite with the girders, was used. The interior of the girder as shown in Figure 6.3 is supported on small longitudinal stringers spanning between floor beams on 7.6 m centers. The bridge uses two planes of cable stays each supported on a 30.5 m high steel pylon, which is a free standing vertical welded steel box 0.9 × 1.2 m in section fixed to the piers with high strength threaded rods anchored in the concrete (see Figure 6.3).

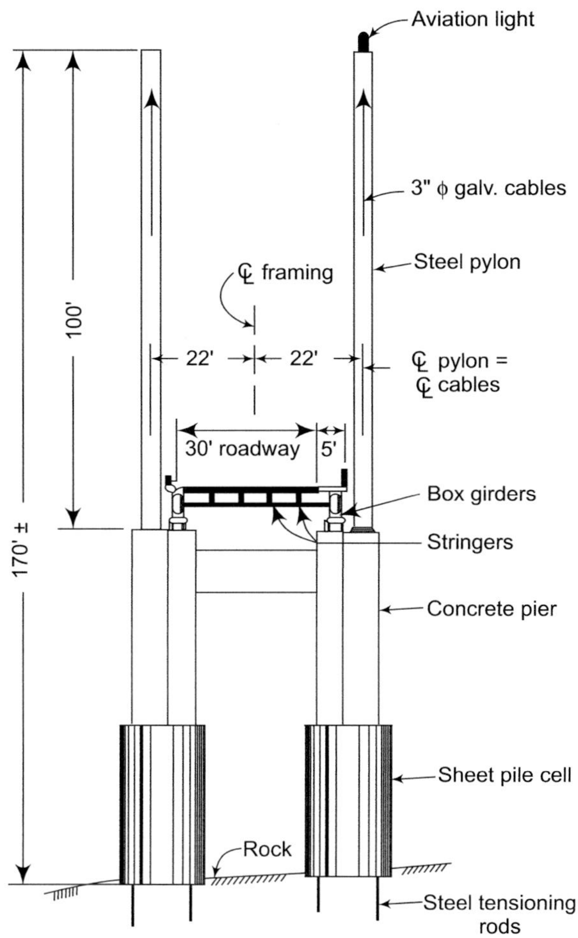

Fig. 6.3 The John O'Connell Memorial Bridge: cross-section of the bridge deck, steel pylons, and concrete piers (Gute, 1973)

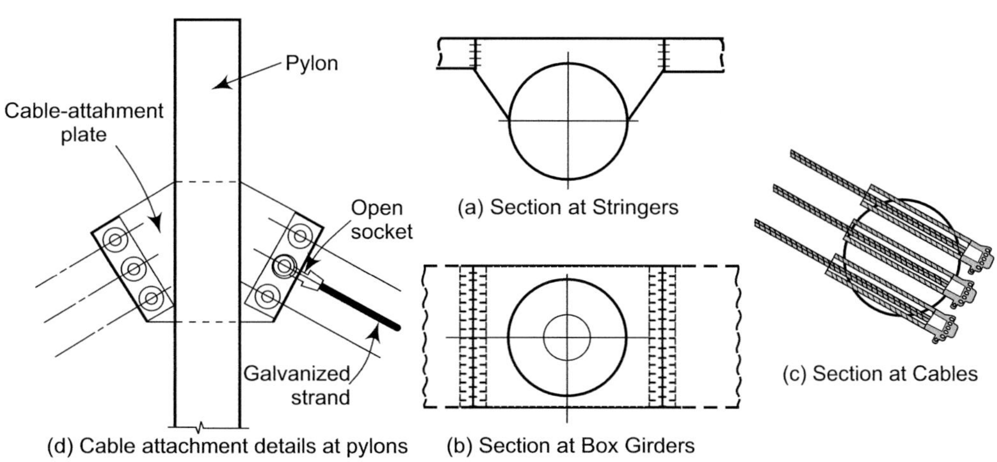

Fig. 6.4 The John O'Connell Memorial Bridge: Miscellaneous details (Gute, 1973)

The plate thicknesses of the pylons range from 38.1 mm at the lower part to 19 mm at the upper part of the pylon. 24 cables are used to transfer the superstructure load to the pylon and balance it. Cables are attached with open sockets to large plates at the tops of the pylons through tubular transverse steel beams at the girder level (See sections a through c in Figure 6.4 and photo in Figure 6.5).

Fig. 6.5 The John O'Connell Memorial Bridge: Photo of the tubular anchor beam (Gute, 1973)

Two land piers and four underwater piers support the bridge. Every pier has a reinforced concrete frame that is about 12.2 m high at the top (see Figure 6.3). Except for the T-shaped piers supporting the pylons, which measure 2.0 × 4.1 m outside, the frame legs are divided into sections measuring 1.2 m by 2.1 m. Short H-piles driven into bedrock support the land piers.

6.1.2 The Ed Hendler Bridge (Pasco-Kennewick)

Seven years after the opening of the John O'Connell Memorial Bridge in Alaska another bridge was opened in Washington State in September 1978. The Ed Hendler Bridge (Pasco-Kennewick) (Figure 6.6) was built on the Columbia River in Washington to connect the central areas of the cities of Pasco and Kennewick. The total length of the crossing is about a kilometer of concrete structure with a suspended portion across the Columbia River that replaced an obsolete 56-year-old, narrow steel truss structure (Grant, 1979).

The bridge spans four lanes of traffic and is 763 m long by 24 m wide. The bridge has a 299 meter main span and two 124 m side spans. A plan and elevation of the structure are shown in Figure 6.7. The bridge girder is made up of 62 sizable, independently precast, prestressed concrete segments. It is suspended over 547 m of length and is structurally continuous throughout. These segments were manufactured in a casting yard on site, loaded onto barges and lifted into place.

The superstructure as shown in Figure 6.8 is a reinforced concrete opened section that consists of two edge trapezoidal box girders that are transversely attached with floor beams every 2.5 m. This system supports the concrete roadway slab. A total of 144 stay cables, attached at 8 m intervals

Fig. 6.6 The John Ed Hendler Bridge (Pasco-Kennewick) Bridge, Washington, US, 1978

along both edges of the concrete girder, are used to support the superstructure. They are arranged in two planes and held at the tops of the two pylons by unique welded steel pylon-head assemblies. Steel wires with a parallel 6.4 mm diameter make up the cables. Each cable has between 73 and 283 wires, depending on the load and location of the cable. The girders and cable anchorages are joined in a monolithic manner. After installation, Portland cement grout is used to fill the gap between the steel wires and the pipe wall, providing corrosion protection for the wire bundles enclosed in polyethylene pipes.

6.1.3 The Hale Boggs Bridge (Lulling)

Studies to assess the viability of crossing the Mississippi River in the delta region close to New Orleans were first conducted in 1970 by the Louisiana Department of Transportation and Development. According to the study, there could be a main span that is between 376 and 640 m long. The applicability of cable-stayed bridges, cantilever trusses, and conventional suspension bridges was compared. In the end, the Department chose to construct a steel cable-stayed bridge to span the Mississippi River close to Luling as part of an interstate route (1-310) that passes close to New Orleans.

The main span of the bridge was set at 376 m making it one of the longest in the world at that time. Side spans of 151 m together with the adjacent approach spans of 79 m were to be continuous forming a main span bridge unit 836 m long. Conventional beam and girder spans were used on the approaches. The bridge (Figure 6.9) is named after the late United States Congressman Hale Boggs. It was opened to traffic on October 6, 1983. The Hale Boggs Bridge was the third major cable-stayed bridge in the United States after the John O'Connell Bridge of Sitka, and the Ed Hendler Bridge in Washington. The Hale Boggs Bridge was the first cable-stayed bridge to be built across the Mississippi River.

The geometric layout of the cable-stayed bridge is displayed in Figure 6.10. All the spans have a constant depth of 4.3 m, which results in a depth-to-span ratio of 1/87. With 1.3 m left and 3 m right shoulders, the bridge accommodates two 3.7 m lanes of traffic in each direction and the total width is 25 m between the parapets. The superstructure comprises two longitudinal steel orthotropic

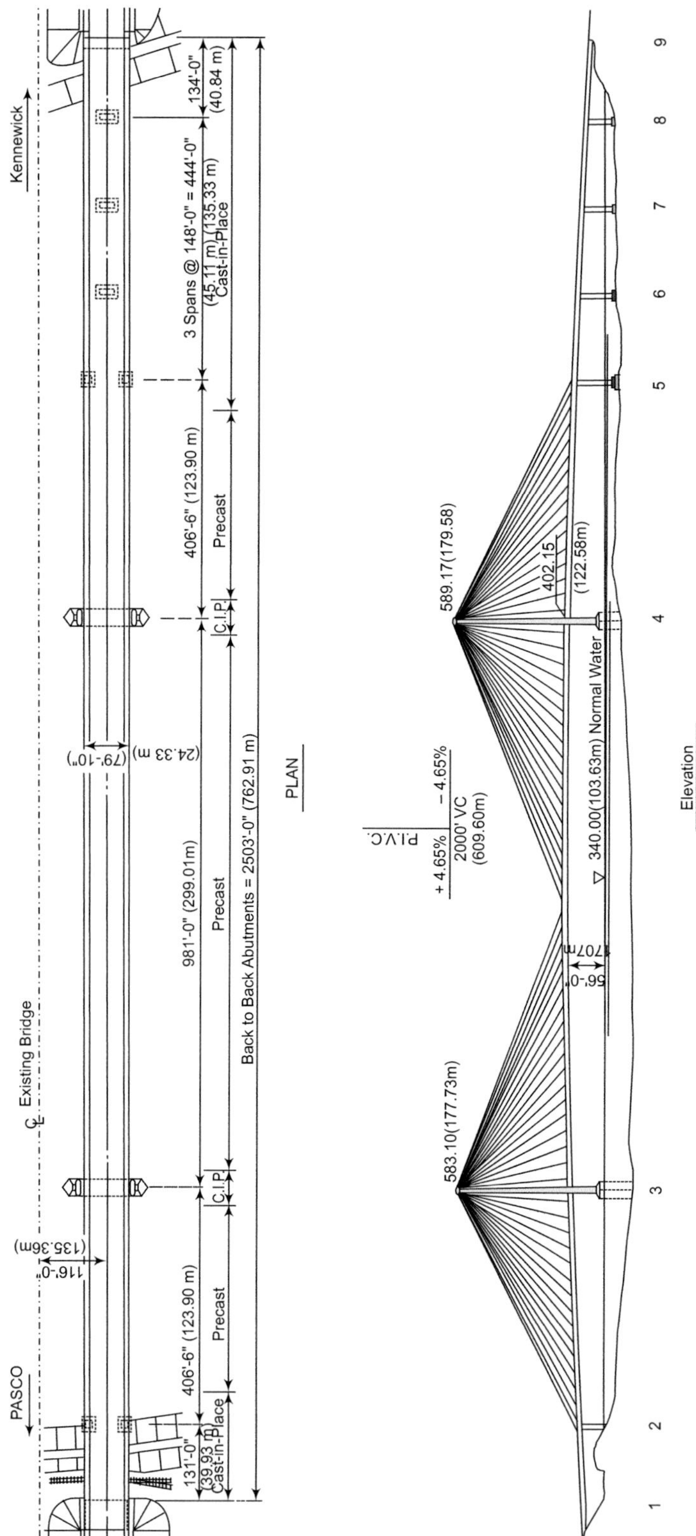

Fig. 6.7 The John Ed Hendler Bridge: plan and elevation

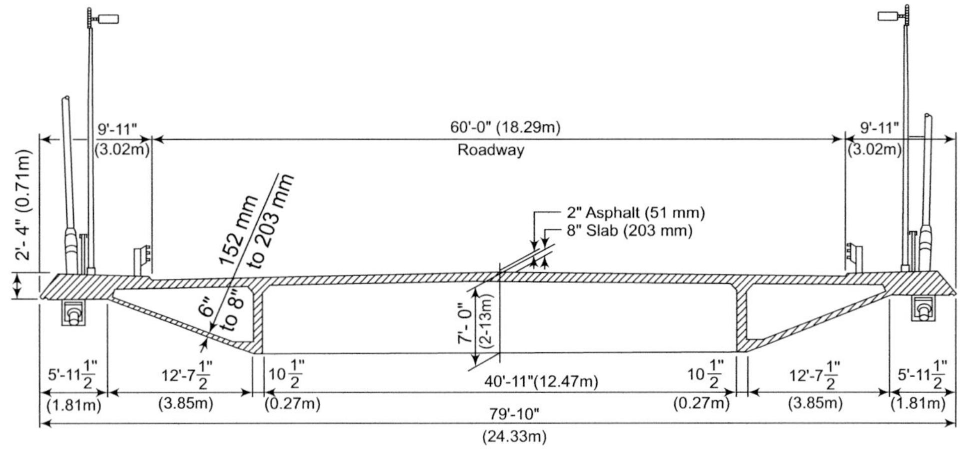

Fig. 6.8 The John Ed Hendler Bridge: cross-section

Fig. 6.9 The Hale Boggs Bridge, Louisiana, US, 1983 (Courtesy of LADOT)

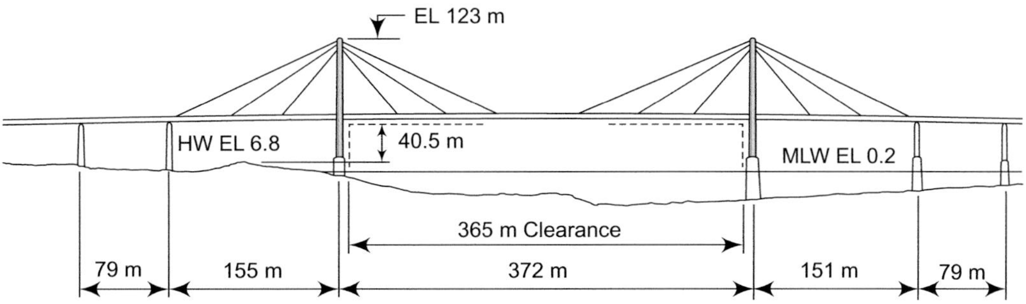

Fig. 6.10 Elevation of the Hale Boggs Bridge (Kealey, 1982)

girders spaced 11.9 m apart (see Figure 6.11a). The transverse floor beams are 0.9 m deep welded plate girders, spaced at 4.6 m and cantilever out at 3.5 m. The deck is an orthotropic plate that acts integrally with the floor beams. The orthotropic deck plate thickness is 11 mm and stiffened in the longitudinal direction by trapezoidal closed ribs that are spaced 0.66 m on center. These ribs are continuous and penetrate through the floor beams. To increase the torsional stiffness of the cross-section, the longitudinal box girders are connected by a full depth plate diaphragm at 13.7 m spacing. In Figure 6.11b, a sloping fairing plate is included to enhance the aerodynamic characteristics of the area between the two pylons. The 64 mm epoxy asphalt wear surface is the final one. For the protection of traffic, steel parapets and median barriers have been installed.

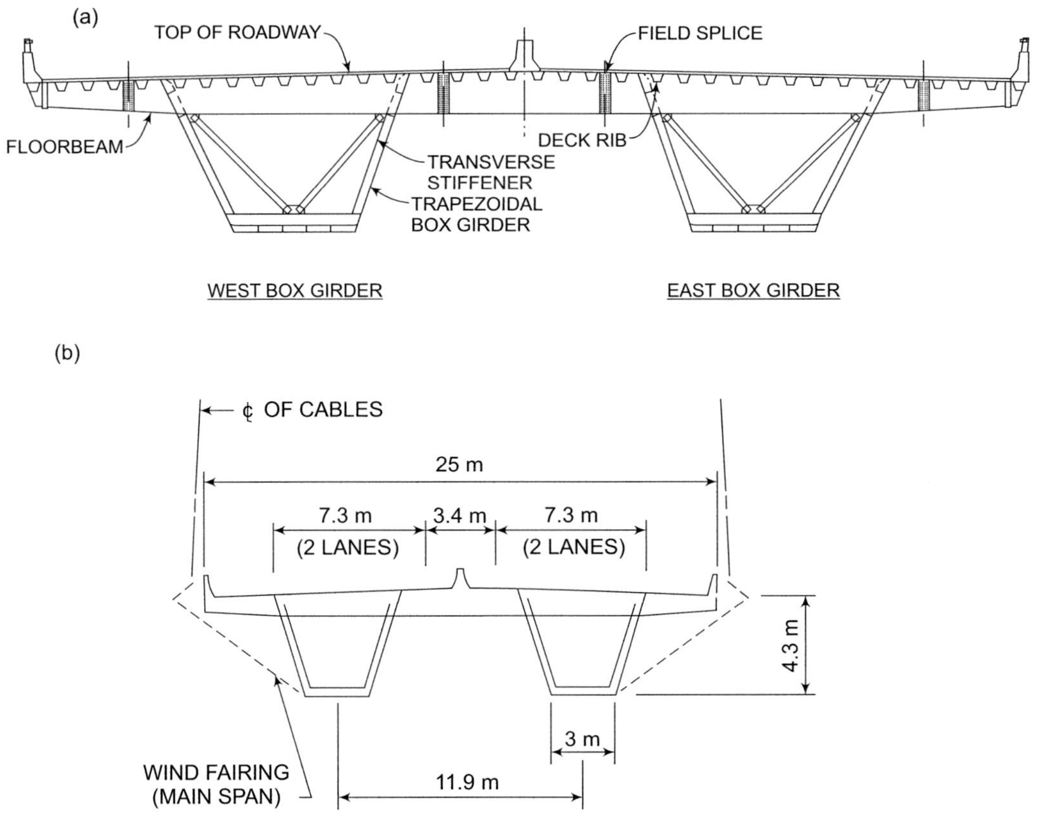

Fig. 6.11 Cross-section of the Hale Boggs Bridge: (a) general layout; and (b) illustrating fairing plate

The two pylons are A shaped with a slight modification. They are about 107 m above the top of the concrete pier. Nine-cell steel boxes that taper in both the longitudinal and transverse directions make up each leg of the pylon. Strong weathering steel has been utilized in the construction since it was decided that the metalwork would be left unpainted to reduce maintenance.

On either side of the bridge roadway, in line with the pylon legs, a plane of cable stays is employed. Each pylon has three sets of stay cables arranged like a fan at the top. Every stay consists of two or four wire strands that are arranged in parallel and consist of 103 and 307, 6.35 mm diameter wires each. The Strands are enclosed in a Polyethylene jacket and Portland cement grout was used to fill the space between the jacket and the cables. The Louisiana Department of Transportation (LADOT) has recently approved a comprehensive program to replace all cables of the bridge that were revealed to be in very poor condition based on the latest inspections. This program also includes rehabilitation of the deck.

6.1.4 The East Huntington Bridge

August 1985 saw the opening of North America's second precast prestressed concrete open-cross-section cable-stayed bridge, the East Huntington Bridge (Figure 6.12) over the Ohio River in West Virginia. It is characterized by one pylon only, main span of 274m and side span of 184 m (Figure 6.13). The main section of the bridge is supported by a 128 m high stay cable pylon and 62 stay cables. Only two expansion joints accommodate longitudinal movements in the girder from Pier S3 to Pier N3. At the bridge pylon, the girder is permitted to move only vertically. The superstructure (Figure 6.14) girder consists of reinforced concrete edge beams, 1.5 m (depth) by 1m (width), and a 200 mm thick roadway slab. The slab is supported by transverse 840 mm deep rolled steel floor beams that are connected to the main longitudinal girders. The concrete pylon comprises hollow, sloping legs that contain elevators for the cable anchorages. The legs terminate at the pylon head which is an I-shaped 27 m high segment. The pylon head houses the cable anchorage chassis, which is shielded by 90 mm-thick, dark-colored precast concrete panels The parallel wire cables comprise 6 mm wires housed in forged steel anchorage assemblies, wrapped in polyethylene pipes, and filled with grout to prevent corrosion in the parallel wire cables. The cables range in length from 61 to 23 m, with each size ranging from 83 to 307 wires (Grant, 1979).

Fig. 6.12 The East Huntington Bridge, West Virginia, US, 1985

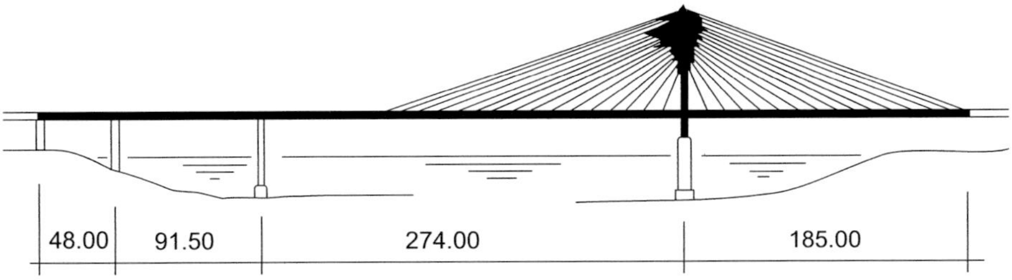

Fig. 6.13 Elevation of East Huntington Bridge (Grant, 1987)

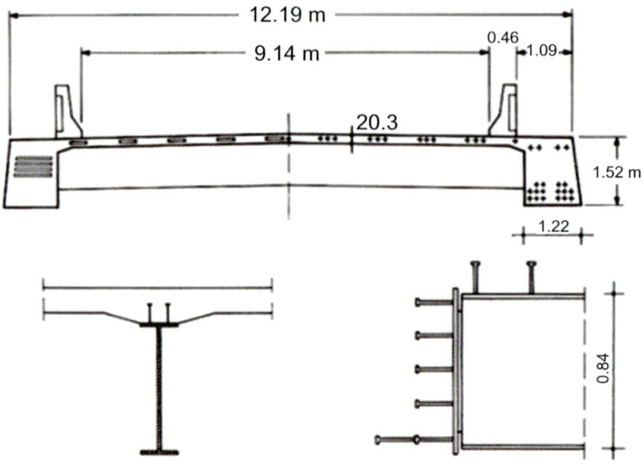

Fig. 6.14 Cross-section of East Huntington Bridge (Grant, 1987)

6.1.5 The Bob Graham Sunshine Skyway Bridge

The Bob Graham Sunshine Skyway Bridge was opened in 1987 and is the second bridge with that name on the site. It replaced a truss bridge that was built in 1954 but was subject to a tragic ship collision accident and FDOT decided to replace it with a modern bridge. With a center span of 365.76 m and flanking spans of 164.59 m each, the cable-stayed bridge has a three-span cable-stayed main span structure (Figure 6.15). This high-level crossing offers 53.35 m of vertical navigation clearance.

Fig. 6.15 The Bob Graham Sunshine Skyway Bridge, Florida US, 1987

The superstructure is made up of a single cell box that is 4.26 m deep and has internal struts and inclined webs to transfer the stay-cable forces throughout the section's depth from the anchorage area at the top slab to the bottom of the girder (Figure 6.16). The main deck box girder is prestressed transversally and vertically to resist local bending due to traffic loads and all shear stresses. Longitudinally, the precast segments are assembled by external prestressing tendons. Together with the axial load created by the component of stay loads, all joints at the top and bottom fibers are maintained under permanent compression under the combined effects of all loads, temperature gradients and long-term redistribution of internal stresses.

A single plane of cables running along the bridge's centerline supports the entire structure. The cables are continuous through the pylon in a fan-like arrangement, with stays angled between 22 and 47 degrees with respect to the deck. At 7.31 m intervals, the cables are fastened to the box girder. The stay-cables are made up of seven-wire low relaxation prestressing strands with a diameter of 60 to 80 mm that are pressure grouted into steel pipes. In order to permanently compress the grout, the cables were overstressed prior to grouting and then released once the grout had hardened. At the deck level, damping devices are installed to regulate vibrations caused by cables. The stays allow for future stay replacement because they are continuous through the pylon, where a double pipe serves as a deviation saddle.

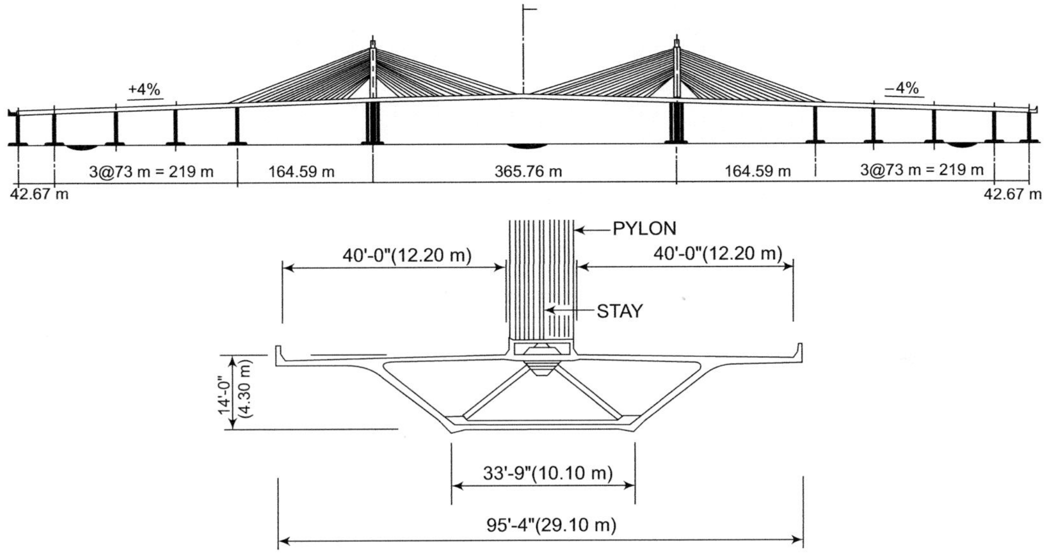

Fig. 6.16 The Bob Graham Sunshine Skyway: Elevation and cross-section (Muller and Tassin, 1987)

Above deck, the pylon is a single shaft; below, it transforms into two elliptical box pier shafts. Rigid connections bind the twin pier shafts, the box girder superstructure, and the pylon together. The balanced cantilever method was used to build the bridge. Two winches fastened to the cantilevers' ends raised the precast segments from the barges into place. In the casting yard, the segments were essentially prestressed in both the transverse and vertical directions. A small amount of longitudinal prestress in the top slab allowed for the cantilever construction to be completed prior to the permanent cables being installed.

6.1.6 The Dame Point Bridge

The Dame Point Bridge (Figure 6.17) is another example of a reinforced concrete open cross-section. However, the main span of this bridge is 97 m longer than the Ed Hendler Bridge (Pasco-Kennewick). The crossing is a 3.2 km-long, high-level bridge structure crossing the navigation channel of the St. Johns River at Jacksonville, Florida. The main river spans of the bridge consist of a 792.6 m long three-span cable-stayed bridge with a center span of 396.3 m and side spans, each 198.15 m long. The structure carries a 6-lane divided highway at a maximum grade of 5 percent. Each of the two divided roadways is 13.26 m wide from curb to curb, consisting of three 3.66 m lanes, a 1.37 m outside shoulder and a 0.91 m inside shoulder. The two roadways are separated by a concrete median barrier that has been designed to be removable and facilitates an additional lane. The length and height of the center span satisfies navigation clearances of 381 m horizontally and 53.4 m vertically.

Fig. 6.17 The Dame Point Bridge, Florida US, 1989 (Courtesy, Jonathan Zander)

The superstructure is an open cross-section like East Huntington deck with a slight difference. The transverse floor beam is prestressed concrete spaced at 5.34 m on centers (Figure 6.18). Where the compression from the cable thrust forces increases, the deck units next to the pylons have a 2.44 m wide reinforced concrete edge girder that varies in depth from 1.52 m to 1.85 m. The post-tensioned transverse floor beams support the deck slab and connect the edge girders. The slab varies in thickness from 22.9 cm to 55.9 cm in the deck units of highest compression. The slab is normally reinforced concrete except it is post-tensioned in areas where the cable thrust forces are not yet distributed over the whole deck width (Loizias and McCabe, 1990).

Two Double-leg pylons with two cable plans were employed. The legs are 29.8 m center to center and extend 92 m above the roadway level. The pylons are of varying solid cross-section above the roadway and of octagonal hollow cross-section below.

The load of the superstructure is transmitted to the pylons via 288 stay cables that use parallel Dywidag threaded bars with a 32 mm diameter. The thick steel pipes, used to house the grade 1040 thread bars, were pressure grouted to withstand any post-grouting loads on the bridge. The range of the steel pipe section is 16.8 cm to 21.9 cm, contingent upon the quantity of contained bars. The cable anchorages are made in such a way that the bar tendons alone can withstand the structure's dead load before grout is applied. Standard Dywidag anchor nuts with an anchor disk and plate are used to secure the bars.

6.1.7 The Talmadge Memorial Bridge

Two years after the Dame Point Bridge in Florida was opened, another bridge, which has similar characteristics was opened in Savannah, Georgia. The Talmadge Memorial Bridge (Figure 6.19) was built on the Savannah River between downtown Savannah, Georgia, and Hutchinson Island. The cable-stayed bridge replaced another bridge that was built in 1953. The bridge was completed in 199. The three-span bridge has side spans of 143.15 m each and a main span of 335.28 m. It offers 56.4 m of vertical navigation clearance.

The deck features an open reinforced concrete cross-section comprising two edge girders attached by a transverse floor beam that supports a reinforced concrete slab. The bridge deck has four traffic lanes and measures 24.54 m wide. The edge girders are 1.37 m deep by 1.37 m wide and connected by the transverse floor beams, which are positioned 8.91 m and 8.61 m apart at the center span and side spans, respectively. They support the slab, which has a thickness of 282 mm. The stay-cables are mounted on H-shaped pylons in a semi-fan configuration. Each pylon has two

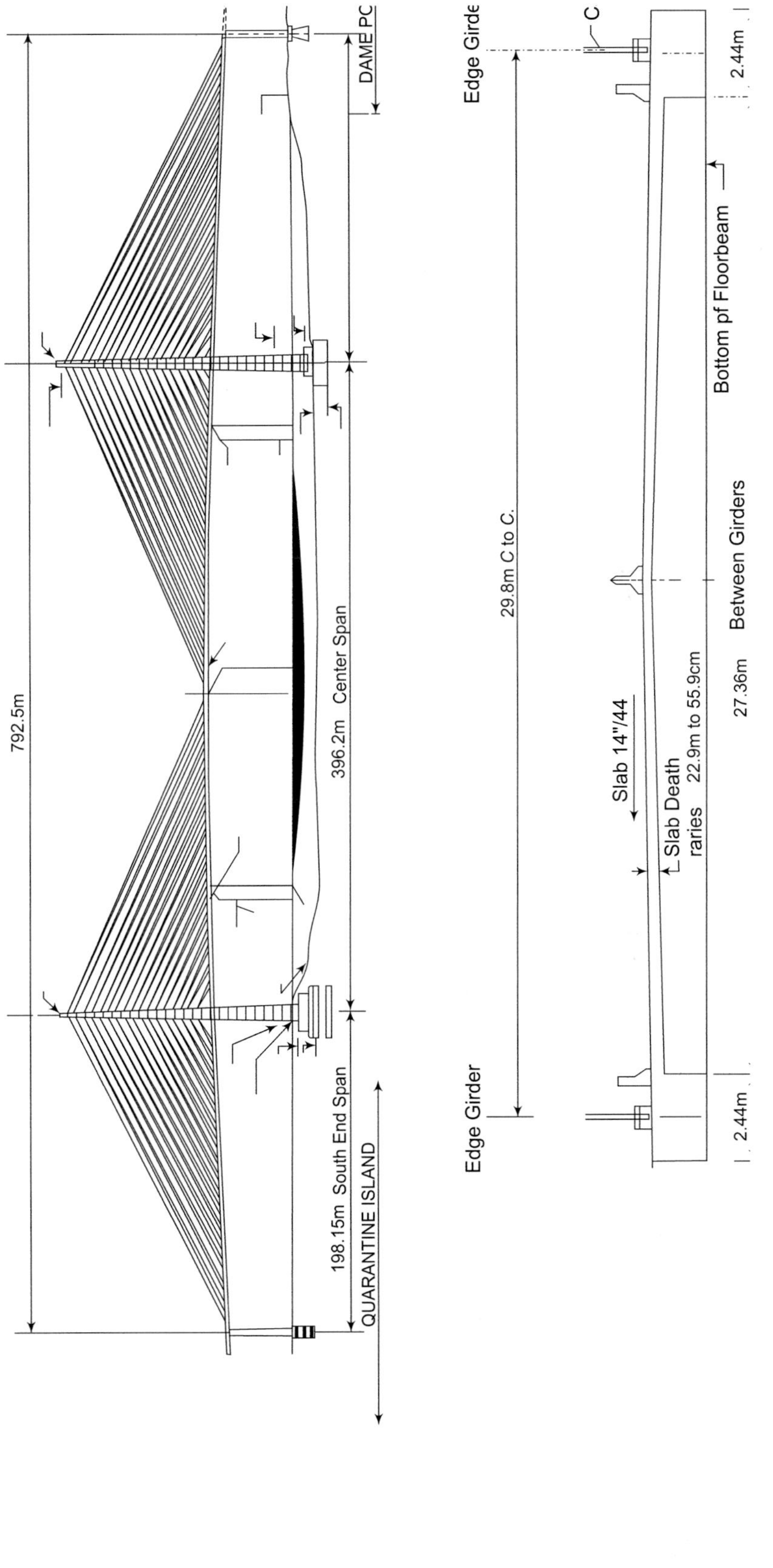

Fig. 6.18 The Dame Point Bridge: elevation and cross-section

Fig. 6.19 The Talmadge Bridge, Georgia US, 1991 (Courtesy, Library of Congress)

struts: a roadway strut and an upper strut. The superstructure, which is suspended by the stay-cables, travels through the pylons and is supported by vertical bearings at the pylon's roadway strut. The stay-cables are made up of seven-wire low relaxation prestressing strands that are parallel and 15 mm in diameter. They are pressure grouted inside a black polyethylene (HDPE) plastic pipe and covered in light-colored PVF tape (Tang, 1995).

6.1.8 The Fred Hartman Bridge

The Fred Hartman Bridge (Figure 6.20) was built between the cities of Baytown and La Porte, Texas to replace the Baytown Tunnel. The tunnel had to be removed when the Houston Ship Channel was deepened to 13.7 m to accommodate larger ships. The bridge was opened to traffic in September 1995.

Fig. 6.20 The Fred Hartman Bridge, Texas, US, 1995 (Courtesy, Bill Word)

The bridge was named after Fred Hartman (1908–1991), the editor of the Baytown Sun Newspaper published in Baytown, Texas. The bridge has four lanes with complete shoulders and a 22 m wide roadway in each direction of travel. The bridge is the first in the United States to have two superstructures running continuously through the entire length of 674.8 m (Figure 6.21). Stay cables are positioned at roughly 15 m intervals in a semi-fan configuration to support the beam. Each superstructure is supported by two cable planes.

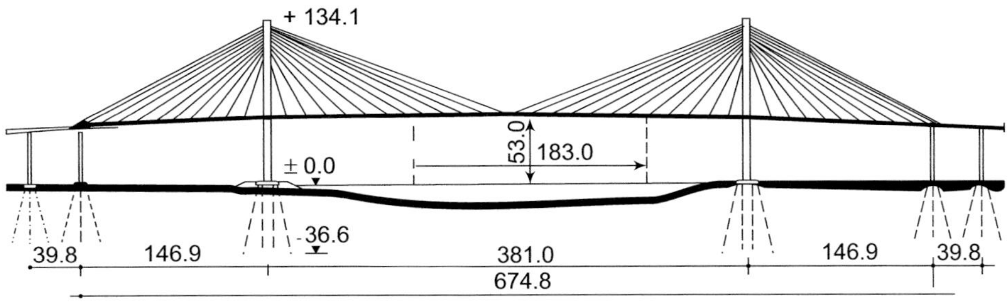

Fig. 6.21 Elevation of the Fred Hartman Bridge (Svensson and Lovett, 1990)

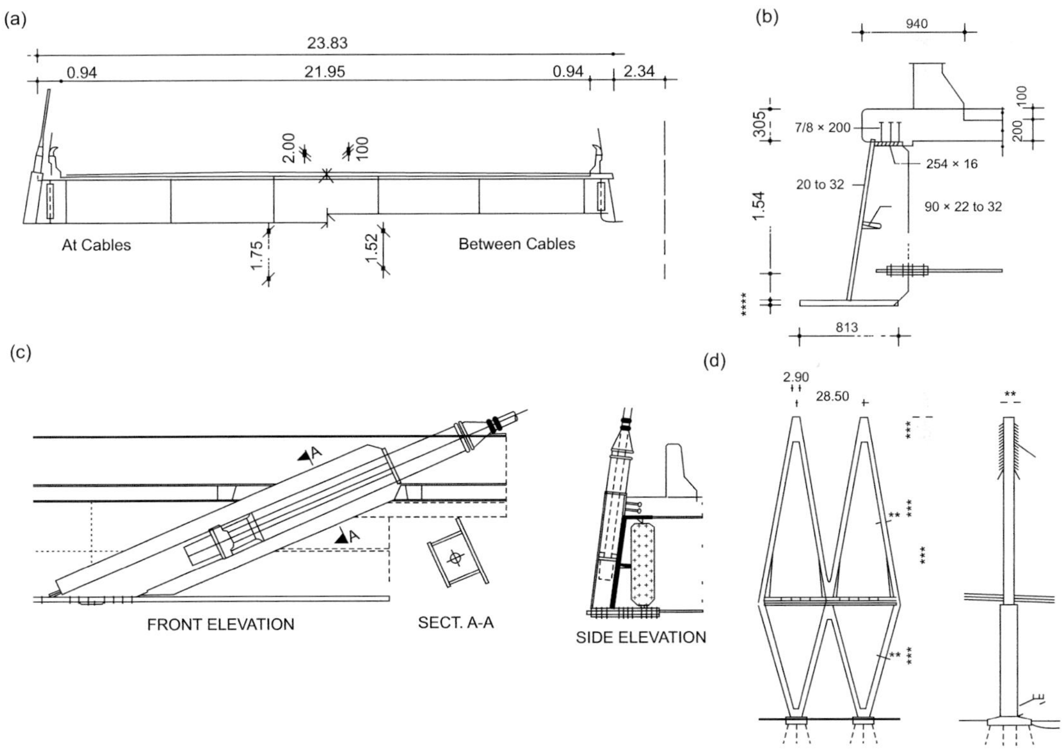

Fig. 6.22 Fred Hartman Bridge miscellaneous details: (a) cross-section; (b) detail of edge girders; (c) stay cable anchorage at girder; and (d) pylons (Svensson and Lovett, 1990)

The bridge has a composite cross-section comprising two longitudinal edge steel plate girders as illustrated in (Figure 6.22a). The girders are composite with a concrete roadway slab with one continuous longitudinal stiffener. The concrete deck is 200 mm thick and has a 100 mm reinforced concrete wearing surface. Shear studs are provided on the main girder and arranged in rows of three with a constant longitudinal spacing of 115 mm. The floor beams (Figure 6.22b) are welded plate girders with 14 mm webs without transverse stiffeners to simplify fabrication. They are connected to the main girders with high-strength bolts. The cables are anchored in welded boxes bolted to the main girders as shown in (Figure 6.22c). The stay cables consist of parallel wires with HiAm (high amplitude fatigue resistance) anchorages in polyethylene pipes, injected with cement grout after installation for corrosion protection.

The pylons are designed to be double diamond shaped. A truss is formed by connecting the two lower A-frames at deck level, which distributes the transverse wind loads to the two foundations. The pylon legs act like cantilevers in a longitudinal direction. The stay cables at the head of the pylons travel through steel pipes that are set into the walls of the pylons (see Figure 6.22d). Each one is securely fastened inside using steel bearing plates that are supported by concrete corbels.

6.1.9 The William H. Natcher Bridge

The William H. Natcher Bridge (Figure 6.23) is another composite cable-stayed Bridge that was built over the Ohio River to carry U.S. Highway 231 and connect Owensboro, Kentucky to Rockport, Indiana. The bridge which was opened on October 21, 2002 has a main span of 366 m and two 152 m side spans each (see Figure 6.24).

Fig. 6.23 The William H Natcher Bridge, Indiana, 2002

Steel floor beams and steel edge girders make up the superstructure. The distance between the floor beams is 4.57 m. Precast concrete pieces with cast-in-place (CIP) infills make up the deck slab. Shear studs attach the slab to the floor beams and edge girders. On the deck slab, a 38-millimeter layer of latex-modified concrete is applied. Every pylon has a trapezoid-shaped head to hold all the cable anchors and two inclined legs that extend above the deck. The two inward-bending pylon legs below the deck are secured in place by a tie strut. With bearings beneath each edge girder, the superstructure is positioned atop the tie strut and travels through each pylon that separates the legs.

The Indiana pylon has expansion type bearings, whereas the Kentucky pylon has fixed bearings. In order to withstand the tension of both the dead load and the live load, the pylon legs are prestressed to a jacking force of 70,300 kN. They are 4.88 × 2.44 m above the roadway and 4.88 × 3.66 m below the tension strut. The bridge is equipped with two semi-fan-shaped cable stay planes that are secured to the pylon and deck respectively.

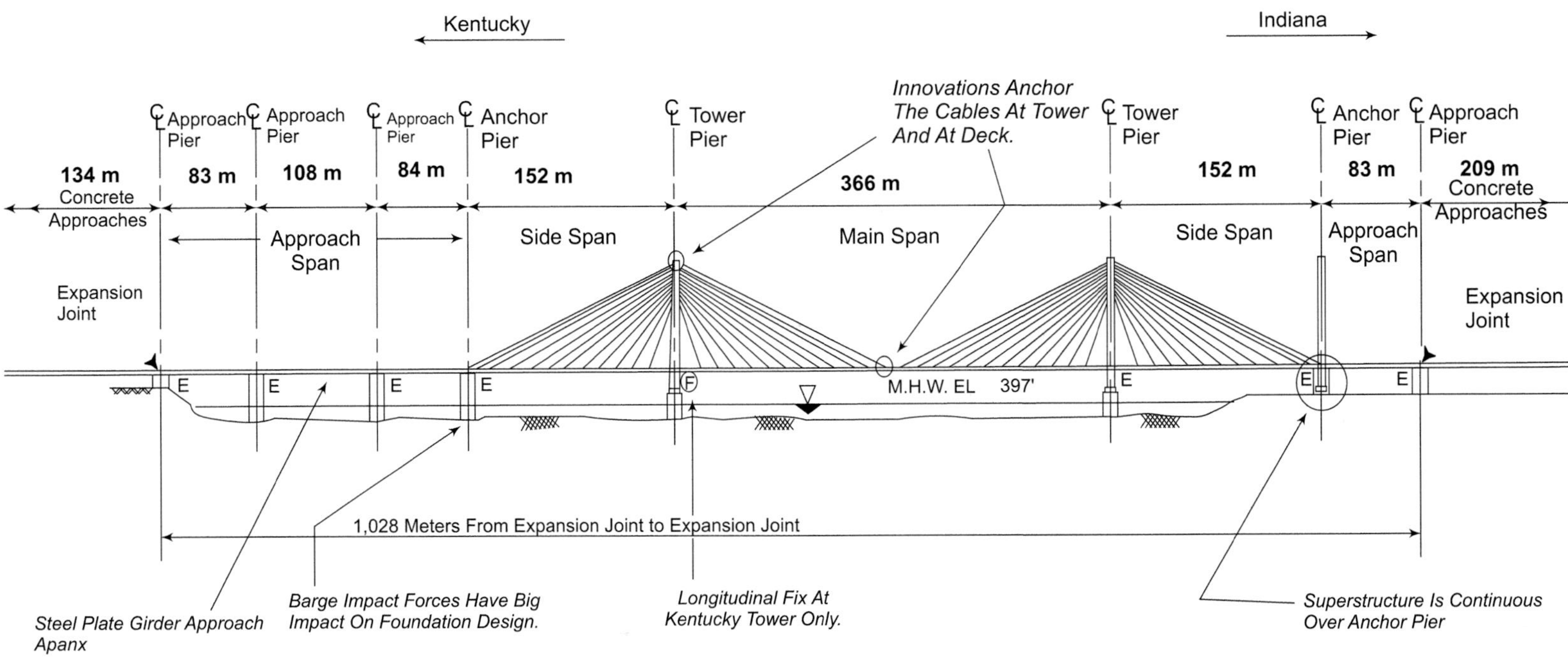

Fig. 6.24 Elevation of the William H Natcher Bridge illustrating boundary conditions (Chandra, and Hsu, 1999)

6.1.10 The Sidney Lanier Bridge

The Sidney Lanier Bridge (Figure 6.25) was built over the Brunswick River in Brunswick, Georgia. It was opened to traffic in 2003 and carries four lanes of U.S. Route 17. It was named after poet Sidney Lanier. It took the place of another movable bridge whose lift span—76.25 m—was judged to be too small for use in navigation in the future. The new bridge is 2372.9 m long (Figure 6.26a). The main spans are cable-stayed structures with two side spans 190.63 m long and a center span 381.25 long. It is intended to have two lanes of traffic going in each direction. The minimum horizontal and vertical clearances for the navigation channel are 122 m and 56.43 m, respectively. This bridge is an open concrete cable-stayed bridge, just like the Talmadge Bridge in Savannah, Georgia. Both bridges feature a thin deck girder and H-pylons. The cables on both have the same vertical clearance of 56.43 and are set up like fans. The Talmadge Bridge, which has a center span of 335.5 m, is somewhat smaller than the others. Also, the Pylons of the Sidney Lanier Bridge are in the waterway and require pier protection while those of Talmadge Bridge are outside the waterway to avoid ship collisions (see Figure 6.25).

Fig. 6.25 The Sidney Lanier Bridge, Georgia, 2003, (Courtesy of Road Traffic)

The superstructure spans 21.5 m between the inside faces of the curbs and 24.71 m from end to end. It is comprised of two longitudinal edge girders and transverse floor beams supporting a concrete slab that is 27.94 cm thick. 1.45 m in width and 1.525 m in depth make up the parallel edge girders. The web of the transverse floor beams, which is 0.53 m wide and varies in depth, is positioned at 8.44 m for the side spans and 8.29 m for the main span. As shown in Figure 6.26b, the floor beams are 1.525 m deep at the edge girders and 2.135 m deep at the deck's centerline, including the slab.

The bridge is supported by two semi-fan stay cable planes. High-density polyethylene ducts are encased with seven-wire strands of Grade 270, measuring 1.524 cm in diameter, to form the stay cables. After the last stressing, cement grout is poured into the duct. Each cable has a different number of strands; the longest and shallowest cables have 50, while the shortest and steepest cables are close to the pylons, with only 18. The top cables are 85.86 m above the deck.

The H-frame pylon was chosen due to its eye-catching design and ease of use. The pylons' height, measured from the base, is 142.435 m. Each pylon is made up of two rectangular, hollow columns that are joined at two points by lower and upper cross ties. The longitudinal dimension of the box section is always 6.405 m. The upper portion of the pylon's leg is 2.745 m wide in the transverse direction. From the deck elevation, this width rises linearly to a base of 7.625 m on top of the foundation pedestal. The construction joints of the pylon legs are spaced approximately 4.88 m apart. The upper cross tie is the hollow rectangular beam that is 5.795 m wide and 4.575 m deep.

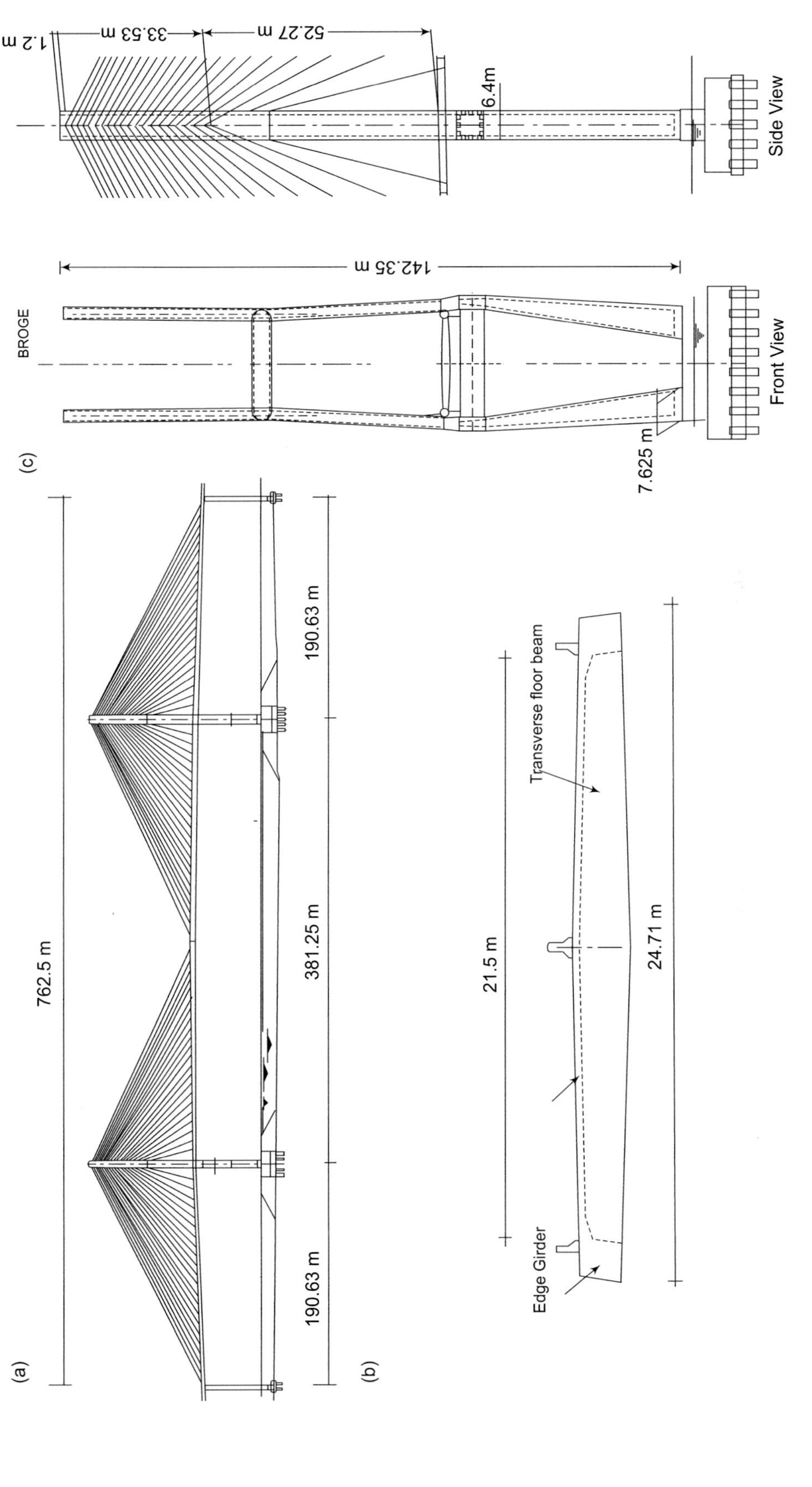

Fig. 6.26 The Sidney Lanier Bridge: (a) elevation, (b) cross-section, and (c) pylon detail

The height above the deck is 38.735 m. At this point, the lowest cable also enters the pylon face. The lower cross tie is 5.49 m deep and 5.795 m wide. It is located 4.27 m below deck level. Both ties are prestressed. A higher prestressing force is needed because the lower cross tie must support the tension force caused by the kink in the pylon leg and service load moments. Although the upper cross tie does not require prestressing because it is under compression, it was prestressed to withstand high reversible bending moments brought on by seismic and wind loads. These loads were also taken into account in the design of the lower cross tie (see Figure 6.26c).

At each pylon, one elastomeric bearing measuring 1.2 m in length, 0.96 m in width, and 0.45 m in thickness supports the deck girder vertically beneath the edge girders. They transfer the girder's vertical load to the pylons and provide partial restraint in the longitudinal direction under live load and wind. A system of longitudinal lock-up devices, or LUDs, joins the girder and pylons. When they experience abrupt seismic movement, they will lock up or freeze. They permit slow movements brought on by creep, shrinkage, and temperature. All movements are restrained by the monolithic connection that joins the edge girders to the anchor piers. Every anchor pier has a 7.32 by 3.05 m hollow rectangular pier shaft. A hammerhead that cantilevers out transversely from the pier shaft is located at the top of it. The deck and the hammerhead are the same width. The loads transferred from the superstructure are carried by pre-stretching, which is oriented transverse to the axis of the bridge. Each main pylon's foundation is made up of forty vertically drilled caissons with a diameter of 1.8 m, spaced 4.575 m apart (Tang et al., 1995).

6.1.11 The Arthur Ravenel Jr. Bridge

Known by another name, the Cooper River Bridge (Figure 6.27), the Arthur Ravenel Jr. Bridge spans the Cooper River in South Carolina, USA, and links the towns of Mount Pleasant and Charleston. Two truss bridges that had deteriorated from neglect were replaced by the bridge, which opened to traffic in July 2005. The old bridges were too low to support big shipping vessels, too narrow for contemporary cars, and had weight restrictions. The bridges' steep gradients, restricted emergency access, central reservations, and one bridge that received a 1995 safety and integrity rating of 4 out of 100 made them unsafe as well. The bridge, which has a main span of 471 m, is the fourth longest in the United States.

Fig. 6.27 The Arthur Ravenel Jr. Bridge, South Carolina, US, 2005, (Courtesy, Skanska)

The main span, high level approaches, ramps, and interchanges were all included in the roughly three-mile-long bridge, which took four years to design and build. Larger shipping vessels can be accommodated by deepening the dredged depth by three m and widening the navigation channel to 305 m thanks to the main span of the bridge. With two 198 m side spans, two 68.75 m anchor spans, and a main span of 471 m, the cable-stayed span has a total suspended span length of 1004.5 m (Figure 6.28). A composite concrete deck with I-shaped steel edge girders and floor beams is used

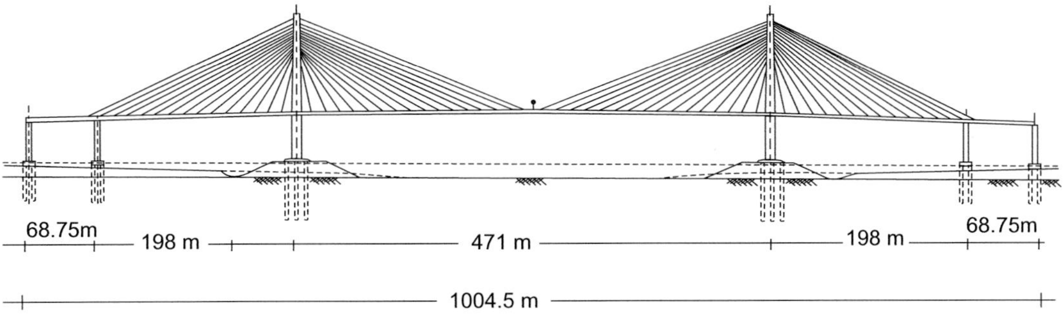

Fig. 6.28 Elevation of the Arthur Ravenel Jr. Bridge

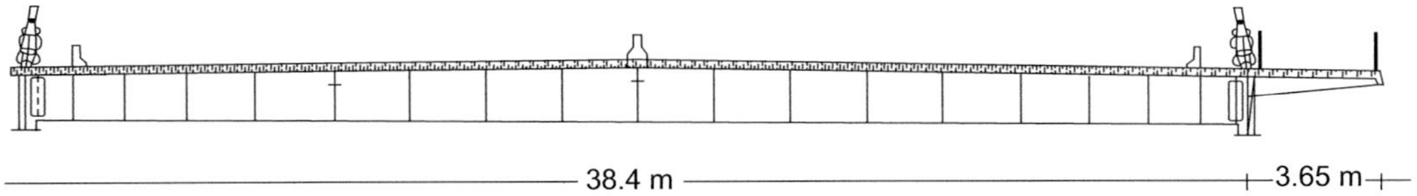

Fig. 6.29 Cross-section of the Arthur Ravenel Jr. Bridge

for the main span. Eight lanes for traffic and a 3.65 m wide bike/walkway are available on the road. As seen in Figure 6.29, the bikeway/walkway is cantilevered outside of the south edge girder.

Two 2 m deep steel I-shaped edge girders and steel floor beams spaced 4.75 m apart make up the superstructure for the suspended main and side spans. A 24 cm concrete deck slab sits atop the beams. The deck consists of five centimeters of latex-modified concrete on the wearing surface and 8,000 psi precast panels with closure strips covering the floor beams and girders. There are eight lanes of traffic on the 38.4 m wide bridge deck in addition to a 3.65 m wide bike and pedestrian walkway. The bridge is supported by two semi-fan-shaped stay cable planes. The loads of the suspended spans are transferred to the pylons by 128 cables, each of which is composed of seven wire strands. A pipe made of high-density polyethylene encloses them. The diamond-shaped pylons are 56.7 m above the water and rest on a series of drilled foundations.

6.1.12 The US Grant Bridge Replacement

Two cable-stayed bridges were opened to traffic in 2006 i.e. US Grant Bridge (Figure 6.30) and the Penobscot Narrows Bridge (Waldo-Hanckoc). Both are examples of replacing aging suspension bridges with new cable-stayed bridges. The original US Grant Bridge suspension bridge had two

Fig. 6.30 The US Grant Bridge, Ohio, US, 2006 (Kumarasena and McCabe, 2008)

steel pylons that extended 213.4 m apart and back spans of 106.7 m. It had no shoulders and just accommodated two travel lanes. Moreover, river pilots of this important navigational corridor, which see the passage of almost 70 000 barges annually, have long expressed concern about safety due to the pylon locations of the 213.4 m main span, which required an angular change in the river navigation path. The original bridge was removed in 2002 to make room for an in-line replacement by a cable-stayed bridge after the Ohio Department of Transportation determined that it was functionally deficient, too expensive to maintain, and impractical to upgrade after nearly eight decades of service. (Kumarasena and McCabe, 2008).

The three-span steel composite cable-stayed bridge is 513.6 m long, with a 266.7 m main span and 140.2 and 106.7 m back spans. This bridge employed two interior longitudinal strut beams at one-third of the deck width in addition to the two edge girders that define a composite deck (see Figure 6.31). Cast-in-place (CIP) concrete pours along the edge girders, floor beams, and strut beams create a composite slab with precast concrete panels that are attached to the steel framing. Because the slab is intended to span between floor beams, strut beams are only functioned to support the longitudinal CIP joint between the precast slab panels during construction but serve no structural purpose in the finished product.

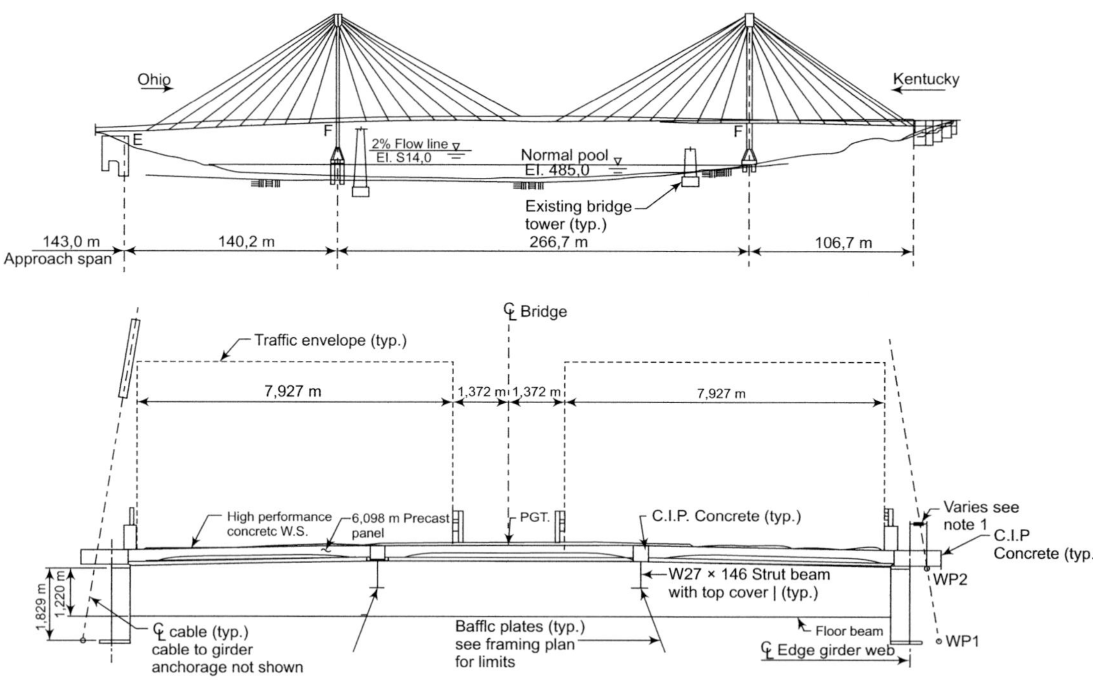

Fig. 6.31 The US Grant Bridge: elevation and cross-section (Kumarasena and McCabe, 2008)

The central pylons of this bridge are slender and consist of a single mast. They rise 57.9 m above the level of the roadway. The foundation constructability was greatly enhanced by the use of compact water-line pylon foundations, which were made possible by the selection of a single leg mast. Another distinguishing feature of this bridge is its pylon head, which eliminated the need of cables directly anchored on the pylon walls and thereby excluded pylon shaft post-tensioning and other complicated details along the length of the pylon stem. The superstructure has significant torsional rigidity due to the two cable planes that are fixed to the central pylon. Along the edge girders, the typical cable spacing is 15.24 m (Figure 6.32). The pylons are supported by compact waterline footings measuring 11 m by 11 m. Four 3 m diameter drilled shafts are socketed roughly 12 m into the rock to support them. The precast I-girder approach spans on the Ohio side, measuring 143.0 m in length, adding 657.0 m to the overall length of the crossing.

Fig. 6.32 The US Grant Bridge: Pylon head details (Kumarasena and McCabe, 2008)

6.1.13 The Penobscot Narrows Bridge

The second bridge that opened in 2006 is the Penobscot Narrows Bridge (Figure 6.33). It links Prospect, Maine, with Verona Island by crossing the Penobscot River on US 1/SR 3. Built at the same location in 1931, the Waldo-Hancock Bridge was an aging steel suspension bridge. It was replaced in an emergency with this cast-in-place cable-stayed concrete structure.

Fig. 6.33 The Penobscot Narrows Bridge, Maine, US, 2006

The length of this bridge's main span originated from foundations in water and placing pylons on land. The outcome is an asymmetrical cable-stayed bridge that has a total length of 646 m with a main span of 354 m. The pylons are 136 m above water. The bridge provides navigation clearance of 41 m (see Figure 6.34).

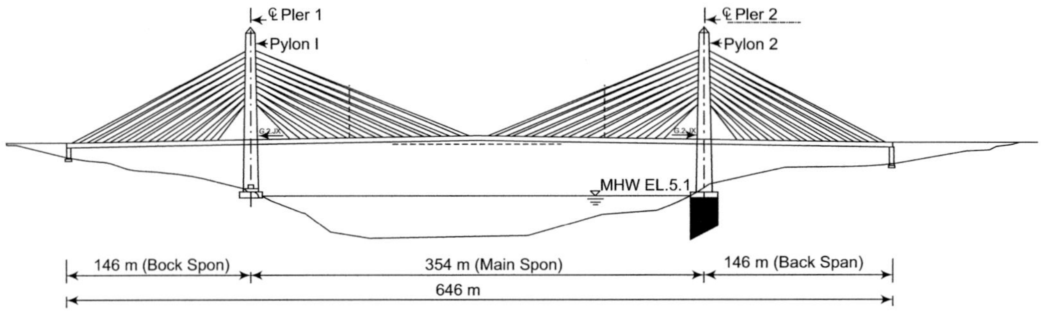

Fig. 6.34 Elevation of the Penobscot Narrows Bridge (Courtesy, Maine DOT)

The span cross section consists of a single trapezoidal concrete box with a deck that is 17.5 m wide and 3.9 m deep. As illustrated in Figure 6.35, the deck arrangement comprises a road featuring a single 3.65 m lane for each direction of traffic, bordered by a 2.1 m shoulder on the right and a 0.3 m shoulder on the left. The cable anchor blocks are located on an elevated 4.4 m median that divides the two travel lanes.

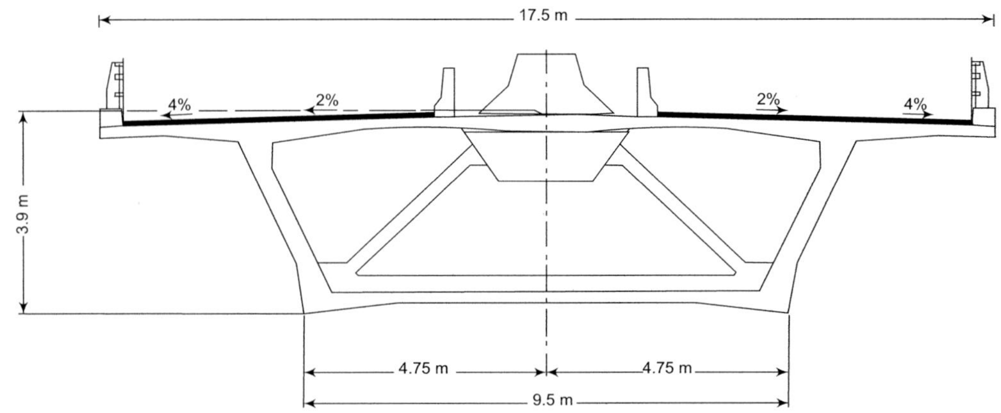

Fig. 6.35 Cross-section of the Penobscot Narrows Bridge (Courtesy, Maine DOT)

Two 131.1 m tall concrete pylons, each with 40 stay cables, support the bridge's spans. The pylons have a constant width across the traffic direction and taper with height in that direction. The bridge is furnished with two semi-fan-shaped cable planes. The cables have lengths between 36.4 and 181.1 m and inclination angles between 22.1 and 52.8 degrees. The cables in the south plane fan are about 13% longer than those in the outer fans because the bridge uses two planes of cables that are anchored to the superstructure's median.

The stay cables are made of un-grouted high-density polyethylene (HDPE) pipe with 41 to 72 loose strands of steel wire encased in it. Even though the number of strands varies for different cable lengths, the HDPE pipe always has a constant diameter of 400 mm and has a double helical fillet on its surface. The 7-wire strands with a diameter of 1.52 cm are filled and coated with epoxy. These continuous strands pass through cradles at each pylon and are attached to anchor blocks at the deck level. Because each strand functions independently, it can be taken out, examined, and replaced. The cradle system of the Penobscot Narrows Bridge is one of its fundamental features. The bridge is among the first to use a cable-stay cradle system, which removes the need for anchorages in the pylons and permits larger stay diameters and, consequently, longer spans. Another unique feature of this bridge is extra emphasis on incorporating new innovative cable protective systems. Four levels of protection are provided to assure redundancy for the stay cable system: epoxy coating on the stay strands; outer layer of HDPE sheathing around the stays system filled with nitrogen gas to purge potential corrosives; and sealed system with monitoring equipment. This proactive approach will allow maintenance crews to identify and address potential concerns in the future, at an early stage. An additional monitoring tool for the Penobscot River Bridge is a series of force monitoring systems on each stay. The force monitoring systems can accurately determine the force within 1% using a portable field laptop unit. Furthermore, six strands in three stays were removed and replaced with carbon fiber strands in June 2007 and were employed to enable monitoring of the carbon fiber material in this special application (Rohleder et al., 2008). An additional feature of this bridge is its observatory, which is reached by elevator from the top of Pier 1, the west pylon. Encased in glass, the bridge observatory is the highest public bridge observatory globally and the only one of its kind in the United States. Figure 6.36 displays a picture of the observatory.

Fig. 6.36 The Penobscot Narrows Bridge: photo bridge observatory

6.1.14 The Veterans' Glass City Skyway Bridge

The Veterans' Glass City Skyway Bridge (Figure 6.37) was opened to traffic on June 24, 2007. The bridge is part of Interstate 280 bridge and runs over the Maumee River in Toledo, Ohio, USA. This landmark cable-stayed bridge incorporates unique structural and aesthetic features. The bridge has a single mast pylon that exists exactly in the center of the channel. All sides of the 60.7 m upper section of the 132.8 m tall single pylon are covered in glass in homage to Toledo's long history as a glass-making hub. Beyond the glass, 384 light-emitting diode (LED) fixtures show over 16 million programmable color combinations at night. The use of a cradle system for the stay cables allows for the main pylon's aesthetic features. One of the few movable bascule bridges still in use on the Interstate system is replaced by this one (Bonzon, 2008).

Fig. 6.37 The Veterans' Glass City Skyway Bridge, Ohio, US, 2006 (Courtesy, FIGG)

The cable-stayed main span crosses the Maumee River with two 187 m spans that flank the main pylon (Figure 6.38). The bridge carries three 3.66 m lanes of highway traffic with two 3.05 m wide shoulders in each direction. The bridge provides about 37.8 m of vertical clearance and 61.0 m of horizontal clearance for the existing navigation channel.

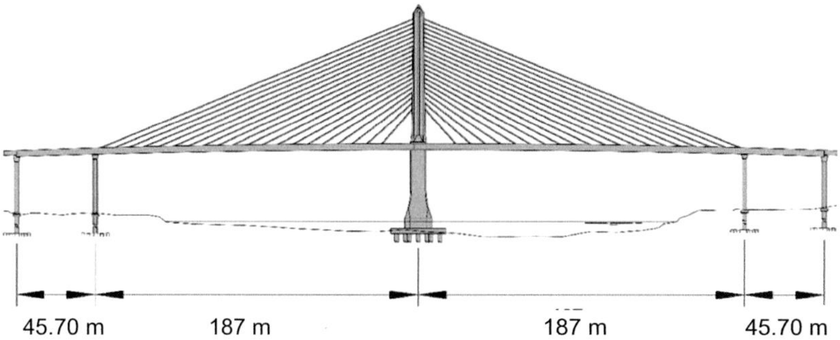

Fig. 6.38 Elevation of the Veterans' Glass City Skyway Bridge (Bonzon, 2008)

Two identical precast segmental trapezoidal box girders make up the superstructure. At the locations of the stay anchorages, they are combined into a rigid frame using a precast concrete triangular frame (Figure 6.39). This triangular component is attached to the neighboring box girders via post-tensioning tendons and concrete closure pours. Traffic loads are transversely transferred from the bridge deck to the stay cables that are anchored in the center by the delta frames. The deck space in between the delta frames is filled in with a post tensioned concrete slab as well.

The pylon of this bridge is one of its main characteristics. It has an octagonal cross-section with unequal sides as shown in Figure 6.40. The two major dimensions vary from 15.8 m longitudinally and 8.5 m transversally at the base to 8.8 m and 4.1 m near the roadway level. A diaphragm that connects the superstructure to the pylon is located at the level of the roadway. The pylon's upper section is 80.7 m above the level of the road. From the footing to the underside of the bridge superstructure, the lower part of the pylon rises 48.4 m. The upper pylon features a distinctive cross-section with a cruciform shape, varying in length from 8.84 to 6.30 m and has a constant width of 4.06 m. The glass panels integrated into the cruciform's four corners have their place thanks to this cross-sectional shape. Maintenance elevators could access the entire height of the upper pylon by passing through the cavities between the glass panels and the solid cross-section. The pylon

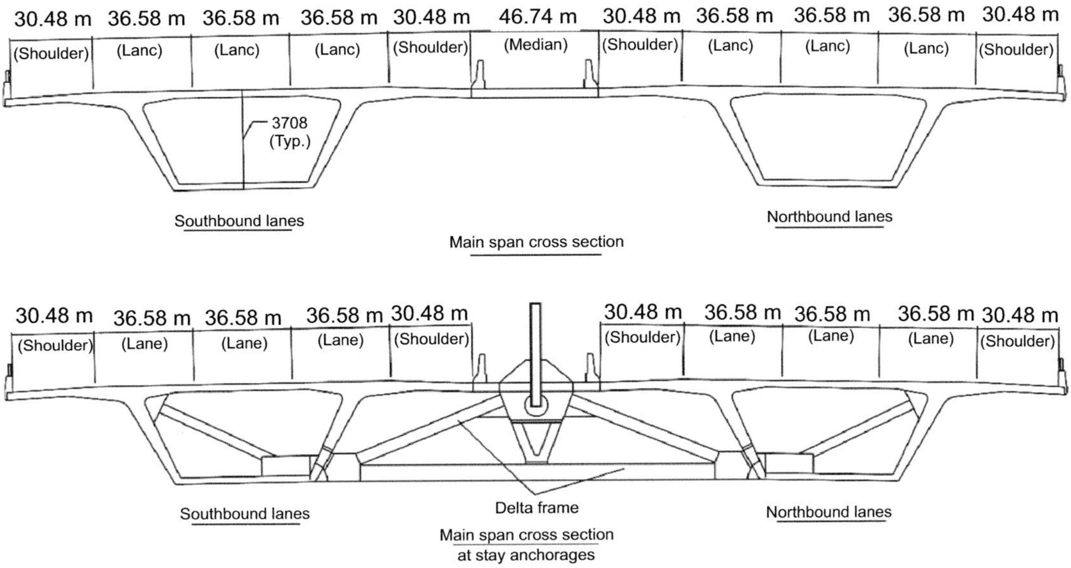

Fig. 6.39 Cross-section of the Veterans' Glass City Skyway Bridge (Bonzon, 2008)

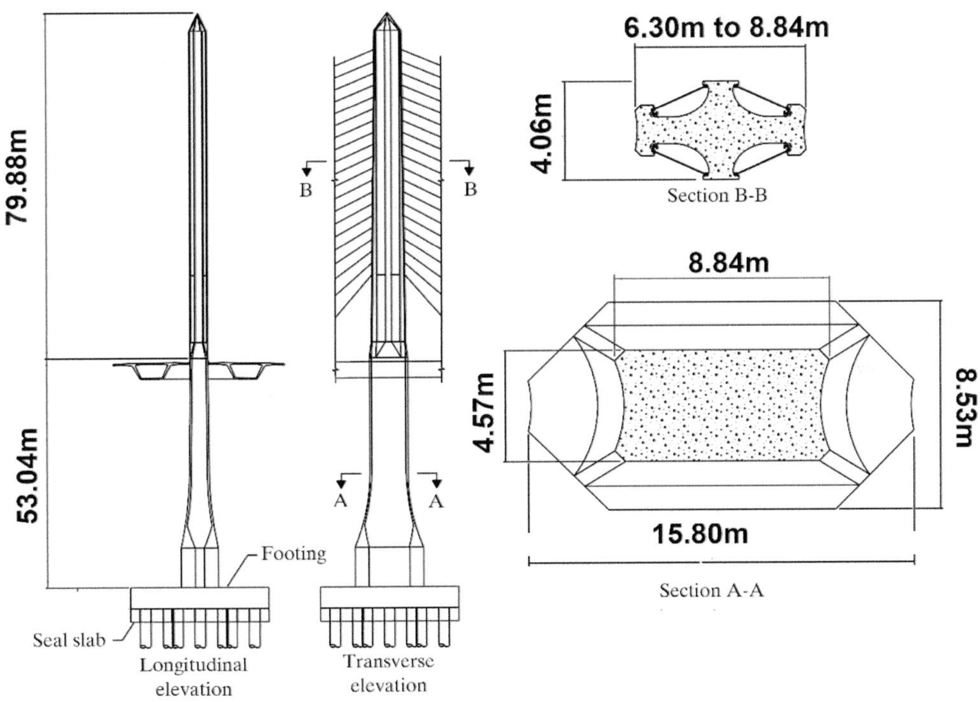

Fig. 6.40 Cross-section of the Veterans' Glass City Skyway Bridge (Bonzon, 2008)

is supported by seventeen 2.44 meter-diameter drilled shafts that are each inserted 4.6 m into the bedrock beneath the river. The drilled shafts are attached to a circular concrete footing that is 31.7 m in diameter and 4.88 m thick.

The superstructure's weight is transferred to the pylon by twenty stay cables arranged in a semi-harp configuration, with deck-level anchorages spaced every 8.53 m. The stays go through the upper pylon at intervals of 2.74 m, arranged in a single plane parallel to the pylon's axis. The stays are connected at the pylons through a cradle connection like those of the Penobscot Narrows Bridge.

6.1.15 The Jesse Brent Memorial Bridge (Greenville Bridge)

The Jesse Brent Memorial Bridge, also known as Greenville Bridge (Figure 6.41), was built over the Mississippi River, and carries US 82 and US 278 between Refuge, Mississippi, and Shives, Arkansas. The bridge replaced the Benjamin G. Humphreys Bridge, a cantilever truss bridge, which had become functionally obsolete. The cable-stayed bridge with its main span of 420 m is ranked number six in the US. Opened to traffic in 2010, this bridge, which has a composite superstructure carries four lanes of traffic (two in each direction), each 3.66 m wide. The bridge has a 3.6 m outside shoulder and a 2.44 m inside shoulder. The total length of the project (bridge, approaches, and new roadway) is about 6 km. The cable-stayed bridge has a main span of 180 m and two side spans 180 m long. The two concrete diamond pylons rise 130 m above water. The Mississippi approach to the bridge includes 905 m of a new roadway and 1952 m of approach structures.

Fig. 6.41 The Jesse Brent Memorial Bridge, Arkansas, US, 2010 (Courtesy, Arkansas Department of Parks and Tourism)

6.1.16 The John James Audubon Bridge

The John James Audubon Bridge (Figure 6.42), was completed and opened in May 2011 to replace the ferry between the communities of New Roads and St. Francisville. The bridge also serves as the only bridge structure on the Mississippi River between Natchez, Mississippi and Baton Rouge, Louisiana. It is named after John James Audubon, an American legend, artist, and ornithologist.

The cable-stayed bridge consists of a continuous 970 m long structure with a 482.5 m main span, 195 m side spans and 49 m flanking spans as shown in Figure 6.43. The main span provides a navigational envelope that is 446 m wide and a 20 m vertical clearance (466 m by 20 m). The bridge is the third longest bridge in the US.

Fig. 6.42 The John James Audubon Bridge, Louisiana, US, 2010 (Courtesy of GEC)

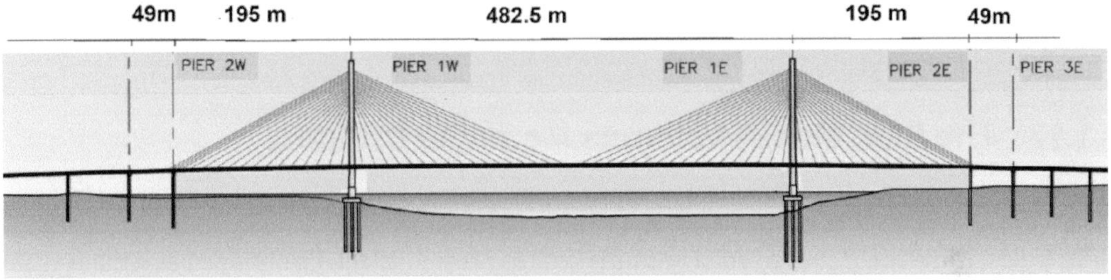

Fig. 6.43 Elevation of the John James Audubon Bridge (Schemmann et al., 2008)

The bridge superstructure is a composite steel section utilizing two 1.83 m deep edge girders that accommodate four traffic lanes as shown in Figure 6.44. The deck is typically 23 m wide. Additional wind fairings are added at the center half of the main span for greater aerodynamic stability. This widens the deck to 26 m. The floor beams span between edge girders at 4.65 m intervals. The deck is suspended from two concrete pylons by a total of 136 stay-cables arranged in a semi-harp configuration in two nearly vertical planes. The stay cables are made of individually greased sheathed and galvanized parallel strands with each stay cable containing two reference strands that can be permanently removed for future inspection. The cable anchorages are spaced apart at 14 m.

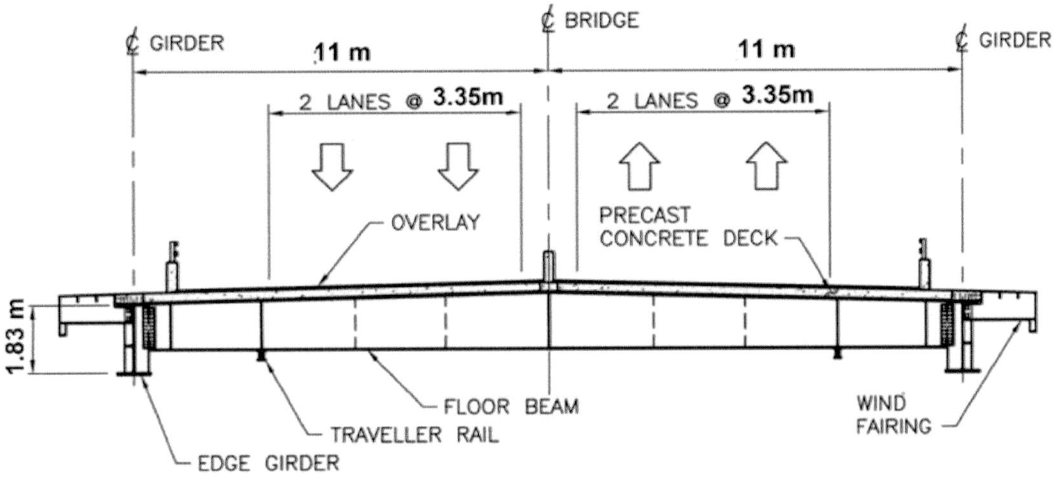

Fig. 6.44 Cross-section of the John James Audubon Bridge (Schemmann et al., 2008)

At Pier 1W/E (refer to Figure 6.43), the edge girders are supported vertically on reinforced elastomeric bearings that are installed on bearing pedestals. They provide Transverse restraint. Pot bearings are used at Piers 2W/E and 3W/E with the pots being transversely guided at Pier 3W/E at the south edge girder. Transverse restraint at Pier 2W/E is provided by shear keys at the floor beam midspan, which engage the pier cap beam. The deck is free sliding at all piers except at 2W, 1W and 1E. Longitudinal deck fixity is asymmetric with restraint provided by fixed linkages at the west pylon and lock-up devices at the east pylon. Longitudinal fixity at 2W is provided to increase the stability of the relatively slender pier columns. Uplift forces from the backstays are resisted by the combined effects of flanking span weights, deck counterweights located over the anchor piers and uplift restraint brackets attached to the anchor piers.

Pylons are H-framed with hollow section legs and cross beams. The pylons piers and anchorage piers are supported on 55 m deep 2.43 m diameter drilled shaft foundations located in the riverbed. Anchor piers at 2W and 2E comprise twin 2.43 m diameter reinforced concrete columns that are extensions of drilled shaft foundations. Struts between the columns are added at water level for ship impact (Schemmann et al., 2012).

6.1.17 The Stan Musial Veterans Memorial Bridge

The Stan Musial Veterans Memorial Bridge (Figure 6.45), also known as the New Mississippi River Bridge, was built over the Mississippi River between St. Clair County, Illinois, and the city of St. Louis, Missouri. It was opened to traffic on February 9, 2014. The bridge has a main span of 457 m which makes it ranked fifth longest bridge in the US. The bridge has a total span of 854 m including a 457 m long main span and two side spans, each 193.78 m long. The super structure is a composite section 26 m wide including two 3.65 m lanes and two shoulders in each direction. The outer shoulders are 3 m wide and the inner shoulders in the vicinity of the median are 1.83 m wide. The bridge can accommodate an additional lane in each direction with this arrangement. The deck is supported by a total of 136 stay-cables arranged in a semi-harp configuration in two vertical planes. The cable stays are anchored to two concrete diamond shaped pylons which reach 133 m above the interstate highway, I-70.

Fig. 6.45 The Stan Musial Veterans Memorial Bridge, Missouri, US, 2014

6.1.18 The Governor Mario M. Cuomo Bridge

The Governor Mario M. Cuomo Bridge, also known as the New Tappan Zee Bridge, is a twin cable-stayed bridge spanning the Hudson River between Tarrytown and Nyack in the State of New

York. The crossing includes two parallel 5 km long bridges crossing the Hudson River. Each of the two bridges comprises a 365.75 m cable-stayed navigation span and157 side spans with common foundations and diverging pylons along a series of 106.7 m continuous steel girder spans (Figure 6.46). The bridges each have 43 total piers, 41 of which are within the river. The cable-stayed main span uses a composite steel cross-section superstructure and provides over 335 m of horizontal and 41.75 m of vertical clearance for ships. The bridge was built to replace the original Tappan Zee Bridge, which had deteriorated over the years. The north span was opened to traffic in August 2017 and the south span was opened in September 2018.

Fig. 6.46 The Governor Mario M. Cuomo Bridge, New York, US, 2018

Each new bridge carries four traffic lanes. The north span carries in addition to the four lanes a shared use path for bicycles and pedestrians (see Figure 6.47b). Furthermore, the crossing has been designed to accommodate the future addition of a rail bridge between the roadway decks (see Figure 6.47c). The cable-stayed main span is supported by 4 semi-V shaped pylons stand inclined at five-degree angles. The height of the pylons is 127.70 m, and they are 7.6 m by 7.9 m at their base. While the pylons taper, they maintain their rectangular shape. All eight pylon legs have symmetric design details. Each composite cross-section deck is supported by 96 stay-cables arranged in a semi-harp configuration in two planes.

6.1.19 The Goethals Bridge

Another cable-stayed bridge that opened in 2018 on the US east coast is the Goethals Bridge (Figure 6.48). This bridge is a twin bridge, just like the Governor Mario M. Cuomo Bridge. It is operated by the Port Authority of New York and New Jersey and spans the Arthur Kill strait to link Elizabeth, New Jersey, and Staten Island, New York. This bridge replaces an old cantilever truss bridge that was constructed in 1928. Two 112 m side spans, which are continuously framed into flanking spans of 58 m in NJ and 48 m in NY, balance the main span of 274.3 m (see Figure 6.49). Together with the gravity loads of the integrally cast counterweight and substructure unit, the flanking spans' structural continuity and dead load eliminate any possibility of uplift, resulting in a redundant and sturdy structural solution. A navigation clearance with a width of 152.4 m and a height of 42.2 m is provided by the central span (Spoth et al., 2019).

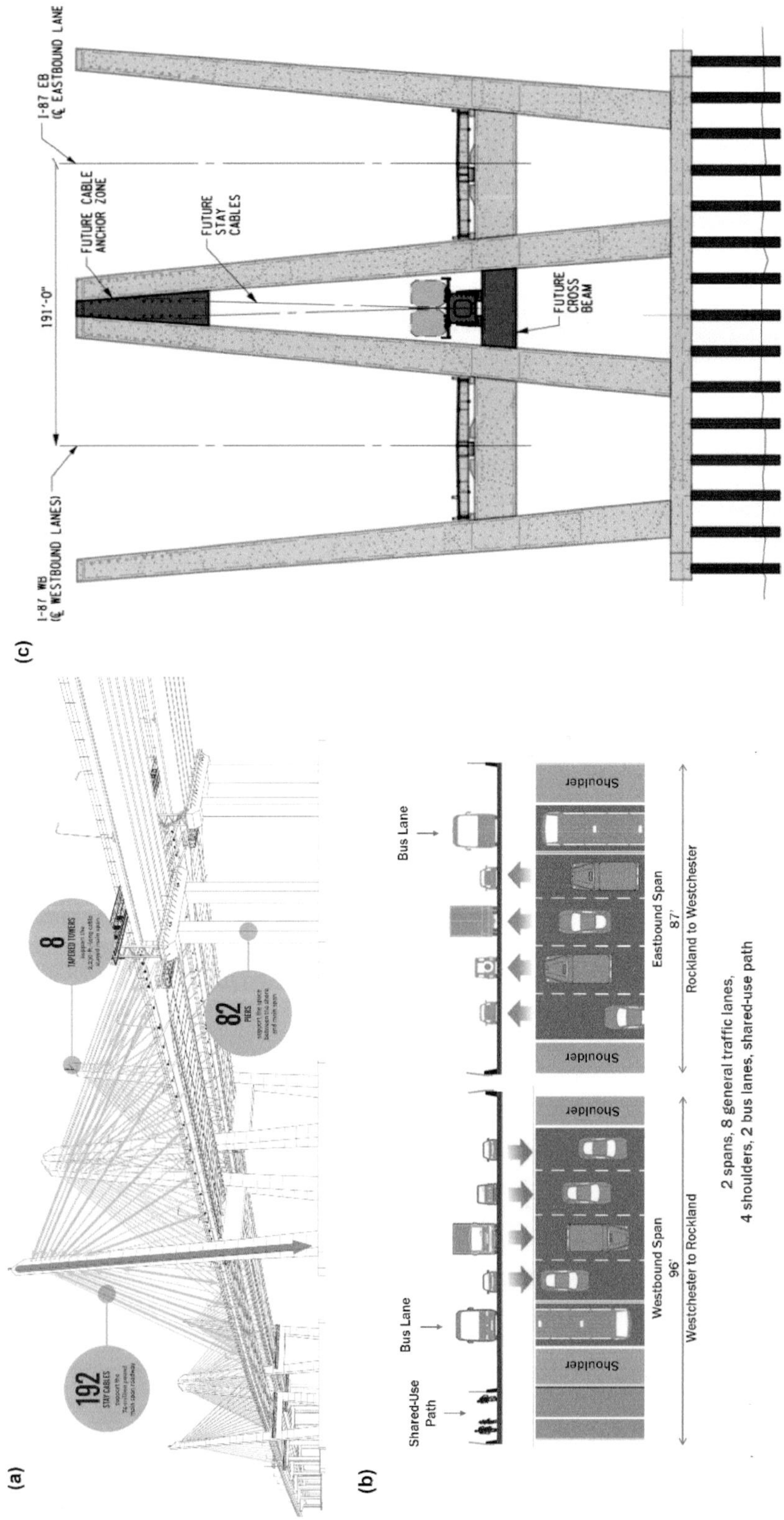

Fig. 6.47 Main characteristics of the Governor Mario M. Cuomo Bridge: (a) elements of the bridge; (b) lanes configuration; and (c) pylons with future rail addition (Courtesy, New York State Thruway Authority)

Fig. 6.48 The Goethals Bridge, New Jersey, US, 2018 (Courtesy, Parsons)

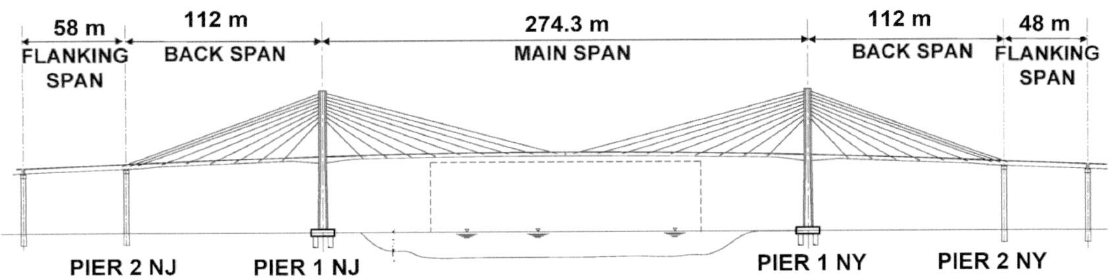

Fig. 6.49 Elevation of the Goethals Bridge (Spoth et al., 2019)

The superstructure (Figure 6.50) is a composite cross-section that comprises two edge longitudinal plate girders that are attached together by floor beams spaced at 4.57 m intervals. There are also two longitudinal trusses added to support the maintenance traveler required by the client. that are spaced at one third and two third the distance between the longitudinal girders. The slab is composed of lightweight precast concrete panels that are designed to be continuous with concrete closure pours over floor beams, edge girders, and in between panels. The slab is supported by a steel framing system. Thereafter, a polyester polymer concrete overlay is applied to the deck panels. The deck is composited with the structural steel framing system due to the closure pours.

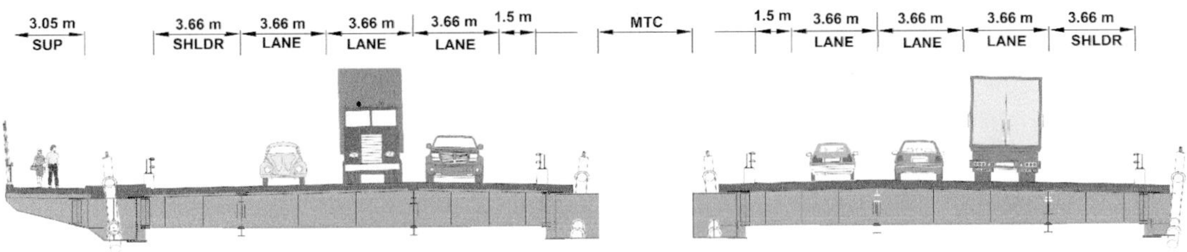

Fig. 6.50 Cross-section of the Goethals Bridge (Spoth et al., 2019)

The height of the pylons was limited to 37 m above the deck level due to the bridge's proximity to Newark Liberty International Airport. This results in a minimum stay cable angle of 15 degrees at the center of the bridge, which is lower than the optimal angle of 23 degrees usually used. This limited the number of stay-cables, hence each composite cross-section deck was supported by a total of 72 stay-cables made of seven-wire strands and arranged in a semi-harp configuration in two planes. The main pylons for each deck are configured as a semi-V in elevation, which was a project requirement. Therefore, the pylon legs were designed with a 5-degree outward lean. A triple set of backstay cables diverge from the top of the pylon and anchor side by side to the edge girder at Pier No. 2, or the anchor pier, in order to counterbalance the main span. The client, Port Authority, has specified major components to provide an extended service life in order to guarantee that the new structure provides exceptional service for future generations. A 150 yr-service life for foundations and substructures and a 100 yr-service life for the rest of the elements except the bearings, which are designed to sustain for 50 years. The necessary service life was verified using a variety of intricate techniques, such as modeling the chloride penetration of concrete, comprehensive permeability testing, material selection, use of sealers and membranes, cautious paint system selection, and a thorough maintenance and operations plan, all of which were documented in a project-specific corrosion protection plan.

6.2 UPCOMING CABLE-STAYED BRIDGES IN THE US

Two cable-stayed bridges are currently under construction and expected to open in 2024. The Corpus Christi Harbor Bridge (Figure 6.51) project in Texas involves the replacement of the current US 181 Harbor Bridge in Corpus Christi, Texas, along with a change in the alignment to interchange between US 181 and Interstate 37. The proposed improvements both to US 181/SH 286 and Harbor Bridge will address structural deficiencies and navigational restrictions of the current bridge, and improve safety, connectivity, and level of service. In total, the project includes the development, design, construction, and maintenance of 10.36 km of bridge and connecting roadway, and demolition of the existing Harbor Bridge. The new bridge comprises six-lanes (3 lanes in each direction, as well as a bike/pedestrian path). The bridge when completed, will be the longest cable-stayed, concrete segmental bridge in North America. It will completely span the ship channel bank-to-bank. It provides a concrete structure with a 170-year design life. With a main span of 506 m, this bridge will be the second longest cable-stayed bridge in the US. The bridge will provide a 62 m vertical navigation clearance. The bridge will be characterized by two identical precast segmental trapezoidal box girders for the superstructure. Two single reinforced concrete mast pylons with

Fig. 6.51 Rendering of the Corpus Christi Harbor Bridge, Texas, US

161 m total height will be used to support the bridge. The stays are aligned in a single plane along the axis of the pylon's 72 dual cables in a semi harp arrangement that will be used to transfer the load of the superstructure to the pylon.

The Gordie Howe International Bridge (Figure 6.52) is a cable-stayed international bridge across the Detroit River, currently under construction. The crossing will connect Detroit in Michigan and Windsor in Canada. The entire project has a total length of approximately 2.5 km.

The new bridge will have two "A"-shaped pylons that will be located on the banks of the Detroit River. The deck will include six lanes for automotive traffic, and a bicycle and walking path. When completed, the Gordie Howe International Bridge will have the longest main span of any cable-stayed bridge in North America at 853 m.

Fig. 6.52 Rendering of the Gordie Howe International Bridge, Michigan, US/Ontario, Canada

6.3 CABLE-STAYED BRIDGES IN CANADA

6.3.1 Hawkshaw Bridge

Canada has a long history in the construction of cable-stayed bridges. Both the Hawkshaw Bridge and the Papineau-Leblanc Bridge are among the first cable-stayed bridges built around the world. Constructed in 1967, the Hawkshaw Bridge (Figure 6.53) is a cable-stayed bridge approximately

Fig. 6.53 The Hawkshaw Bridge, New Brunswick, Canada, 1967

332 m long. It has a main span of 217.34 m and north and south spans of 57.74 m each. There are two lanes, one for each traffic direction. The bridge superstructure is composed of two steel girders running across the river and connected transversely by floor beams that work integrally with an orthotropic steel deck supported on top. The two 32.96 m tall steel pylons are on concrete piers. The cable stays are made up of six strands, each with a diameter of 62 mm.

6.3.2 The Papineau-Leblanc Bridge

Similar to the Hawkshaw Bridge, the Papineau-Leblanc Bridge (Figure 6.54) was among the first cable-stayed spans in North America. It links Laval and Montreal, Quebec, and crosses the Rivière des Prairies. It is the first to combine a torsional stiff center box girder with a single vertical plane. It has two equal side spans of 90 m each, and a center span of 241 m (Figure 6.55 a). With a depth-to-span ratio of 1/68, the superstructure is made up of a two-cell rectangular box girder that is 10.4 m wide and 13.6 m deep. There are transverse floor beams, 0.76 m deep, spaced at 4.57 m, and cantilevers at approximately 8.4 m on each side of the box made integrally with the orthotropic steel slab on top. The orthotropic slab and transverse floor beams are supported by diagonal struts.

Fig. 6.54 The Papineau-Leblanc Bridge, Quebec, Canada, 1969

The Pylons consist of a single box steel section that tapers from 1.8 m by 1.8 m at the base to 1.5 m by 1.5 m at the top, and they rise 38.4 m above the deck. A 500 mm thick ASTM A441 high-strength steel plate is used to make them. Sliding bearings support the girder, which is fixed rigidly to each pylon. A load of 4082 tons is supported by each pylon, which weighs 127 tons. (Demers and Simonsen,1971).

6.3.3 Alex Fraser Bridge

The Alex Fraser Bridge, also known as the Annacis Bridge, (Figure 6.56) connects Richmond and New Westminster with the North Delta in Vancouver, British Columbia across the Fraser River. With a main span of 465 m, this bridge was the longest cable-stayed bridge in the world when it opened on September 22, 1986.

The bridge has two side spans of 182.75 m each and two approach spans of 50 m each in addition to its 465 m main span (see Figure 6.57). It provides a 58.4 m navigation clearance above High Water. The bridge carries 6 lanes of highway traffic. The pylons rise 154.3 m above the top of its foundation. The superstructure is of the composite type consisting of a structural steel skeleton consisting of 2.1 m deep twin edge I girders and transverse floor beams that are tapered 1.6 m to 1.8m deep and spaced every 4.5 m. The floor beams support a composite precast concrete deck which is overlaid with a concrete cast in place. The superstructure is supported by 192 cable stays

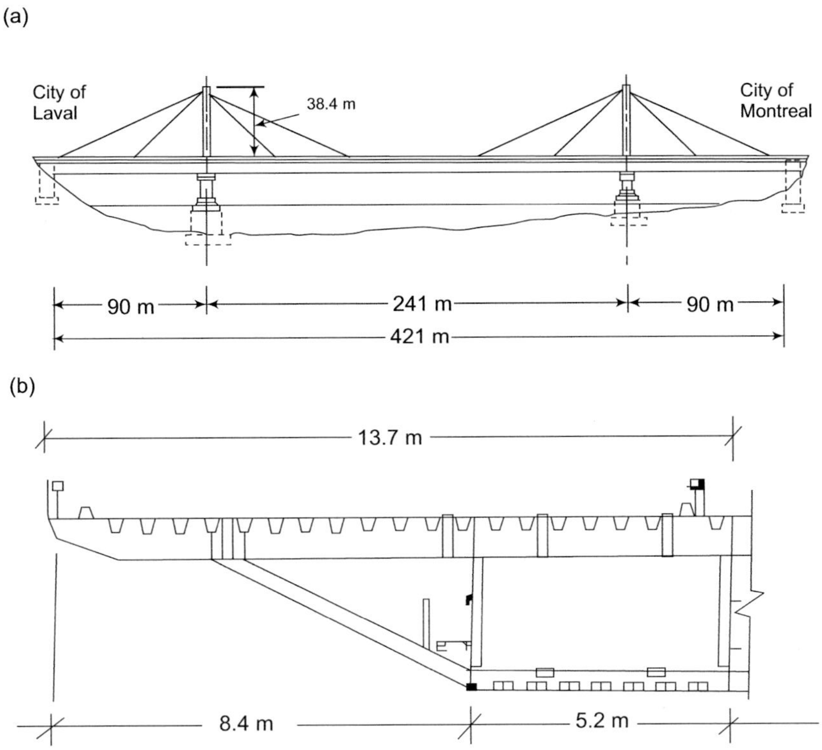

Fig. 6.55 Elevation and cross-section of the Papineau-Leblanc Bridge (Demers and Simonsen,1971)

Fig. 6.56 The Alex Fraser Bridge, Vancouver, Canada, 1986

that are arranged in a semi fan configuration. Stay lengths range from 49.5 m to 237.5 m and their diameters range from 80 mm to 130 mm. The 7 mm diameter galvanized wire strands that make up cable strands are covered in a black polyethylene sheath. Each cable has a cast steel socket on both ends that is filled with zinc. Cables end at the tie beams in the pylons where arrangements are made for jacking and adjustment. The reinforced concrete bents and pylons are designed to behave ductile during seismic activity. Every foundation is supported by steel piles (Taylor and Simonsen, 1984).

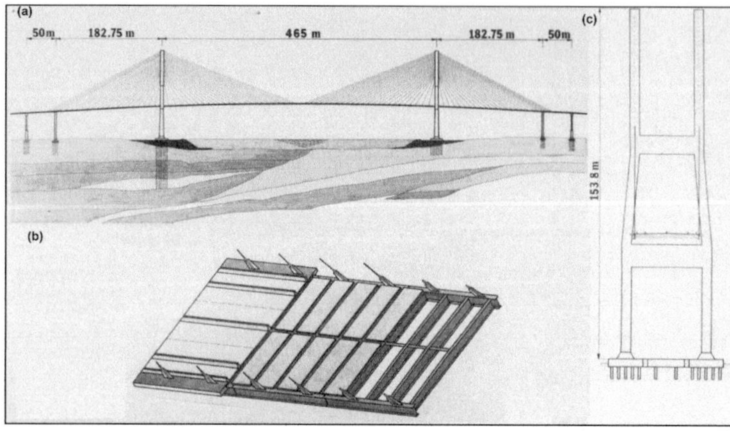

Fig. 6.57 The Alex Fraser Bridge: (a) elevation; (b) isometric view of deck; and (c) pylon

Four years later another cable-stayed bridge was opened in the same region but not for the highway; rather it carries the SkyTrain medium-capacity rapid transit system in the Metro Vancouver area across the Fraser River (Figure 6.58). The bridge has two 123 m tall pylons and has a navigation clearance 45 m above the Fraser River and valley. The main span is 340 m, and the total length is 616 m. The bridge is the longest cable-stayed transit-only bridge in the world. It is designed by Buckland and Taylor, the same firm that designed the Alex Fraser Bridge, and used the same structural system for the superstructure, nevertheless this bridge used a diamond shaped pylon.

Fig. 6.58 The Skybridge, Vancouver, Canada, 1990

6.3.4 Pitt River Bridge

One more cable-stayed bridge was opened in British Columbia in 2009. The Pitt River Bridge (Figure 6.59) spans the Pitt River between Port Coquitlam and Pitt Meadows in British Columbia, Canada. The bridge is part of Highway 7, carrying Lougheed Highway across the river. The bridge has a main span of 190 m and two side spans of 95 m. This bridge is 40 m wide and carries eight lanes of traffic. Because of the sizable deck area, this bridge was designed with three planes of cables.

Fig. 6.59 The Pitt River Bridge, British Columbia, Canada, 2009 (Courtesy, SYSTRA)

Around thirty cables with a diameter of 15.7 mm and seven wire strands were needed for each side span, while about 60 strands were needed for the main spans. A parallel harp arrangement was created by aligning the stays. The conventional composite steel and concrete deck served as the superstructure. It was composed of precast concrete panels that were 175 mm deep and covered in a 50 mm thick low permeability concrete overlay. The primary longitudinal beams are 1.5 m in depth. The floor beams are 4.5 m apart and have a depth of 1 meter. For the pylons, a solid rectangular section filled with 40MPa concrete was utilized. The middle leg was typically 5 m by 2 m in cross-section, while the outer pylon legs measured 5 m by 1.8 m. The section above the deck tapered from 5 m to 3 m, and the cables were secured by going through the pylon and coming to an end on the other face. The pylons are connected at deck level by a solid concrete diaphragm; however, they are not connected above deck (Tassin and Hall, 2009).

6.3.5 Olivier-Charbonneau Bridge

The Olivier-Charbonneau Bridge was opened to traffic in 2011 (Figure 6.60). It connects Laval's St. François district and Montreal through a crossing on Rivière des Prairies. The length of the Main Bridge is roughly 1,200 m. The crossing is made up of two 24 m continuous concrete girder approach spans, a seven-span continuous constant-depth steel girder structure with internal spans of 96 m, and a cable-stayed bridge with a 115-280-115 m span arrangement. Six traffic lanes and a multipurpose path with a width of 3 m are present on the main bridge of Olivier-Charbonneau.

The composite cross-section of the superstructure is made up of 2 m deep steel edge girders, both of which are composite with a 240-millimeter-thick concrete slab, and floor beams spaced at 4.5 m. The concrete slab is made of full-depth precast panels. Placed in a grid four panels wide, each panel spans across floor beams in the longitudinal direction (see Figure 6.61 b).

Two planes of cable stays, each with eighty stay cables organized in a semi-fan shape, support the main span cable-stayed superstructure and fasten onto the edge girders at intervals of 13.5 m. The cables consist of several 15.7 mm diameter galvanized seven-wire strands, with their number ranging from 29 to 84. Cables come in 180 mm and 250 mm sizes. The strands are shielded from corrosion by their galvanized coating. Moreover, a firmly extruded High-Density Polyethylene (HDPE) sheathing and petroleum wax were used to fill the gaps between the wires. The constructed bundled strands are further shielded from corrosion by enclosure in an exterior HDPE stay pipe.

Fig. 6.60 The Olivier-Charbonneau Bridge, Laval-Montreal, Canada,2011 (Courtesy, Parsons)

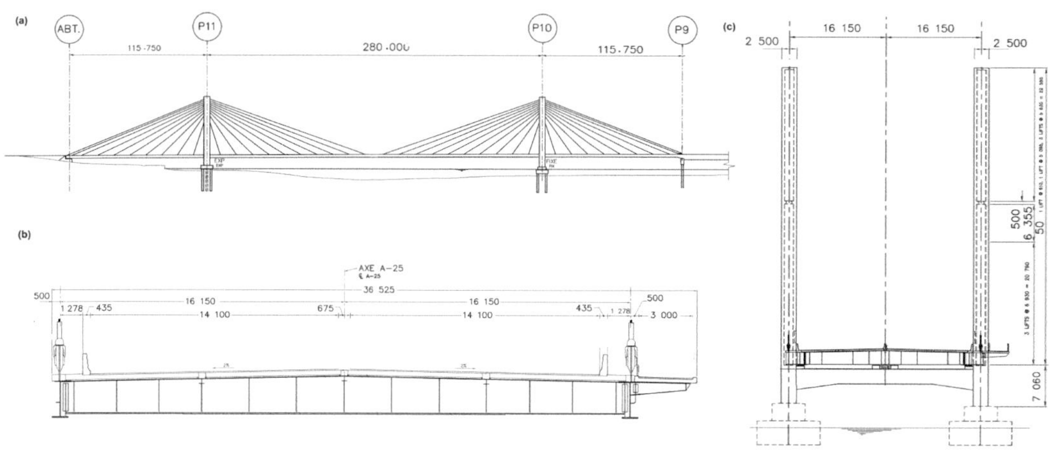

Fig. 6.61 The Olivier-Charbonneau Bridge: (a) elevation; (b) cross-section; and (c) pylons (Courtesy of Parsons)

The pylon legs (Figure 6.61 c) comprise a rectangular solid section 5 m by 2.5 m up to the level of the deck. They have hollow cross-sections above the deck with wall thicknesses 600 mm and 500 mm in the longitudinal and transverse directions respectively. The upper segment of each pylon leg comprises tube guides that are spaced every 400 mm and allow the cable stays to enter through and get anchored to the pylon head.

6.3.6 The Deh Cho Bridge

The Deh Cho Bridge (Figure 6.62) is a 1.1 km-long cable-stayed bridge across a 1.6 km span of the Mackenzie River on the Yellowknife Highway (Highway 3) near Fort Providence, Northwest Territories. The bridge, which was opened on November 30, 2012, provides a strategic connection for northern residents with the rest of Canada, and helps create economic development opportunities in the North.

Fig. 6.62 The Deh Cho Bridge, Northwest Canada, 2012

Two vertical equilateral trusses make up the superstructure, and they are joined at the top and bottom chord levels by Chevron cross bracings. The precast concrete deck of the bridge, which is 11.3 m wide and 235 mm thick, functions as a composite structure to support two lanes of traffic. the superstructure's total depth along the bridge's whole length is 4.5 m. With a total length of 1,045 m, the new jointless superstructure has a span arrangement of 90 m, 3 × 112.5 m, 190 m (navigation channel), and 90 m (see Figure 6.63). The span to depth ratio of the bridge is 42. At each pylon head, four groups of three stays each, arranged in two cable planes, are secured with cast steel sockets that are connected by pins. Steel truss outrigger systems are used to anchor the stays, which are locked coil cables with a diameter of 100 mm, at the centers of the back spans and the third point of the main span.

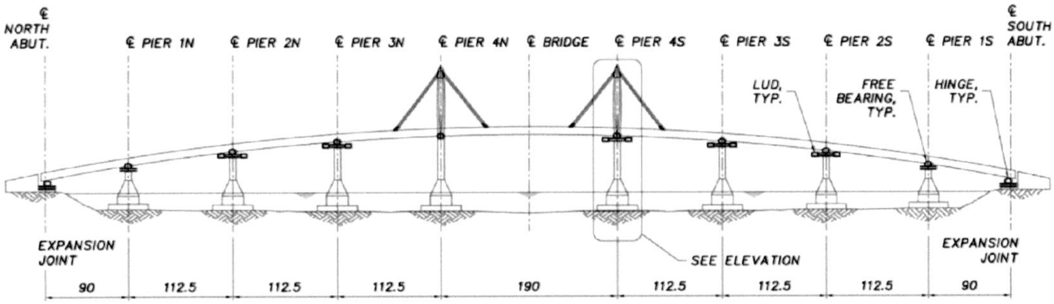

Fig. 6.63 Elevation of the Deh Cho Bridge (Courtesy of Infinity)

The bridge is furnished with two steel A-pylons (refer to Figure 6.64), each of which is supported by two spherical bearings that permit the pylon to move in its longitudinal direction. Disk bearings at the piers and abutments are generally used in the structural system, with bearing guides in the transverse bridge direction, but longitudinal movements are permitted due to temperature variations. The superstructure is longitudinally restrained only at Pier 4 North. Lock-up Devices (LUD) are used at all other piers, with the exception of the piers closest to each abutment. Temperature displacements are permitted by the Lock-up Devices without creating restraining effects; however, in the event of longitudinal impact forces resulting from seismic loads or gusty winds, the devices lock and rigidly connect the superstructure to the piers, allowing load sharing amongst engaged piers. Every pier is made up of an upper steel head and a lower solid concrete cone that is reinforced with an outer steel shell. The steel head is composed of two legs that are inclined, a base, and a tie-beam that joins the legs. At the point where the pier connects to the foundation, the steel head and the lower concrete cone are joined. There are high-strength bars that are post-tensioned to guarantee a tight connection at all times. The Deh Cho Bridge's eight piers are supported by concrete spread footings that are inserted using cofferdams into the Mackenzie Riverbed (Schueller and Singh, 2012)

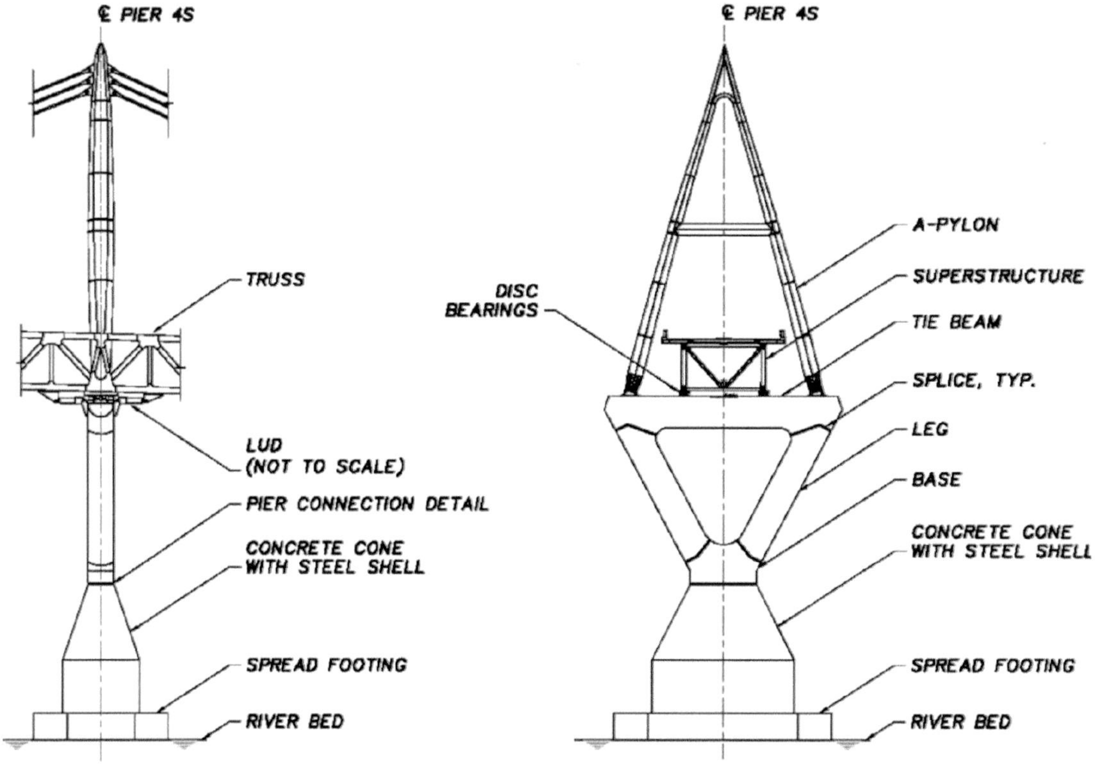

Fig. 6.64 The Deh Cho Bridge: Pylon and superstructure (Courtesy, Infinity)

6.3.7 Port Mann Bridge

The Port Mann Bridge (Figure 6.65) was completed in 2015. It replaces a steel arch bridge that spanned the Fraser River, connecting Coquitlam to Surrey in British Columbia in the Vancouver metro area. The entire crossing includes the new, ten-lane cable-stayed bridge, and approaches both Surrey and Coquitlam as part of British Columbia's Gateway Program to address Greater Vancouver's transportation needs.

The cable-stayed bridge consists of two pylons, 10 traffic lanes, one pedestrian sidewalk, and a total of 288 stay cables. The superstructure is a twin composite section deck as shown in Figure 6.66.

Fig. 6.65 The Port Mann Bridge, Vancouver, Canada, 2015

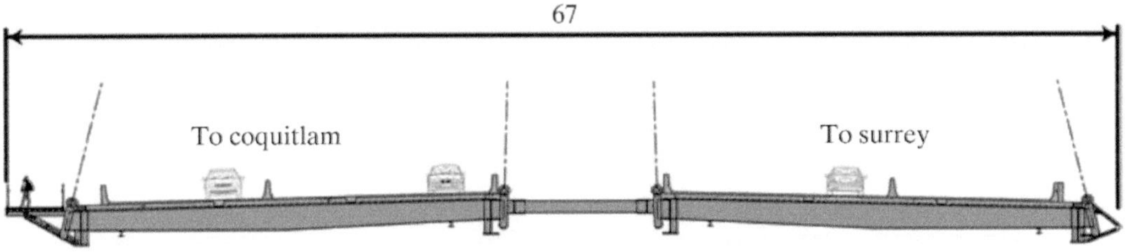

Fig. 6.66 Cross-section of the Port Mann Bridge (Courtesy, T.Y. Lin)

6.3.8 The Nipigon River Bridge

In the same year another cable-stayed bridge was opened to traffic in Ontario. The Nipigon River Bridge (Figure 6.67) Carries Highway 11 and Highway 17 and is designated as part of the Trans-Canada Highway, across the Nipigon River near Nipigon, Ontario. The bridge has two spans, each measuring 112.8 m and 139 m, and four lanes. The bridge is distinguished by a single pier line made

Fig. 6.67 The Nipigon River Bridge, Ontario, Canada, 2015

up of three independent pylons joined at the foundation and the deck below. There is a 13 m wide divided highway carried by the superstructure. This comprises two lanes of 3.75 m, a 3.0 m shoulder on the north side, and a 2.5 m shoulder on the south side. Three planes of stay cables, with eleven cables on each side of the pylon for each plane, support the deck structure. The deck is a composite cross-section made up of three transverse floor beams spaced 4.5 m apart and three plate girders that are longitudinally aligned with the three planes of the stay cables. The 225 mm precast concrete road deck is supported by the floor beams. H-piled foundations support the pylons and abutments.

6.3.9 Samuel De Champlain Bridge

In June 2019, Montreal celebrated the opening of the Samuel De Champlain Bridge over the Saint Lawrence River (Figure 6.68). This new Samuel De Champlain Bridge Corridor Project includes the cable-stayed bridge. It was decided that building a new bridge would be more advantageous for the area both financially and socially, given the continuously rising maintenance costs associated with the current truss bridge.

Fig. 6.68 The Samuel De Champlain Bridge, Quebec, Canada, 2019

The west approach structure, the cable-stayed bridge, and the east approach structure are the three separate superstructures that make up the bridge, and they are all supported by steel and concrete piers. Up to four lanes of traffic in each direction, a central transit corridor for potential light-rail service, and a multiuse pathway for pedestrians and cyclists are all features of the bridge.

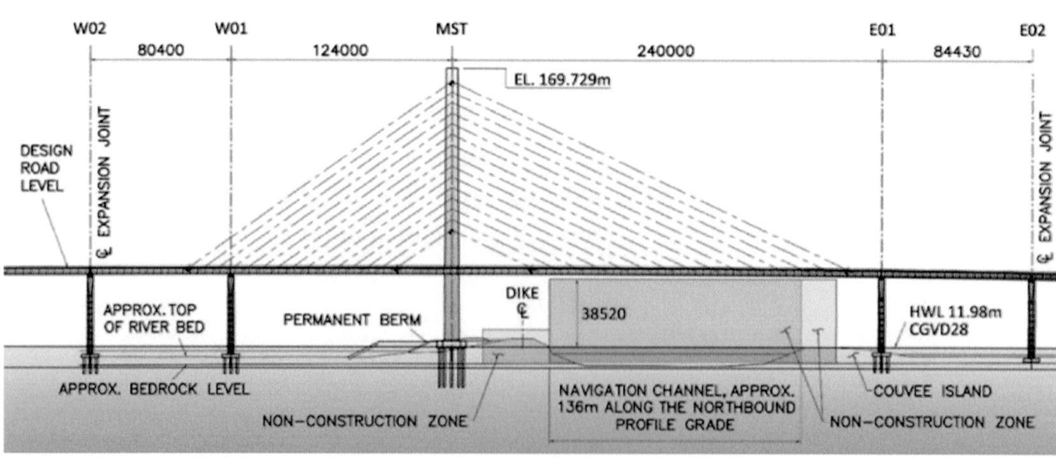

Fig. 6.69 Elevation of the Samuel De Champlain Bridge (Nader, 2019)

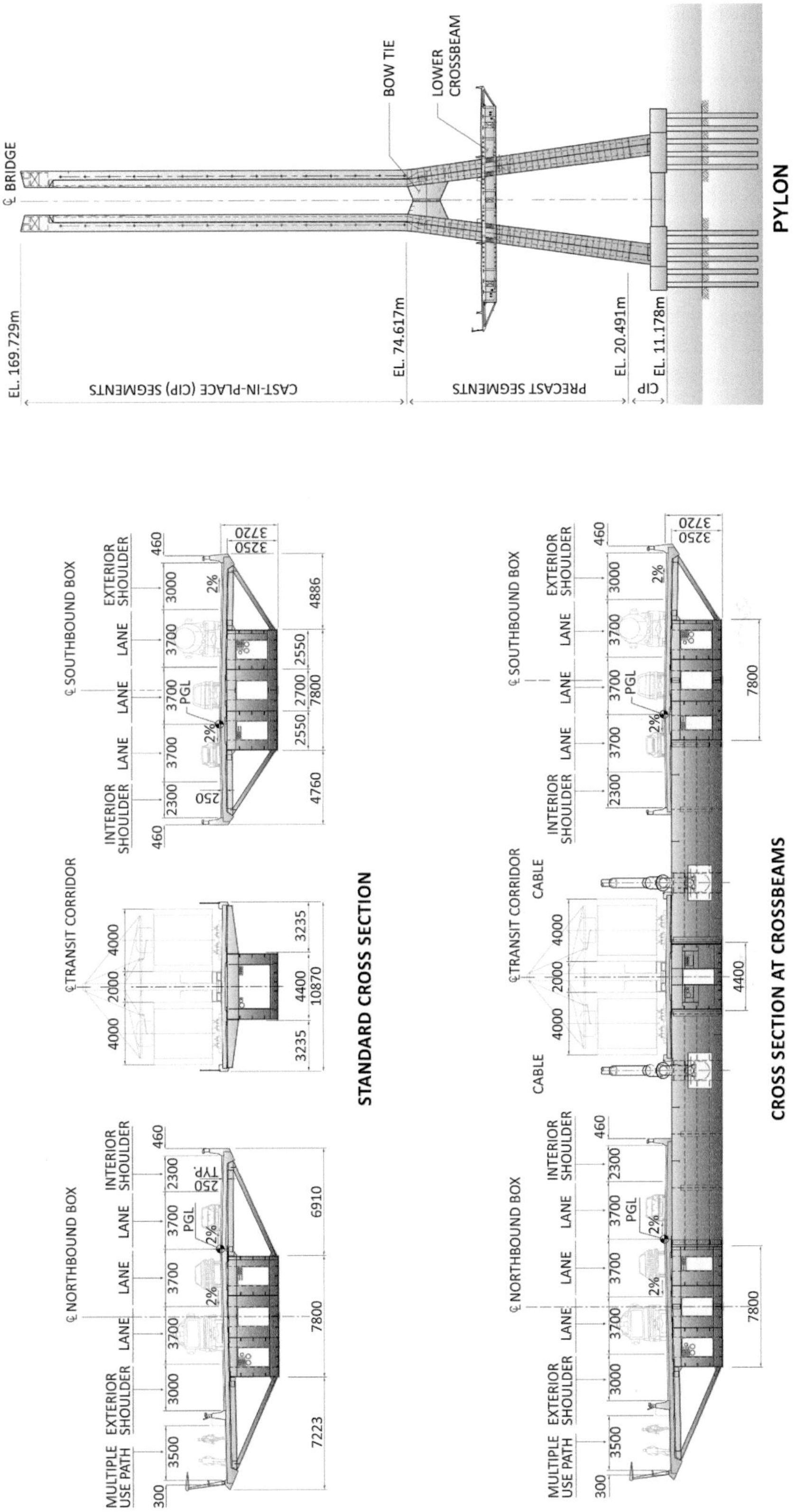

Fig. 6.70 Cross-sections and pylon of the Samuel De Champlain Bridge (Nader, 2019)

With spans of 124 m and 240 m, the single pylon cable stayed bridge is a two-span construction (Figure 6.69).

The northbound, southbound, and center transit corridors are supported by three longitudinal orthotropic box girders that make up the superstructure (See Figure 6.70). Precast deck panels and lightweight composite girders make up the structure. The weight of the girders is transmitted to the stay cables via the cross beam since the three girders are joined by crossbeams at each pair of stay cables. To achieve overall balance at the pylon, concrete counterweights were used in the shorter back span due to unbalanced spans and stay arrangement. Concrete that was poured inside the steel girder compartments served as the counterweights. The 127 7-wire strands that make up the stay cables are low-relaxation, ASTM A416 (ASTM 2018) Grade 1860 compliant, and each strand has a minimum breaking strength of 279 kN guaranteed. The cables have undergone hot-dip galvanization. Every strand was covered with high-density polyethylene (HDPE) sheathing after being waxed. Each stay's strands were inserted into an HDPE stay pipe.

Fig. 6.71 Samuel De Champlain Bridge**:** Photo illustrate the crossbeam from underneath the pylon

The pylon consists of two shafts built of precast and cast-in-place concrete segments on a CIP footing with piles. The pylon shafts are hollow and are connected by two cross beams. The lower crossbeam is framed into the superstructure and supports about 60 m of the back span and main span superstructures. Architecturally, the lower cross beam functions as a cross-passage between the three longitudinal girders; as a center for the coordination and distribution of utility lines in the superstructure and the pylon; as the chief elevator service landing; and as a base station for the under-bridge maintenance gantry (see photo in Figure 6.71). The lower portions of the shafts up to the upper cross beam are sloped at 1:7 from the vertical, while the upper portions are vertical, free-standing, and support the stay-cable anchorages (Nader, 2019).

References

Bonzon, W.S., The I-280 Veterans' Glass City Skyway: New Landmark Cable-Stayed Bridge, Ohio, *Structural Engineering International*, Volume 18, Issue 1, 43–48, 2008.

Chandra, V. and Hsu, R., The Innovative William Natcher Cable-Stayed Bridge, *IABSE Conference-Cable-Stayed Bridges-Past, Present and Future,* Malmo, Sweden, 1999.

Demers, T.G. and Simonsen, O.F., Montreal Boasts Cable-Stayed Bridge, *Civil Engineering-ASCE*, pp 59–63, August 1971.

Grant, A., The Pasco-Kennewick Intercity Bridge, *PCI Journal*, Volume:24, Issue 3, pp 90–109, May-June 1979.

Grant, A., Design and Construction of the East Huntington Bridge, *PCI Journal*, Volume 32, Issue 1, pp 20–29, May-June 1979.

Gute, W., First Vehicular Cable-Stayed Bridge in the U.S., *Civil Engineering-ASCE*, pp 50–55, November 1973.

Kealey, T.R., Mississippi River Bridge, Luling (Louisiana, USA), *IABSE STRUCTURES C-20/82*, pp 4–5, 1982.

Kumarasena, S. and McCabe, R, US Grant Bridge Replacement, *Structural Engineering International*, Volume 8, Issue 1, 56–60, 2008.

Loizias, M.P. and McCabe, R.J., Design and Construction of the Dame Point Concrete Cable-stayed Bridge in Jacksonville, Florida, *Proceedings of the International Bridge Conference*, IBC-90-61, Pittsburgh, 1990.

Muller, J. and Tassin, D., Design principles and construction methods of the Sunshine Skyway Bridge, *IABSE Symposium*, Paris, Versailles, pp 53–58, 1987.

Nader, M., Accelerated Bridge Construction of the New Samuel De Champlain Bridge, *Journal of Bridge Engineering*, Volume 25, Issue 2, 2020.

Rohleder, W.J., Tang, B., Doe, T.A., Grace, N. and Burgess, C.J., Carbon Fiber-Reinforced Polymer Strand Application on Cable-Stayed Bridge, Penobscot Narrows, Maine, Journal of the Transportation Research Board, NO. 2050 pp 169–176.

Schemmann, A.G., Bergman, D.W. and Shafer, G., John James Audubon Bridge-Design-Build Delivery of the Longest Span Cable-stayed Bridge in the United States, *Proceedings of the International Bridge Conference*, IBC-90-61, Pittsburgh, 2008.

Schueller, M. and Singh, P.R., Design and Construction of the Deh Cho Bridge Challenges, Innovation, and Opportunities, Proceedings of the Conference of the Transportation Association of Canada Fredericton, New Brunswick, 2012

Spoth, T., Condell, S. and Yuwei, S., The New Goethals Bridge, *Structural Engineering International*, Volume 29, Issue 1, 94–100, 2019.

Svensson, H.S. and Lovett, T.G., The twin Cable-Stayed Composite Bridge at Baytown, Texas, *IABSE Symposium*, Brussels, pp 317–322, 1990.

Tang, M-C, Talmadge Memorial Bridge, Savannah, Georgia, *Structural Engineering International*, 5:1, 15–16, 1995.

Tang, M-C, Jang, D., Lee, H., Sidney Lanier Cable-Stayed Bridge, Georgia, *Proceedings of the International Bridge Conference*, IBC-99-16, Pittsburgh, 1999.

Tassin, D. and Hall, C. Pitt Progress, Bd &e Issue 54 pp 32–34 2009

Taylor, P.R. and Simonsen, O.F., Annacis Island Bridge, *IABSE 12th Congress, Vancouver, BC*, pp 1086–1087, 1984.

CHAPTER 7

Development of Modern Cable-stayed Bridges in China and Japan

7.1 DEVELOPMENT OF MODERN CABLE-STAYED BRIDGES IN CHINA

The design and construction of cable-stayed bridges in China initiated in the early seventies of the last century. The first bridge was built in 1975 with a span length of 54 m (Xiang, 1999). Some concrete cable-stayed bridges were built in the period 1975-1982. The Jinan Bridge built over the Yellow River in 1982 had a main span of 220 m, the longest span achieved at that time. 19 more cable-stayed bridges were built between 1982 and 1990 in 12 provinces of China. The main span-length was raised to 260 m for the Yonghe Bridge, and 288 m for the Dongying Bridge, which were the only cable-stayed bridges with a steel deck at that time. China built ten more bridges between 1990 and 2000, with many spans beyond 400 m. Also, this period was characterized by different technologies employed for building the deck. A composite cross-section was used for three bridges in Shanghai.They were, the Nanpu Bridge built in 1991 with a main span 423 m, the Yangpu Bridge, built in 1994 with main span of 602 m, and the Xupu Bridge built in 1996. The 2nd Wuhan Bridge in Hubei; the Tongling Bridge in Anhui, the 2nd Chongqing Bridge all were built in 1995 on the Yangtze River employing prestressed concrete technology for the deck. Two more bridges were built in Hong Kong. The Kap Shui Mun Bridge, with a main span of 430 m was opened to traffic in 1997 and used an orthotropic steel deck; and the Ting Kau Bridge, completed in 1998 used a composite cross section.

7.1.1 The Second Nanjing Bridges across Yangtze River

A prominent evolution in the construction of cable-stayed bridges started in China in the early 2000s. In 2001 and 2005, two more cable-stayed bridges were completed in Nanjing, China and broke the records as both are in the list of the top thirty longest cable-stayed bridges in the world; the list contains 21 more Chinese bridges. The 1238 m long Second Nanjing cable-stayed bridge (Figure 7.1) is made up of five continuous steel box girder spans covered in a 50 mm epoxy asphalt concrete pavement. The spans are positioned as follows: 58.5 m + 246.5 m + 628 m + 246.5 m + 58.5 m. Each side span has one additional pier (Liu and Z.X., 2004). As shown in Figure 7.2, the superstructure is an orthotropic box steel girder that is 38.2 m wide and 3.5 m deep. The cables are set up in a semi-fan form, with 20 pairs of cables spaced 15 m apart in each fan.

Fig. 7.1 The Second Nanjing Bridge, China, 2001

At 195.41 m in height, the reinforced concrete cable pylon creates a distinct, inverted "Y"-shaped structure. A dual-wall steel cofferdam and a footing with 21 drilled shafts serve as the cable pylon's foundation. A 1500 t ballast weight is applied to the supplement pier in order to counteract uplift forces on the transition and piers. Throughout the trapezoid shape, the ballast weight is dispersed longitudinally.

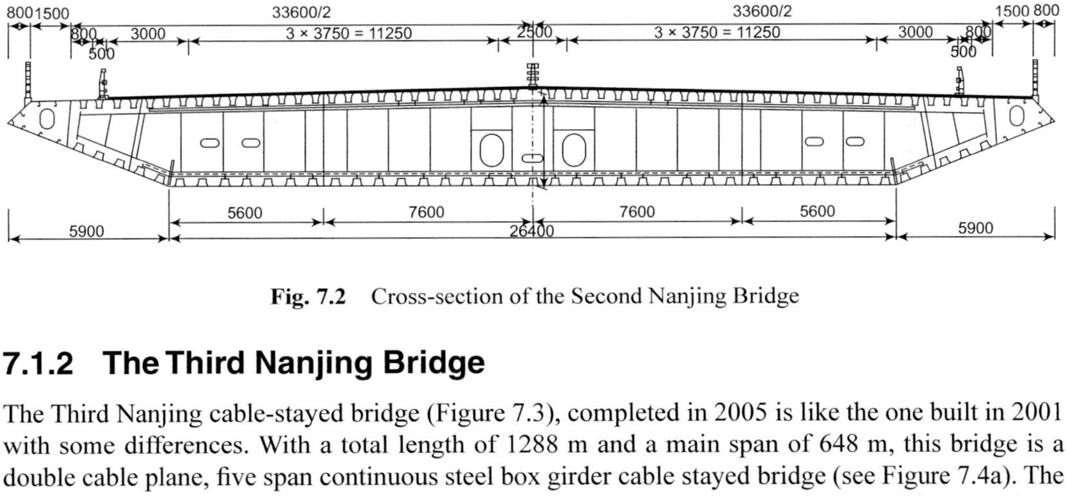

Fig. 7.2 Cross-section of the Second Nanjing Bridge

7.1.2 The Third Nanjing Bridge

The Third Nanjing cable-stayed bridge (Figure 7.3), completed in 2005 is like the one built in 2001 with some differences. With a total length of 1288 m and a main span of 648 m, this bridge is a double cable plane, five span continuous steel box girder cable stayed bridge (see Figure 7.4a). The girder is divided into 89 general segments, 15 m in length (Yue Li et al., 2017). The bridge's pylons, which have an inverted V shape, are 215 m tall and have 720-meter-radius lateral round curves on either side. Every pylon has four beams, two of which are steel structures and one of which is a reinforced concrete structure at the bottom. (Figure 5.216c). The steel girder is 1288m in length, 37.2 m in width and 3.2 m in height (Figure 7.4 b).

Fig. 7.3 The Third Nanjing Bridge, China, 2005 (Courtesy, Mageba)

7.1.3 Sutong Bridge

Sutong Bridge across Yangtze River (Figure 7.5) came next in 2008. With its 1088 m main span, it set itself as the longest cable-stayed bridge in the world for almost twelve years until the Hutong Bridge broke the record. The Sutong Bridge crosses the Yangtze River approximately 100 km upstream from Shanghai connecting the cities of Suzhou and Nantong located on the southern and northern banks respectively. The total length of this link is about 8.2 km of bridge structures.

The cable-stayed bridge is a seven span which has a span arrangement of 100 m + 100 m + 300 m + 1088 m + 300 m + 100 m + 100 m = 2088 m (Figure 7.6). An orthotropic steel box girder makes up the superstructure cross-section of the bridge. With the wind fairing, the overall width is 41.0 m, allowing for 8 dual traffic lanes. As shown in Figure 7.7, the cross-section depth is 4.0 m. Closed steel ribs are used to stiffen the steel box longitudinally. Transverse plate diaphragms are offered with smaller distances down to 5.27 m locally and a typical distance of 4.0 m around the two pylons. The structural steel's typical yield strengths are 345 MPa and 370 MPa (Miao et al., 2006).

The cable stay systems are made of parallel wire strands consisting of 7 mm wires, each with a cross-sectional area of 38.48 mm. The stay cables are arranged in double inclined cable planes with a standard spacing of 16 m in the central span and 12 m near the ends of the back spans along the girder. The cables have a nominal tensile strength of 1,770 MPa. The longest backstay has a maximum cable size of 158 mm, while the main span stays close to the pylons have a minimum cable size of 111 mm. The longest cable weighs 59 tons and is approximately 577 m long.

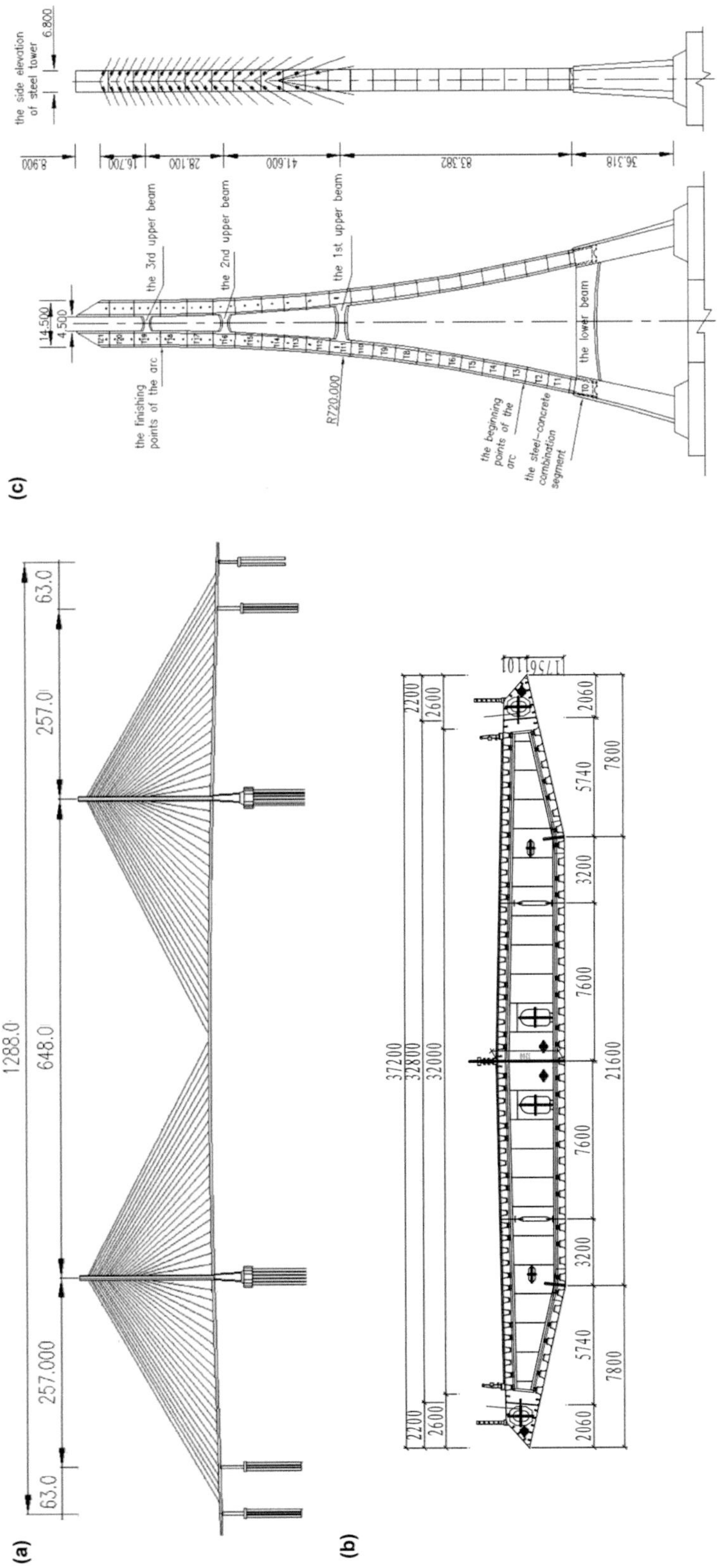

Fig. 7.4 The Third Nanjing Bridge: (a) elevation; (b) cross-section; and (c) pylon

Fig. 7.5 Sutong Bridge, China, 2008 (Courtesy, AECOM)

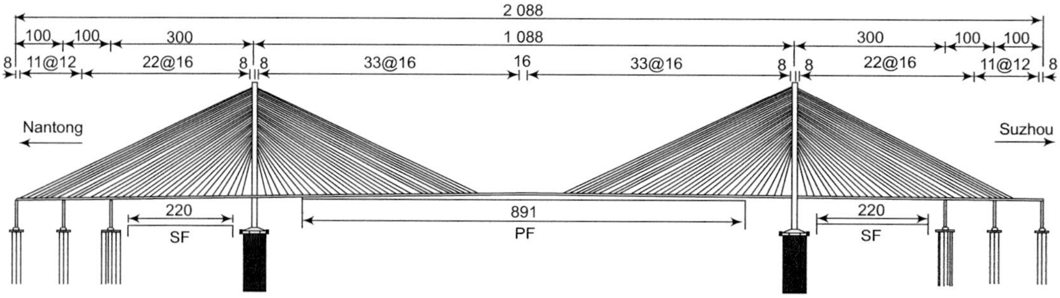

Fig. 7.6 Elevation of Sutong Bridge, (Miao et al., 2006)

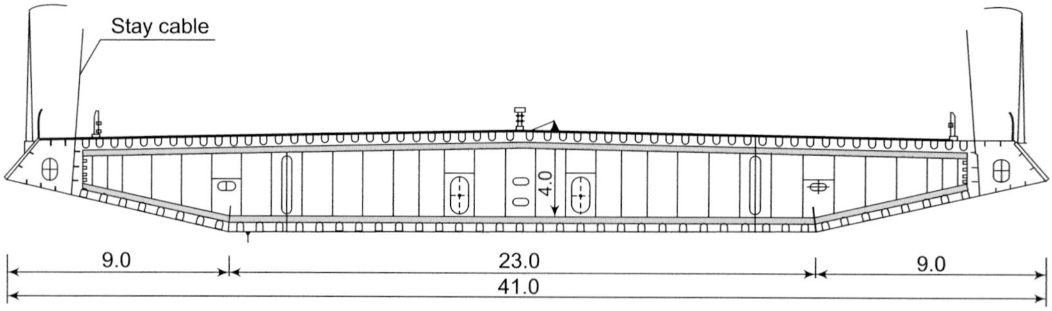

Fig. 7.7 Cross-section of Sutong Bridge, (Miao et al., 2006)

The concrete inverted Y-shaped pylons are approximately 300 m high (Figure 7.8). Shear studs at the top of the pylon secure the steel boxes containing the stay-cables to the concrete. A cable anchorage steel box's largest segment weighs roughly 36 tons. The full floating structural system and fixed structural system were not feasible due to wind and temperature. Therefore, viscous dampers were installed as a semi-floating system. Viscous dampers dissipate seismic energy but at the same time introduce small restrictions to the slow movements caused by temperature, traffic, and static wind. Each pylon has four dampers in total. Before the displacement restriction devices are turned on, there can be a relative movement of 750 mm between the girder and the pylon, which is allowed before the displacement restriction equipment is activated. Each of the four dampers at a single pylon has a linear stiffness of 100 MN/m of movement if a relative movement beyond 750 mm occurs.

The pylon or piers are supported by bored friction piles with an approximate diameter of 2.65 m. Each pylon is supported by 131 piles. Piers' foundations range from 19 piles per pier to 36 piles per pier. The pile lengths vary between 108 and 116 m.

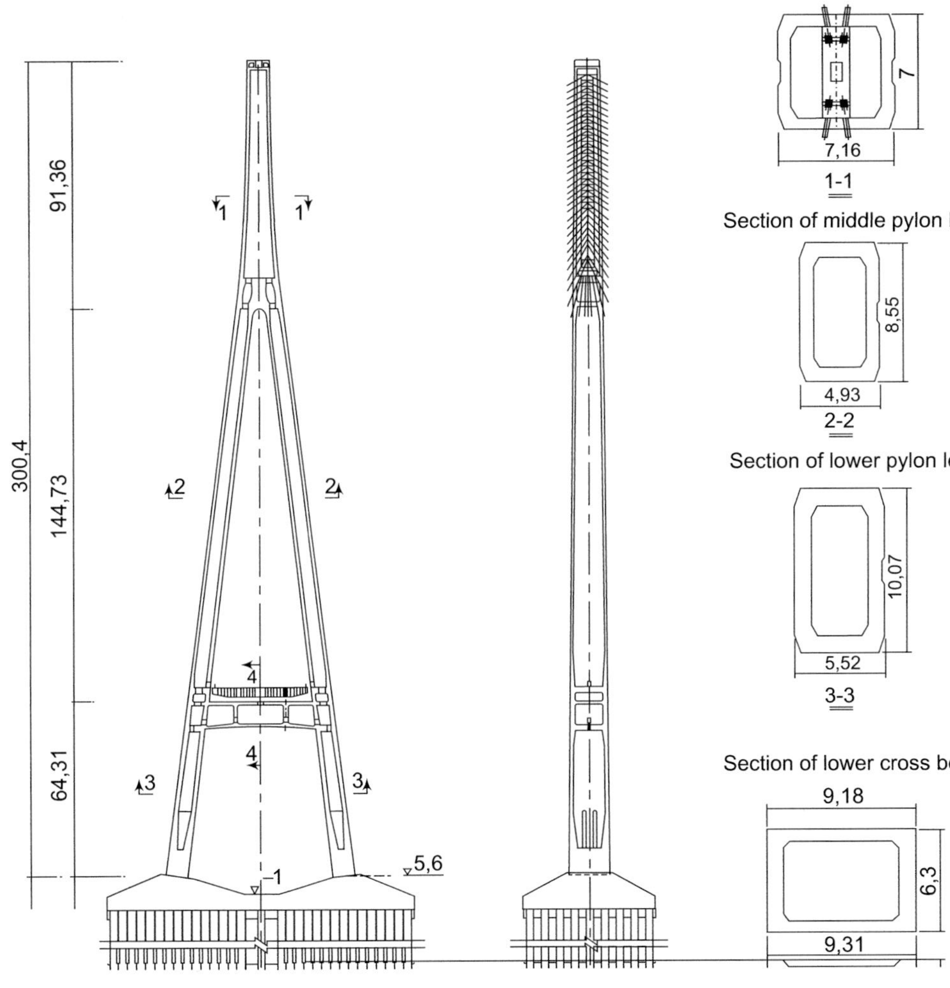

Fig. 7.8 Pylons of Sutong Bridge (You et al., 2008)

7.1.4 Stonecutters Bridge

Stonecutters Bridge (Figure 7.9) opened a year later in Honk Kong. With a 1018 m main span it could not break the record by Sutong Bridge, nevertheless, it keeps its ranking as the fourth longest cable-stayed bridge in the world. It is a section of Route 8 (formerly known as Route 9), an east-west expressway that connects the cities of West Kowloon and Lantau Island with Hong Kong International Airport (Hansen et al., 2004).

Fig. 7.9 Stonecutters Bridge, Honk Kong, China, 2009 (Hansen et al., 2004)

The bridge is 1596 m long overall. Its main span extends 1018 m across Rambler Channel, with four back spans measuring 79.75, 70, 70, and 69.25 m on each side. The two stay cable planes have a semi-fan arrangement and are anchored at the deck's outer edges, spaced 10 m apart for the back spans and 18 m apart for the main span (see Figure 7.10).

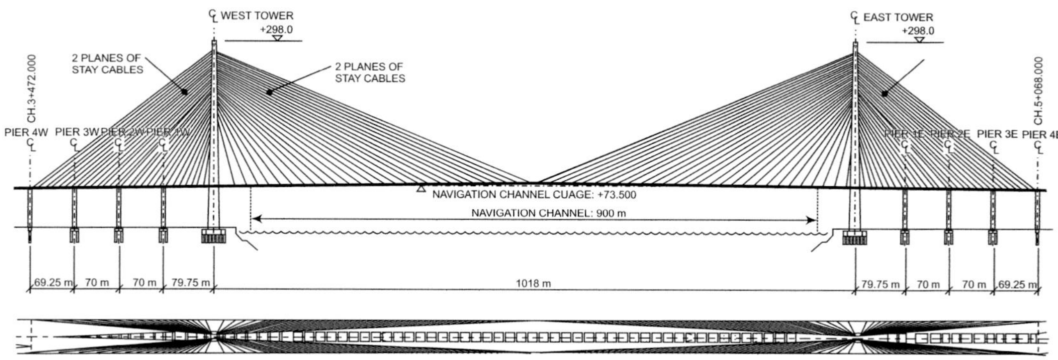

Fig. 7.10 Elevation and plan of Stonecutters Bridge (Hansen et al., 2004)

The bridge deck is a twin box-girder, steel in the main span and prestressed concrete in the back spans. The interface between steel and concrete is located 49.75 m into the back spans. The two longitudinal girders are connected by cross girders. The alignment of the main span is straight and the deck width is constant. However, the back spans must accommodate widening roadways at the western end and a curved alignment with superelevation at the eastern end (see Figure 7.11).

The freestanding pylons have a circular cross-section in concrete up to 170.5 m above the ground. Typically, couplers connect the Ø50 mm vertical reinforcement. Stainless-steel bars make up the outer reinforcement layer, which increases the structure's resilience in the harsh marine environment. The upper part is a composite made of the inner concrete wall and stainless-steel skin. Shear studs are used to connect the skin to the concrete wall in order to achieve composite action. A glazed steel structure is arranged on the top 5 m and serves as both a structural lighting element and a place to store maintenance equipment. One of Stonecutters Bridge's most recognizable visual elements is the tall, circular, freestanding pylons with a metallic upper portion. The deck is solidly

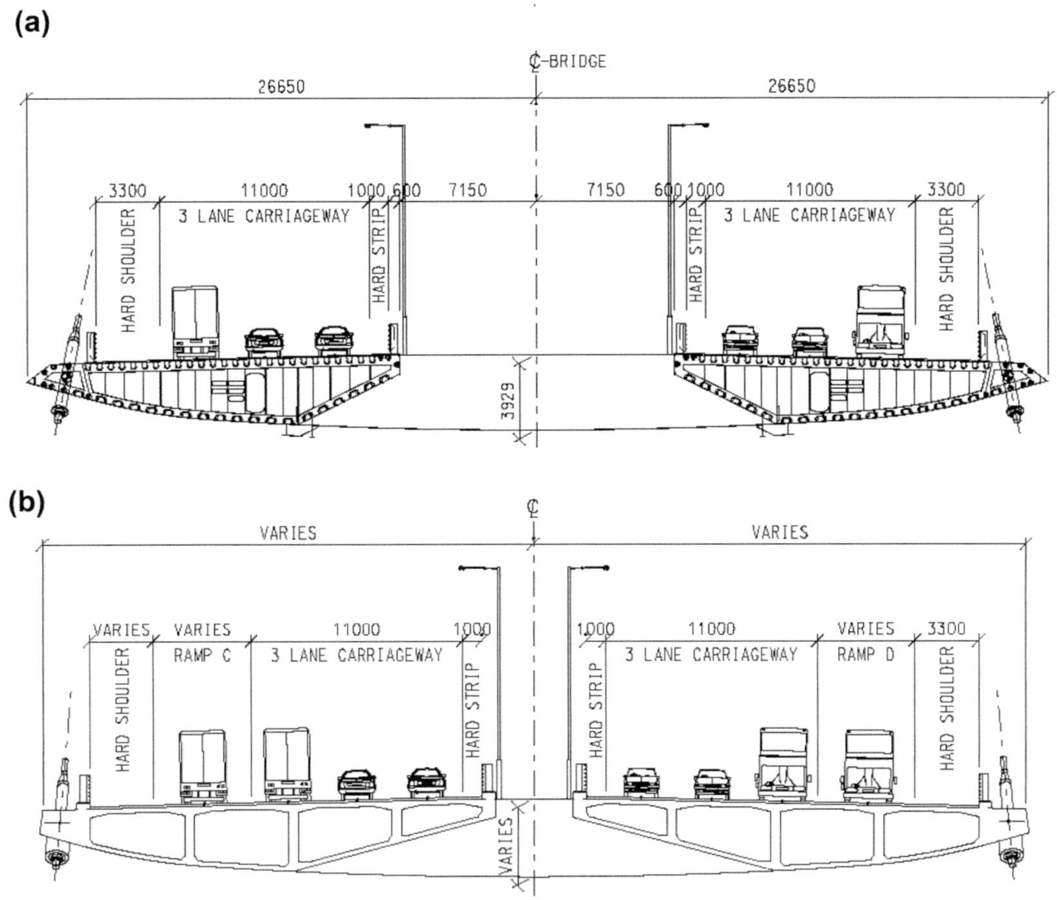

Fig. 7.11 Cross-section of Stonecutters Bridge: (a) steel deck of main span; and (b) concrete deck at back spans (Hansen et al., 2004)

attached to the piers in the rear spans. The back span piers are monolithically connected to the deck and the three intermediate piers are single columns (refer to Figure 7.12). The back span piers and bearings on the pylons limit the bridge deck's lateral movement. The back span piers and the hydraulic buffers at the pylons provide longitudinal direction restraint. The buffers are configured so that movements resulting from static actions, like temperature and mean wind, can occur while they fuse and limit movement under seismic shaking and strong winds.

7.1.5 Hutong Yangtze River Bridge

In the period from 2009 to 2020 China built dozens of long main span cable-stayed bridges. The list includes but is not limited to three cable-stayed bridges in 2010: the 926 m Edong Yangtze River Bridge; the 816 m Jingyue Yangtze River Bridge; and the 708 m Minpu Bridge, Shanghai. Two more giant cable-stayed bridges were completed in 2013: the 780 m Zhangzhou Xiamen Bridge across the Jiulong River and the 818 m Jiujiang Yangtze River Expressway Bridge. The 630 m Tongling Road-Rail Bridge was opened in 2015. Three more bridges were opened in 2016: the 638 m Wangdong Bridge; the 720 m Duge Bridge in Liupanshui and the 800 m Yachi River Bridge in Guizhou. The 800m Second Wuhu Bridge was opened in 2017. Three cable-stayed bridges were completed in 2019: the 820 m Shishou Yangtze River Bridge; the 828 m Chizhou Yangtze River Bridge; and the 920 m Jiayu Yangtze River Bridge in Xianning.

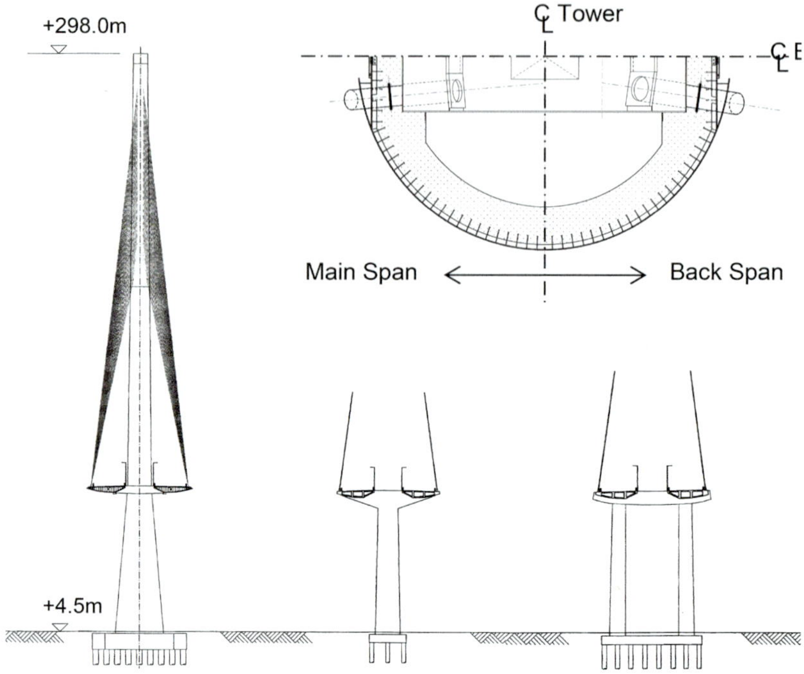

Fig. 7.12 Pylon and piers of Stonecutters Bridge (Hansen et al., 2004)

Construction of Hutong (HuSuTong) Yangtze River Bridge (Figure 7.13) started in 2014 and six years later the bridge was opened in July 2020. It links Tongzhou District, Nantong, in the north with Zhangjiagang City, Suzhou, in the south. Figure 7.14 depicts the span layout of the double-pylon steel truss girder cable-stayed bridge, which has a main span of 1092 m and a span layout of (142 + 462 + 1092 + 462 + 142) = 2300 m.

Fig. 7.13 Hutong Yangtze River Bridge, China, 2020

The width of the bridge deck is 35 m, and its height is 16 m. The upper portion carries a six-lane highway, and the lower portion carries a four-lane railway track. The structural system comprises three main longitudinal trusses. The upper deck carrying the roadway is an orthotropic steel slab that is supported by transverse floor beams that act integral with the orthotropic slab plate and sit on top of two transverse trusses spaced every 14 m and transfer their loads to the verticals of the main longitudinal trusses. The lowermost railway deck is a box structure made of orthotropic steel. As shown in Figure 7.15, a concrete bridge deck composite section is used in the highway bridge deck within 252 m of the side spans in order to counteract the auxiliary pier supports' negative reaction force.

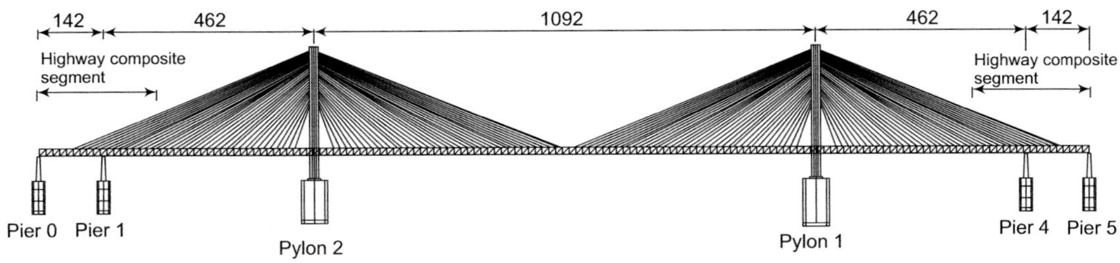

Fig. 7.14 Elevation of Hutong Yangtze River Bridge (Lu and Li, 2018)

The stay cables of the bridge were set as a three-cable plane structure corresponding to the three-piece truss structure of the main girders. A total of 432 parallel wire cables are used to support the suspended spans. Cable sizes range from a minimum of 110 mm for the back stays near the pylons to a maximum of 150 mm for the longest backstay. The cables are spaced on the upper part of the pylons and are equally spaced at the bridge deck on the side spans and the main span at equally spaced distances of 14 m intervals. The pylons are diamond-shaped and divided into three parts depending on the geometry. The heights of the upper, middle and lower parts of the pylons are 131, 140 and 54 m, respectively, as shown in Figure 7.16. Caisson foundations were adopted for the pylons with plane length and width 86.9 and 58.7 m, respectively.

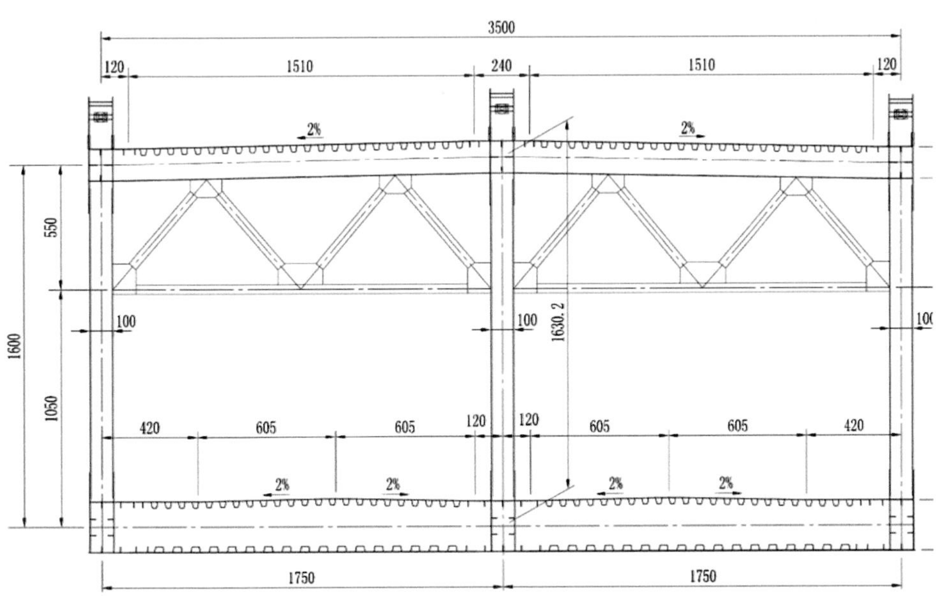

Fig. 7.15 Cross-section of Hutong Yangtze River Bridge (Lu and Li, 2018)

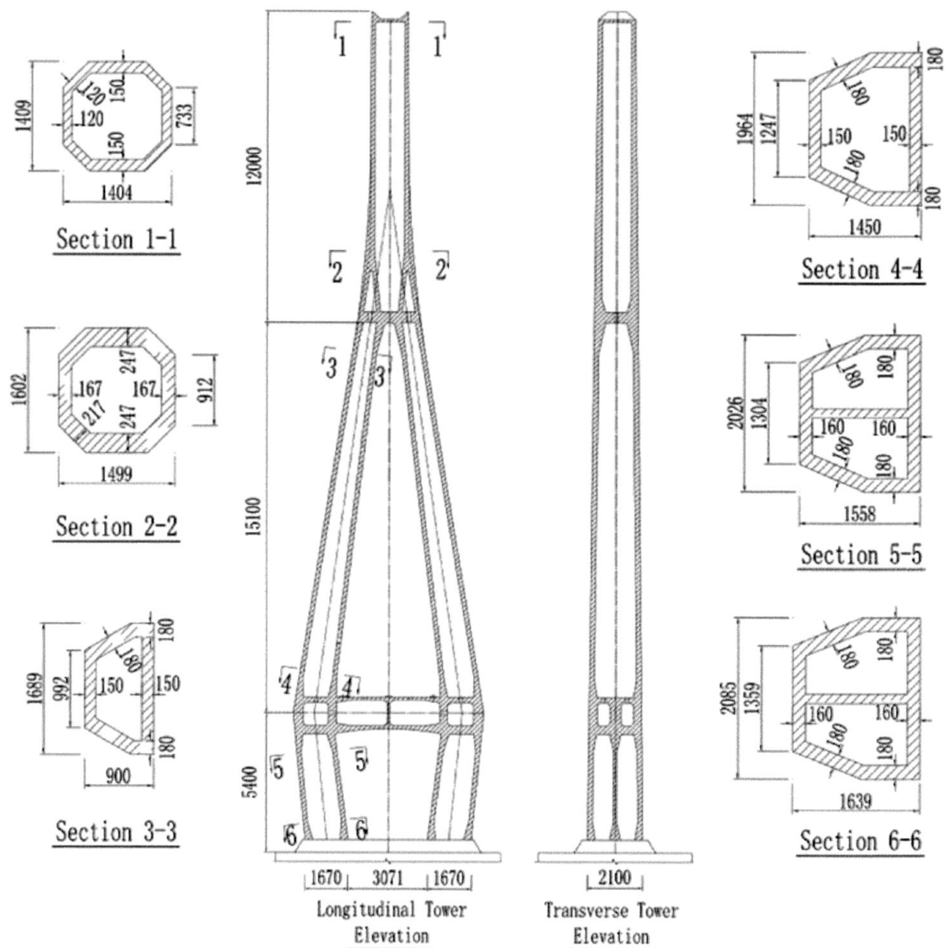

Fig. 7.16 Pylons of Hutong Yangtze River Bridge (Lu and Li, 2018)

7.2 JAPANESE EXPERIENCE WITH CABLE-STAYED BRIDGES

Construction of cable-stayed bridges started in Japan in the late sixties of the last century. The Onomichi Bridge, which opened in 1968, is regarded as Japan's first cable-stayed bridge. It links the islands of Honshu and Mukai-Lima, totaling 385 m in length. There is a 215 m center span and 85 m side span. The orthotropic plate deck that spans the two longitudinal plate girders, each measuring 3.2 m deep and positioned 10.2 m apart, makes up the superstructure. This bridge has a depth-to-span ratio of 1/67. The steel pylons are designed like a portal frame.

7.2.1 The Toyosato Bridge

The Toyosato Bridge (Figure 7.17) in Osaka is one of the early cable-stayed bridges of Japan. Completed in 1970, this bridge has a total length of 377 m with a 216 m main span and two 80.5 m side spans. The 19.10 m deck carries four lanes of traffic and two bicycle paths. The superstructure is a trapezoidal box section, 10.5 m wide at the top flange and 7 m wide at the bottom flange. The depth is 3.0 m, about 1/72 of the span. The orthotropic deck is supported by transverse cross beams at 1.8 m on centers which cantilever out 7 m from box 4 to produce a total deck width of 28 m. The A shaped pylons are 35.0 m above the piers. The cable stays are arranged in a single span configuration consisting of eight cables anchored to the transverse beams of the steel pylons. Upper stays comprise 16 strands of 154 wires each. Lower stays consist of 12 strands of 127 wires each. The wire diameter is approximately 5 mm. Each strand is fabricated with parallel wires. Upper and lower stays have diameters of approximately 280 and 220 mm respectively. In addition to the zinc coating, the stays are covered by a synthetic resin wrapping and are made continuous over saddles in the pylon. Two rocker bearings transfer the load of each pylon to its supporting reinforced concrete piers. They are hinged longitudinally and fixed transversally to the bridge. There are concrete shear blocks installed on each pier that restrain the transverse motion of the bridge under seismic loads.

Fig. 7.17 The Toyosato Bridge, Japan, 1970

7.2.2 Arakawa-Ohashi Bridge

One year later, another cable-stayed bridge was opened in Tokyo. The Arakawa-Ohashi Bridge (Figure 7.18). The bridge has two 60 m side spans in addition to its 160 m main span. There are four traffic lanes on the deck. Two outside plate girders placed 4.5 m apart from the box's web and a central box girder measuring 4 m wide make up the superstructure. The orthotropic deck's overall width is 17.9 m. The cross section has a depth of 2.4 m and a depth-to-span ratio of 1/67. The superstructure as well as the 34.0 m pylon are all made of steel. The cable stays are arranged in a single span configuration consisting of eight cables.

Fig. 7.18 The Arakawa-Ohashi Bridge, Japan, 1971

7.2.3 Rokko Bridge

Construction of the Rokko Bridge (Figure 7.19) was completed in 1976. It connects Rokko Island and Kobe and has a center span of 220 m with two side spans, each measuring 90 m. It was intended to be a six-lane, double-decker roadway bridge.

Fig. 7.19 The Rokko Bridge, Japan, 1976

The superstructure comprises two edged Warren trusses that are designed such that they are continuous over the three spans and supported by semi-fan arranged cables. The longitudinal truss's upper and lower steel deck plates are fastened to its chord members to function as a part of its flange and help support its bending moments. The deck plates are supported by cross-girders, which are placed every 2.5 m. The cross-girders are supported as simple girders on the chord members and are rigidly attached to the steel deck. The cables consist of parallel wire strands that include 217 wires with a diameter of 5 mm in each strand. A 2 mm thick plastic wrapping made of polyethylene film reinforced with glass fibers shields the cables from corrosion. The pylons are H shaped. Each leg has a cross-section that measures 2.0 by 2.2 m and has a maximum plate thickness of 32 mm. To allow base rotation in the longitudinal direction while preventing it in the transverse direction, the pylons are simply supported on the shoes and fixed at intermediate piers (Figure 7.20).

7.2.4 The Katsushika Harp Bridge

1988 saw opening of the Katsushika Harp Bridge (Figure 7.21) in Tokyo. Located amidst the Yotsugi entry/exit on the Central Circular Route and the Hirai-Ohashi entry/exit, it was built over the separation levee where the Ayase River divides into the Arakawa and Nakagawa Valleys.

The total length of the bridge is 455 m with a central span of 220 m. This bridge has a unique feature as the pylons are not equal in height as shown in Figure 7.22.

7.2.5 Hitsuishijima and Iwakurojima Bridges

The Great Seto Bridge, which connects Okayama and Kagawa territories in Japan by means of a network of double-deck bridges spanning five small islands in the Seto Inland Sea, was constructed in 1988 and included the two bridges (Figure 7.23). The bridges support one railroad track in each direction on the lower deck and two lanes of traffic on the upper deck in each direction.

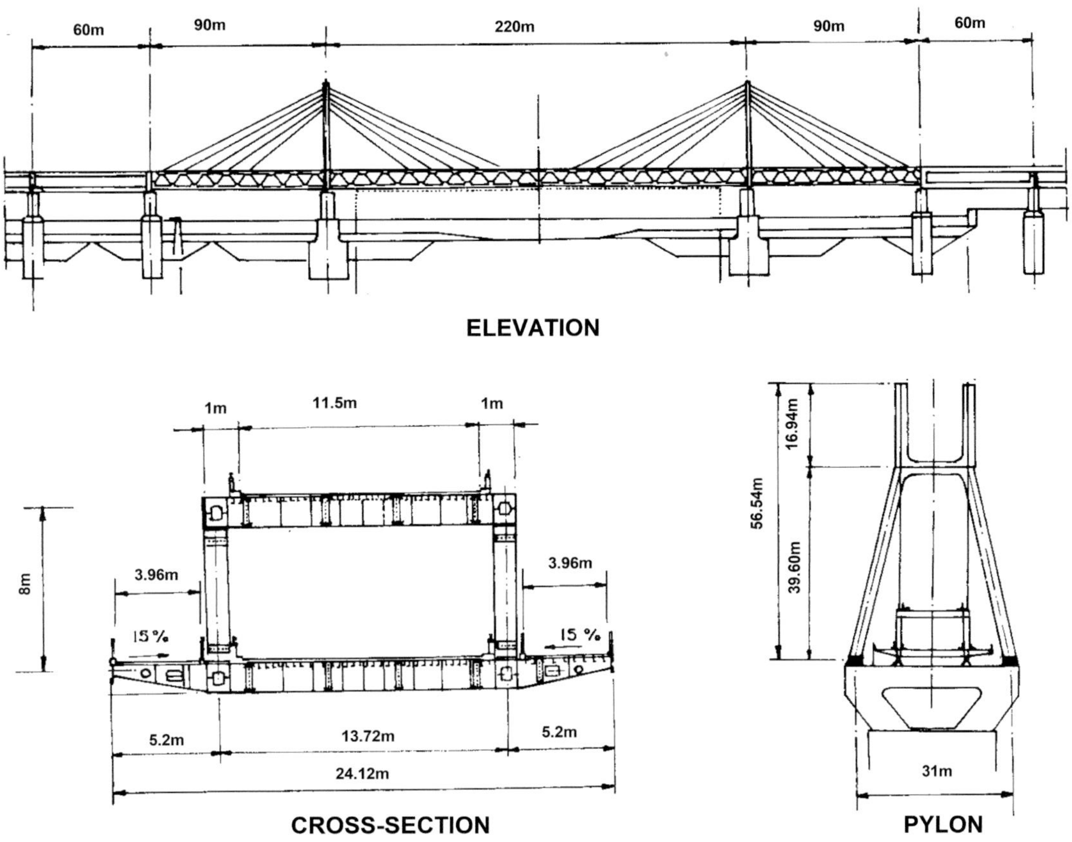

Fig. 7.20 General Configuration of the Rokko Bridge

Fig. 7.21 The Katsushika Harp Bridge, Japan, 1985 (Courtesy of Shutoko)

Each bridge has a main span of 420 m and two side spans of 185 m. The superstructure is 27.5 m wide and composed of two longitudinal Warren trusses with 13.9 m high vertical members. The upper and lower deck plates are supported on cross girders that are spaced at equal intervals. The superstructure is supported by stay cables made of parallel wire strands using 7 mm diameter steel wires anchored to HiAm sockets. The H shaped pylons are rigid steel frames that rise 136 m on top of their supporting piers. The bridges are provided with spring shoes and shock absorbers that are installed at the truss ends to control the longitudinal movements due to seismic loads.

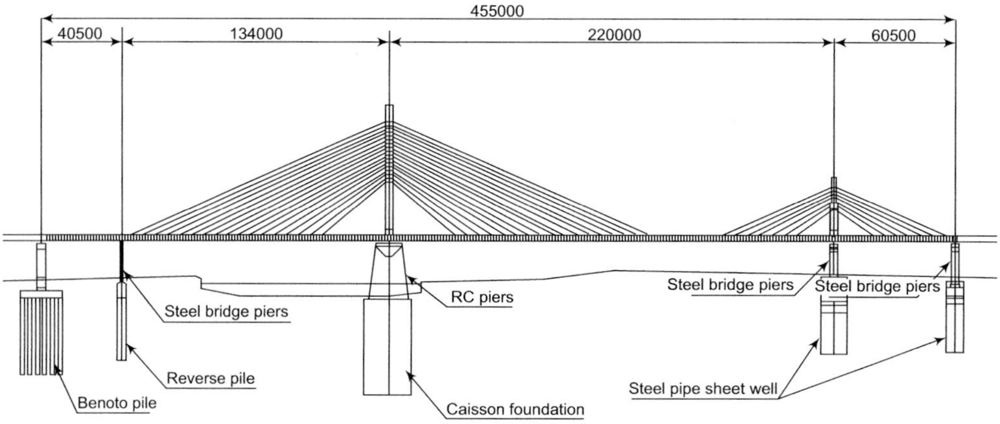

Fig. 7.22 Elevation of the Katsushika Harp Bridge, Japan

Fig. 7.23 Hitsuishijima and Iwakurojima Bridges, Japan, 1988

7.2.6 Yokohama Bridge

The Yokohama Bridge (Figure 7.24) was added to Japan's infrastructure network in 1989. The bridge is part of the Bayshore Route of the Shuto Expressway connecting Honmoku futo and Daikoku futo. The 860 m bridge crosses Tokyo Bay with a main span of 460 m. With a total span of 860 m (200, 460, and 200 m), the bridge is a continuous, double-deck, three-span cable-stayed structure. The girder is made up of a double-decked steel truss box, with the lower deck serving the two-lane national route and the upper deck serving the six-lane Yokohama Expressway Bay shore route. Each of the bridge's two 172-meter-tall H-shaped pylons are welded together as a single, monolithic piece. The pylon on the Daikoku side is provided with a lookout lounge at the lower transverse beam level and one can enjoy the view of the Yokohama Port from a height of 50 m above sea level.

The superstructure comprises two main longitudinal trusses 12.0 m deep. The width of the superstructure is 31.0 m centerline to centerline of the longitudinal trusses that are connected to each other at each vertical element by transverse diaphragms at the upper deck and cross beams at the lower deck. There is a bottom lateral bracing system that is also included at the lower deck. Wind fairings are included and cantilevers extend out another 4.60 m on both sides. As shown in

Fig. 7.24 The Yokohama Bridge, Japan, 1989

Figure 7.25, the upper deck is a 3.0 m deep steel orthotropic box section stiffened by longitudinal ribs and transverse diaphragms. The lower deck for each direction comprises a steel orthotropic plate supported by five longitudinal stringers that transfer the loads to the horizontal cross beam, which is supported directly by the longitudinal trusses through a rigid connection. The stay cable system is demi harped and comprises two planes of stay cables. 88 dual stay cables are transferring the superstructure load to the pylons. The steel pylons rise 172 m on top of their supporting foundations and are fixed firmly to them. The boundary conditions for the structure are illustrated in Figure (7.26). Rocker links are provided at the pylons and piers. They are composed of steel and have circular bearings at both ends and a solid main body in the center. Steel rods are inserted into the bearings through an inner ring and bolted to two circular steel plates on either side of the main body (Figure 7.27). They are intended to be elements of tension. Girder longitudinal movement is made possible by the multi-rotational pin connections at both ends, which function similarly to a pendulum system. The girder is suspended across the pylons and piers in a transverse direction. Wind tongues on the pier-girder and pylon-girder connections limit the transverse movements of girders. To allow for slight relative motion, there is a tiny space between the wind tongues and the girder. The wind shoes are located at the interface of the girder connections with the wind tongue. In essence, they are sliding bearings with a stainless-steel wind tongue side and a Teflon polytetrafluoroethylene (PTFE) surface on the girder side. The maximum transverse load that side bearings and wind shoes can withstand for pier–girder and pylon–girder connections, respectively, is 27,654 and 48,346 kN.

7.2.7 Aomori Bay Bridge

The Aomori Bay Bridge (Figure 7.28) was completely opened to traffic in July 1994. It is a prestressed concrete cable-stayed bridge in Aomori. The bridge is 498 m long, with a center span of 240 m, and a width of 25 m. In order to link port facilities that were divided by a railroad station, it was constructed as a port road bridge. It is a conspicuous part of Aomori's skyline. The bridge carries two lanes in each direction and two sidewalks.

The main girder has a 3-cell box in cross section made of prestressed concrete (PC) that is 25 m wide and 3.5 to 2.5 m deep (see Figure 7.29). Because the stay cables are arranged in a single plane and the cable tension adjustment must be done inside the main girder, the 3-cell box-girder increases torsional rigidity. To divide the cable tension among the four webs, diaphragms are installed in the main girder's cable anchoring sections. A reinforced concrete (RC) structure with a diamond shape serves as the primary pylon. In order to decrease dead load and enhance aesthetics, the main pylon was made slender using high strength concrete with a characteristic strength of 60 MPa. The stay cable arrangement is a single plane fan system with two parallel cables in one stay. The pylons are supported by large concrete reinforced caissons.

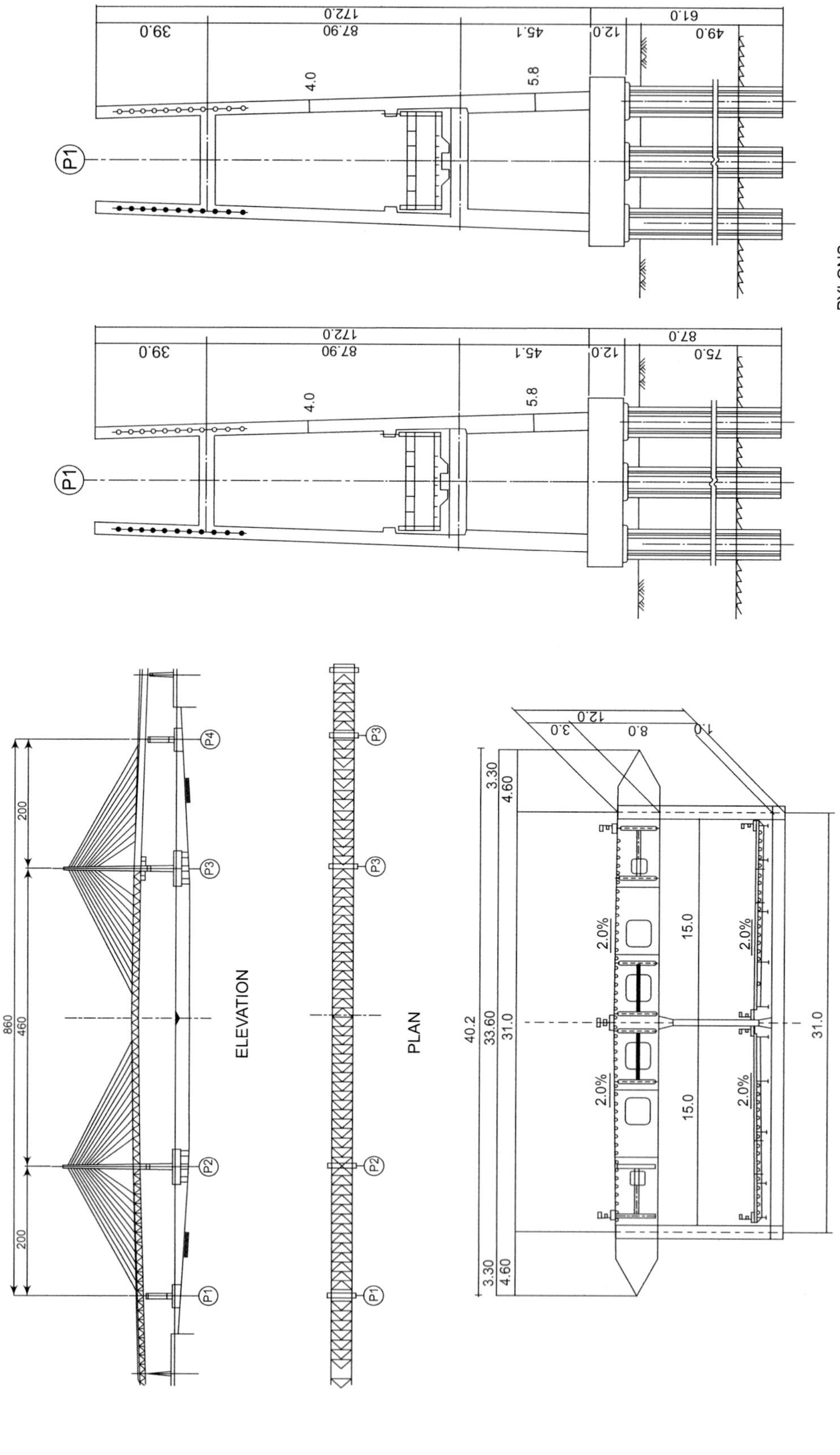

Fig. 7.25 General configuration of the Yokohama Bridge (Wada et al., 2010)

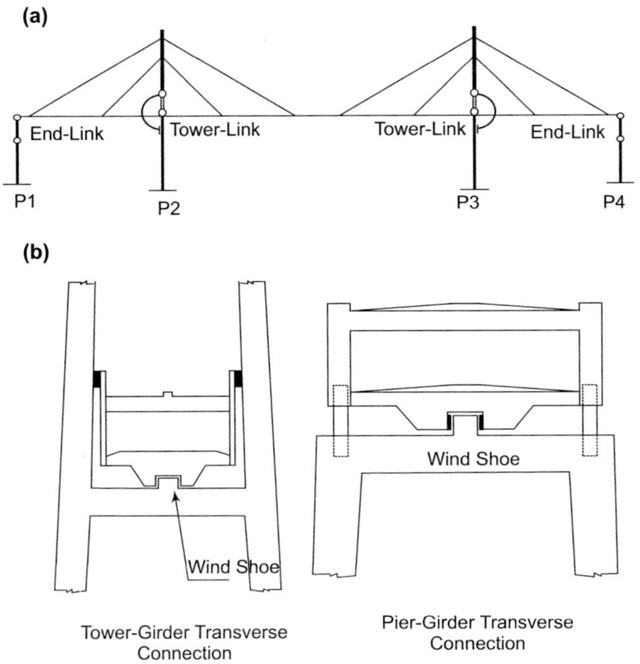

Fig. 7.26 The Yokohama Bridge: (a) boundary conditions at Pylons and piers of ; and (b) boundary conditions details (Siringoringo et al., 2014)

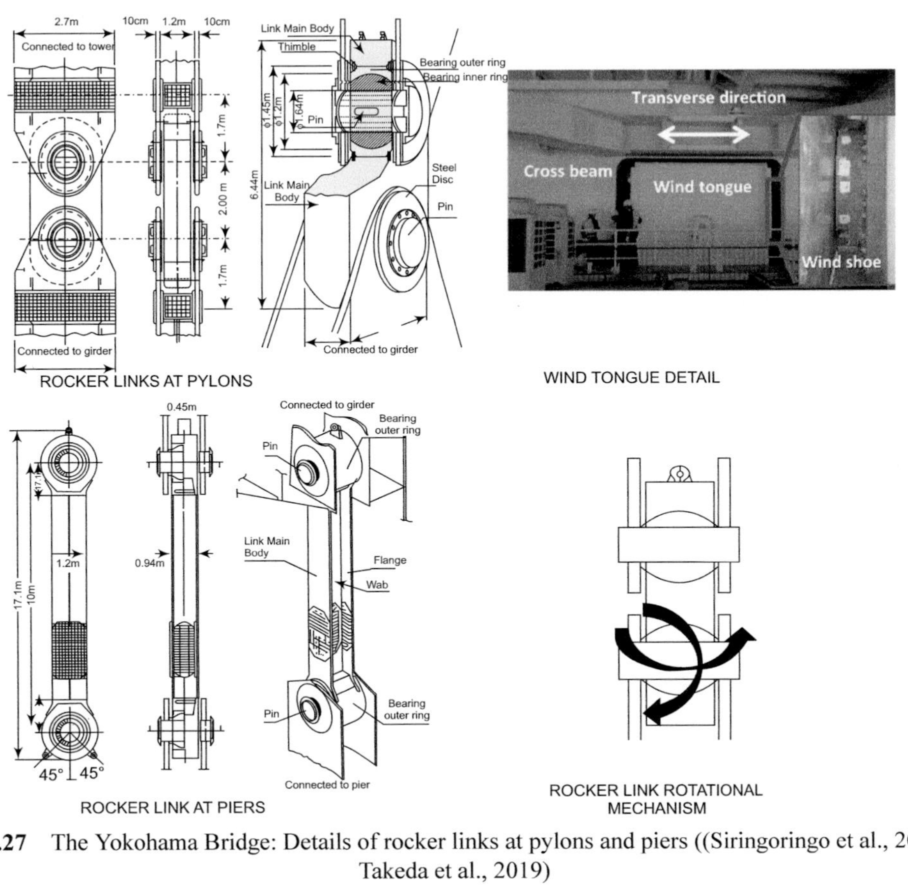

Fig. 7.27 The Yokohama Bridge: Details of rocker links at pylons and piers ((Siringoringo et al., 2014 & Takeda et al., 2019)

Fig. 7.28 The Aomori Bay Bridge, Japan, 1994

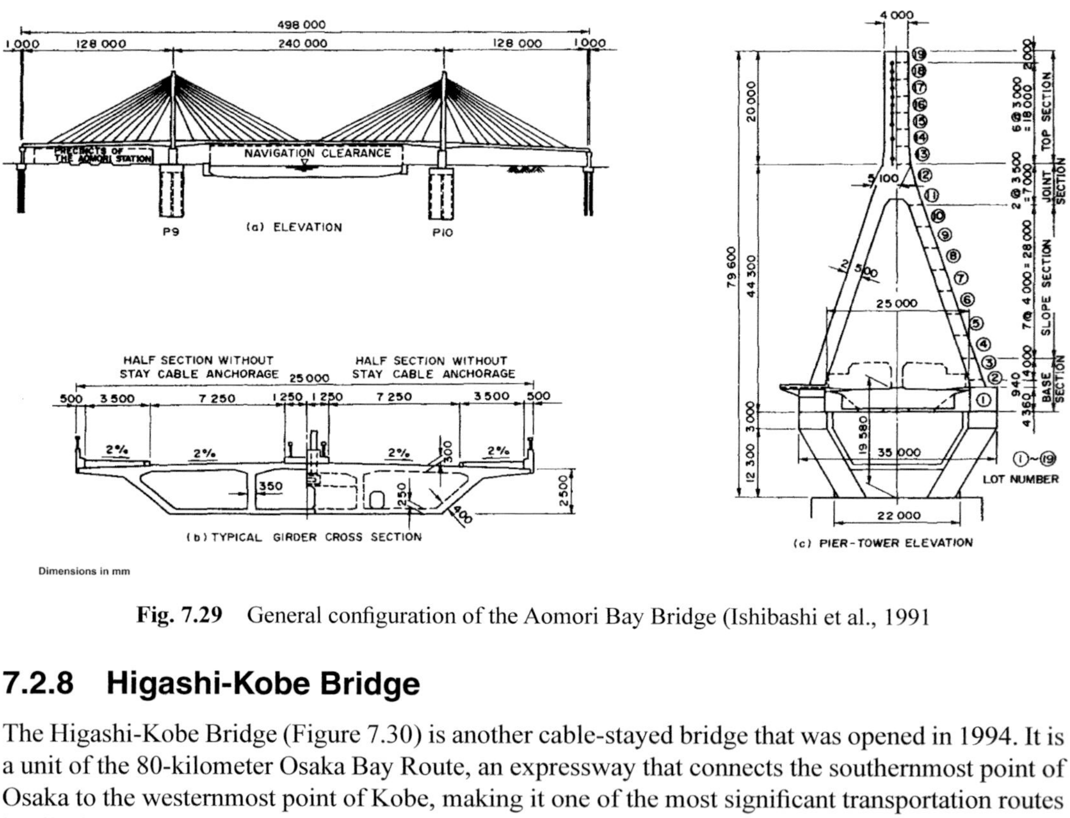

Fig. 7.29 General configuration of the Aomori Bay Bridge (Ishibashi et al., 1991

7.2.8 Higashi-Kobe Bridge

The Higashi-Kobe Bridge (Figure 7.30) is another cable-stayed bridge that was opened in 1994. It is a unit of the 80-kilometer Osaka Bay Route, an expressway that connects the southernmost point of Osaka to the westernmost point of Kobe, making it one of the most significant transportation routes in all of Japan. The bridge is 885 m long in total. The side spans are each 200 m in length, while the center span is 485 m.

The bridge is an 885 m long, double-decked, three-span cable-stayed construction. There are three traffic lanes on each deck. The superstructure is made of two Warren trusses and a steel orthotropic slab structure with 12 mm thick deck plates for the deck that are supported by transverse and longitudinal ribs spaced 70 cm and 3 m apart, respectively. The deck cross-section is displayed in Figure 7.31. With the primary girder's longitudinal displacements primarily limited by the cables, it can move longitudinally on all supports.

Fig. 7.30 The Higashi-Kobe Bridge, Japan, 1994

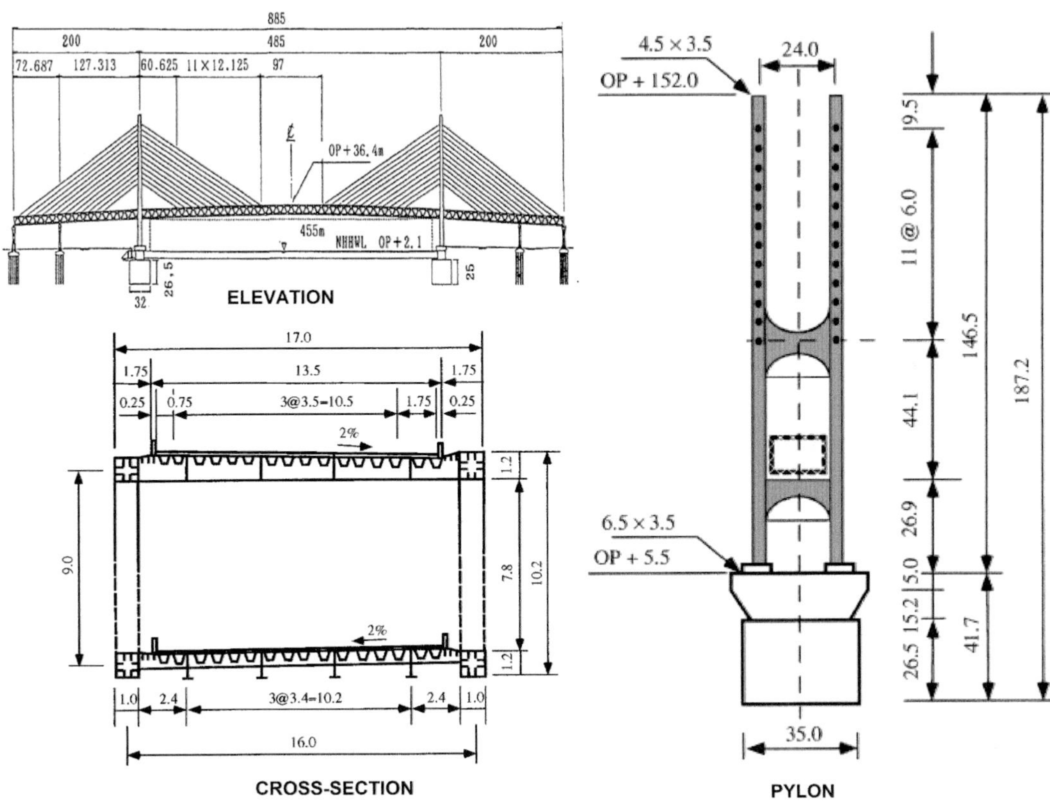

Fig. 7.31 General configuration of the Higashi-Kobe Bridge (Ganev et al., 1998)

The superstructure load is transferred to the pylons by 96 stay cables that are arranged in a harp format. The truss chord members enclose the cable anchors. Every cable comprises 24 to 301 7 mm-diameter wires in it, depending on its location. Polyethylene tubes are used to encase them to prevent corrosion. The bridge's H-shaped pylons are 146.5 m high. Each pylon column has a cross-section that is 3.5 m × 6.5 m at the base and 3.5 m × 4.5 m at the top. There are 24 m in horizontal space between the two legs. Rectangular-cross-section hollow steel pieces are welded together to form the columns' structure.

Pneumatic caissons measuring 35 m (W), 32 m (L), and 26.5 m (H) are used as the pylon foundations. Each caisson has six rows of six cells each with partition walls inside it. The floating caisson method was used to build the foundations. Pile foundations support the secondary piers at the side spans.

7.2.9 Tsurumi Tsubasa Bridge

The Tsurumi Tsubasa Bridge (Figure 7.32) opened in 1994, just like the Higashi-Kobe Bridge. Built over the Tsurumi Fairway, which runs between Yokohama's Daikoku-Futo Wharf and Ogishima Island, the bridge is 1020 m long, single-plane, three-span continuous cable-stayed with a 510-meter center span.

Fig. 7.32 The Tsurumi Tsubasa Bridge, Japan, 1994

As shown in Figure (7.33), the girder's total width, including fairings, is 38 m. Because the bridge has a single-plane, cable-stayed deck with extremely long spans, it has a five-cell box section with a depth of four m to enhance torsional rigidity. With 68 stay cables supporting the superstructure, the cables were set up in the shape of a semi-fan. The cables are galvanized wire bundled into non-grout-type parallel wire strands (PWS). A maximum of 4997 mm-diameter polyethylene-coated wires make up each cable. The pylon height measured from the top of the caisson foundation is 180 m. With a trapezoidal cross section, the upper pylon is diamond shaped. The lower pylon begins at the upper caisson slab and comprises a steel console, which supports the pylon leg. The purpose of the steel-frame reinforced concrete structure, which is located approximately 20 m above the upper slab of the caisson foundation smoothly transfers the load from the pylon to the caisson foundation.

7.2.10 Ikara Bridge

The Ikara Bridge (Figure 7.34) was opened in 1996 to connect Nagashima Island to Ikara Island. It is a prestressed concrete cable-stayed bridge with a main span of 260 m and total length of 670 m distributed as: 41.950 m – 43.100 m – 120 m – 260 m – 120 m – 43.100 m – 41.952 m. It has a vertical navigation clearance of 18 m and accommodates two traffic lanes and one sidewalk.

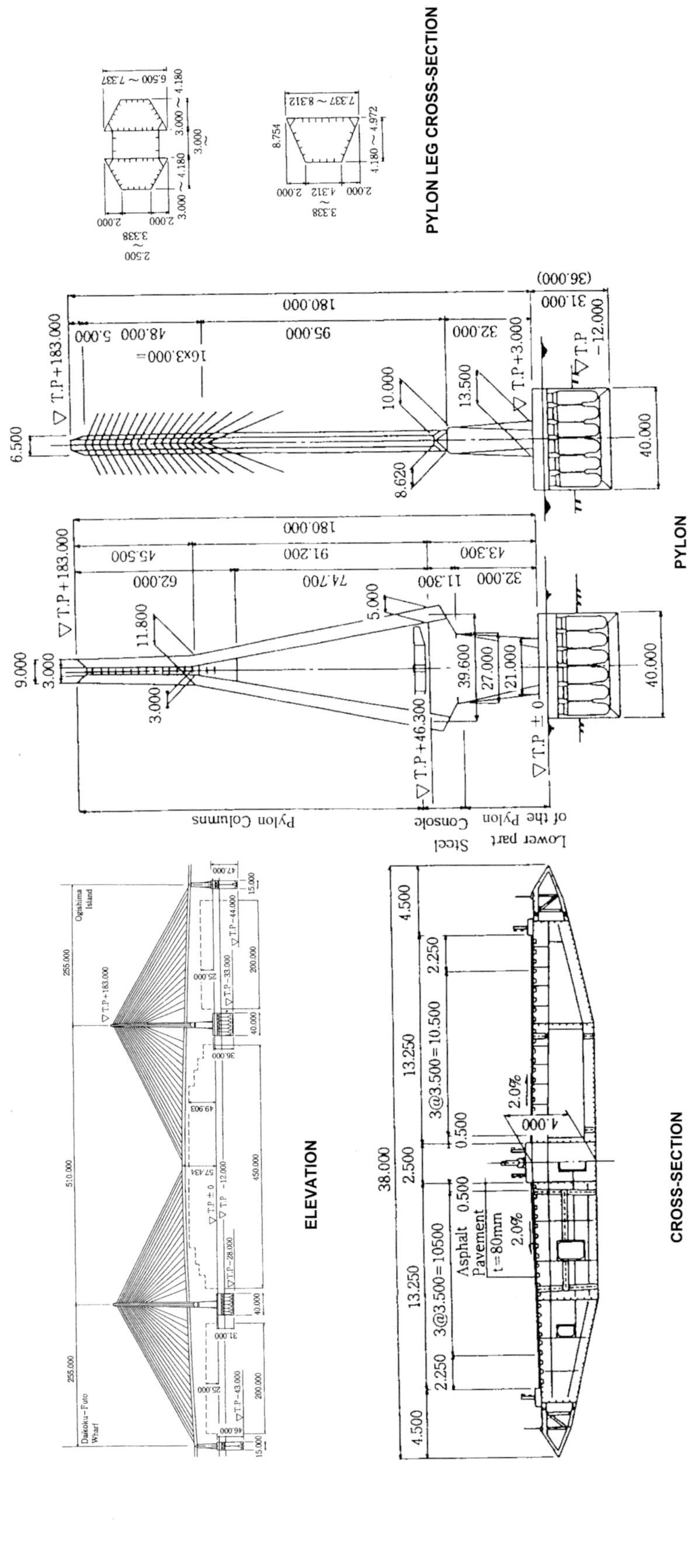

Fig. 7.33 General configuration of the Tsurumi Tsubasa Bridge (Enomoto et al., 1994)

Fig. 7.34 The Ikara Bridge, Japan, 1996

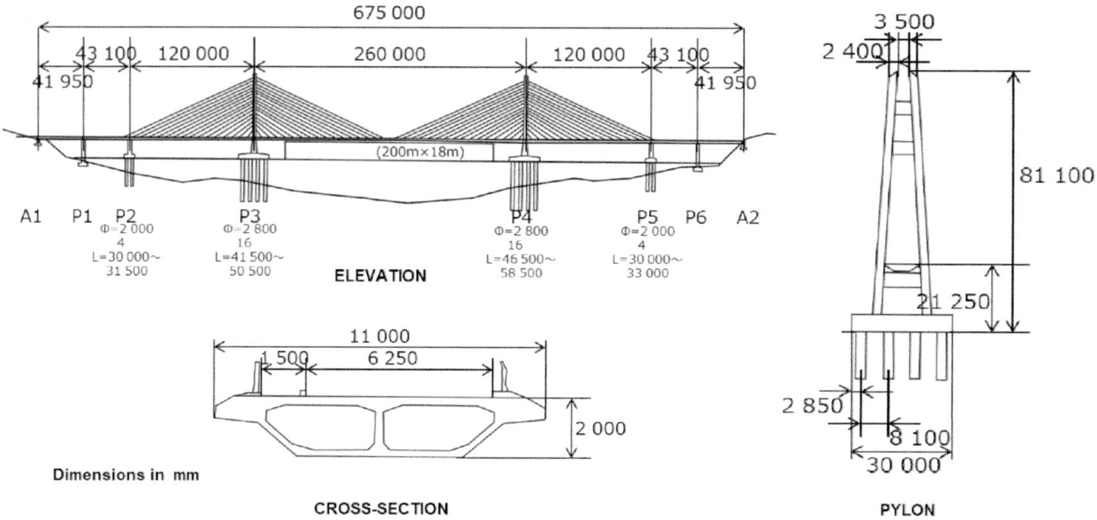

Fig. 7.35 General configuration of the Ikara Bridge, Japan, 1996

The superstructure as shown in Figure 7.35 is a two cell prestressed concrete trapezoidal box girder with a deck width of 11 m and 2.0 m depth. The reinforced concrete pylons rise 81.10 m above the foundation and the bridge has two planes of stay cables. The cables are arranged in a harp configuration. 128 cables are supporting the superstructure and transfer its load to the pylons. The pylons are supported by groups of 16 drilled shafts, 2.8 m diameter.

7.2.11 The Cable-Stayed Meiko Grand Bridges

The Meiko Grand Bridges (Figure 7.36) are three cable-stayed bridges that form a connecting link over the Nagoya Port. The bridges are part of the national highway Route 302 (Nagoya Loop-2) and the Ise Bay Highway that link the Tomei and Meishin Expressways. These bridges are: (i) the West Bridge (Meiko-Nishi Bridge); (ii) The Central Bridge (Meiko-Chou Bridge); and (iii) the East Bridge (Meiko-Higashi Bridge). From west to east. center span lengths of the bridges are 405, 590 and 440 m. respectively.

Fig. 7.36 The Meiko Grand Bridges, Japan, 1998

The Meiko West Bridge is composed of two adjacent cable-stayed bridges 50 m apart. The first bridge was completed in 1985 and the second one was opened in 1998, each carrying three lanes. Each bridge is a steel cable-stayed bridge with three spans of (175 + 405 + 175 m). The Central Bridge has a total length of 1170 m including a 590 m main span and two side 290 m spans. The East bridge is 700 m long including a 410 m main span and two 145 m side spans. Elevations of the three bridges are illustrated in Figure 7.37.

All main girders have an orthotropic steel deck and a streamlined, trapezoidal box cross section with triangular fairings (Figure 7.38). For the West Bridge, the box girders are around 2.8 m deep, while for the other two bridges, they are 3.5 m deep. As a result, the Central Bridge's girder is so shallow that its depth-to-span length ratio is 1:170 and its depth-to-width ratio is 1:17. The girder and the pylon are connected by longitudinal prestressing cables, which produce an elastic constraint and an improved damping effect, in order to lessen the forces caused by earthquakes that act on the pylon to limit the girder's excessive longitudinal movement. Owing to the East Bridge's comparatively short side spans in relation to its main span, an orthotropic steel deck supporting the side spans' reinforced concrete deck slab serves as a counterweight. Additionally, this arrangement helps to keep the deck plate from buckling locally.

All three of the Meiko bridges use a semi-fan, two-plane cable arrangement. The strands were manufactured in the shop with a built-in corrosion protection which made grouting unnecessary at the site. An extruded polyethylene jacket covers a bundle of 7 mm diameter zinc-galvanized wires that provide corrosion protection. The stay cables on the Meiko Central Bridge have a maximum diameter of 153 mm. They are made up of 397 steel wires and have a 10 mm thick polyethylene sheath covering. A total of 163 to 379 cables with a diameter of 5 mm were used for the West Bridge.

The Central and East bridges were built with diamond-shaped steel pylons that had an A-frame above the deck and angled legs beneath it to minimize the foundation size (see Figure 7.39). As illustrated, the pylons of the West Bridge, on the other hand, are A-shaped. The Central Bridge's pylon shafts took on an octagonal cross section, whereas the East Bridge's original design included a horizontal strut between the pylon top and road deck.

7.2.12 Tatara Bridge

The Tatara Bridge was opened in 1999. It is part of the Nishiseto Expressway. The expressway is a series of roads and bridges that is one of the three routes of the Honshū-Shikoku Bridge Project connecting the islands of Honshū and Shikoku across the Seto Inland Sea in Japan. The bridge is 30.6 m wide carrying a four-lane highway as well as additional lanes for bicycles, motor bikes and

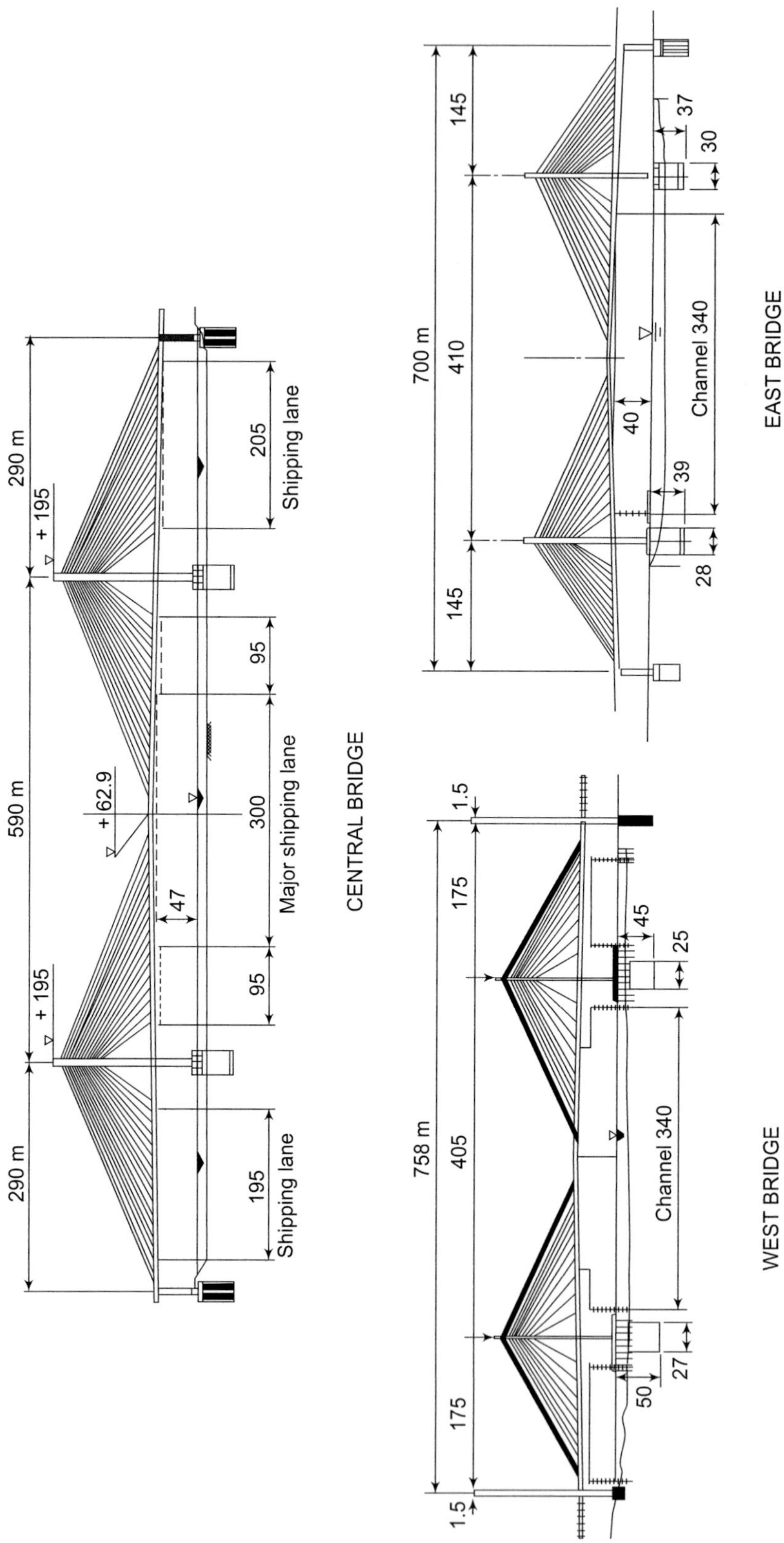

Fig. 7.37 Elevations of the Meiko Grand Bridges (Ito, 1998)

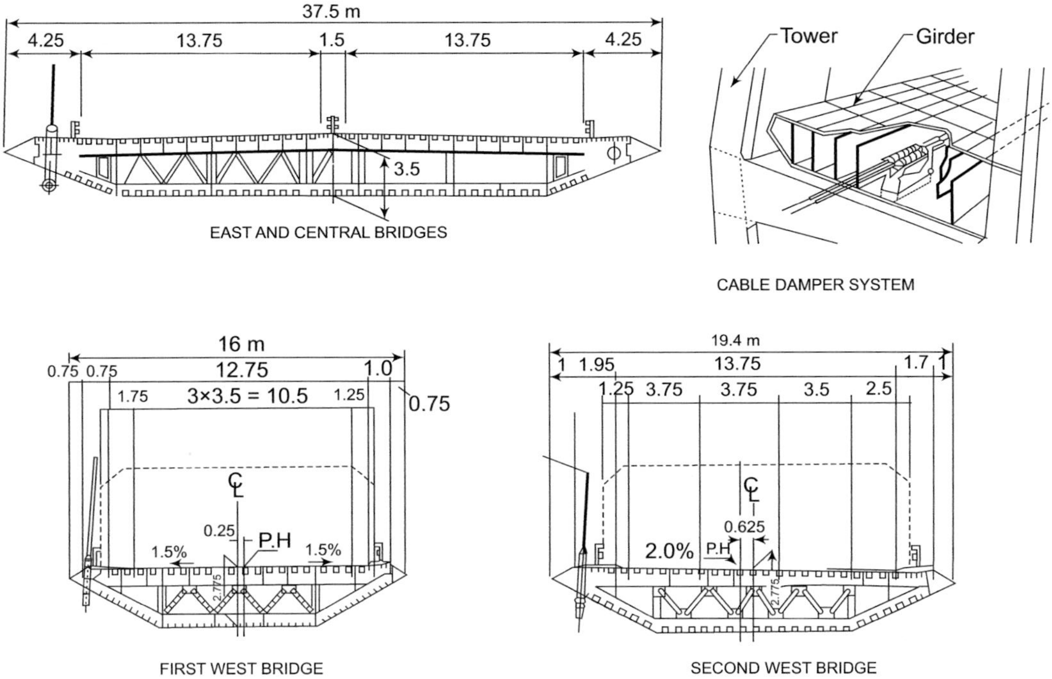

Fig. 7.38 Cross-sections details of the Meiko Grand Bridges (Ito, 1998)

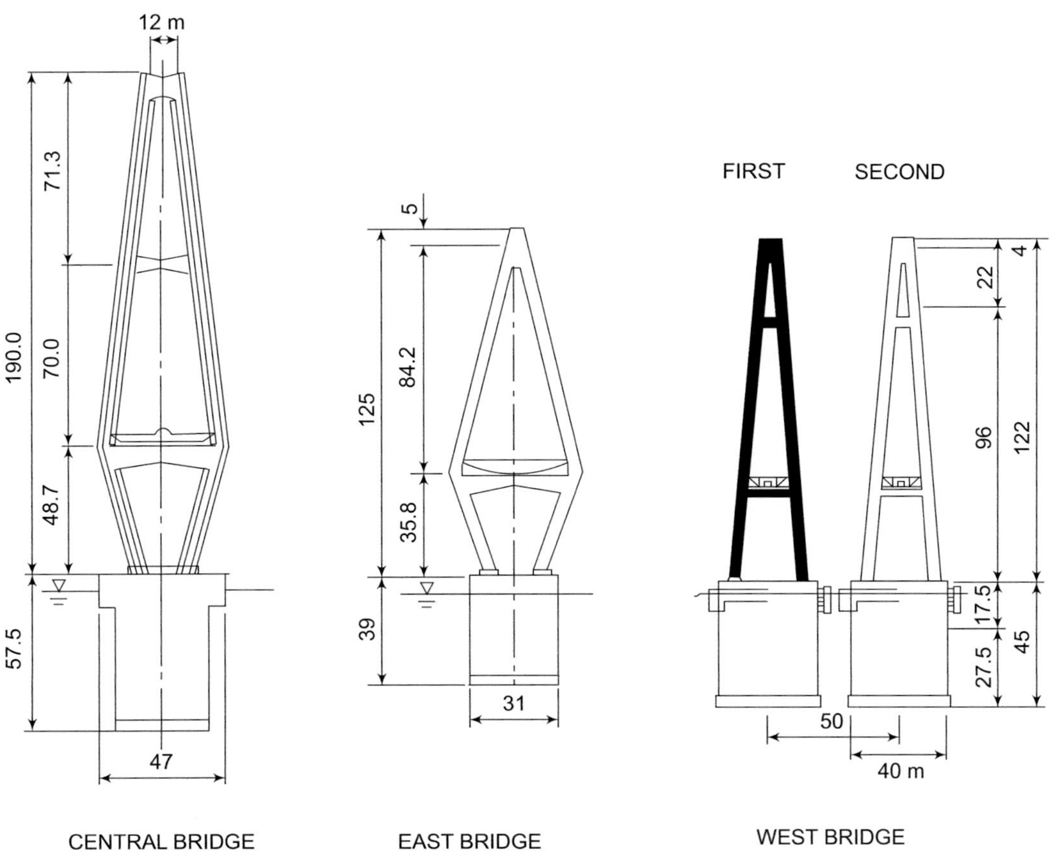

Fig. 7.39 Pylons of the Meiko Grand Bridges (Ito, 1998)

Fig. 7.40 The Tatara Bridge, Japan, 1999

pedestrians. With a main span of 890 m and a total length of 1480 m, Tatara Bridge is the eighth longest cable-stayed bridge in the world (Figure 7.40). The bridge as shown in Figure 7.41 has a total length of 1480 m, distributed as: 50 m – 50 m – 170 m – 890 m – 270 m – 50 m respectively. The superstructure of the main span is a steel box girder 2.7 m deep with fairings. The girder is made of prestressed concrete. These substantial PC girder sections strengthen the bridge's rigidity by acting as counterweights against elevating forces. The girders are held up at the pylons by fixed hearings that prevent lateral movement and elastomeric bearings that permit vertical movement.

The bridge has two-plane and semi-fan shaped stay-cables arrangements. The semi-parallel wire strands used to create the 168 cables are prefabricated in a shop, covered with polyethylene tubes, and bundled together. The ends of the strands are fixed by sockets that are highly resistant to bending fatigue. The steel pylons have an inverted Y shape and stand 223 m tall (Saeiki et al., 1998). The cross-section of the main pylon is a cross-shaped section with corners cut for high-speed wind stability (see Figure 7.41).

7.2.13 Megami Ohashi Bridge

The Megami Ohashi Bridge (Venus Wing) (Figure 7.42) connects the southern and western parts of Nagasaki City, which is divided by Nagasaki Port, in the shortest possible distance, easing the chronic traffic congestion in the city center and boosting the industry, economy, and culture of the entire region. It was opened to traffic in December 2005. As shown in Figure 7.43, the main span is 480 m and the side spans are 200 m. It was completed on December 11, 2005. The steel box girder with an orthotropic steel deck and fairing is 31.1 m wide and 2.717 m deep. The H-shaped steel pylons are 170 m high. To suppress vortex-induced oscillations, two tuned mass dampers (TMD) are installed at the top of each column (the N columns and S columns) of the main pylons. A total of 104 cables, arranged in a dual plane configuration, are designed to transfer the superstructure loads to the pylons. They are made of parallel wire strands, precoated and encased in polyethylene tubes.

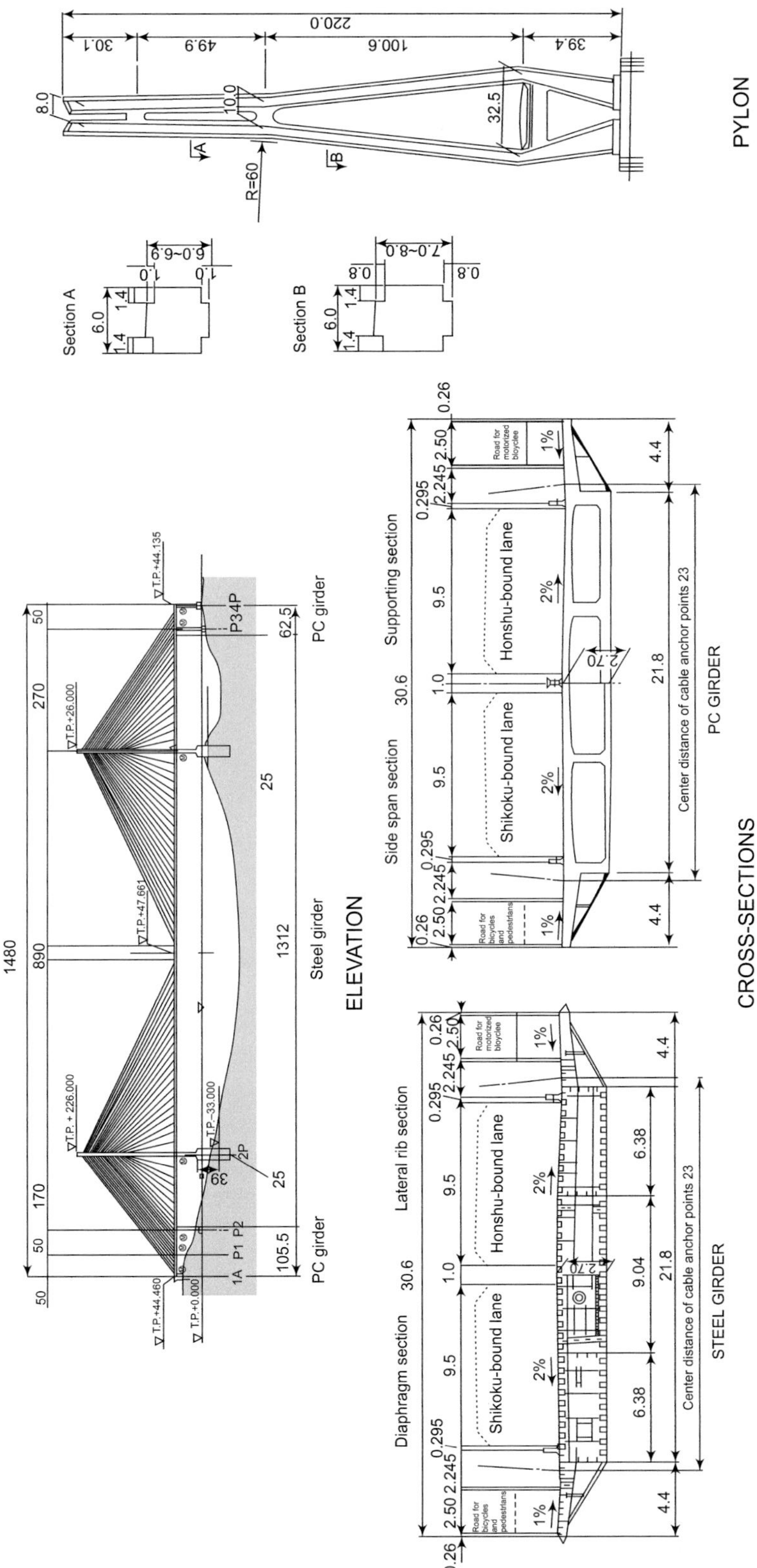

Fig. 7.41 General configuration of the Tatara Bridge (Yabuno et al., 2003)

Fig. 7.42 Megami Ohashi Bridge Japan, 2005

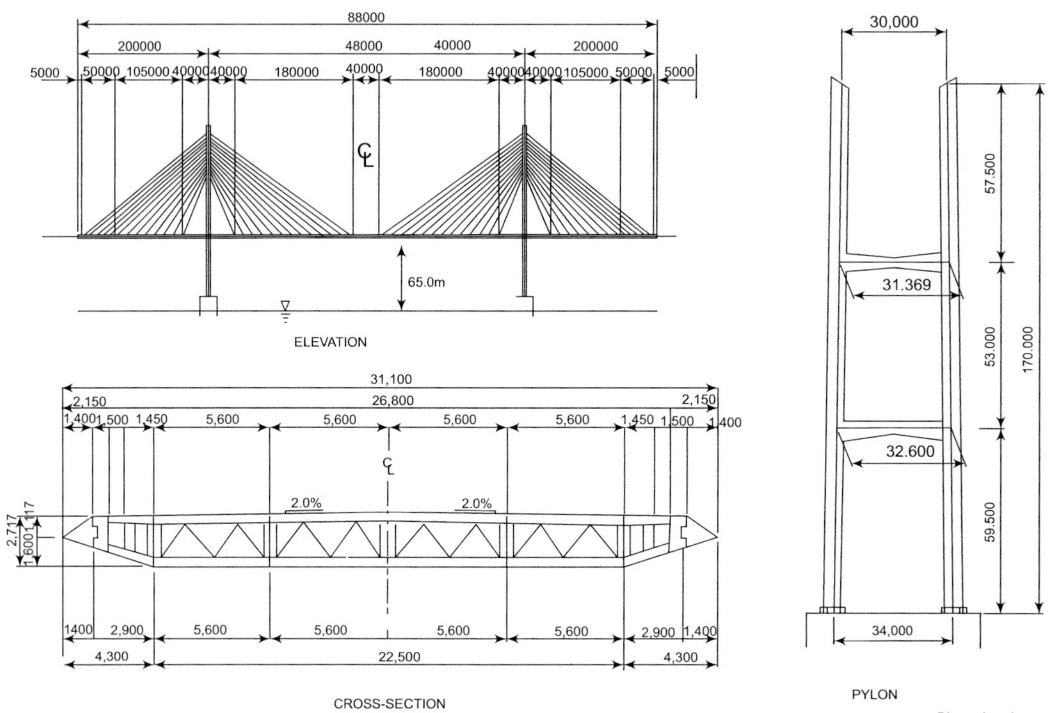

Fig. 7.43 General configuration of Megami Ohashi Bridge

7.2.14 Shin-Minato Bridge

Shinminato Bridge (Figure 7.44) is in Imizu, Toyama Prefecture, and was recently opened in September 2012. It is the largest cable-stayed bridge on the Sea of Japan coast and links the east and west districts around the Toyama Shinminato Harbor entrance. The bridge accommodates a horizontal clearance of 270 m and a vertical clearance of 47m.

The bridge is a continuous structure with five spans. The center span of the bridge is 360 m, and its overall length is 600 m (Figure 7.45). A three-cell PC box girder is used for each side span, while the twin-box type steel girder with an orthogonally stiffened steel deck makes up the central portion of the main girder, which is about 351.5 m long between the pylons.

Fig. 7.44 Shinminato Bridge, Japan, 2012

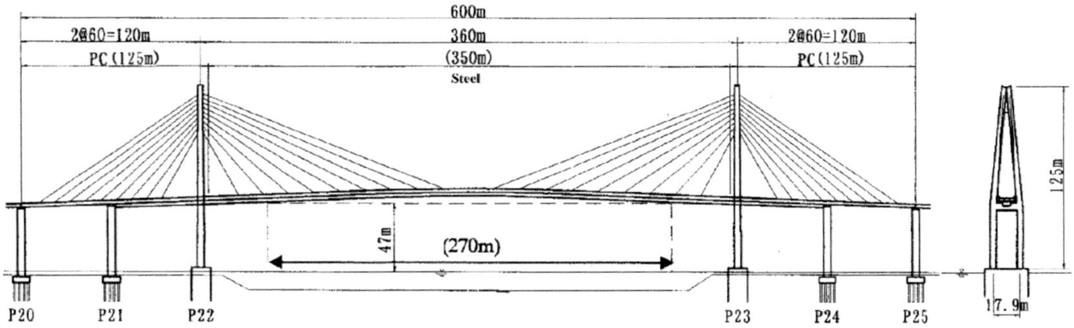

Fig. 7.45 Elevation of Shinminato Bridge

Due to the relatively short side span length, a hybrid system consisting of steel and concrete with distinct unit weights was utilized. As a result, when compared to alternatives using only steel and concrete, the most cost-effective solution was found. The girder section's two edges had fairings installed on them. Wind tunnel tests were used in experimental research to determine its shape. The steel and concrete girders were connected at a distance of approximately 5 m from the center span pylon. Over a 1.5 m length, the transition girder transfers stress between steel and concrete girders using headed studs for vertical shear and a bearing system with prestressed PC wires for a normal force. A two-plane, semi-fan-style cable arrangement suspends the girder. There are 109 to 211 wires with a 7 mm diameter in each cable.

Underneath the steel deck plate in the girder, a pedestrian and bicycle path are installed between two boxes. The installation of elevators at pier numbers 21 and 24 (refer to Figure 7.44) facilitates convenient access to this girder passage. An A-shaped pylon is 127 m above sea level. A steel pylon with a lower weight was chosen because the site has extremely poor soil conditions. From the base to deck level, the A-shaped pylon's cross-section is rectangular. The pylon's cross-section has bevels at each corner above deck level. A decision was made to use the section with corner bevels in order to avoid critical oscillations after wind tunnel tests.

Pneumatic caissons serve as the foundation for the main pylons because they can carry enough weight in the load bearing stratum, which is located about 40 m below sea level. The RC pier foundations were intended to be cast-in-place concrete piles with a diameter of 1200 mm.

References

Enomoto, M., Morikawa, H., Takano, H., Ogasawara, M. and Hayashi, H et al., Design and Construction of Tsurumi Tsubasa Bridge Superstructure, *Fourth International Bridge Engineering Conference*, pp 249–258, 1994.

Hansen, K., Hauge, L., and Hussain, N, Stonecutters Bridge-Detailed Design, *fib Symposium Avignon, France*, April 26–28, 2004.

Ganev, T., Yamazaki, F., Ishizaki, H. and Kitazawa, M., Response Analysis of The Higashi-Kobe Bridge and Surrounding Soil in the 1995 Hyogoken-Nabbu Earthquake, Earthquake Engineering and Structural Dynamics, Volume 27, pp 557–576, (1998).

Ishibashi,T., Fujjita, K., Fujimori, S., Ishihara, S., Suehiro, T. et al., Design and Construction of Aomori Bay Bridge, *IABSE Symposium*, Leningrad, pp 271–276, 1991.

Ito, M., The Cable-Stayed Meiko Grand Bridges, Nagoya, *Structural Engineering International*, Volume 8, Issue 3, pp 168–171, 1998.

Liu, L. and Z, X., Second Nanjing Cable-Stayed Bridge, *Structural Engineering International*, Volume 14, Issue 1, 34–36, 2004.

Li, Y., Li, Z. and Wu, Q., Experiment and Calculation Method of the Dynamic Response of Deep-Water Bridge in Earthquake, *Latin America Journal of Solids and Structures*, Volume 14, pp 2518–2533, 2017.

Lu, L. and Li, J., Longitudinal Vibration and Its Suppression of a Railway Cable-Stayed Bridge Under Vehicular Loads, *International Journal of Structural Stability and Dynamics,* Volume 18, No. 4, 2018.

Miao,J. Pei, Minshan, Zhng, X. and Xiao, R., Global Analysis of the SuTong Cable-stayed Bridge. Journal of Highway and Transportation Research and Development, VOL.1, No. 1 pp 51–55, 2006

Saeki, S., Okukawa, A. and Ohashi, H. The Honshu-Shikoku Bridges, Structural Engineering International, 8:1, 10–15 1998

Siringoringo, D.M., Fujino, Y. and Namikawa, K, SeismicResponseAnalyses of the Yokohama Bay Cable-Stayed Bridge in the 2011 Great East Japan Earthquake, *Journal of Bridge Engineering,* Volume 19, No. 8, 2014.

Takeda, T., Mizutani, T., Nagayama, T. and Fujino, Y., Reproduction of Cable-Stayed Bridge Seismic Responses Involving Tower–Girder Pounding and Damage Process Estimation for Large Earthquakes, *Journal of Bridge Engineering*, Volume 24, No. 2, 2019.

Wada, K., Tomita, N. and Takano, H., Construction of the Yokohama Bay bridge superstructure, *IABSE Symposium*, Leningrad, pp 177–182, 1991.

Xiang, H., Retrospect & Prospect of cable-stayed bridges in China, IABSE Conference, Malmo, 1999

Yabuno, M., Fujiwara, T., Sumi, K., Nose, T. and Suzuki, M. et al. (2003). Design of Tatara bridge. *IHI Engineering Review*, 36(2), 40–56.

You, Q., He, P., Dong, X., Zhang, X. and Wu, S.S. et al, Sutong Bridge – The Longest Cable-Stayed Bridge in the World, *Structural Engineering International*, Volume 18, Issue 4, pp 390–395, 2008.

CHAPTER 8

Evolution of Cable-Stayed Bridges in Asia and Africa

8.1 MODERN CABLE-STAYED BRIDGES IN ASIA

8.1.1 South Korea

A number of significant cable-stayed bridges, including the first Jindo Bridge (70 + 344 + 70 m) and the Dolsan Bridge (85 + 280 + 85 m) built in 1984, were part of South Korea's first generation of sea-crossing cable-stayed bridges. The 1994 collapse of the Sungsu Truss Bridge brought to attention the importance of bridge maintenance programs and their requirement to keep similar accidents from happening in the future. As a result, stricter guidelines for bridge management and operational programs were released by the government in 1995. These guidelines included field measurements, load capacity testing, instrumentation, and systematic visual inspection. Also, a new era of structural health monitoring started and implemented on existing bridges like Jindo Bridge to evaluate its structural health. A new generation of cable-stayed bridge sea-crossings started between 2000 and 2019, which include several large cable-stayed bridge structures. Table 8.1 summarizes some of the major bridges built during this period. Three of these bridges are selected and discussed in detail.

Table 8.1 Major Cable-Stayed Bridges in South Korea 2000-2019

Bridge Name	*Year*	*Type*	*Total Length of project (m)*	*Main Span (m)*
Seohae	2000	Composite	7,310	470
Youngheung	2001	Steel Box Girder	1,250	240
Second Jindo	2005	Steel Box Girder	484	344
Machang	2008	Composite	1,700	400
Shinwando	2009	Steel Box Girder	430	200
Incheon	2009	Steel Box Girder	1,480	800
Busan-Geoje (I)	2010	3-phlon composite	1,650	230
Busan-Geoje (II)	2010	2-pylon composite	1,870	475
Bukhang	2011	Composite	1,114	540
Second Dolsan	2012	Prestressed Concrete	464	230
Chilsan	2019	Prestressed Concrete	820	320

8.1.1.1 Seo-hae Bridge

Completed in November 2000, the Seo-hae Bridge (Figure 8.1) is a segment of the newly constructed six-lane west Coastal Highway. It includes three distinct bridge types—the cable-stayed, balanced long span concrete box girder, and segmental short span concrete box girder bridge types—and crosses Asan Bay about 65 km south of Seoul. The bridge spans two abutments and is 7.3 km long; has 100 piers of support. the main span of 470 meters that is cable-stayed. traverses the main channel, which eventually serves as the Asan Port's entrance.

Fig. 8.1 The Seo-hae Bridge, South Korea, 2000

The bridge is a 5-span steel composite cable-stayed bridge with a span distribution of 60 m – 200 m – 470 m – 200 m – 60 m (Kim and Cho, 2001). The bridge is 34 m wide and carries 6 lanes and two side lanes. The superstructure is a composite section comprising 2 main 2.8 deep edge girders and two internal longitudinal stingers for the purpose of supporting the precast slab. The transverse floor beams are located every 4.1 m. The deck consists of 26 cm thick precast units. There are 144 cable stays transferring the service loads to the pylons. They are arranged in a semi-fan configuration in two parallel vertical planes and made of 7 parallel wire strands and protected with a high-density polyethylene sheath. The stays are anchored to the deck at distances of 12.3 m. Each cable consists of 37 to 91 wires depending on its location and the amount of load it is transferring. The pylons are H-type reinforced concrete. The heights of the north and south pylons above the foundations are 179 and 182.26 m respectively. Each pylon is provided with three crossbeams (Figure 8.2).

8.1.1.2 Incheon Bridge

The Incheon Bridge project (Figure 8.3), a 13 km toll bridge, connects the Incheon International Airport to the south of the city with an expressway link across the Incheon Port entrance. The expressway enters the Incheon Port by crossing the main shipping channel. As a result, a cable stay bridge with an 800 m main span was built as part of the project. Low level prestressed concrete viaducts form the approaches to the main bridge.

The main span of the bridge is 800 meters consisting of two flanking spans 80 m each and two side spans 260 meters each. To ensure aerodynamic stability and minimize self-weight, a steel box girder with an orthotropic deck was chosen (Figure 8.4). To increase the torsional stiffness of the entire bridge, reverse Y-shaped concrete pylons were used in conjunction with two-sided cable stays spaced 15 meters apart at the exterior girders. A semi-fan type cable arrangement was chosen and to lessen cable oscillation and static wind loading, comparatively small-diametered parallel wire cables were used. The end piers are likewise two separate hollow section twin columns connected by a crossbeam on the top, measuring 55.8 meters in height, and the supplementary piers are two separate hollow section twin columns measuring 58.1 meters in height. To counteract the uplift forces in

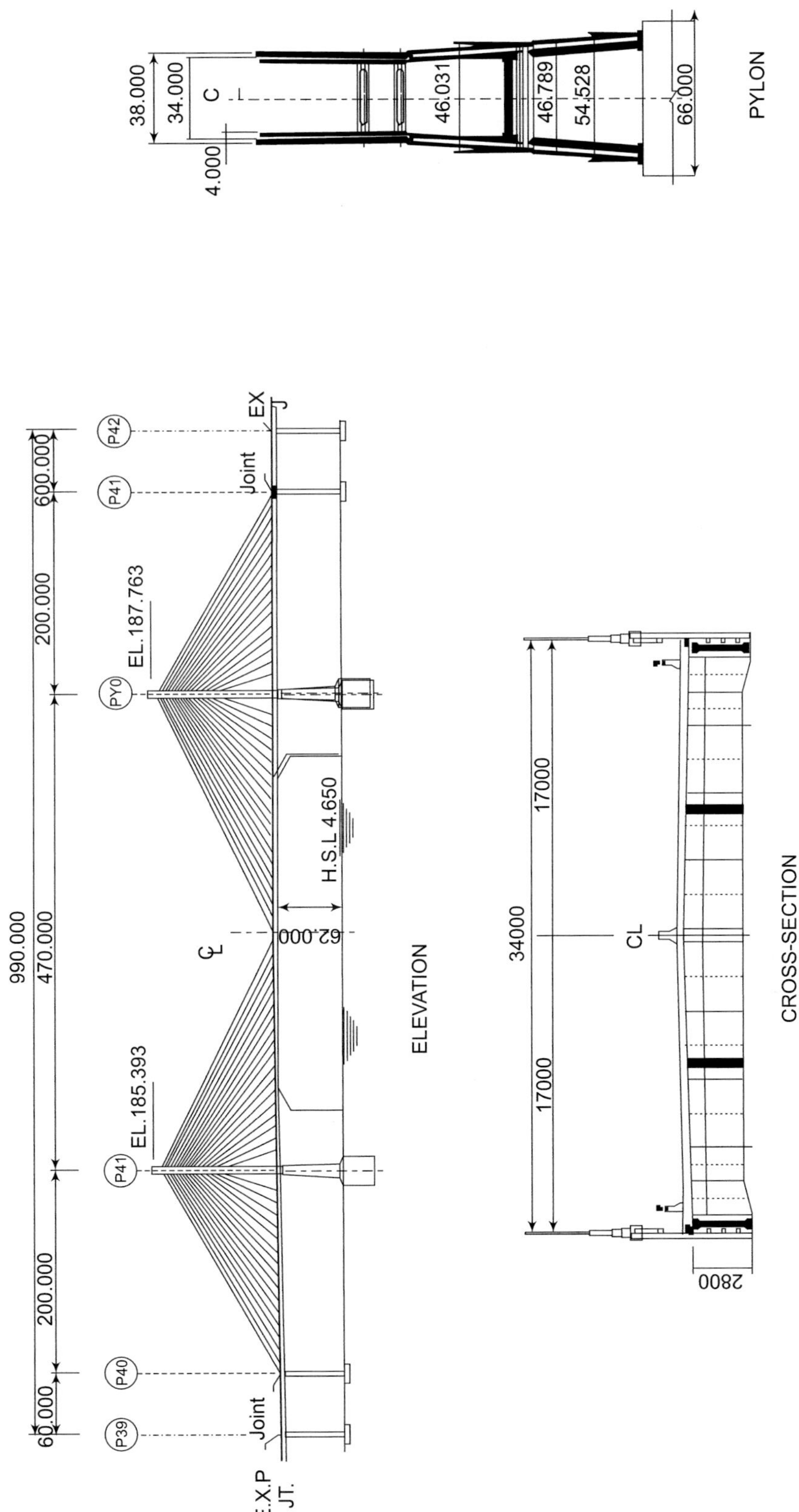

Fig. 8.2 General Configuration of the Seo-hae Bridge (Kim & Cho, 2001)

Fig. 8.3 The Incheon Bridge, South Korea, 2009

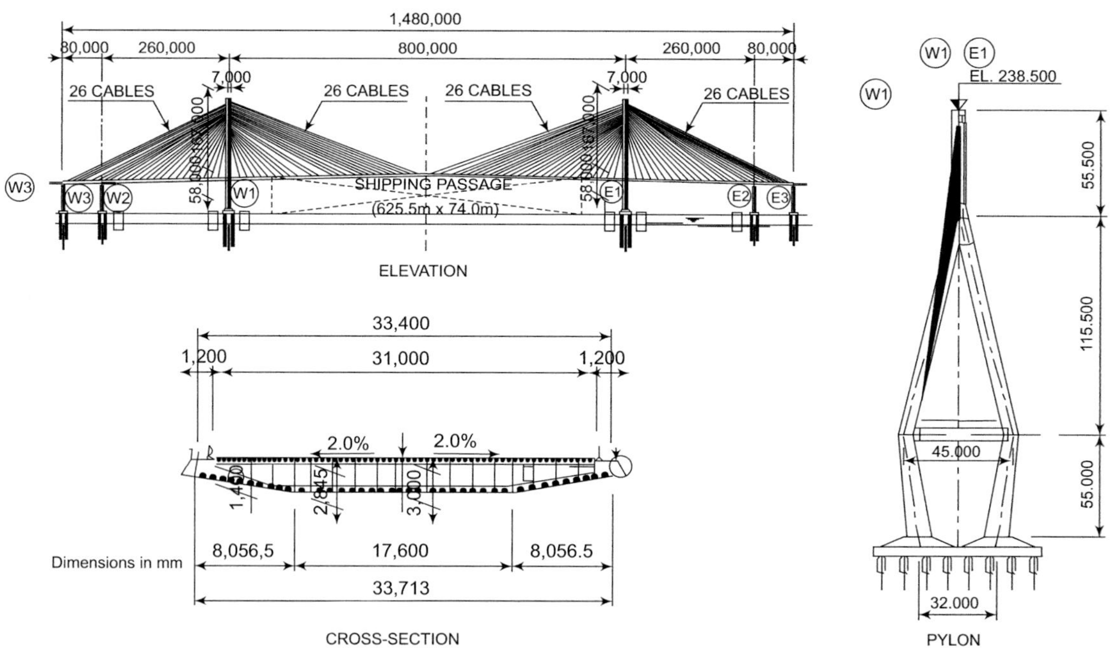

Fig. 8.4 General arrangement of the Incheon Bridge (Ha et al., 2012)

these piers, counterweights and tie-down cables are both installed. The multi-column drilled shifts, or pylon foundations, are made up of 24 piles with a diameter of 3.0 meters.

8.1.1.3 Busan-Geoje Bridge

Completed in December 2010, the Busan–Geoje Fixed Link is a significant infrastructure link located in the southeast region of Korea to connect Metro Busan City and Geoje Island. It is 8.2 km long and consists of three main parts: two cable-stayed bridges (Lot 1 and Lot 2) and an immersed tunnel (Lot 3) as illustrated in Figure 8.5.

Lot 1 includes the three-pylon cable-stayed bridge. The multi-span composite steel and concrete viaducts that connect Geoje Island to Jeo Island span a total of 1650 meters over the length of the two islands. With 108 m of side spans and 230 m of main spans, the cable-stayed bridge has a total length of 676 m (Figure 8.6).

Fig. 8.5 General arrangement of the Busan-Geoje project, South Korea, 2010 (Jeong and Kim, 2012)

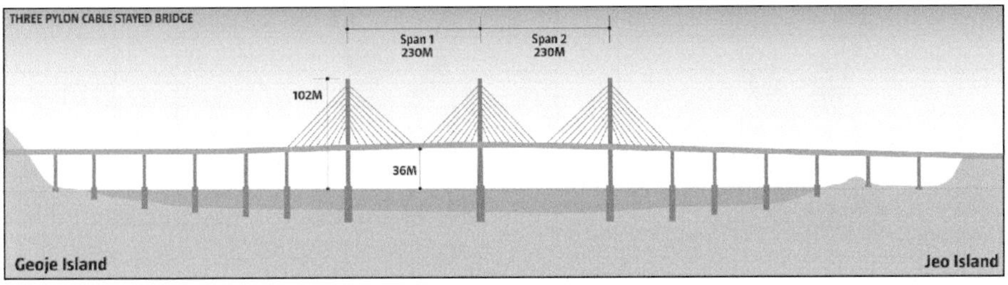

Fig. 8.6 Busan-Geoje Bridge Lot 1

Section 2 of Lot 2 connects Jeo Island and Jungjuk Island through the two-pylon cable-stayed bridge. The 1870 m long bridge is made up of two sections: the main cable-stayed bridge, which is 919 m long, and steel-concrete composite viaduct approach spans, which can reach a maximum length of 90 m (Figure 8.7). Having a 475 m main span and 222 m side spans, it offers 435 m of horizontal clearance and 52 m of vertical navigation.

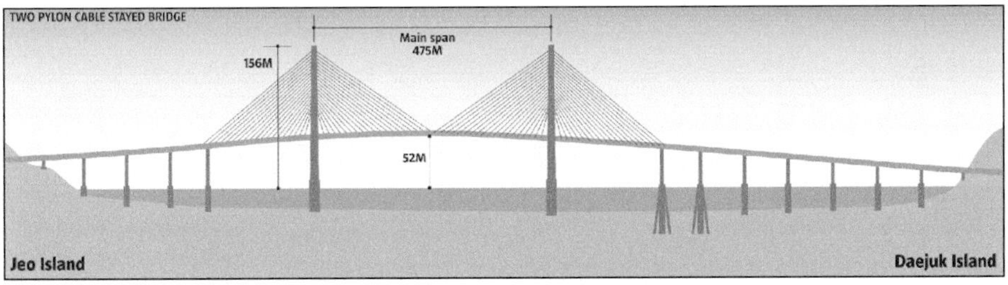

Fig. 8.7 Busan-Geoje Bridge Lot 2

The superstructure comprises steel–concrete composite plate girders with 12 m long × 24 m wide steel segments connected by bolted splices. Prefabricated panels are joined by concrete in situ connections to form the concrete deck slab. The design called for the lower pylon legs to taper inwards, as seen in Figure 8.8, allowing the foundations' plan area to be significantly reduced. The open caisson foundations are situated in up to thirty meters of water and are built on competent rock. Construction was accomplished using the balanced cantilever method.

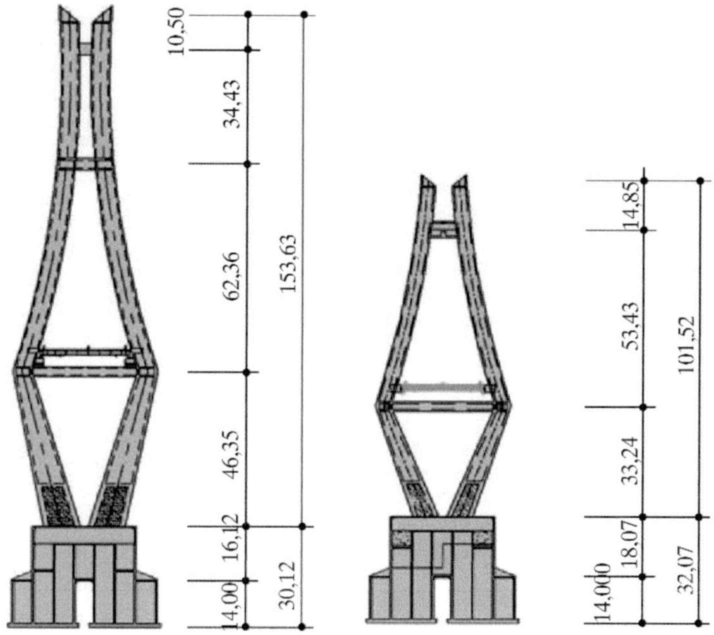

Fig. 8.8 Busan-Geoje Bridge pylons (Jeong and Kim, 2012)

8.1.1.4 Second Dolsan Bridge

The Second Dolsan Bridge, which was opened in 2012 is the main bridge on the Yeosu Road between Woodoo-Ri and Jongwa-Dong, linking the Yeosu port with Dolsan Island across the South Sea of Korea (Figure 8.9). The main objective of the new link was to shorten the First Dolsan Bridge's route and lessen traffic. Additionally, it served as a crucial infrastructure link during Yeosu's 2012 Expo. It is the first bridge in South Korea constructed with a superstructure made of reinforced concrete.

Fig. 8.9 Second Dolsan Bridge, South Korea, 2012 (Courtesy, Park Dae-Yong)

As shown in Figure 8.10, the 744 m Second Dolsan Bridge is made up of an approach bridge and a main bridge that is cable-stayed. The 280-meter approach viaduct has eight 35-meter-long spans made of cast-in-place slabs and post-tension beams. The 464 m long cable-stayed bridge is made up of two 117 m long side spans and a 230 m main span. The bridge provides a 120 × 22 m navigation channel and carries four lanes of traffic (Park & Lee, 2012).

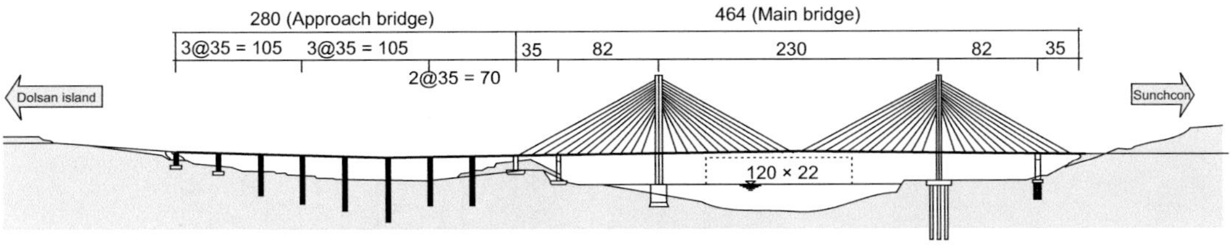

Fig. 8.10 Elevation of Second Dolsan Bridge (Park & Lee, 2012)

The superstructure is an open-cross-section made of reinforced concrete with two edge beams spaced 24 meters apart, measuring roughly 1.5 meters deep, 1.81 meters wide, and 2.49 meters wide. The deck slab is 24.2 m wide and has a 0.25 m thickness at each edge beam (Figure 8.11). The edge beams are connected transversely every 4.5 m by a floorbeam. The deck can move freely longitudinally, but it is transversely restrained at the pylons. However, at the piers, the deck is free to move longitudinally and transversely. The superstructure load is transferred to the pylons by 104 stay cables. They are set up in a semi-fan configuration in two planes. Sizes of the cables range from 23 to 49 multistrands of 15.7 mm. A strand has a tensile strength of 1860 Mpa.

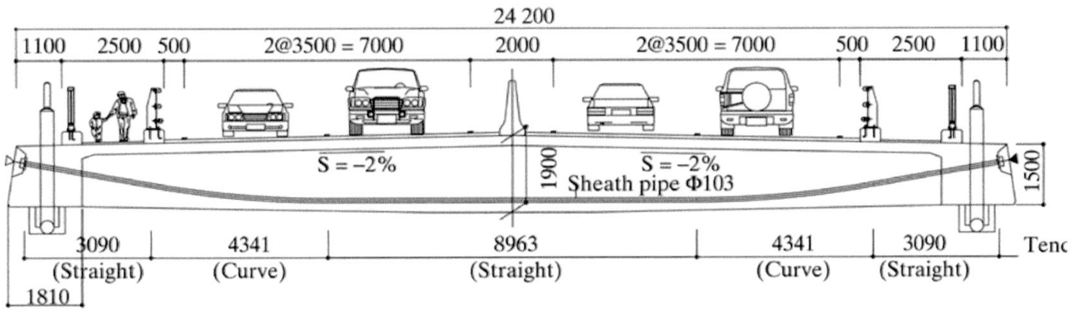

Fig. 8.11 Cross-section of Second Dolsan Bridge (Park & Lee, 2012)

The "H"-shaped pylons are approximately 90 meters high, while the crossbeam is roughly 60 meters high (Figure 8.12). The cross-section of the pylon is a hollow box with 0.6 m thick walls. The pylons' legs are vertical from 56.6 to 90 meters and sloped at an angle of 85° from the foundation to a height of 56.6 meters. The pylon's upper crossbeam is prestressed and features a 3.5 × 3.2 m hollow box section with 0.6 m thick walls. Piers can be found at the ends of the side span. Every cross-section of a pier is a 3.0 × 3.0 m hollow box with 0.6 m thick walls. For the Dolsan and Suncheon pylons, respectively, two 12.5 m diameter cylindrical caissons with 0.8 m thick walls and fifteen 2.4 m diameter cast-in-place concrete piles were utilized. Bridge elements are designed with a 40 MPa cylinder strength as standard.

8.1.2 Vietnam

As a developing economy, Vietnam's need for roads and bridges has grown critically. Therefore, in recent years, Vietnam has seen a rapidly expanding transport infrastructure especially long span bridges due to the geographical features of the country with many rivers and straits. Several large cable-stayed bridges have been built in Vietnam during the last two decades. The list includes but not limited to Bai Chay Bridge in 2006; Phu My Bridge in 2009; Can Tho Bridge in 2010; Nhat Tan Bridge in 2015; and the Vam Cong Bridge in 2019.

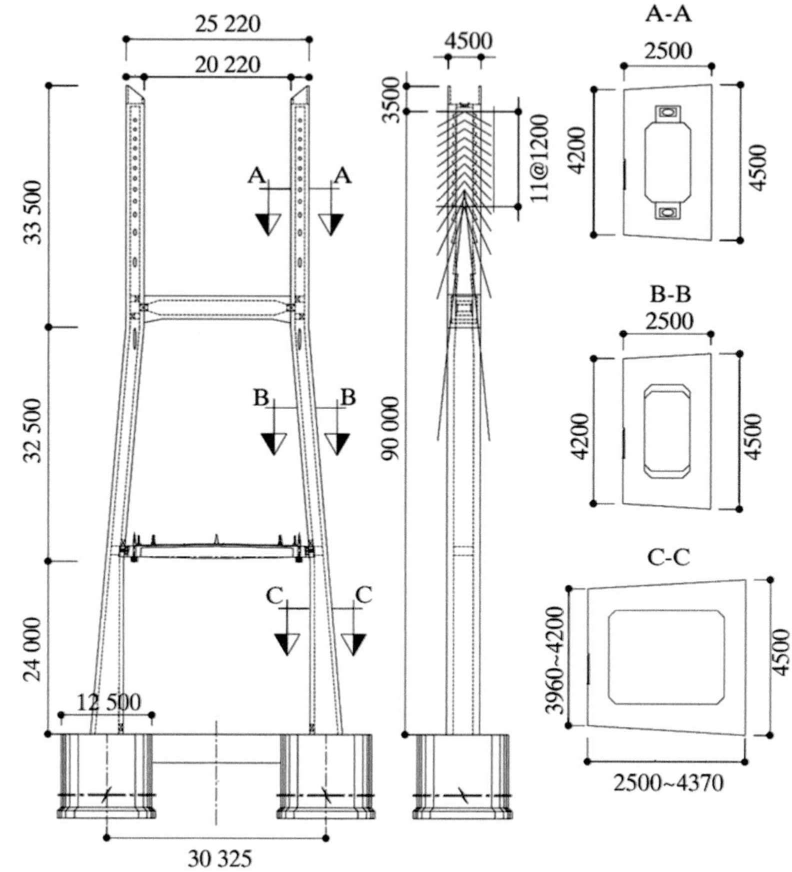

Fig. 8.12 Pylons of Second Dolsan Bridge (Park & Lee, 2012)

8.1.2.1 Bai Chay Bridge

The Bai Chay Bridge (Figure 8.13) is part of the Vietnam National Highway No. 18, which connects the International Airport of the capital city, Hanoi, with the town of Mong Cai on the Chinese border. The bridge crosses the Ha Long Bay Strait and has a total length of 903 m distributed as 35 m; 86 m – 129.5 m – 435 m – 129.5 – 86 m (Figure 8.14). The bridge as a single plane cable-stayed prestressed concrete bridge has one of the longest main spans in the world of this kind. The bridge provides a navigation clearance 200 m in width and 50 m in height. The bridge carries two lanes of traffic as well as additional lanes for motor bikes in each direction.

Fig. 8.13 Bai Chay Bridge, Vietnam, 2006

The superstructure is a reinforced concrete box girder 3.72 m deep with one central cable plan and a width of 25.3 m (Figure 8.14). A steel pipe bracing inside the box girder was provided as shown in Figure 8.14. Steel pipe bracings were installed in two different configurations at 3.25-meter intervals: square pipe (250 × 250 × 16 mm) and round pipe (267.4 × 9.3 mm). At the anchorage locations of the square steel pipe brace, prestressing tendons measuring 12 × 15.2 mm were inserted to withstand the tensile force exerted by the stay cables. 56 stay cables total—28 on each side of the main pylon—are fitted inside colored HDPE ducts and have between 35 and 71 strands (15.7 mm) per stay cable. The pylons have different hollow sections and are 9.5 meters above the deck level. Within every pylon, there are permanently installed Tuned Liquid Dampers. A fiber-reinforced plastic (FRP) box, 80 mm high, filled with water, makes up each tuned liquid damper (TLD). 344 dampers were installed per pylon. Vibration testing on the bridge showed that the TLD decreased the pylon's dynamic acceleration response by 58%, which in turn decreased the pylon's maximum deflection at its top. The main pylon pier is supported by a large caisson foundation that was constructed by the pneumatic method, where compressed air pressurizes the inside of the working chamber to compensate for the water pressure encountered during excavation (Nakamura et al., 2007).

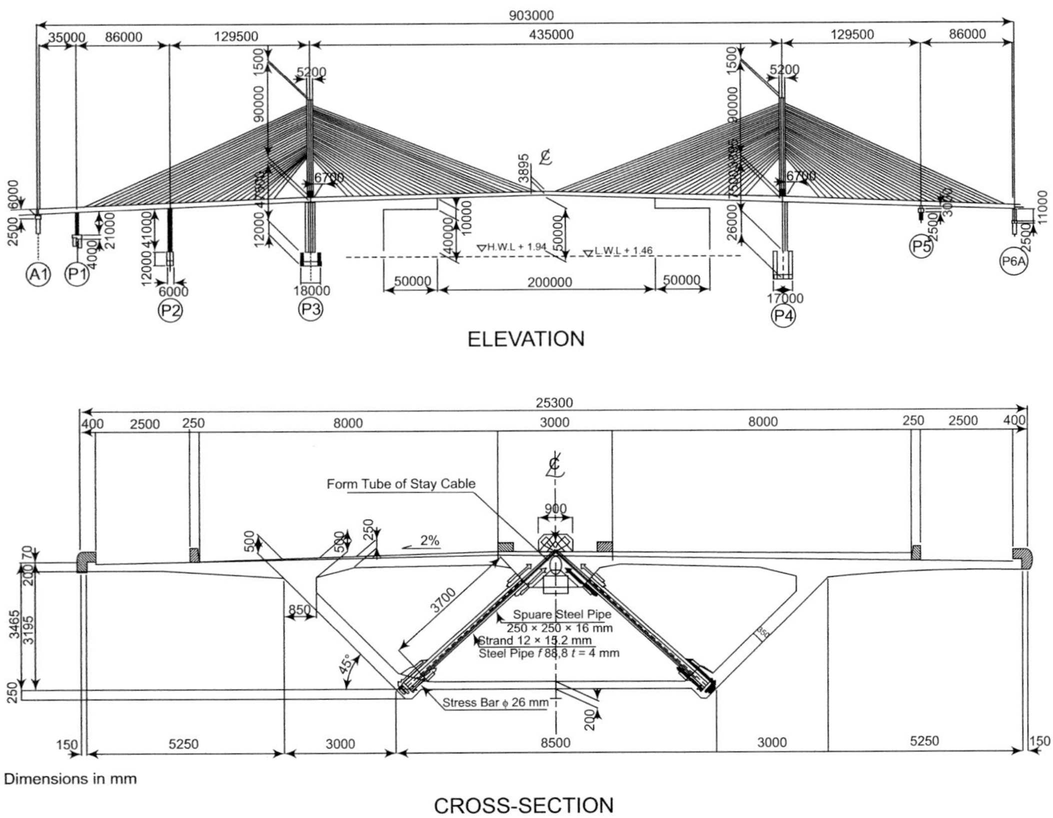

Fig. 8.14 General arrangement of the Bai Chay Bridge (Nakamura et al., 2007)

8.1.2.2 Phu My Bridge

Ho Chi Minh City (HCMC)'s PhuMy Bridge connects Districts 7 and 2 across the Saigon River (Figure 8.15). The bridge is a part of the new ring road that is being built around HCMC and is anticipated to be a major transportation route in the future that connects the central and northern regions of Vietnam with the southern Mekong Delta. The central span of the cable-stayed bridge measures 380 meters, while each of the back spans measures 162.50 meters (Figure 8.16). The pylons' overall height is 93 meters above deck level and 134.50 meters above the levels of the pile caps. At high tide, a minimum of 45 meters of vertical clearance is available for river traffic across a 250-meter wide area in the middle of the bridge.

Fig. 8.15 Phu My Bridge, Vietnam, 2009

The main span deck is 27 meters wide overall, with two vehicle and truck lanes, three traffic lanes in each direction, a dedicated motorcycle lane, and pedestrian footways (Figure 8.17). Two longitudinal concrete girders connected by transverse prestressed concrete cross girders spaced five meters apart make up the reinforced concrete open cross-section superstructure. Both longitudinally and transversely prestressed, the side spans have a solid concrete cross-section.

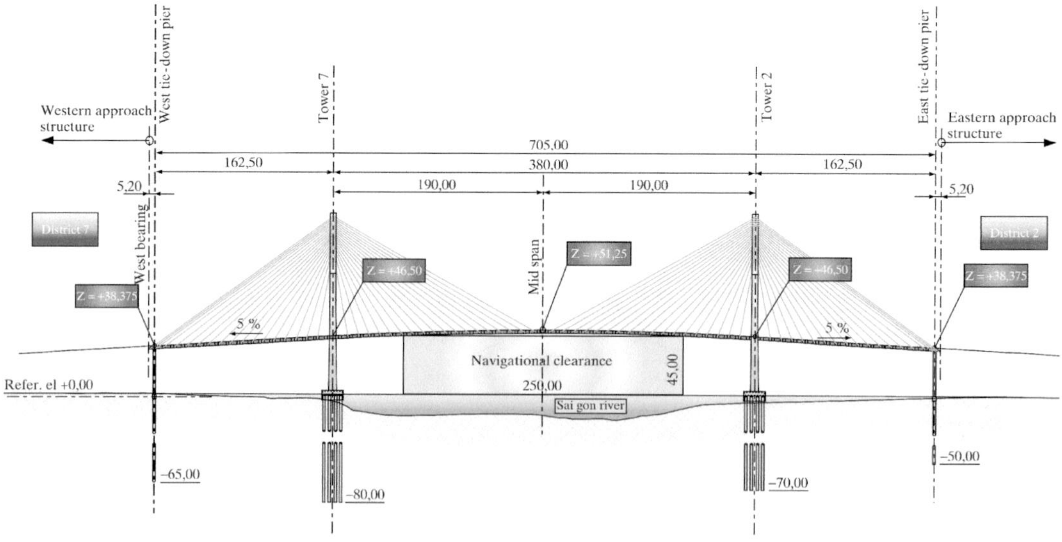

Fig. 8.16 Elevation of Phu My Bridge (Moir et al., 2010)

Stay cables are arranged in two vertical planes. At intervals of 10 meters, the stay cables are attached to the longitudinal edge beams. At the tiedown pier, the final three stay cables on the back span are connected.

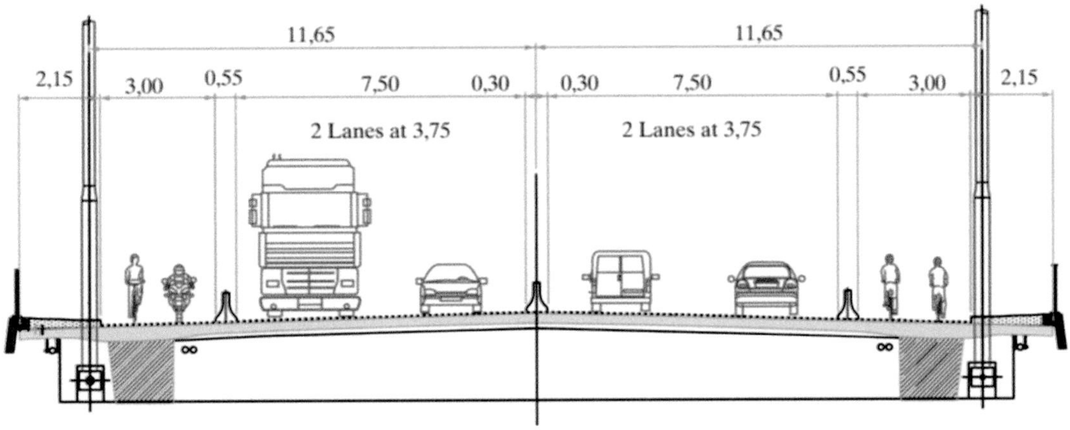

Fig. 8.17 Cross-section of Phu My Bridge (Moir et al., 2010)

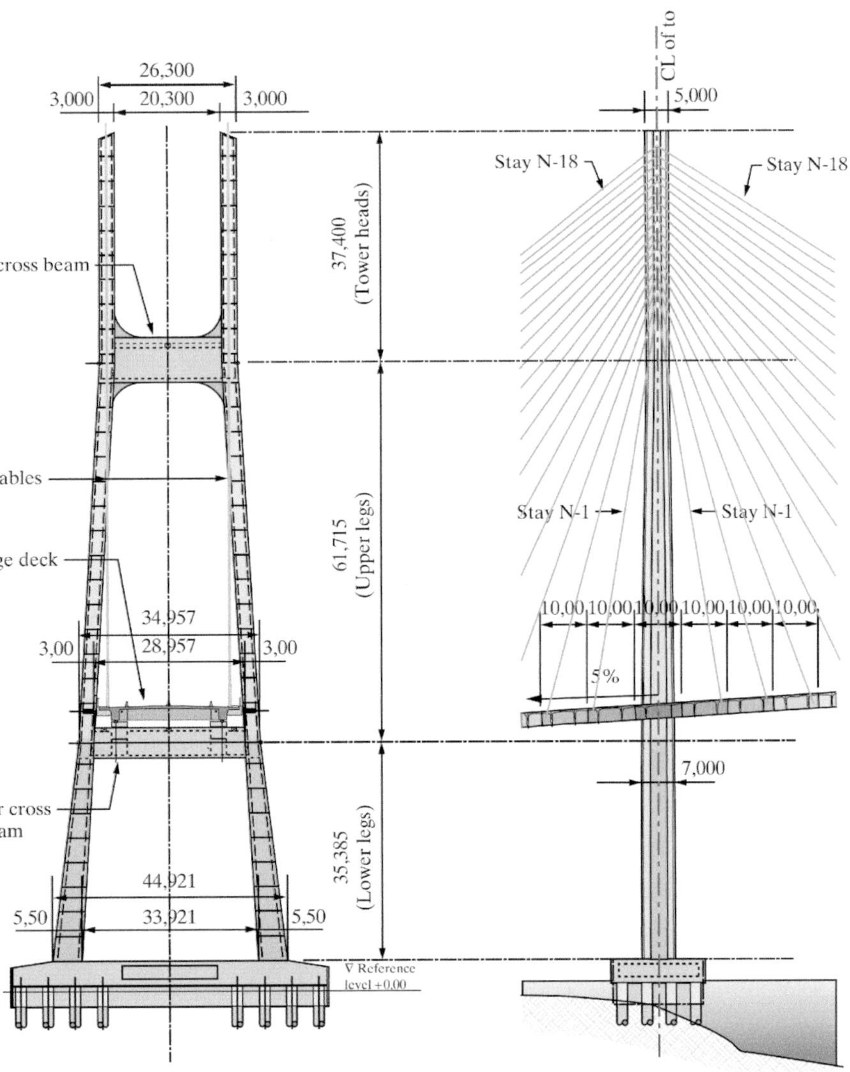

Fig. 8.18 Pylons of Phu My Bridge (Moir et al., 2010)

The H-shaped concrete pylons are shown in Figure 8.18. Each leg is a box section with external dimensions ranging from 5.5 m × 7 m to 3 m × 5 m. The stay anchorages are located in a vertical plane inside the upper legs. Two crossbeams, the lower of which was precast before being lifted into its final position, connect the legs. Every major bridge pylon is supported by two sets of fourteen 2.05-meter-diameter piles that can reach a depth of 80 meters. At the end of every cable-stayed back span are twin prestressed rectangular concrete columns known as the tie-down piers. Large-diameter bored piles support the tie-down piers (Moir et al., 2018).

8.1.2.3 Can Tho Bridge

The Can Tho Bridge (Figure 8.19) crosses the Hau (Bassac) River, the largest distributary of the Mekong River. The bridge is located in Binh Minh District, Vinh Long Province, about 170 km south of Ho Chi Minh City. The four-lane bridge has two 2.75 m wide walkways. The bridge deck is 26 m wide and made of pre-stressed concrete. Span lengths are 2 × 40 m – 150 m – 550 m – 150 m – 2 × 40 m while the total pylon height is 175.3 m. With a 550 m main span, Can Tho Bridge is the longest-spanning cable-stayed bridge in Vietnam and all of South-East Asia. Conclusion of the construction of the bridge was delayed due to the collapse of a 90-meter approach ramp during construction.

Fig. 8.19 Can Tho Bridge, Vietnam, 2010 (Courtesy of Bui Thuy Dao Nguyen)

8.1.2.4 Nhat Tan Bridge

Located within Hanoi's second City Ring Road, Nhat Tan Bridge (Figure 8.20) reduces the driving distance between Noii International Airport and the city. The bridge is 8.9 kilometers long overall; the bridge itself is 3.755 meters long, and the approach structures that lead to it on both sides total 5.18 kilometers in length.

Fig. 8.20 Nhat Tan Bridge, Vietnam, 2015

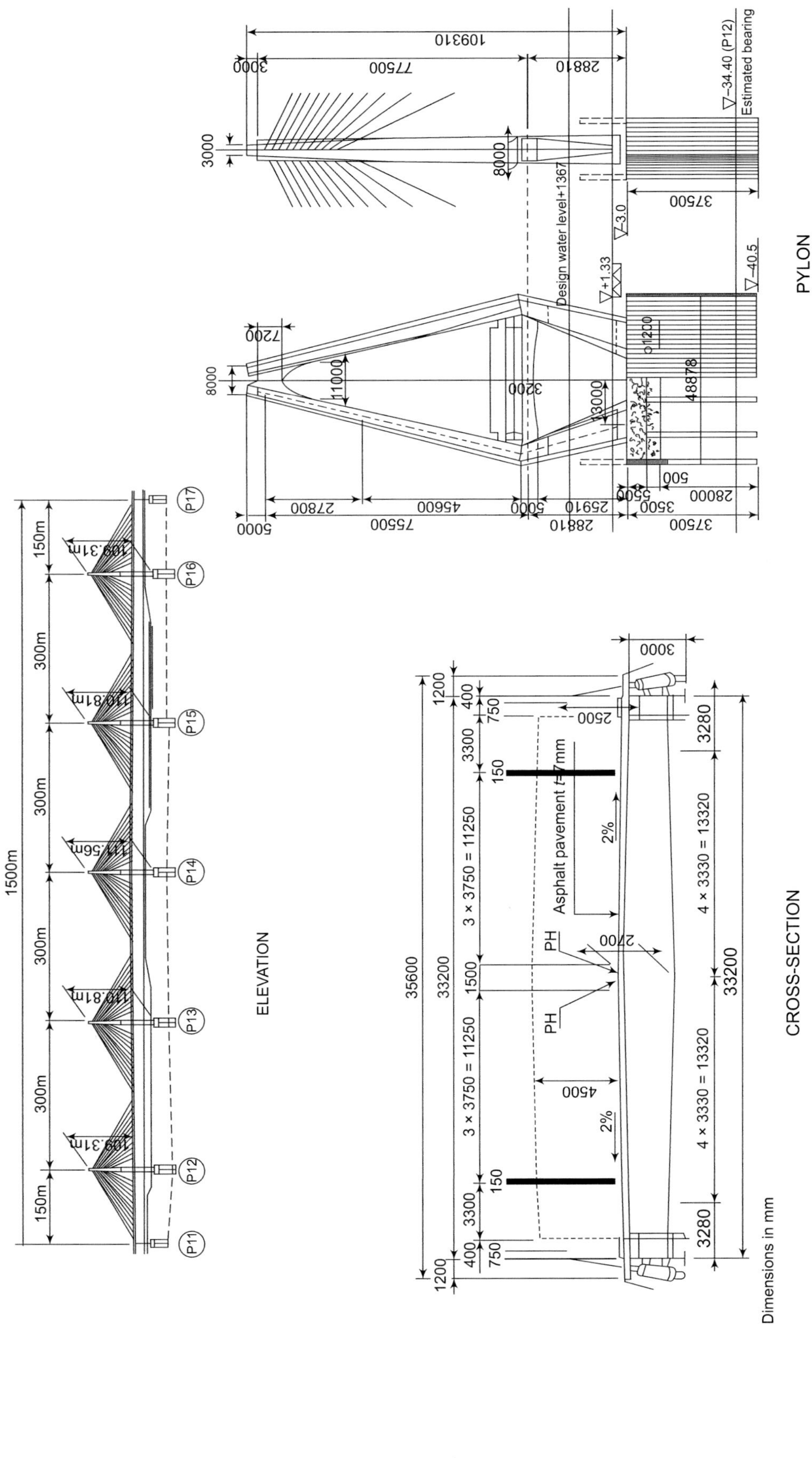

Fig. 8.21 General configuration of Nhat Tan Bridge, Vietnam, 2015 (Swit et al., 2016)

The 1500 m long cable-stayed section is made up of six spans, measuring 150 m, 4 × 300 m, and 150 m. The width measures 34.6 meters, encompassing four lanes for vehicle traffic, two bus lanes, two motorcycle lanes, and two walkways for pedestrians. The superstructure is a composite cross-section comprising two edge steel I girders 3 m deep connected transversely with floor beams as shown in Figure 8.21. Ten longitudinal steel stringers are installed for the purpose of supporting the precast concrete slabs and achieving the composite cross-section. The slabs are designed continuously through concrete closure pours over the floor beams, edge girders, and between panels over the stringers.

The bridge has two planes of stay cables. With a diameter of 7 mm and a tensile stress of up to 1770 Mpa, cables are composed of 121, 151, 163, and 313 strands. High-Density Polyethylene (HDPE) is used to coat the cables in protection; it is impervious to wind and UV light. Moreover, the coating has grooves built into it to absorb the energy created when raindrops hit each other during a heavy downpour. An inverted V forms the shape of the reinforced concrete pylons. The highest point is 111 meters above the foundation and 74 meters above the bridge deck. The main pylon's cross-section transforms at the level of the steel girders from a regular hexagon at the base to a heptagon. (Swit et al., 2016).

8.1.2.5 Vam Cong Bridge

The Vàm Cống Bridge (Figure 8.22) runs 2.97 km along the Hau River in southern Vietnam, linking Lap Vo District in Dong Thap Province and Thot Not District in Can Tho City. The six-lane bridge was opened in May 2019. Its cable-stayed bridge section is designed as a 3-span cable-stayed bridge (870 m) with a central span of 450 m and is distributed as: 210 m + 450 m + 210 m = 870 m (Figure 8.23). It is the second-longest cable-stayed steel bridge in Vietnam, and allows travel at a maximum speed of 80 km/h.

Fig. 8.22 Vam Cong Bridge, Vietnam, 2019

The superstructure is a composite section with two 2 m deep edge steel plate girders 25.8 m apart and connected transversely with a floor beam every 4.0 m. The bridge has two planes of stay cables arranged in a semi-fan configuration. Cables are anchored at 12 m intervals on the upper surfaces of the outer edge girders. Cables are made of 15.7 mm galvanized high tensile 7-steel strands. The number of strands in each cable varies according to its location and ranges from 30 to 73 strands per cable.

The shortest cable is 41.9 m at the pylon between the side spans and the longest is 234.7 m. The pylons are H-shaped resting on a 2.5 m diameter bored pile foundation. Its height is 130.9 m from the pile cap to the top of the pylon and 110 m from the deck to the top of the pylon, respectively. The concrete strength is 50 Mpa.

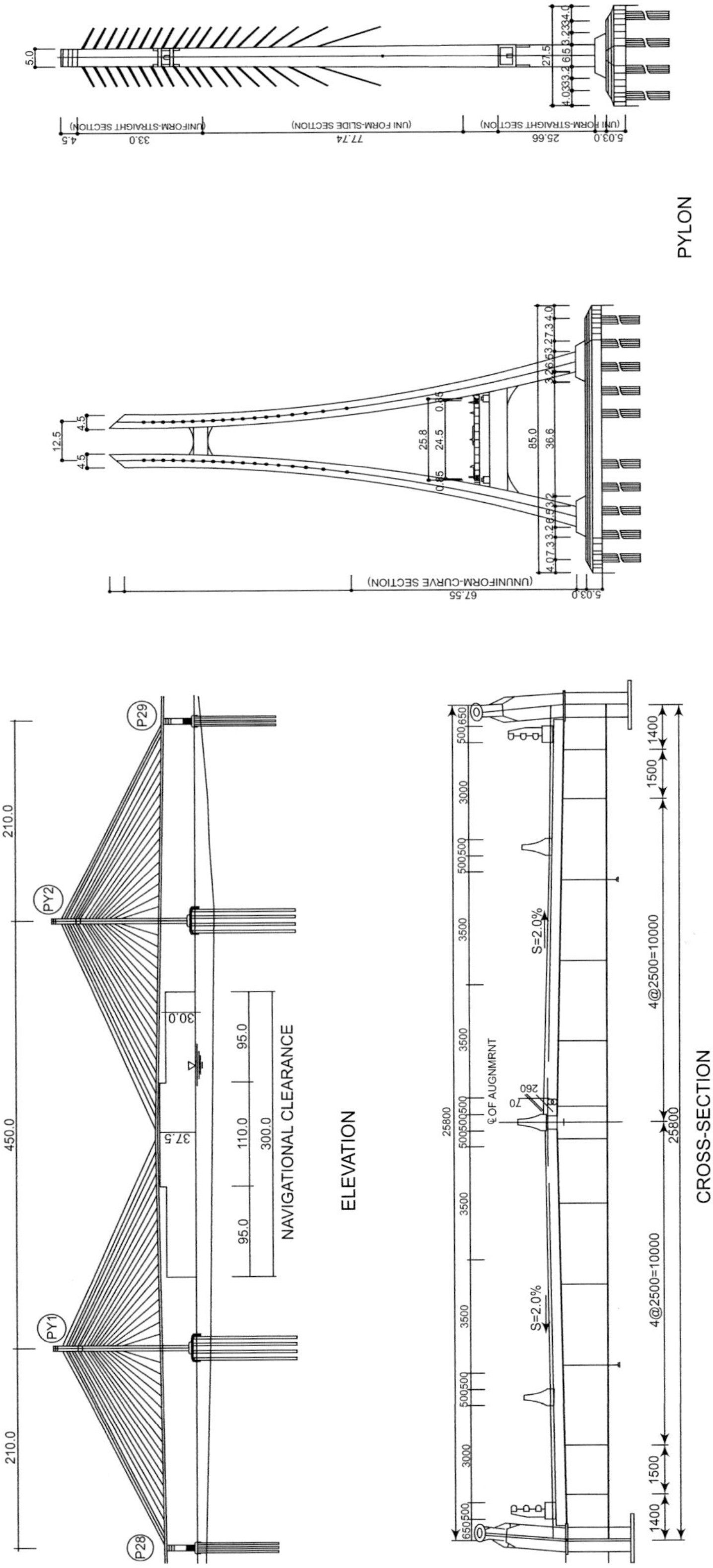

Fig. 8.23 General configuration of Vam Cong Bridge (Lee et al., 2013)

8.1.3 Thailand

Several giant cable-stayed bridge projects were completed in Thailand during the last two decades most importantly, Rama VIII Bridge in 2002; Industrial Ring Road Bridge (Bhumibol Bridge) in 2006 and Kanchanapiske Bridge in 2007. These bridges are outlined in this section.

8.1.3.1 Rama VIII Bridge

Bangkok's Rama VIII Cable-stayed Bridge spans the Chao Phraya River (Figure 8.24). It is a crucial component of the ongoing efforts to enhance traffic flow throughout the Bangkok metropolitan area and a vital link in the heart of Bangkok.

Fig. 8.24 Rama VIII Bridge, Thailand, 2002

Overlooking the Chao Phraya River, the main bridge is a 475-meter-long cable-stayed structure (Figure 8.25) over the 100 m wide navigation channel. The bridge offers a 10.4 m vertical clearance with only one supporting pylon. The length of the main span is 300 m. Two 50 m back spans and a 75 m anchor span make up the land span. With two traffic lanes on either side and a 5.3 m wide promenade for bicyclists and pedestrians, the main span deck spans 29.2 meters. The superstructure is a composite cross-section. It comprises two 1.6 m deep longitudinal edge plate girders 23 m apart connected transversely by 1.3 m deep floor beams spaced at 4.5 m intervals.

Precast panels and cast-in-place infill strips over the floor beam and girder flanges make up the longitudinally spanning concrete deck slab over the floor beams. The central 10 m wide by 2.5 m deep post-tensioned concrete box girder spine, which supports post-tensioned transverse ribs separated at 5.4 m centers, makes up the two 50 m long rear spans (see Figure 8.25). The thickness of the box web is 600 mm at the midspans and 750 mm close to the supports. In a similar vein, the thickness of the bottom slab ranges from 400 to 300 mm. Between the post-tensioned box webs, the transverse ribs are 380 mm thick, reducing to 300 mm in the cantilevers. Every transverse rib position on the center box's webs is vertically post-tensioned. The anchorages for the back stays are located within a post-tensioned central box that is 10 meters wide and 9 meters deep, making up the 75-meter anchor span. Ten 19 × 12 mm straight tendons, mostly found in the bottom slab, are used to longitudinally post-tension the center box during the construction phase. The deck and transverse ribs resemble those in the back spans (Figure 8.25). Reinforced concrete makes up the 3 m deep central longitudinal diaphragm beam to which the stays are attached. There are transverse fin walls that are 380 mm thick at each transverse rib position. The self-weight of the anchor span and the ballast concrete, which is gradually inserted into the bottom of the box as the backstays are constructed, resist the uplift caused by the stays. The fin walls have vertical post-tensioning to transfer the weight of the ballast and the box to the longitudinal central beam that anchors the stays.

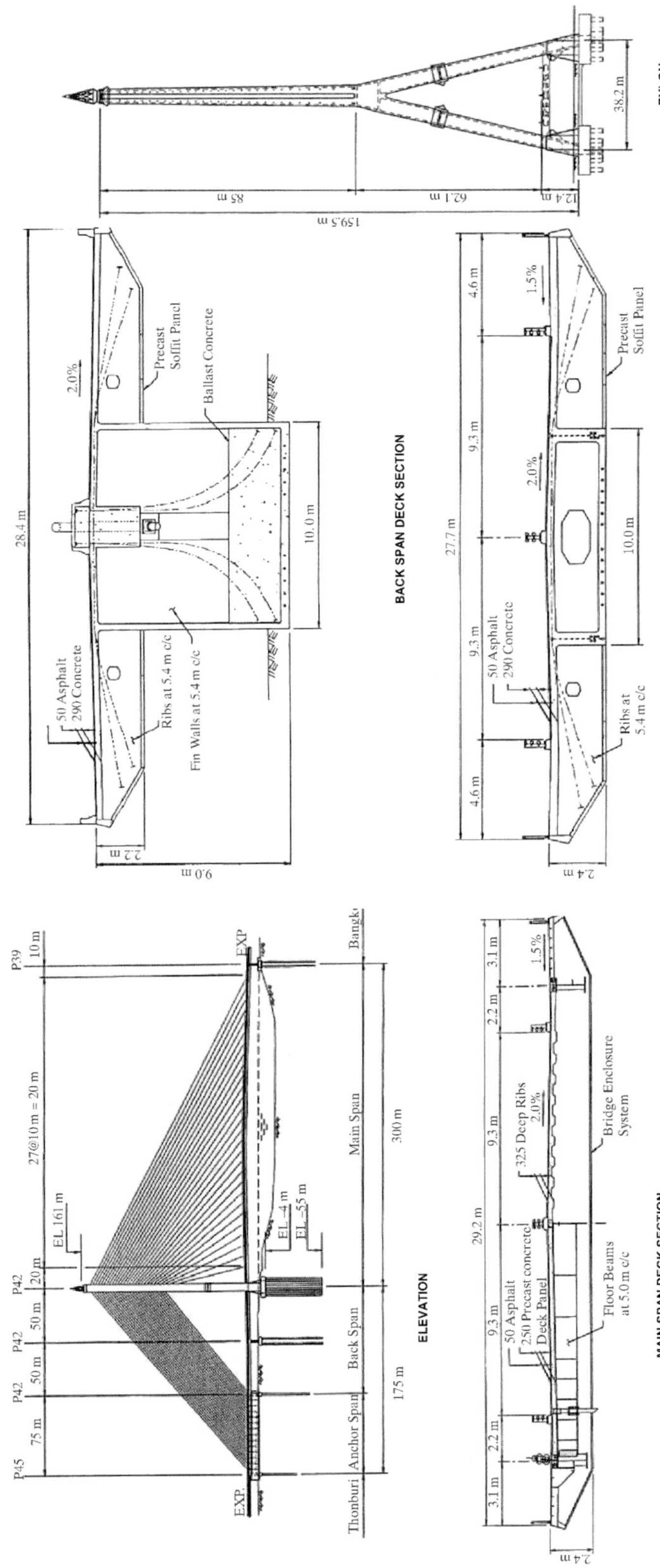

Fig. 8.25 General arrangement of Rama VIII Bridge (Nankorn et al., 2002)

Two semi-fan-shaped stay planes that are anchored on either side of the deck section support the main span. The open main span deck portion is given torsional stiffness by the two planes. The stays in the land span are grouped into a single center plane that resembles a harp. Seven wire prestressing strands rated at 1770 MPa are used in the cable stay system. The primary span stays close to the pylon have a minimum of 11 strands, while the longest backstay has a maximum of 65 strands. The 160-meter-tall concrete main pylon is shaped like an inverted Y. The pylon and its legs consist of a straightforward rectangular box segment that is meticulously crafted to resemble a jump. The longitudinal width of the box portion is 7 m at all times. The pylon box section's transverse width increases from 5 m at the top to 7.5 m at the connection with the legs, where there is a considerable requirement for lateral bending.

The legs that are below the taper widen to a minimum of 4 m at the top of the pile cap, 5.4 m at the connection with the pylon. To reduce force transfer at the joint between the pylon and the legs, two diaphragms are used. This section of the pylon's reinforcement was designed and detailed using finite element and strut and tie models. Under each pylon leg, there are two sets of eighteen 55-meter-deep piles that support the main pylon (Nankorn et al., 2002).

8.1.3.2 Industrial Ring Road Bridge

Completed in 2006, the Industrial Ring Road Project (Figure 8.26) features two cable-stayed bridge crossings over Bangkok's Chao Phraya River, with navigation spans of 398 and 326 meters. Three approach structures make up the project, and the primary north-south road is connected to the west approach through a central peninsula interchange. The location of the interchange between the two main bridge structures placed strict geometric restrictions on how the cable-stay back spans were arranged as well as the interchange itself.

Fig. 8.26 Industrial Ring Bridge, Thailand, 2006 (Khamdee et al., 2009)

The four primary longitudinal girders and the variable depth cross-girders, spaced four meters apart, define the composite cross-sections that make up the main navigation span superstructures (Figure 8.27). The deck is is 35.9-m wide fabrication of 12-meter-long deck modules that were erected in a cantilever in line with the stay spacing and served as the foundation for the design. Seven wire-galvanized strands with a diameter of 15.7 mm make up the stay system. Prestressed concrete was used to build the back spans. There are differences in the number of strands in each cable, ranging from 40 to a maximum of 90. The gold-colored HDPE protective sheath features double helical ribbing to enhance the aerodynamic performance of the stays, specifically their reaction to vortex shedding. The diamond-shaped pylons are heavily chamfered to articulate light and shade in the intense sunlight. They are hollow and dimensioned for maximum visual effect. The bases of the pylons measure 6.9 × 20.2 m. The pylon legs were further reduced to 2.65 m × 5.12 m above deck level, and from 8.0 m × 6.83 m below deck level to 3.1 m × 6.2 m. The average wall thickness is 0.70 m. The outer dimensions of the stem anchorage zone measure 5.0 × 5.0 m, giving it an octagonal shape. The pylon was intended to be built with a jump form system, with 6.0 m panels as standard heights. The main bridge pylons' form is complemented by the Y-shaped piers of the back span supports. Every foundation is made up of bored piles with a 1.5 m diameter and a 60 m depth (Khamdee et al., 2009).

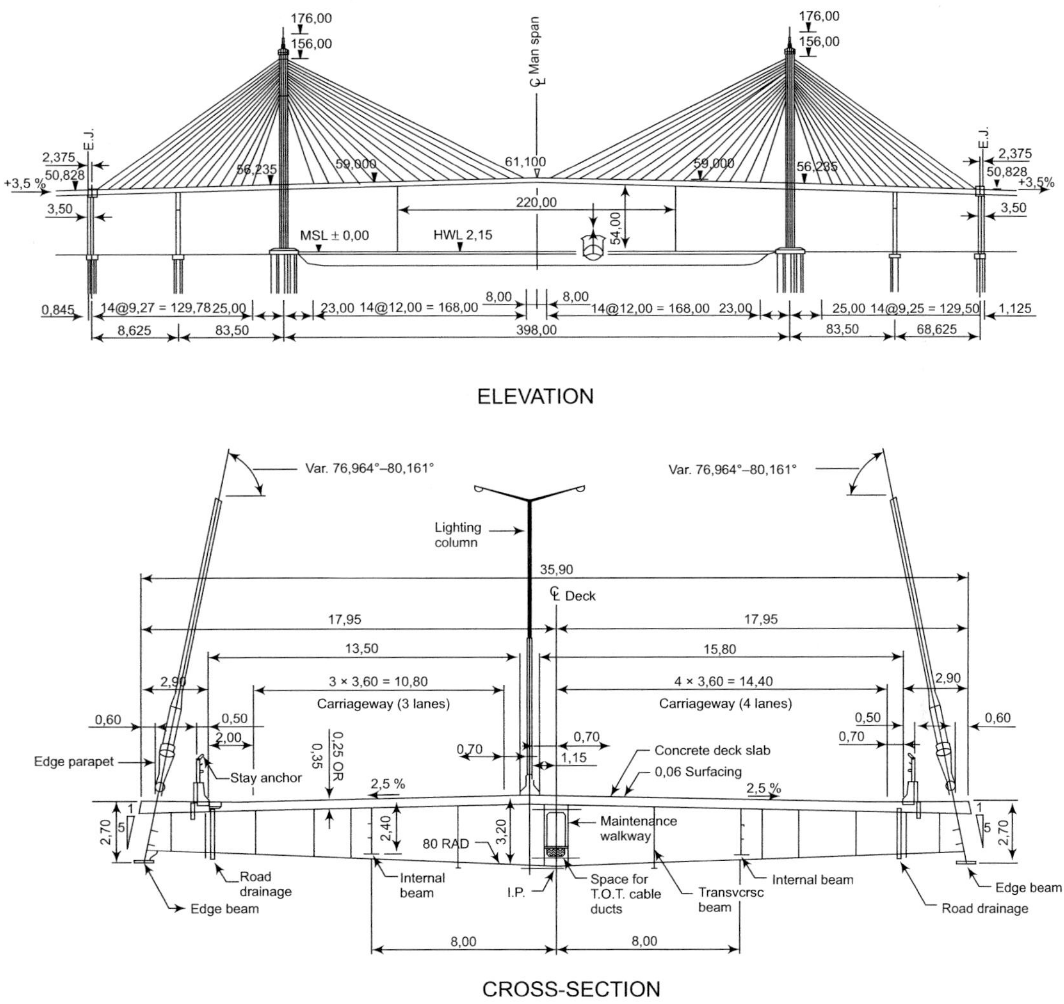

Fig. 8.27 General arrangement of Industrial Ring Bridge (Khamdee et al., 2009)

8.1.3.3 Kanchanapisek Bridge

The Southern Outer Bangkok Ring Road Project's primary component, the Kanchanapisek Bridge (Figure 8.28), spans the Chao Phraya River. With two side spans and a main span that is 500 meters long, the bridge is currently the longest span across a river, measuring 951 meters in total length (Figure 8.29).

Fig. 8.28 Kanchanapisek Bridge, Thailand, 2007 (Jomvinya et al., 2009)

In order to accommodate six lanes for traffic, the deck is designed as a composite section with a steel member 3.3 meters deep, a total width of 36.7 meters, and a reinforced concrete slab 260 to 310 millimeters thick, covered by 50 millimeters of asphaltic concrete. The deck's steel grade complies with AASHTO M 270 Grades 50 and 50W. The longitudinal beam and floor beam have an I-shape and are composed of steel Grade 50W (refer to Figure 8.29).

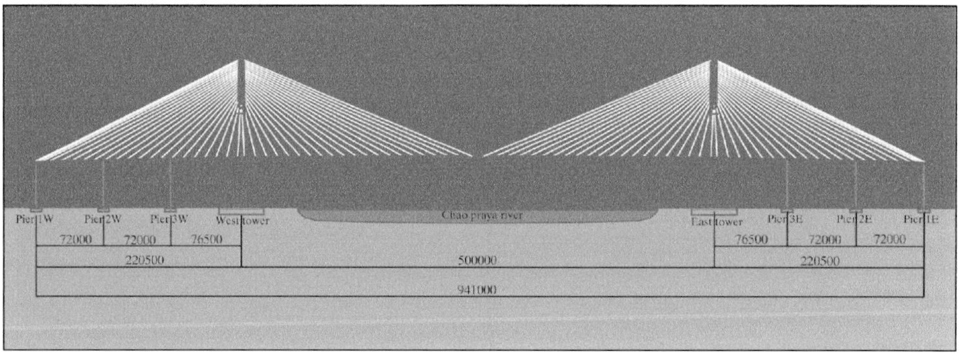

Fig. 8.29 Elevation of Kanchanapisek Bridge (Jomvinya and Vicat., 2009)

The individual parallel-galvanized 15.7 mm diameter strands that make up the stayed cables have a guaranteed ultimate tensile strength of 1860 Mpa. With the addition of petroleum wax and HDPE coating, the strands are triple protected against corrosion (galvanization wax and HDPE). 168 cables hold up the superstructure. There are between 23 and 72 strands in cables near the pylon and in the middle of the main span.

The A-shaped reinforced concrete pylon has a hollow core and a total height of 184.64 meters. It is divided into three sections: the lower leg of the pylon is 49.44 meters above the ground, the upper leg is 61.50 meters, and the top portion is 76.70 meters, which is the upper anchorage zone, or pylon head (Figure 8.31). The spire is constructed of gold-covered stainless steel. A total of fifty cast-in-place bored piles with a diameter of two meters support each 24-meter-long, 24-meter-wide, and 4-meter-deep pylon pile cap. A prestressed tie beam connecting the footings beneath each leg balance the tension force caused by the inclination of the pylon legs. Three anchor piers support each side span, allowing for precise control over the main span's moment and deflection. The purpose of the three counterweighted anchor piers on each side is to secure the stayed cables. The anchor piers are permanently under tension, as they are transferred to the pile cap, which measures 28.6 m in width, 12 m in length, and 3.5 m in thickness. The pile cap is supported on eight bored piles that are 64 m deep.

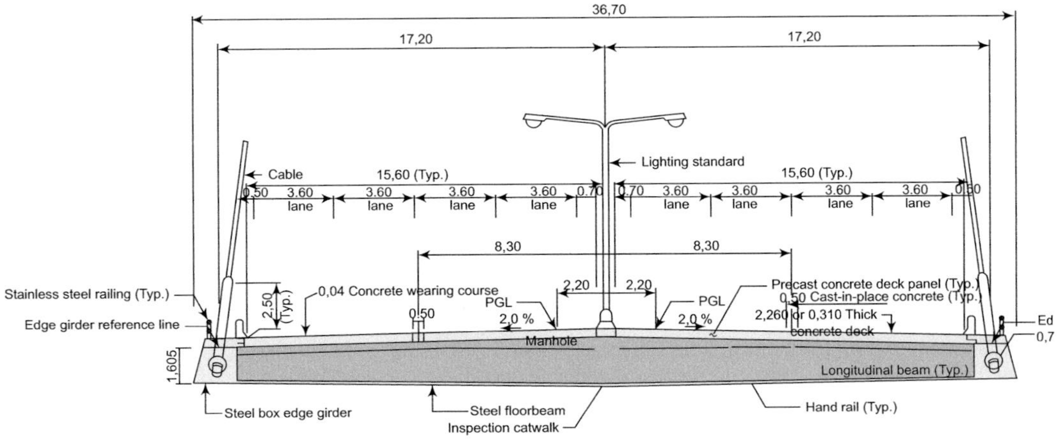

Fig. 8.30 Cross-section of Kanchanapisek Bridge (Jomvinya and Vicat, 2009)

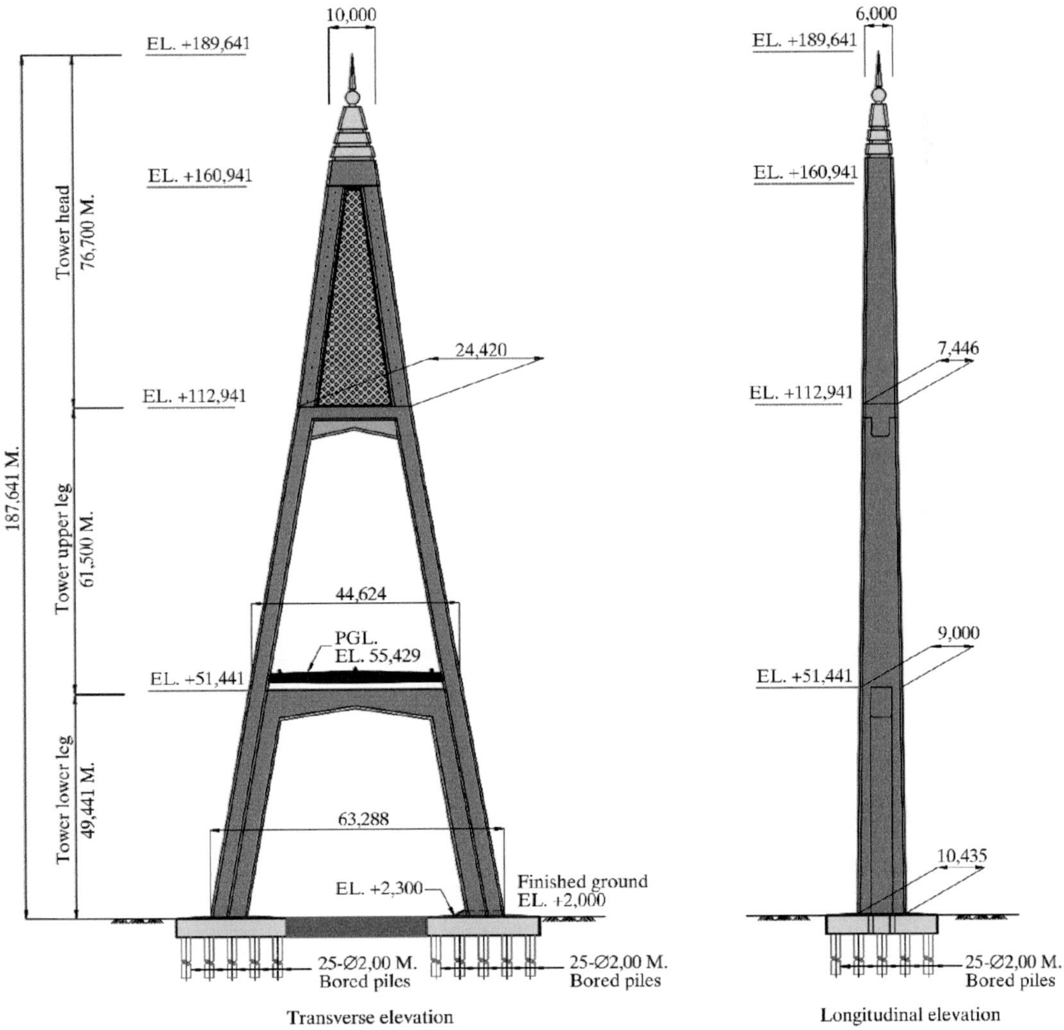

Fig. 8.31 Pylon of Kanchanapisek Bridge (Jomvinya and Vicat, 2009)

8.1.4 Malaysia

Malaysia joined the list of countries that implemented cable-stayed bridges in 1985 when the Penang Bridge was opened as part of the 13.5 km E36 Highway in the State of Penang which connects Perai on the mainland side of the state with Gelugor on the island, crossing the Penang Strait. The 225 m cable-stayed bridge has a prestressed concrete open cross-section. Several cable-stayed bridges were completed early this century. Table 8.2 outlines the significant cable-stayed bridges built in the country.

Table 8.2 Significant Cable-Stayed Bridges in Malaysia

Bridge name	*Year*	*Type*	*Total Length of project (m)*	*Main Span (m)*
Penang Bridge	1985	Prestressed concrete open cross-section	13,500	225
Seri Saujana Bridge	2002	Prestressed concrete double box girder archcable-staved	300	300
Seri Wawasan Bridge	2003	Prestressed concrete	240	169
Sungai Muar Bridge	2004	Prestressed concrete box girder	632	132
Sungai Prai Bridge	2004	Prestressed concrete box girder	485	185
Sultan Abdul Halim Muadzam Shah Bridge	2006	Prestressed concrete open cross-section	22,500	240
Sungai Johor Bridge	2006	Composite steel box girder	1,708	500

Geographically some of these bridges share the same region. The Penang Bridge and the Sultan Abd Halim Muadzam link Penang Island to the mainland over the Malacca Strait and are about 7.7 km apart. The Sungai Prai is located not far from the Penang Bridge at the estuary of Prai River on the strait. Similarly, Seri Saujana Bridge and Seri Wawasan Bridge are located over the Putrajaya Lake in Putrajaya about 3 km apart. It can be noticed that all these bridges except the Sungai Johor Bridge are prestressed concrete with a superstructure that is either an open section or box girder. The Sungai Johor Bridge is characterized by its unique composite cross-section comprising a torsionally stiff steel box girder and a concrete deck slab that is used to transfer the compressive axial forces and connected to the upper plate of the box by shear studs.

Both the Penang Bridge and the Sultan Abdul Halim Muadzm Bridge have a similar design. Also, Sungai Muar Bridge and Sungai Prai Bridge have similar designs except for minor changes. Therefore, the discussion herein is limited to the Seri Saujana Bridge, Sungai Muar Bridge, and Sultan Abdul Halim Muadzam Bridge.

8.1.4.1 Seri Saujana Bridge

The Seri Saujana Bridge is unique in the sense that it combines two bridge types, the arch and cable-stayed, leading to a spectacular landmark with sculptural quality (Figure 8.32). The bridge is in Putrajaya, the new Administrative Center of the Federal Government of Malaysia. This bridge is also known as Bridge 8.

The structure is a combination of two overhead inclined arches and a two-pylon cable-stayed bridge with a main span of 300 m. The bridge deck as shown in Figure 8.33 is 32 m wide. It consists of a multi-cell box girder consisting of a central box that houses the anchorages of the stay cables

Fig. 8.32 Seri Saujana Bridge, Malaysia, 2002

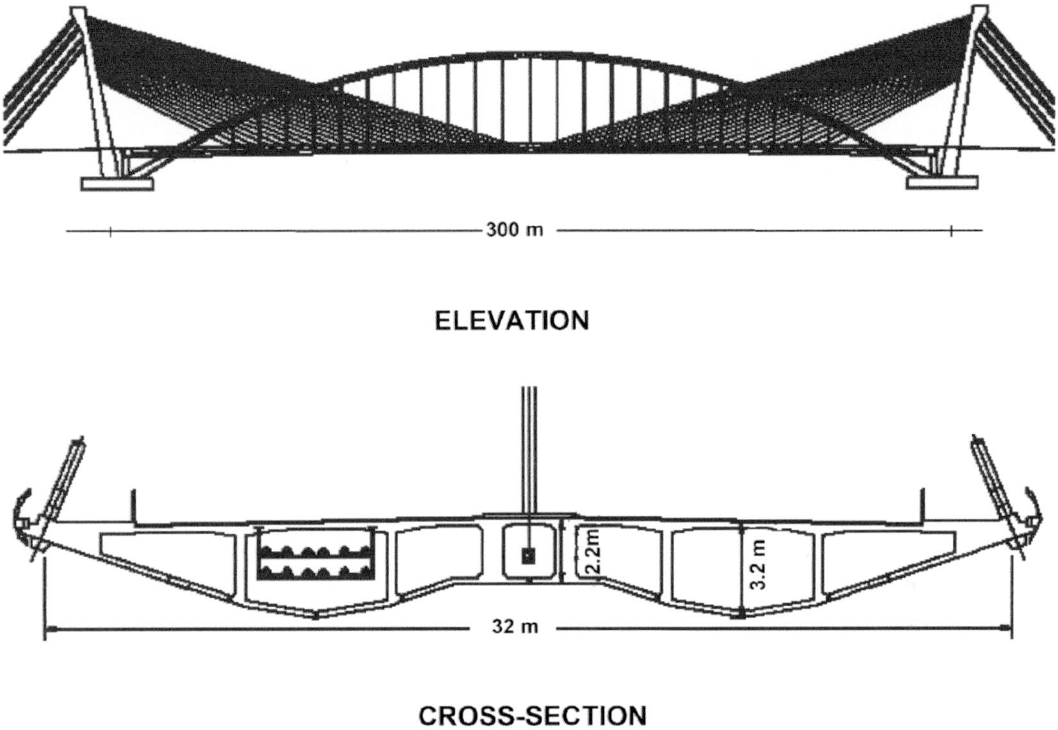

Fig. 8.33 General arrangement of Seri Saujana Bridge

and two wing trapezoidal boxes carrying the roadway. Hence the superstructure is suspended by the central one plane of cable stays and the inclined hangers at both sides.

The deck serves as a tie-beam for the arch because it is prestressed in both the longitudinal and lateral directions. Wing bending resulting from differential load situations caused by the combined action of the arch and stay systems is covered by lateral prestressing cables in a transverse direction. The bridge deck is continuous without expansion joints.

The arches consist of 2.20 m diameter cold-rolled steel tubes. Their 34-meter height above the deck is supported by K-bracings, which are likewise made of steel pipes. One plane of cables hangs the bridge in the center. The bridge deck is supported by twenty backstay cables in two spread planes that are fixed externally by steel trumpets, and twenty front stay cables. The sword-shaped pylons are 73 meters high and 78 degrees slanted backward. PE coated strands with seven wires make up the cables. Four to six threaded prestressing bars with coupling nuts make up the inclined hangers.

After the tensioning work was finished, the steel pipe casing was welded and filled with grout. (Klein and Yamout, 2003).

8.1.4.2 Sungai Muar Bridge

Sungai Muar Bridge (Figure 8.34) crosses the Muar River, 40 km south of Melaka. It simplifies the north-south Melaka/Muar links, by connecting directly to the Bunga interchange.

Fig. 8.34 Sungai Muar Bridge, Malaysia, 2004

The total length of the structure is 632 m, subdivided in the central cable-stayed span of 132 m and two access ramps of eight spans of 32 m (24 m for the banks) (Figure 8.35). The structure carries two lanes of 3.50 m, one shoulder of 1.7 m in each direction and a 2.0 m central median.

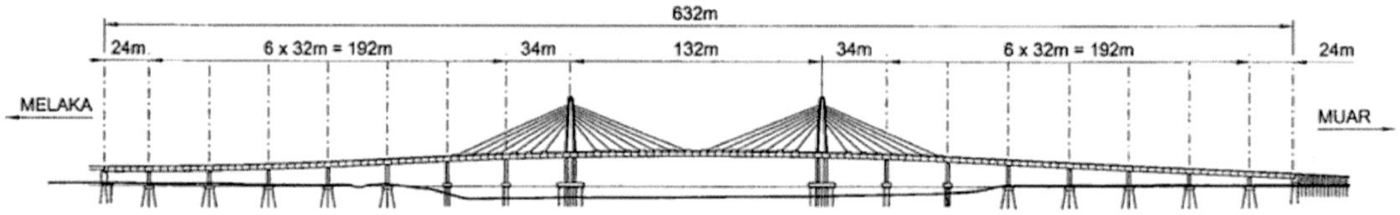

Fig. 8.35 Elevation of Sungai Muar Bridge (Tanis et al., 2003)

The superstructure is a trapezoidal concrete box girder with an upper width of 21.4 m and a constant height of 2.5 m (axial) as illustrated in Figure 8.36. It is made structurally continuous over the entire length of the structure. The cross-section is prestressed in the longitudinal direction with 19K15 Freyssinet tendons and the upper slab is prestressed transversely with 5T15 Freyssinet flat cables at every interval of 0.80 m.

The central span is attached to the pylons by 2 × 7 axial stays. The spacing between the cable anchorages is 8 m on the deck and 1.40 m on the pylon. A connection is made with the pylon through a design saddle consisting of a multitube complex of separate individual tubes. Each stay cable strand runs in one central span anchor to the corresponding anchor side span, anchoring in the pylon occurs in the saddle by friction between the strand (HDPE sheath) and the tube (aluminum) of the saddle strand.

The pylons are 30 m high above the deck with dimensions as shown in Figure 8.37. The dimensions in the upper portion where the saddles exist are: a web of 850 mm thick, suitable for

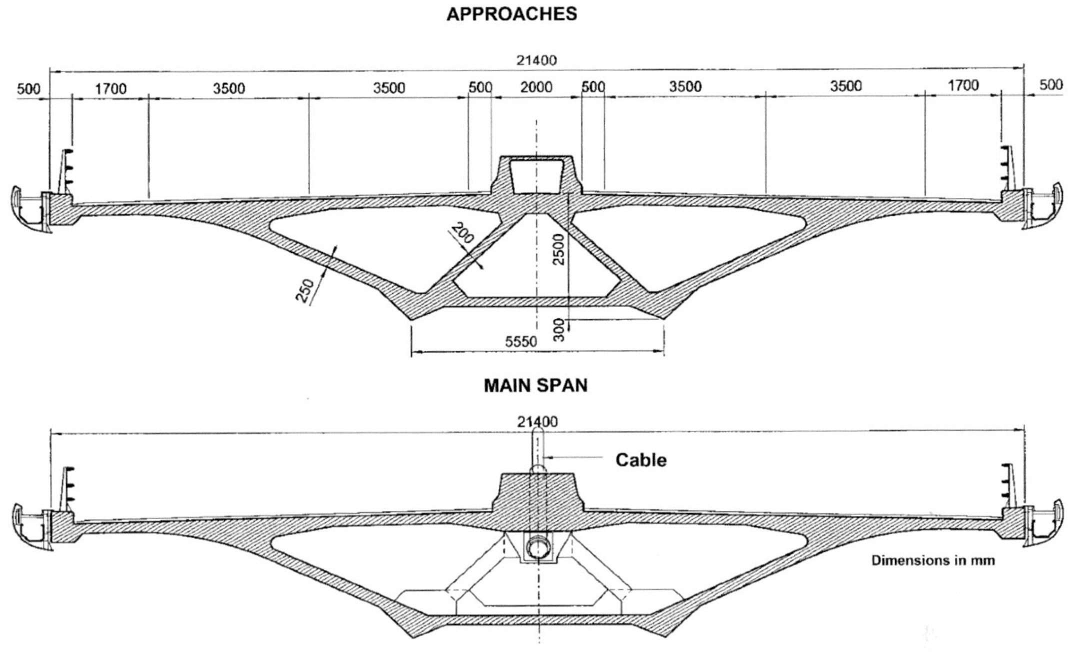

Fig. 8.36 Cross-section of Sungai Muar Bridge (Tanis et al., 2003)

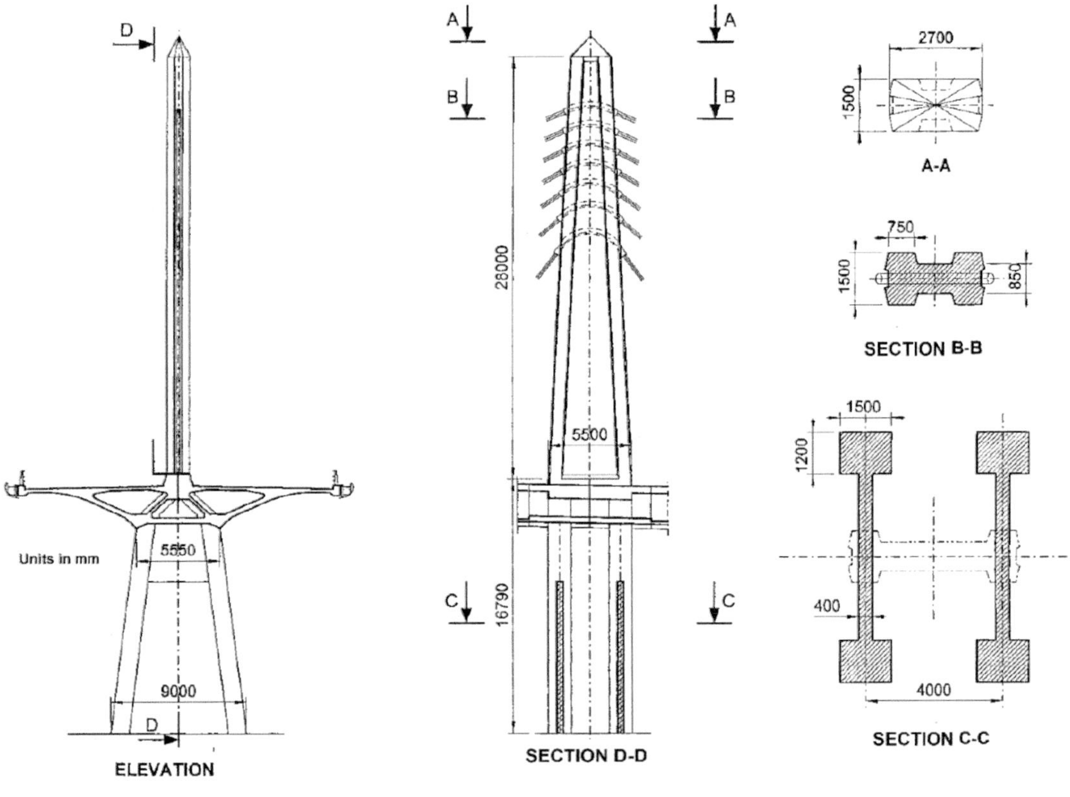

Fig. 8.37 Cross-section of Sungai Muar Bridge (Tanis et al., 2003)

diameter – 400 mm – tubular saddles thickened at the ends, in the horizontal direction, at 1.50 m. Vertically, the width of the pylon along the axis of the bridge varies depending on the curved length of the saddles: the general appearance of the pylon is therefore a "sword" with width varying from 5.30 m to 2.70 m. Under the deck, the pylon is extended below through four posts to the level of the pile foundations. Each pair of posts is attached transversely by a rigid diaphragm.

8.1.4.3 Sultan Abdul Halim Muadzam Shah Bridge

The Sultan Abdul Halim Muadzam Shah Bridge (Figure 8.38) connects the south of Penang Island and the peninsula of Malaysia. It has a total length 22.5 km off which about 16.5 km is the marine bridge.

Fig. 8.38 Sultan Abdul Halim Muadzam Shah Bridge, Malaysia, 2006

The bridge deck carries dual roadways comprising 4 traffic lanes and 2 motorcycle lanes and has a width of 29.8 m at typical deck sections. The main bridge is a two-pylon three-span prestressed concrete cable-stayed bridge, with a span arrangement of (117.5 + 240 + 117.5) m (see Figure 8.39). The pylons and the main girder are monolithic. The superstructure is a prestressed concrete open cross-section, which is 34.6 m wide, comprising top slab, transverse diaphragms and two edge beams. The pylons are H-form pylons, and the stay cables are formed of parallel strands and are arranged in 2 semi-fan cable planes. Each pylon column carries 18 pairs of stay cables that are anchored by the deviation saddles in the pylons and anchor blisters in the main girder. The foundations consist of large diameter bored piles, varying from 2.3m to 2.0 m in diameter.

8.1.5 Indonesia

Indonesia covers thousands of islands, the population however is concentrated on six major islands: Sumatra, Java, Kalimantan, Sulawesi, Riau, and Papua New Guinea. To keep up with the economic growth, Indonesia has also adopted advanced techniques in bridge design and construction reflected in numerous prestressed concrete cable-supported structures for crossing big rivers as well as for urban infrastructure developments. Bridges with spans more than 60 m are considered nonstandard and follow advanced methods of design. Selection of a proper bridge is tied to its location and the needs of local traffic. Cable-stayed bridges in particular, have gained widespread popularity in Indonesia since the late nineties and more bridges are opened every year. Table 8.3 outlines the significant cable-stayed bridges that were built in Indonesia during the last twenty years. The Suramadu Bridge that has the longest main span is discussed in detail herein.

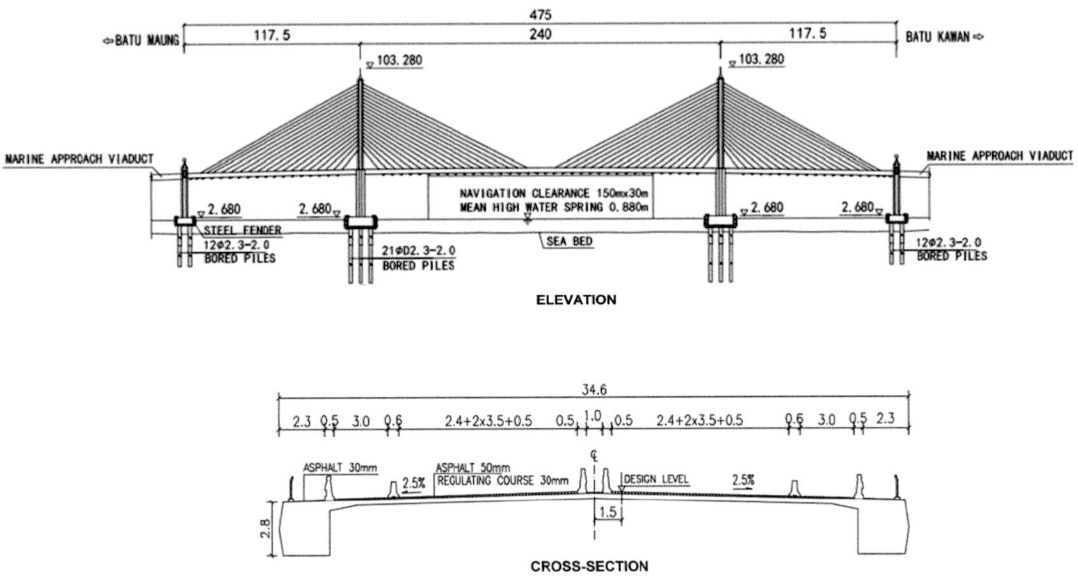

Fig. 8.39 General arrangement of Sultan Abdul Halim Muadzam Shah Bridge (Man et al., 2018)

Table 8.3 Significant Cable-stayed bridges in Indonesia

Bridge name	*Location*	*Year*	*Type*	*Total Length of project (m)*	*Main Span (m)*	*Number of pylons*	*Number of cable planes*
Teuku Fisabilillah Bridge	Riau Isles	1998	Prestressed concrete open cross-section	642	350	2	2
Pasupati Bridge	West Java	2005	Prestressed concrete box girder	2,282	106	1	1
Grand Wisata Overpass	West Java	2007	Prestressed concrete	81	81	1	2
Siak Indrapura Bridge	Riau	2007	Prestressed concrete open cross-section	1,196	200	2	2
Suramadu Bridge	East Java	2009	Composite	5,438	434	2	2
Sukarno Bridge	North Sulawesi	2013	Prestressed concrete box girder	622	120	1	2
Siak IV Bridge	Riau	2014	Composite	699	156	1	2
Galalapoka Bridge	Maluku	2014	Prestressed concrete open cross-section	1,065	150	2	2
Jambatan Mahkota	East Kalimantan	2015	Prestressed concrete open cross-section	680	340	2	2

8.1.5.1 Suramadu Bridge

The Suramadu Bridge (Figure 8.40) is located in the Indonesian province of East Java, in the northern region. It links Madura Island and Surabaya by crossing the Madura Strait. With twin tower pylons, twin cable planes, and a steel-concrete composite beam, the main bridge is a cable-stayed structure. As shown in Figure 8.41, the span arrangement is 192 + 434 + 192 m = 818 m. A prestressed concrete continuous beam with box section and span length of 40 + 7 × 80 + 40 m = 640 m makes up the approach bridge on each side. The approach and main bridges are linked to the V-pier.

Fig. 8.40 Suramadu Bridge, Indonesia, 2009

The main bridge is a floating structure. Side piers are only provided with vertical bearings. Longitudinal earthquake-resistance dampers are positioned at pylon towers to restrict movement along the bridge. At the bridge end, concrete stoppers are placed on the V-pier to stop transverse movement, and rubber positive blocks are positioned between the main girder and the pylon shaft.

The superstructure is a composite cross-section comprising two 2.8 m deep steel box main girders at the edges and two inner stringers as shown in Figure 8.41. Floor beams spaced at 4 m are attaching the girders and transfer the slab load to them. High strength bolts are used for connections between main girders, floor beams and main girders, and stringers and floor beams Transverse dimension for the outsides of the box webs is 2.3 m. The floor beam and stringer, along with the web and bottom flanges, are all made of welded I-sections. Prefabricated panels with a thickness of 250 mm are assembled into the concrete deck slab using concrete in situ connections. Throughout the entire bridge, transverse prestressing tendons are positioned in the concrete deck slab. Longitudinal prestressing tendons are also arranged in the deck slab at the bridge end and mid span where the horizontal components of the stay cables are small. (Consortium of Chinese Contractors, 2005).

The main span is suspended by two planes of stay cables arranged in a semi-fan configuration. 68 stay cables in total are suspending the main span. In order to anchor the stay cables, which are tensioned at the pylon end and fixed at the main girder end, cold cast anchor devices are used. The cables are anchored at intervals of 12 meters on the main girder and 2.2 meters at the pylon. High-density polyethylene is extruded into two layers, one of which is colored and the other is black, to create cables that are composed of galvanized parallel steel strands with a diameter of 7 mm. Steel wires typically have a strength of 1670 MPa.

The pylon includes the seat, lower pylon shafts, mid pylon shafts and upper pylon shafts. The total height is 141.331 m. There are three transverse beams installed on each pylon as shown in Figure 8.42. Pylon shafts and transverse beams are all in the hollow box section. Each pylon is supported by 56,2.4 m diameter bored piles that are encased by steel pipes. The pile cap is 57.2 m, 34m along the bridge, and 6 m in thickness.

Each side pier of the main bridge has two sets of earthquake-resistant steel spherical bearings arranged beneath the main girders; one set is movable in two directions, while the other is movable longitudinally and fixed transversely. The transversely fixed bearing has a horizontal supporting capacity of 6000 kN and a vertical supporting capacity of 8000 kN. The bearing can move up to ±60 cm along the bridge, and it can rotate up to 0.02 rad. Underneath the main girders at each pylon are two longitudinal viscous dampers. The main parameters of the damper are exponent of velocity, $\alpha = 0.4$; damping coefficient, C = 3000 kN $(m/s)^{-0.4}$; normal damping force, F = 2400 kN; length of stroke (mm) ± 600; minimum safety factor of damping force = 1.5. The main bridge expansion joint at the bridge end is the Maurer swivel-joist expansion joint.

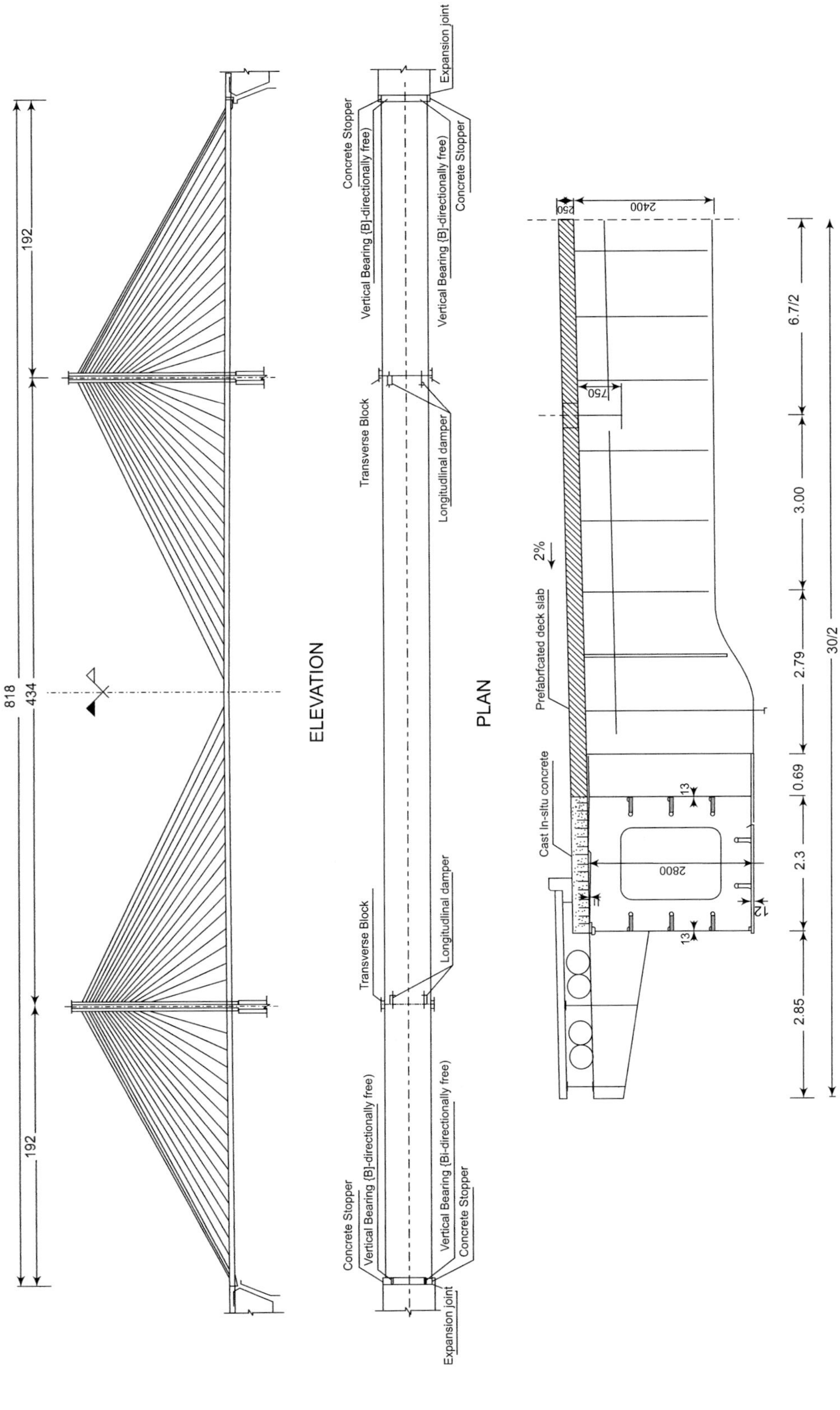

Fig. 8.41 General arrangement of Suramadu Bridge (Courtesy of CCC)

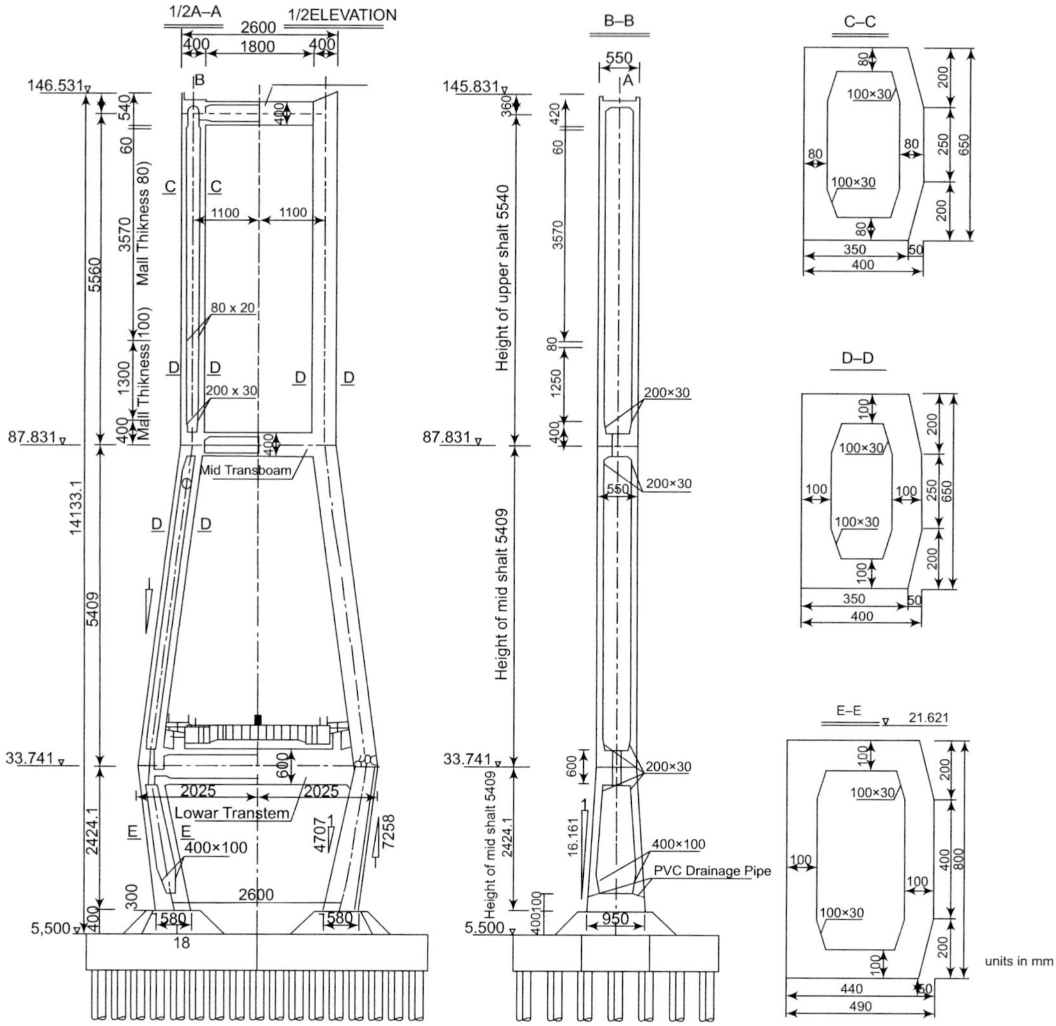

Fig. 8.42 Pylon of Suramadu Bridge (Courtesy, CCC)

8.1.6 India

Roads and bridges in India have attained significant growth in the last few decades. Cable-stayed bridges were built as means of river crossings such as the Second Hooghly Bridge on Hooghly River, the Kota Chambal Bridge on Chambal River, the signature bridge on the Yamuna River and more. Cable-stayed bridges were also built and employed as overpasses in large cities such as the Krishnarajapuram Bridge in Bengaluru and the Ram Jhula-Twin Bridges in Nagpur. Recently India opened a cable-stayed bridge as part of the Konkan railway. Table 8.4 outlines the significant cable-stayed bridges in India. The second Hooghly Bridge, Rajiv Gandhi Sealink project, and the Signature Bridge at New Delhi are discussed in detail.

Table 8.4 Significant Cable-stayed bridges in India

Bridge Name	*Year*	*Type*	*Total Length of Project (m)*	*Main Span (m)*	*Number of Pylons*	*Number of cable planes*
Akkar Bridge	1988	Prestressed concrete open cross-section	154	77	1	2
Second Hooghly Bridge	1992	Composite	823	452	2	2
Krishnarajapuram Bridge	2000	Prestressed concrete	151	106	1	2
Naini Bridge	2004	Prestressed concrete	610	260	2	2
Rajiv Gandhi Sealink	2010	Prestressed concrete box girder	4,700	500	1	2
Worli Twin Bridges	2010	Prestressed concrete box girder	4,700	150	2 per bridge	2 per bridge
Basholi Bridge	2015	Prestressed concrete box girder	592	350	2	2
Kota Chambal	2017	Prestressed concrete box girder	1,500	350	2	1
Signature Bridge	2018	Composite	575	251	1	2 at main span and 1 at back span
Ram Jhula-Twin Bridges	2019	Prestressed concrete box girder	180	90	1 per bridge	2 per bridge
Anji Khad Bridge	2021	Steel Truss	437	290	1	2

8.1.6.1 The Second Hooghly River Bridge

The urban area of Calcutta, in the coastal state of West Bengal, eastern India, is home to the Second Hooghly River Bridge (Figure 8.43). There are two end spans of 182.88 meters each on this cable-stayed bridge, which has a main span of 457.20 meters. A median strip measuring 1.7 meters divides the two 12.3-meter-wide three-lane roadways on either side of the bridge, which has 2.5-meter-wide walkways. The main span of the bridge has a navigational clearance of 33.87 meters. Through the bridge, Indian National Highways 2 and 6 from Delhi and Bombay/Madras can directly connect to the Calcutta Port. Figure 8.44 depicts the bridge's general layout.

Fig. 8.43 Second Hooghly River Bridge, India, 1992

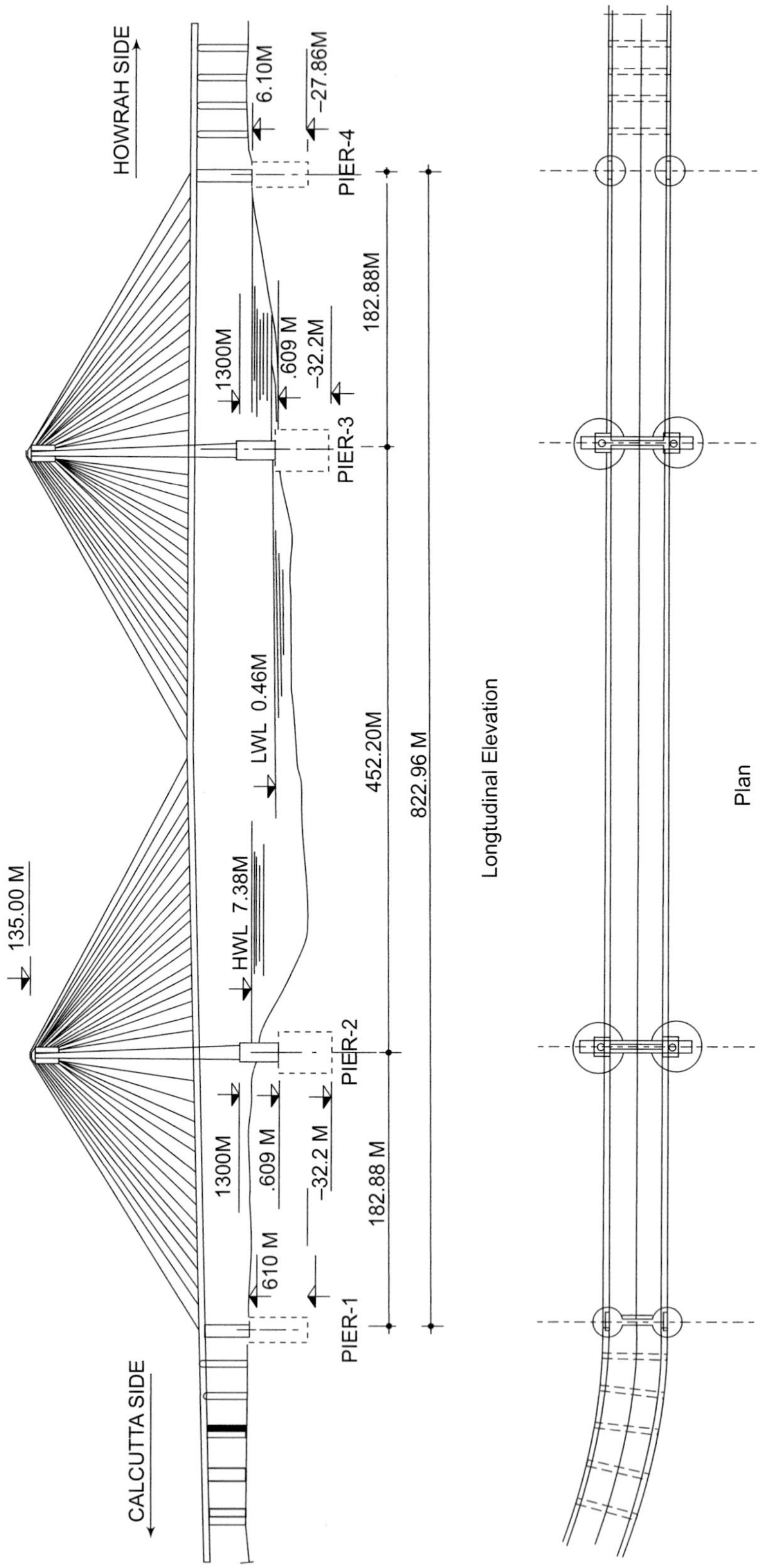

Fig. 8.44 General arrangement of Second Hooghly River Bridge (Gupta, 1991)

The superstructure is composite consisting of two main longitudinal steel girders 2 m in depth, with steel cross girders spaced at 4.1 m intervals. Stay cables are configured in two semi-fan configurations. Cables are made of parallel-wire bundles of 7 mm wires.

The pylons are lateral, rigid frames that have portals at the level of the pylon head and beneath the deck. The bolted steel box girders that make up the pylon units have legs that measure 4 × 4 m at the bottom and 4 × 3 m at the top. The box girder walls consist of two steel plates that are up to 25 mm thick. A number of prestressing rods, each measuring 4 meters in length, firmly anchor the pylon legs to the piers. Twin circular caissons, with diameters varying from 23.8 m for the Howrah pylon to 8.0 m and 12.0 m for the piers, support the main pylons and the anchor piers. A pair of perpendicular parallel diaphragms brace the caissons of the main pylons.

8.1.6.2 Rajiv Gandhi Sea Link

The eight-lane, 4.7 km long Rajiv Gandhi Sea Link (Figure 8.45) is an offshore road bridge located west of the Mumbai Peninsula in Mumbai, India. The districts of Bandra in the north and Worli in the south can travel across Mahim Bay more quickly thanks to it. The two separate twin bridges that make up the project each have two navigation spans and four lanes of traffic. With a typical span of 50 meters and six continuous spans, the Sea Link is made up of viaduct modules (north approach, link viaduct between cable-stayed bridges, and south approach) that are supported by disc and pot bearings. The cable-stayed bridges are the Bandra and the Worli cable-stayed twin bridges. A 4-lane link bridge that connects to the mainland is also located at the southern end. The general layout of the bridge is shown in Figure 8.46.

Fig. 8.45 Rajiv Gandhi Sea Link, India, 2010

The two 250 m main spans of each of the two Bandra cable-stayed bridges are supported by a pylon consisting of four solid legs and two end transition spans, which are 50 m on either side. Each bridge has two semi-fan-shaped stay planes that are spaced six meters apart and attached to the deck's edges. The stays' top anchorages are situated in the upper vertical section of the pylon, inside a steel chassis (Meyer et al., 2011).

The 150 m main spans of each of the two Worli cable-stayed bridges are bounded at both ends by two 50 m conventional approach spans. The superstructure loads are transferred to the pylons by two planes of stay cables. They are placed 6.0 meters apart along the bridge deck in a semi-fan configuration. The pylon is composed of two solid legs that are inclined outward beneath the deck, connected by a cross beam near the top end, and tied together through the deck and inward

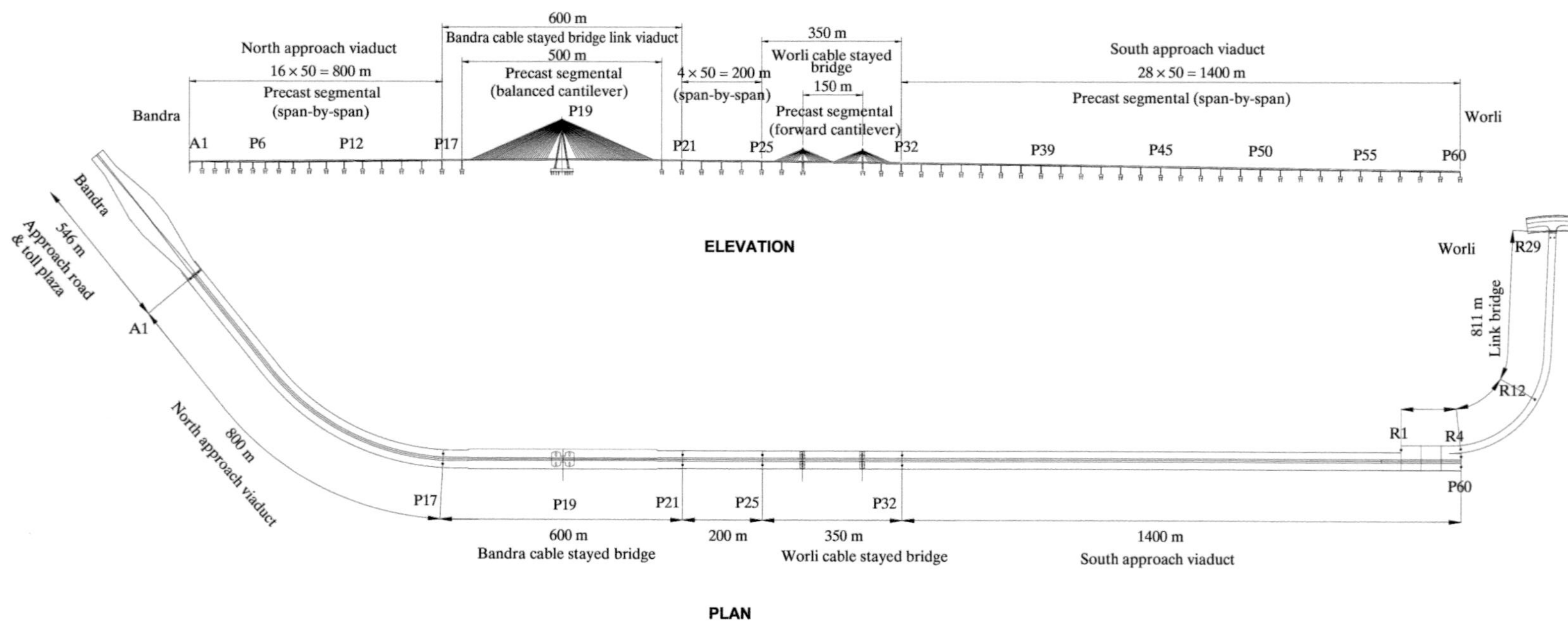

Fig. 8.46 General arrangement of Rajiv Gandhi Sea Link (Meyer et al., 2011)

above it. The bridge superstructure for the two structures is composed of 3 cells. It is a prestressed concrete trapezoidal section 22.7 m wide and 3.0 m deep for each bound as shown in Figure 8.47. For durability, high-performance concrete with a cube strength of 60 MPa, with micro silica and fly ash, was used. For both cable-stayed bridges, the deck is monolithically tied into the pylon (Meyer et al., 2011).

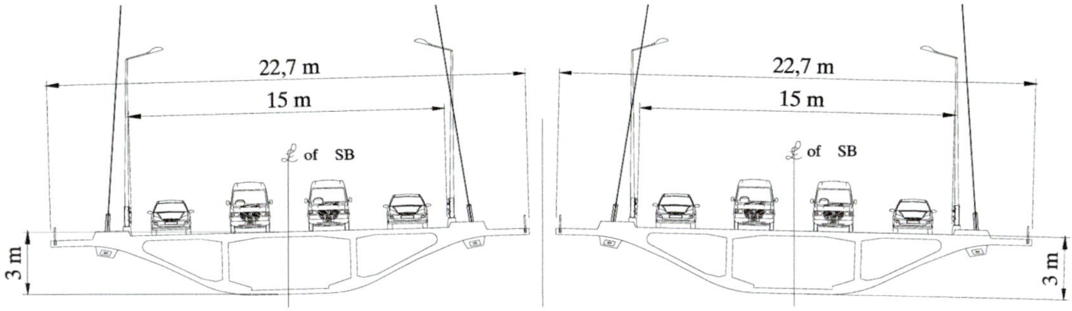

Fig. 8.47 Cross-section of Rajiv Gandhi Sea Link (Meyer et al., 2011)

8.1.6.3 The Signature Cable-Stayed Bridge

Designated as a symbolic representation of the emerging India, the Signature cable-stayed bridge (Figure 8.48) spans the Yamuna River at Wazirabad, New Delhi. With an overall length of 675 meters and a main span of 251 meters, the bridge is an asymmetric cable-stayed construction. Four lanes per direction make up its eight lanes. It is around 35 meters wide. A gateway connecting the Wazirabad region with New Delhi is formed by the 150-meter-tall steel pylon situated towards the eastern shore.

Fig. 8.48 Signature cable-stayed bridge, India, 2018

The deck supports two 14.0 m wide four-lane roads that are divided by a concrete crash barrier, along with two lateral emergency pathways, for a total width of 35.2 m (Figure 8.49). Two edge I-shaped longitudinal main girders and I-shaped cross girders spaced 4.5 meters apart make up the cross-section. A third central main girder is placed to facilitate distribution of heavy live loads. Two cable planes, spaced 13.5 meters apart, support the deck. At a distance of 13.5 meters from their dead end, the cables are directly anchored to the outer main girder webs. Parallel wire strands measuring 15.7 mm make up the cables. The number of strands per cable varies depending on the location; at the main span, it can range from 55 to 123, and for each of the backstays, it can be upto 127. Corrosion protection was applied using hot dip galvanized wires and individually coated strands covered by an outer Polyethylene-pipe.

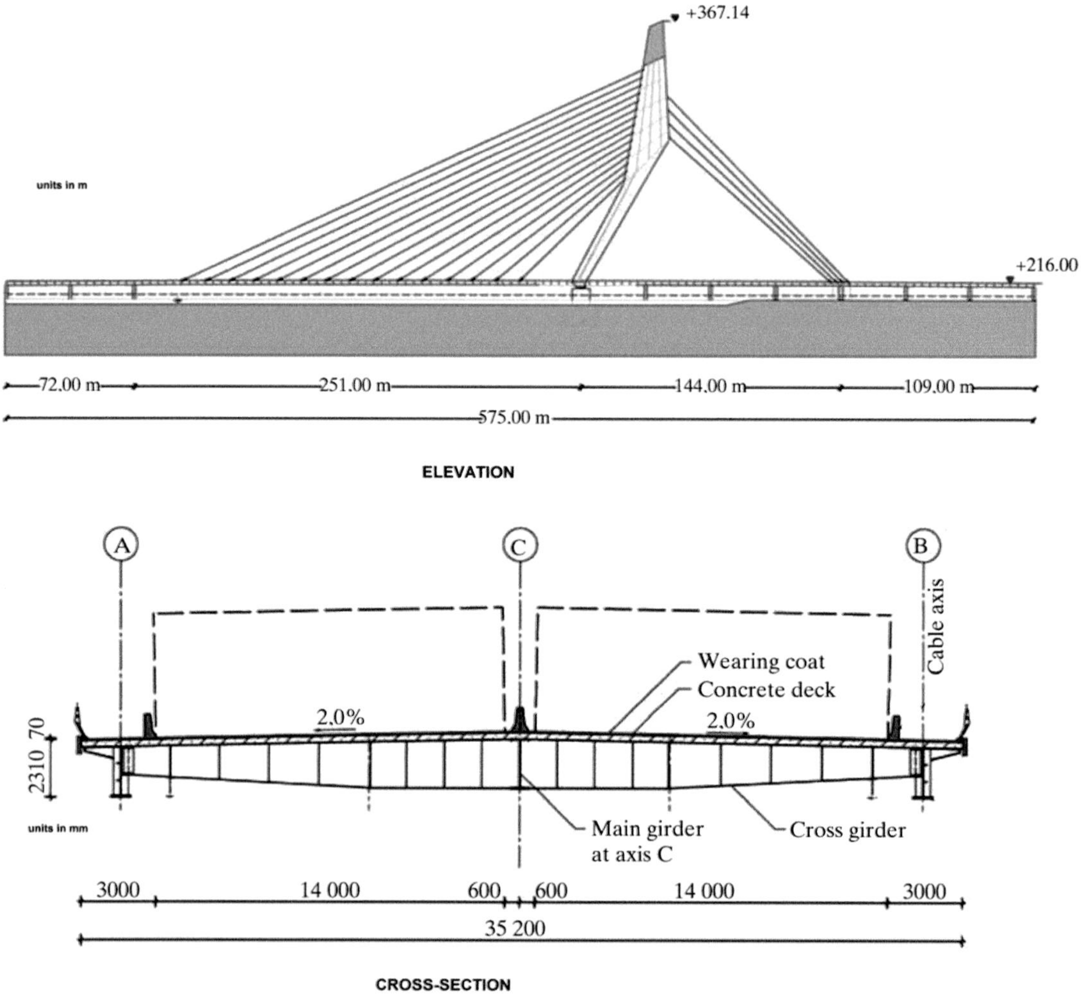

Fig. 8.49 General arrangement of the Signature cable-stayed bridge

The pylon consists of two legs made of steel boxes with a hollow box-section stiffened by internal bracings and merged into one upper pylon zone where the cables are anchored. The pylon is monolithically connected to the deck. The deck's concrete can accommodate the longitudinal horizontal force generated by the pylon. A robust steel cross tie connecting the two legs at the level of the cross girders carries the horizontal forces in the transverse direction. To avoid bending into the substructure, the pylon legs are supported by large spherical bearings. Each bearing has a maximum vertical force transmission capacity of 170 MN.

8.1.7 Turkey

Turkey recently inaugurated three significant cable-stayed bridges: the Golden Horn Metro Bridge in 2014; Nissibi Euphrates Bridge in 2015; and Agin Bridge in 2015.

8.1.7.1 Golden Horn Metro Bridge

The Golden Horn Metro Crossing Bridge (Figure 8.50) connects the South-West part of Istanbul to Taksim Square and Atatürk Airport. 8.1.7.3. The bridge, which is a segment of M2 Metro, is made up of a main cable-stayed bridge with a span configuration of 90 m + 180 m + 90 m, two approach viaducts, and a swing bridge. An orthotropic steel deck with three cells makes up the superstructure. The deck is 13.7 m wide and sits approximately 17 m above sea level. The steel pylon is 54 meters above the deck. A metro station serving line M2 is located on the main span of the cable-stayed bridge.

Fig. 8.50 Golden Horn Metro Crossing Bridge, Turkey, 2014

8.1.7.2 Nissibi Euphrates Bridge

The 80th km of the Adıyaman-Diyarbakır highway in Turkey is home to the Nissibi bridge (Figure 8.51). Southeast Anatolia's Euphrates River is crossed by the bridge over the reservoir of the Atatürk Dam.

Fig. 8.51 Nissibi Euphrates Bridge, Turkey, 2015

There are 610 meters of bridge in total. An orthotropic steel box section measuring 380 meters in length, 26.5 meters in width, and 2.70 meters in depth makes up the 400 meters main span between the two pylons. At either end of the main span, a prestressed concrete box section measuring 10 meters is rigidly fastened. The two 105 m side spans have a prestressed concrete cross-section. The deck is carried by two planes of 20 cables arranged in a semi-fan configuration. Cables are made of a 7 wire, 15.24 mm galvanized strand and the cable sizes vary depending on the location and force in the stay cable. For added corrosion protection, each individual strand is housed in a PE tube that has been filled with grease. An external HDPE pipe encloses the entire bundle of threads.

The two inverted Y pylons are 97.78 meters tall structurally, measured from the top of the footing to the top of the pylon. Concrete reinforced pylons are used. They are composed of composite materials, with the steel anchor chassis located at the top near the cable stay anchors. The steel box that serves as the core is surrounded by an outer shell made of reinforced concrete. The typical pier foundation is built on a 13.3 m × 6 m × 1.5 m spread footing, and the abutment foundation is made up of a 28.7 m × 12.2 m × 1.5 m spread footing. The pylon foundations are made up of rectangular spread footings measuring 50 m × 20 m × 5 m (Bayraktar et al., 2017).

8.1.7.3 Ağın Bridge

Ağın Bridge (Figure 8.52) spans Lake Keban in Elazig Province in eastern Turkey. It links the town of Agin to the city of Elazig. It replaced the historic Karamagara Bridge, which was dismantled in 1974 due to the construction of the Keban Dam. For a period of about 40 years the town was accessible only by ferry boats across the lake. The bridge has a main span of 280 m and back spans of 120 m each, and thus an overall length of 520 m. The two steel pylons reach 55 m above deck level. The main deck is composed of an orthotropic steel box section with a width of 13 m. The deck is supported by friction pendulum bearings for seismic isolation.

Fig. 8.52 Ağın Bridge, Turkey, 2015

Other significant cable-stayed bridges in Asia are listed in Table 8.5. It is observed that while the cable stayed bridge concept was developed in Germany these structures are built extensively in North America. However, the renaissance in its construction and application took place in Asia, which is reflected in the large number of different kinds of cable-stayed bridges with long main spans (400 m-1092 m) in several countries of Asia

8.2 CABLE STAYED BRIDGES IN AFRICA

Africa is the second Continent that applied cable-stayed bridges. The Wadi Kuf Bridge (Figure 8.53) inaugurated in 1971 South of Benghazi, Libya , designed by the internationally acclaimed engineer Riccardo Morandi is considered one of the first concrete cable-stayed bridges in the world. In fact, this bridge held the world record for concrete bridges between 1972 and 1977. It has a main span of 282 m, two 97.5 m side spans and stiff A shaped pylons integrated with V-piers. This bridge is characterized by single post-tensioned concrete forestays , backstays and beams. The beams at each plane are built monolithic with the pylons. The superstructure is a single-cell box girder that varies from 4 m to 7 m at the pylons. The box is 7.4 m wide and forms a 13m deck with its cantilever

Table 8.5 Other Significant Cable-stayed bridges in Asia

Bridge Name	*Country*	*Year*	*Type*	*Total Length of Project (m)*	*Main Span (m)*	*Number of Pylons*	*Number of Cable Planes*
Wadi Leban Bridge	Saudi Arabia	1997	Prestressed concrete box girder	763	405	2	3
Ligang Bridge	Taiwan	2000	Hybrid (steel in main deck and concrete side spans)	510	330	1	1
Aung Zeya	Myanmar	2000	Steel truss with concrete deck	1,154	300	2	2
Maha Bandula	Myanmar	2000	Steel truss with concrete deck	1,110	130	2	2
Wadi Abdoun Bridge	Jordan	2000	Prestressed concrete open cross-section	417	132	2	2
Macapagal Bridge	Philippines	2007	Composite	908	360	1	2
Ettchad Melli Bridge	Iran	2007	Prestressed concrete	113	60	1	2
Keppel Bay Bridge	Singapore	2008	Composite	250	150	1	1
New Taipei Bridge	Taiwan	2010	Composite	1,100	200	1	2
Lali Bridge	Iran	2010	Composite	455	255	2	2
Neak Loeung Bridge	Cambodia	2015	Prestressed concrete	640	340	2	2
Al Emarah City Bridge	Iraq	2017	Composite	168	72	1	2
Basra Bridge	Iraq	2018	Composite	295	150	2	1
Cebu–Cordova Link	Philippines	2020	Prestressed concrete box girder	8,900	390	2	1
Brunei Channel Bridge	Brunei	2020	Prestressed concrete open cross-section	290	145	1	2
Brunei Eastern Channel Bridge	Brunei	2020	Prestressed concrete open cross-section	520	260	2	2

Fig. 8.53 Wadi Kuf Bridge, Libya, 1971

flanges. There is a 55 m long simply supported drop-in center portion of the main span comprising three double-T beams (Podolny, 1973).

The Kwanza River Bridge was built in Angola in 1975 with a main span of 260 m. and It has undergone substantial rehabilitation recently. It is also considered one of the first cable-stayed bridges worldwide. A list of significant cable-stayed bridges in Africa is displayed in Table 8.6. The Suez Canal Bridge and the Mohammed VI Bridgeare are discussed in detail herein.

8.2.1 Suez Canal Bridge

Qantara is a city about 50 kilometers south of the Mediterranean in Egypt where the Suez Canal Bridge (Figure 8.54) is located. As part of the global Northern Coastal Highway, which links the nations of Africa, Asia, and Europe that border the Mediterranean Sea, it is a crucial link. The Suez Canal Bridge Project spans 9 km in total, with approximately 3.9 km of bridges. The portion that crosses the Suez Canal is a cable-stayed bridge with two lanes in each direction and a maximum vertical grade of 3.3% to allow for efficient traffic flow. The main bridge (Figure 8.55) is a steel cable-stayed bridge with a total length of 730 m, central span of 404 m, and two side spans of 163 m (Chodai, 1997). The bridge provides a horizontal navigational clearance of 384 m and a very high vertical clearance of 70 m.

Table 8.6 Significant Cable-stayed bridges in Africa

Bridge Name	*Country*	*Year*	*Type*	*Total Length of Project (m)*	*Main Span (m)*	*Number of pylons*	*Number of Cable Planes*
Wadi Kuf Bridge	Libya	1971	Prestressed concrete box girder	475	280	2	2
Kwanza River Bridge	Angola	1975	Composite	420	260	2	2
Wadi Dib Bridge	Algeria	1998	Steel	502	280	2	2
Suez Canal Bridge	Egypt	2001	Steel orthotropic box girder	3,900	404	2	2
Aswan Bridge	Egypt	2002	Prestressed concrete box girder	977	250	2	1
El Mek Nimir Bridge	Sudan	2007	Composite	642	80	2	2
Lekki-Ikoyi Link Bridge	Nigeria	2013	Prestressed concrete box girder	600	125	1	2
Saleh Bey Viaduct	Algeria	2014	Prestressed concrete box girder	756	259	2	1
Mohammed VI Bridge	Morocco	2016	Prestressed concrete open cross-section	954	376	2	2
Sidi Maarouf Bridge	Morocco	2018	Composite steel box girder	285	136	1	1
Source of the Nile Bridge	Uganda	2018	Prestressed concrete box girder	525	290	1	1
Rod El Farag (Tahya Misr) Bridge	Egypt	2019	Hybrid	540	300	2	4

With four traffic lanes, 0.8-meter sidewalks, and a 1.2-meter median strip, the superstructure is a single-cell trapezoidal steel box girder that is 20.8 meters wide. The depth is roughly 2.6 meters in the center and 1.2 meters at the edges. The upper and lower decks are composed of 16 mm and 11 mm steel plates, respectively, and are orthotropic plates that are strengthened by longitudinal ribs.

Fig. 8.54 Suez Canal Bridge, Egypt, 2001

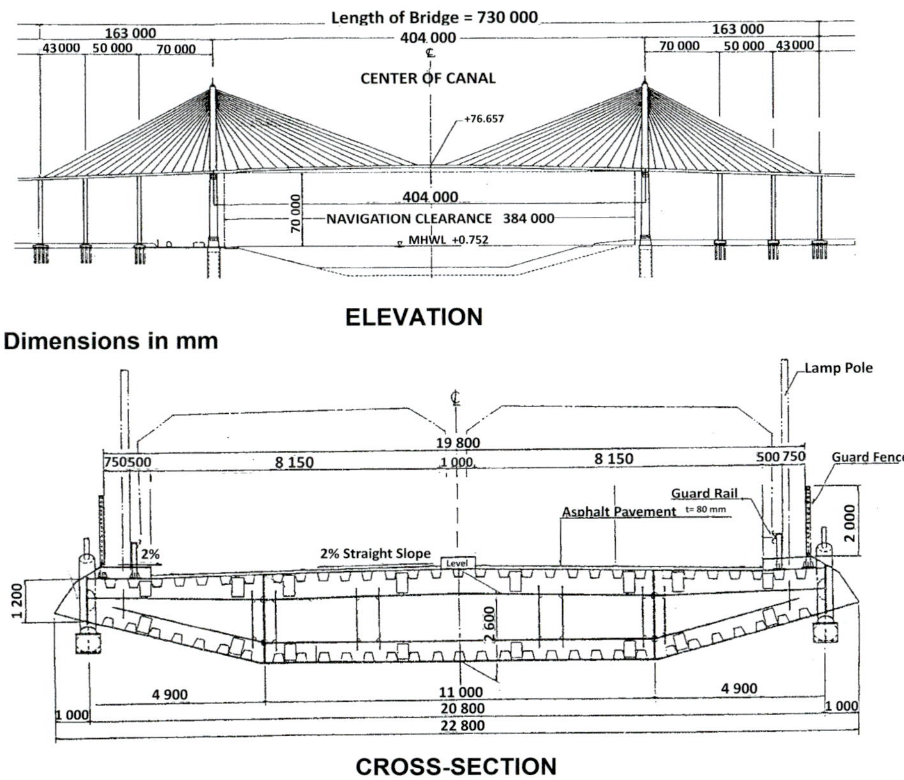

Fig. 8.55 Elevation of Suez Canal Bridge (Courtesy of MTC, Egypt)

As illustrated in Figure 8.55, to reinforce the section, solid steel cross diaphragms with a 10 mm thickness are added at the locations of the cables and midway between them. The cable-stayed bridge has two planes of cables located at the edges of the superstructure. The bridge has a total of 128 cable stays distributed in a semi-fan arrangement. The typical anchorage spacing along the superstructure is 10.50 m. A closer anchorage spacing (3.50 m) is used for the cable stays over the anchorage piers. The stays use epoxy-coated 15 mm - 7 wire strands with 1800 MPa ultimate strength. The number of strands per stay varies from 15 to 38 depending on the location of the stay.

The H-shaped R.C. concrete pylons are roughly 160 meters high. The pylon cross-section is a variable box section with base dimensions of 7.6 m × 7.8 m × 0.7 m. It gradually reduces with an average slope of 1:35 to reach top level dimensions of 2.5 m × 4.5 m × 0.5 m. (see Figure 8.56).

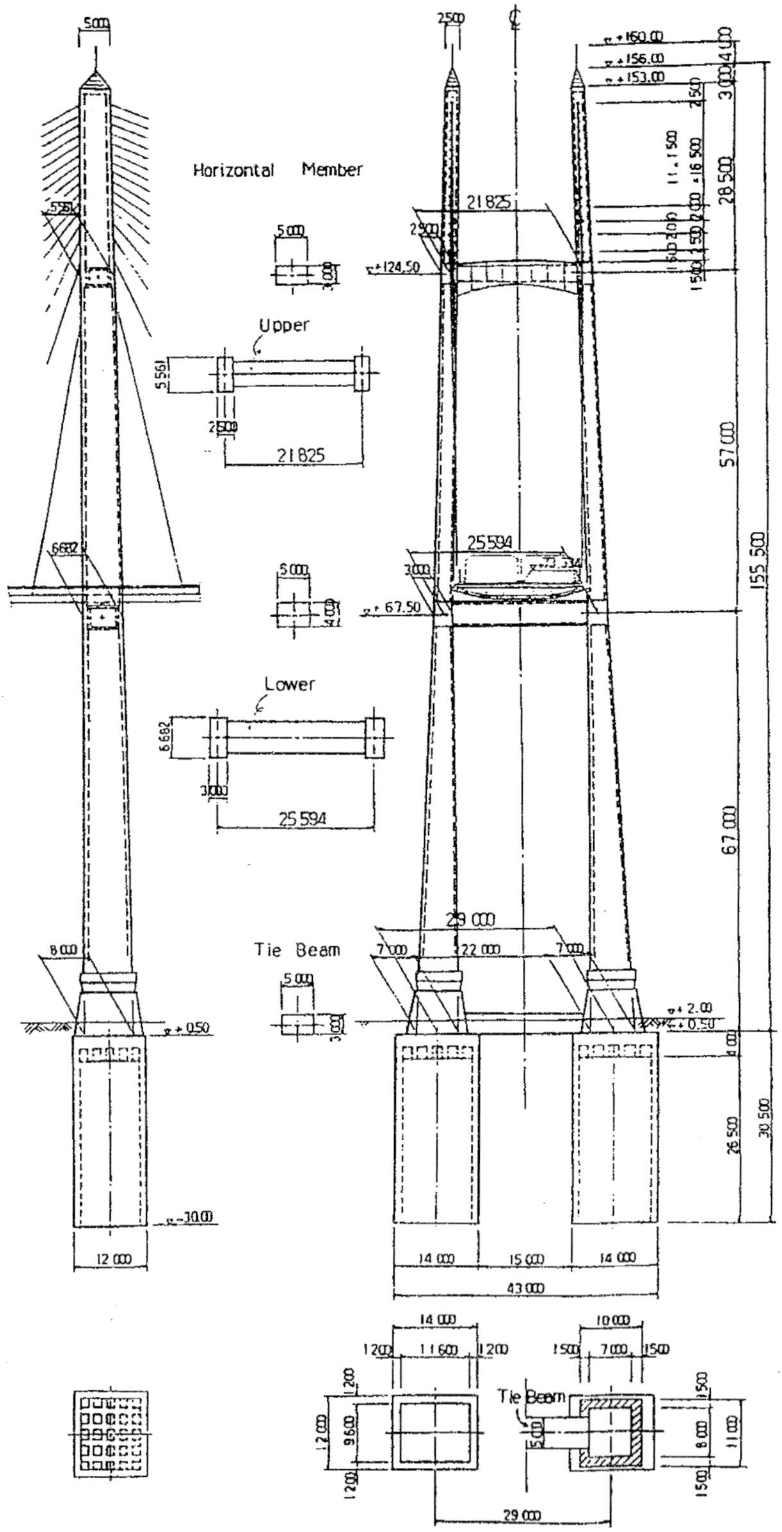

Fig. 8.56 Pylon of Suez Canal Bridge (Courtesy, MTC, Egypt)

The Mohammed VI Bridge

The Mohammed VI Bridge (Figure 8.57) was inaugurated in 2016 and spans across a section of desert connecting Rabat to the city of Salé. The 950 m long bridge is considered the longest concrete open cross-section cable-stayed bridge in Africa and the second longest African cable-stayed bridge after the Suez Canal Bridge. The bridge includes six lanes of vehicle traffic and spans between two 200 m high arched pylons that symbolize the new gates to the cities of Rabat and Salé. The bridge provides 100 m vertical clearance above the Bou Regreg River.

Fig. 8.57 The Mohammed VI Bridge, Morocco, 2016

The longitudinal axis is oriented North-South. With a total length of 951.66 m between abutment axes, the structure is broken down into two parts: a 742 m cable-stayed viaduct with two pylons, including a 376 m-long main span and two 183m-long side spans from the abutment C0 to the abutment P3 as shown in Figure 8.58; and a viaduct consisting of five spans, from pier P3 to abutment C8 and made of 208.86 m long prestressed concrete girders. The deck is simply supported on the abutments longitudinally and restrained transversely by seismic stops. The superstructure, as shown in Figure 8.59, is 30.4 m out-to-out and 25.1 m between the inside faces of the curbs. It consists of a concrete slab 25 cm thick supported by two longitudinal edge girders and transverse floor beams. The parallel edge girders are 2.2 m wide by 2.0 m deep. The transverse floor beams are made of steel and spaced at 4 m. The deck is longitudinally and transversely prestressed.

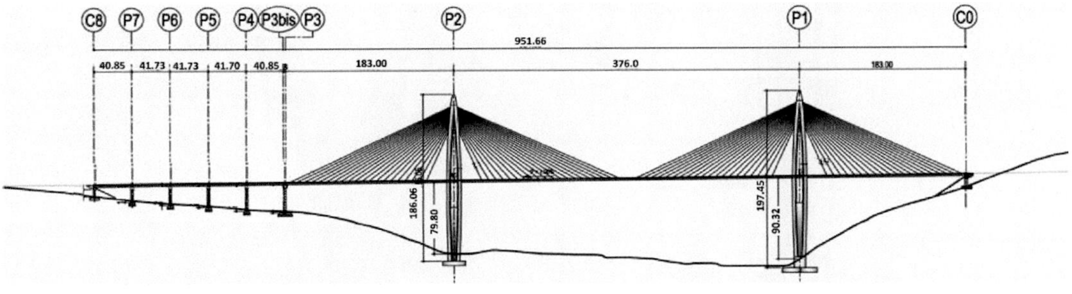

Fig. 8.58 Elevation of the Mohammed VI Bridge (Boujemaoui et al., 2011)

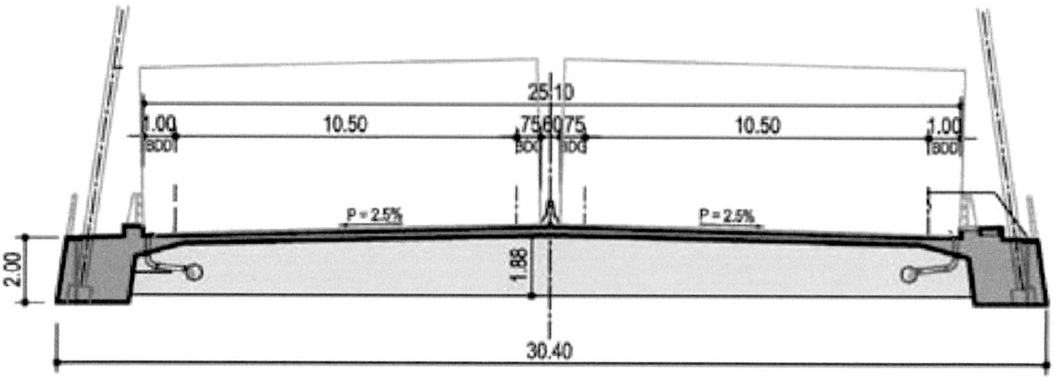

Fig. 8.59 Elevation of the Mohammed VI Bridge (Boujemaoui et al., 2011)

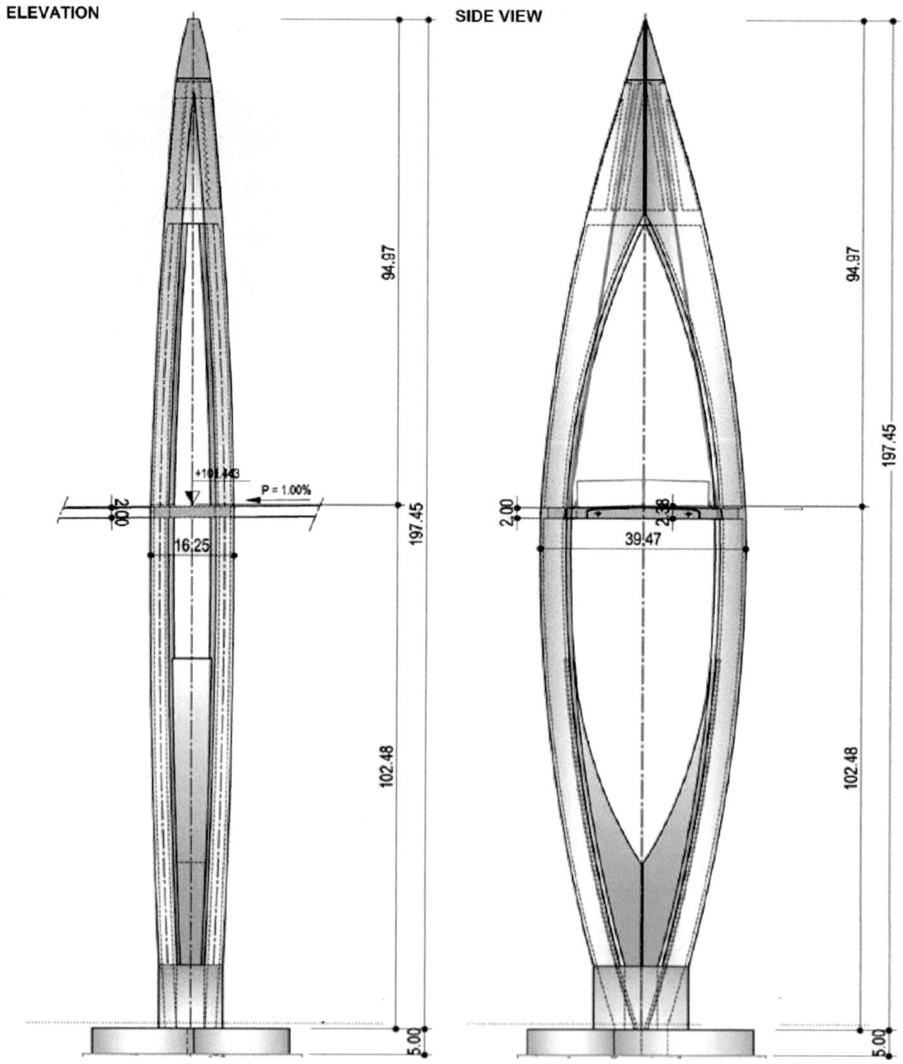

Fig. 8.60 Pylon of the Mohammed VI Bridge (Boujemaoui et al., 2011)

The stay cables are arranged in a semi-fan two planes configuration. The superstructure is supported via 40 pairs of parallel multi-strand stay cables spaced every 8 m at their anchorage in the deck. The stays are made of 15.7 mm diameter 7 wires parallel wire strands, galvanized, sheathed, and raincoated. Damping of the stays is intended to obtain a logarithmic decrement of 4%. With a wind of 15 m/s, the vibration amplitude should not exceed 10 cm.

Each of the reinforced concrete pylons constitutes four concrete legs (box) joining head and foot as shown in Figure 8.60. In the upper part of the anchoring zone stay cables, the pylon is made up of two thick concrete boxes, and thin veils to ensure the architectural continuity and overall appearance. The junction between the four legs and the anchor boxes is made through the floor. The pylon-deck junction is secured by installing spacers in prestressed concrete connecting the four legs and thus resuming the thrusts related to the curvature of the pylon. In the lower part, the four legs of the pylon are connected via reinforced concrete walls.

References

Bayraktar, A., Turker, T., Tadla, J., Kursun, A. and Erdis, A. et al., Static and dynamic field load testing of the long span Nissibi cable-stayed bridge, *Soil Dynamics and Earthquake Engineering*, Volume 94, pp 135–157.

Boujemaoui, M., Eloualidi, M., Vadon, H., Belanger, E., Latallerie, G. et al. Design of the Guarded Viaduct on Oued Bouregreg (in French), *TRAVAUX* Number 879, pp 106–114, 2011.

Chodai Co., LTD, *The Detailed Design Study on the Project for Construction of the Suez Canal Bridge in Egypt*, Report for the Ministry of Transport and Communications, The Government of the Arab Republic of Egypt, 1997.

Consortium of Chinese Contractors, *Indonesian Suramadu Bridge Detailed Design of Main Bridge,* 2005.

Gupta, S.P., The Second Hooghly River Bridge, Calcutta, *Structural Engineering International*, Volume 1, No. 3, pp 7–9, 1991.

Ha, S., Kim, J., Hwang, C., Shin, H., Kim, M. et al., Nonlinear Analysis of Incheon Bridge Considering Time-Dependent Behavior of Concrete Pylon (in Korean), *Journal of the Korean Society of Computing and Structural Engineering*, Vol. 25, No. 2, 2012.

Jeong, S. and Kim, J., The Immersed Tunnel and Bridges of Busan–Geoje Fixed Link, *Structural Engineering International*, Volume 22, Issue 1, pp 20–25, 2012.

Jomvinya, K. and Vicat, E., Kanchanapisek Bridge Over the Chao Phraya River, Thailand, *Structural Engineering International*, Volume 19, No. 1, pp 58–62, 2009.

Khamdee, S., Farquhar, D. and Jordet, E., Design of Industrial Ring Road Project, Bangkok, Thailand, *Structural Engineering International*, Volume 19, No. 1, pp 48–52, 2009.

Kim, W. and Cho, K., Conceptual Design of the Seo-hae Bridge, *Structural Engineering International*, Volume 11, No. 1, pp 18–21, 2001.

Klein, P. and Yamout, M., Cable-Stayed Arch Bridge, Putrajaya, Kuala Lumpur, Malaysia, *Structural Engineering International*, Volume 13, No. 3, pp 196–199, 2003.

Lee, Y., Kang, J., Bae, S., Yun, Y., Lho, B et al., Design of Vam Cong Cable Stayed Bridge in Vietnam, Journal of Korean Recycled Construction Resources (in Korean), Volume 1, No., 2, pp 121–127, 2013.

Man, H., Bo, L., Dong, P. and Man, C., Design of Main Bridge of Second Penang Bridge in Malaysia, *International Journal of Transportation Engineering and Technology,* Volume 4, No. 2, pp 35–42, 2018.

Meyer, M., Sharma, D. and Paradkar, D., Rajiv Gandhi Sealink–Mumbai, India, *Structural Engineering International*, Volume 21, No. 1, pp 94–98, 2011.

Moir, G., Edmonds, C., Walser, P., and Romberg, M., Construction Engineering of Phu My Cable-Stayed Bridge, Vietnam, *Structural Engineering International*, Volume 20, No. 3, pp 331–337, 2010.

Nakamura, T., Tsuchida, K., Ohno, H. and Nagamoto, N., Bai Chay Bridge, Vietnam, *Structural Engineering International*, Volume 17, No. 3, pp 210–213, 2007.

Nanakorn, C., Chitnuyanondh, L., Limsuwan, E., Rama VIII Bridge, Bangkok, Thailand, *Structural Engineering International*, Volume 12, No. 3, pp 158–160, 2002.

Park, D. and Lee, B., Second Dolsan Bridge: A Cable-Stayed Bridge with a Concrete Edge Girder, *Structural Engineering International*, Volume 22, No. 1, pp 44–48, 2012.

Podolny, W., CABLE-Stayed Bridges of Prestressed Concrete, PCI Journal, Volume 18, No. 1 pp 68–79, 1973.

Schlaich, M., Subbarao, H. and Korian, J., A Signature Cable-Stayed Bridge in India—The Yamuna Bridge at Wazirabad in New Delhi, *Structural Engineering International*, Volume 23, No. 1, pp 14–17, 2013.

Swit, G., Krampikowska, A., Chinh, L. and Adamczak, A., Nhat Tan Bridge-The Biggest Cable-Stayed Bridge in Vietnam, *Procedia Engineering*, volume 161, pp 666–673.

Tanis, J., Fremont, G. and Vivien Noel, The cable-stayed bridge of Muar in Malaysia (in French), *TRAVAUX* Number 793, pp 71–75, 2003.

CHAPTER 9

Experience of Latin America with Cable-stayed Bridges

9.1 CABLE STAYED BRIDGES IN LATIN AMERICA

In this section some of the signature bridges from South America in addition to Mexico, Central America, and the Caribbean Islands are discussed. Since covering all bridges in each country may require another book, significant bridges of Latin America will only be outlined.

9.1.1 Mexico

9.1.1.1 The Tampico Bridge

The Tampico Bridge I (Figure 9.1) is one of the first cable-stayed bridges built in Latin America. Completed in 1988, this bridge crosses the Panuco River in Tampico City (Gulf of Mexico). The total length of the entire crossing including approach viaducts is 1543 m. The cable-stayed bridge has a main span of 360 m, two side spans of 70 m and an additional two spans at each side for load balancing each is 70 m (Figure 9.2a). The bridge has an 18 m deck which accommodates four lanes of traffic.

Fig. 9.1 The Tampico Bridge, Mexico, 1988

The central 293.50 m section of this hybrid cable-stayed bridge is supported by an orthotopic steel girder, while the remaining main span and lateral short spans are supported by prestressed concrete girders (Figure 9.2.b). The stays consist of 44 stay cables arranged in a semi-fan axial suspension configuration. The stays consist of 30 to 60 galvanized 7-wire, 15 mm strands that are contained within a HDPE duct that has been injected with petroleum wax. Prior to construction, the stays were prefabricated on the deck. Two pylons in the shape of an inverted Y rise 123.5 meters above the foundation of the bridge.

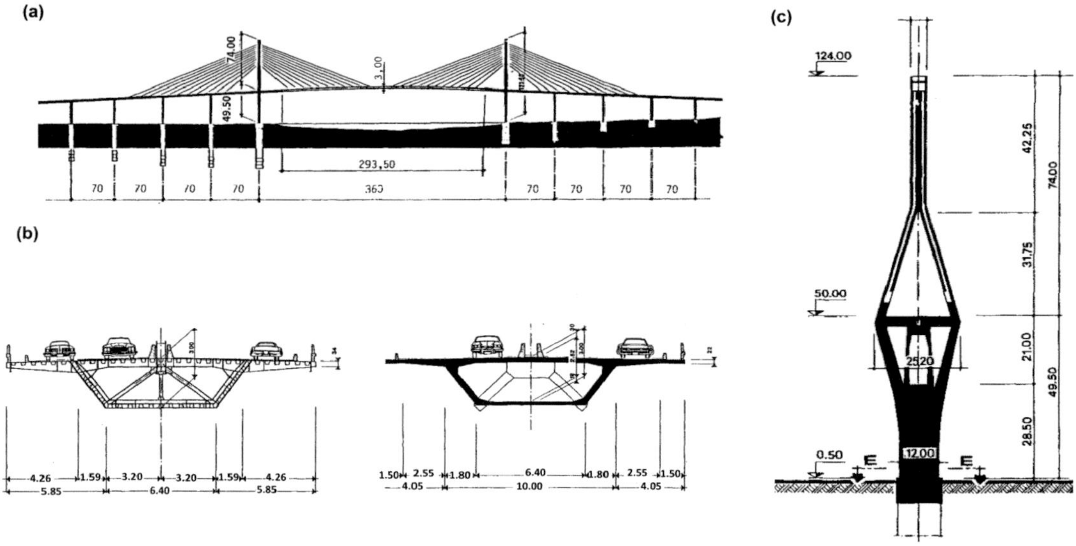

Fig. 9.2 The Tampico Bridge: (a) elevation; (b) cross-sections; and (c) pylon (Zambrano et al., 1990)

9.1.1.2 Mezcala Bridge

The Mezcala Bridge (Figure 9.3) was put into service in 1993. It is located at 221 kilometers off the Autopista del Sol, which connects the city of Cuernavaca with Acapulco. The bridge has a total length of 911 m made up of 6 spans ranging from 39 m to 311 m (Figure 9.4). The superstructure is a composite cross-section comprising two steel I plate girders at the two edges and connected by

Fig. 9.3 The Mezcala Bridge, Mexico, 1993

floor beams that support the reinforced concrete slab. The bridge has 140 stay cables distributed in 12 semi-harps. Due to site conditions. the extreme pylons, which are stabilized by classical back stays are shorter than the central one. The composite deck passes freely through the two legs of each pylon and is simply supported on the pier, which constitutes the lower part of the pylon.

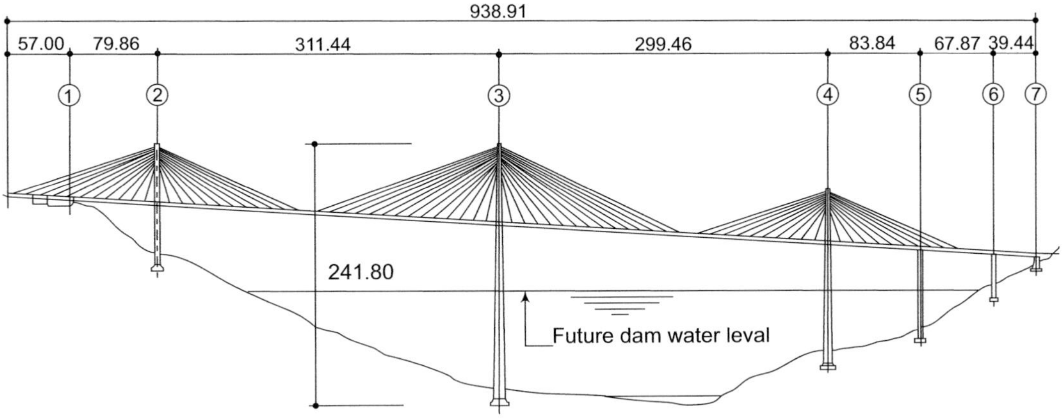

Fig. 9.4 Elevation of the Mezcala Bridge, Mexico, 1993

9.1.1.3 The Puente de la Unidad Bridge

During the last two decades several cable-stayed bridges were constructed in Mexico.

The Puente de la Unidad Bridge was opened to traffic in 2003. It crosses the Santa Catarina River and connects the cities of Monterey and San Pedro Garza García in the Mexican state of Nuevo León. It has a main span of 185 m and a total length of 304 m. Its design as shown in Figure 9.5 has adopted the Calatrava design for the Alamillo Bridge in Spain. The Matute Remus Bridge with a total length of 930 m was opened to traffic eight years later in 2011 in Guadalajara.

Fig. 9.5 The Puente de la Unidad Bridge, Mexico, 2003

9.1.1.4 The Baluarte Bridge

The Baluarte Bridge (Figure 9.6) is one of the most significant cable-stayed bridges that were opened in Mexico during the last ten years. It is located between the municipalities of Concordia in Sinaloa and Pueblo Nuevo in Durango, along the Durango–Mazatlán highway, Mexico 40D and crosses a deep canyon with a depth of 390 m. It was opened to traffic in 2012. The total length of the bridge is 1.124 m, with a main span of 520 m (the longest ever built in Mexico) and two lateral spans of 250 and 354 m.

Fig. 9.6 The Baluarte Bridge, Mexico, 2012

The main span has a composite section with an approximate deck width of 19.76 m, and the side spans have prestressed concrete box sections with 22 m of deck width, approximately (Figure 9.7). Cables are regularly spread along 884 m of the deck. A total of 76 cables are used with a semi fan layout in two planes. Cables are anchored to the pylons and the deck. The bridge is supported by 8 reinforced concrete frame piers, two pylons of the diamond type, and two abutments at the ends. The height of the piers ranges from 40 to 140 m, approximately. The tallest pylon has a height of 165 m, while the other one has a height of 147 m.

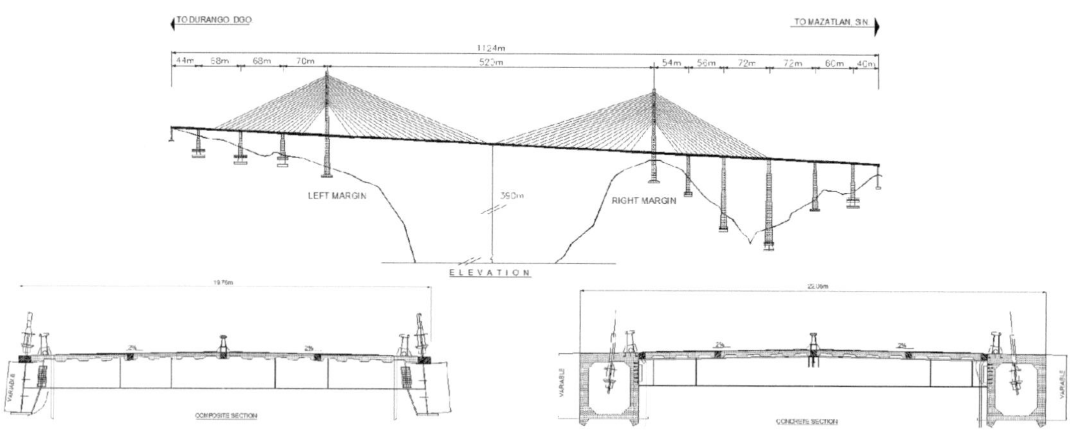

Fig. 9.7 Elevation and cross-sections of the Baluarte Bridge (Gomez et al., 2010)

9.1.1.5 Vidalta Bridge

A year later the Vidalta Bridge was opened in Mexico City. The bridge (Figure 9.8) crosses a valley of great dimensions in Mexico City to give access to a complex of houses located at one of its edges. The valley is a natural space within the city with a vegetation of trees and shrubs, which must be protected as much as possible. Therefore, the first condition that the project had to meet was to alter it as little as possible. Only one support could be placed on it as close as possible to the buildings. The position of this support has given rise to a bridge with two spans of 60 m and 180 m. However, the pronounced difference between the two spans was the main problem that has been raised in this project, due to the difficulty in balancing the forces that the stay cables generate in the pylon.

Two solutions that complement each other were implemented in the design. First, an inclined pylon which divides the deck into two stretches of 78.5 m and 161.5 m was used (Figure 9.9). Second, concrete is used to build the side span, which thus becomes a counterweight, whereas the main span is lightweight since it is made of steel. This way load balancing was achieved.

Fig. 9.8 The Puente Vidalta Bridge, Mexico, 2013 (Courtesy, CFCSL)

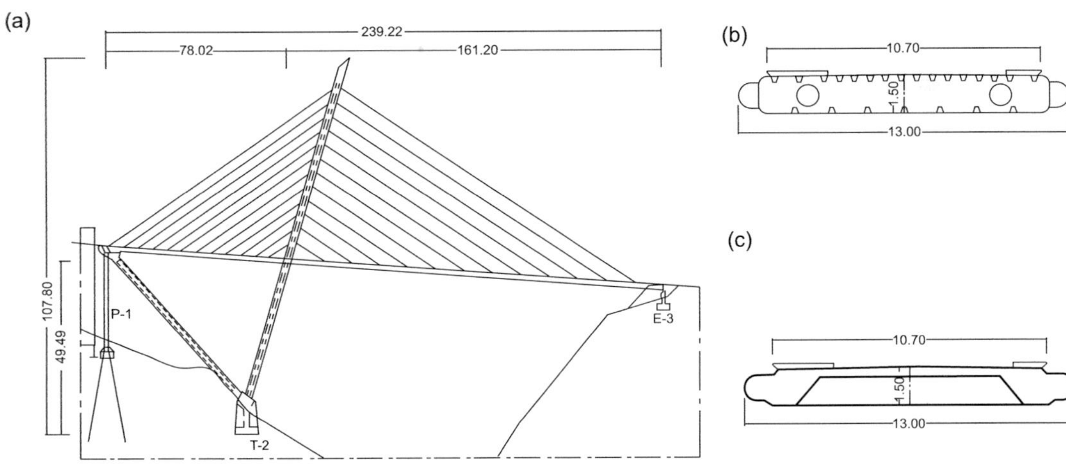

Fig. 9.9 The Puente Vidalta Bridge: (a) elevation; (b) steel cross-section main span; and (c) concrete cross-section side span (Courtesy, CFCSL)

9.1.2 Caribbean

9.1.2.1 Jesus Izcoa Moure Bridge

The Jesus Izcoa Moure Bridge (Figure 9.10) is the first cable-stayed bridge built in Puerto Rico. It was opened to traffic in 2008 and connects the cities of Toa Alta and Naranjito, in Puerto Rico by the Puerto Rico Highway 5. The bridge carries four lanes with a 29.2 m wide deck. The whole crossing (including approach spans) is 703 meters long. The total cable stayed bridge length is 315 meters with a main span of 157.5 m over La Plata River. The cable stayed portion is supported by two hollow diamond-shaped pylons and 96 stays.

9.1.2.2 Mauricio Báez Bridge

The Mauricio Báez Bridge (Figure 9.11) was opened to traffic in January 2007 in the Dominican Republic. It is located near San Pedro de Macorís, in the east of the country, about 40 km east of

Fig. 9.10 The Jesus Izcoa Moure Bridge, Puerto Rico, 2008

Fig. 9.11 The Mauricio Báez Bridge, Dominican, 2007

the capital city of Santo Domingo. The bridge carries six lanes and has a total length of 613.8 m with a main span of 390 m. The superstructure is a composite section, and the side spans are all reinforced concrete structures. The pylons are diamond shaped with inclined legs made of concrete while the top shaft is made of steel. Since the bridge can be exposed to tropical hurricanes the deck is streamlined by aluminum aerodynamic fairings to get the required stability. The bridge is provided with seismic dampers that link the deck to the substructures to withstand the earthquake action.

9.1.3 Panama

9.1.3.1 Centennial Bridge (Puente Centenario)

Panama has three major crossings on the Panama Canal. The Centennial Bridge (Figure 9.12) is the second bridge on the Panama Canal. It was built to relieve traffic on the Bridge of the Americas, the first Panama Canal bridge, and to replace it as the carrier of the Pan-American Highway. The bridge which was opened in 2004 is a single plane cable-stayed bridge with a total length of 1052 m and spans from 60 – 60 – 66 – 200 – 420 – 200 – 46 = 1052 m (Figure 9.13); the ratio of the side opening to central opening is therefore 0.48. The 34.1 m wide bridge deck made from prestressed concrete is 80 m above the water level and consists of a trapezoidal box girder and wide cantilevers. The height of the pylons (top cable) above the bridge deck is 94 m; this corresponds to 0.22 times the main opening. The fan-shaped arrangement of cables has a constant spacing of 6.0 m on the superstructure and 1.5 m at the pylon. The superstructure and pylons are monolithically tied with each other. Therefore, the bearings on all piers can only move in the longitudinal direction.

Fig. 9.12 Centennial Bridge, Panama, 2004

The superstructure is made of a trapezoidal box girder which is 4.5 m high. The bottom width is 8 m and the top portion is 16.2 m wide and has 8.95 m cantilevers on both sides. The girder is post-tensioned in both the longitudinal and transverse directions (Figure 9.14). The roadway slab is supported by the main girders and the transverse diaphragms, which are spaced every 6 m. Interior stiffening steel struts were provided at the locations of the cable anchorage to the deck to transfer the vertical components of cable loads to the web.

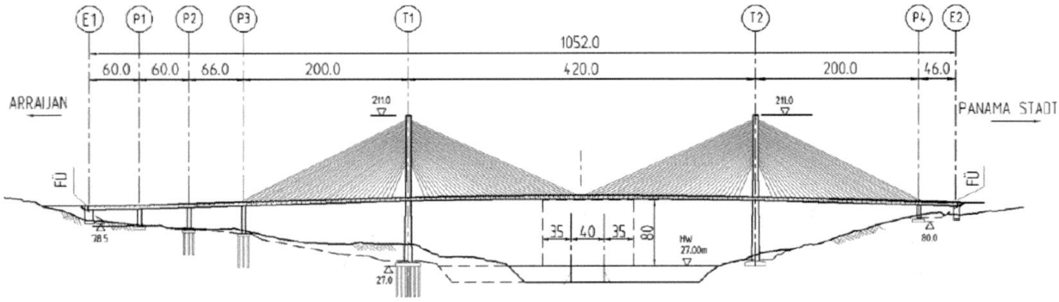

Fig. 9.13 Elevation of the Centennial Bridge (Saul et al., 2005)

The pylons are made of hollow sections of reinforced concrete as shown in Figure 9.15. The upper segment is provided with a steel box which is composite with the section for cable anchorage. The cables consist of 43 to 80 parallel monostrands 15 mm in diameter with a tensile strength of 1820 N/mm^2 after galvanizing. The monostrands are protected against corrosion as they are galvanized, protected by a 1 mm thick PE pipe, and a grease filling. Beyond that, the entire cable is covered with a PE tube.

9.1.3.2 The Atlantic Bridge

The Atlantic Bridge (Figure 9.16) or the Third Panama Crossing crosses the Panama Strait on the eastern coast of Panama. Completed in 2019, it is the third bridge over the canal. It features a cable-stayed bridge with a main span of 530 m and two side spans of 230 m (Figure 9.17). The east and west approaches are 1074 m and 756 m long, respectively. The bridge provides 365.5 m and 75 m horizontal and vertical navigation clearances.

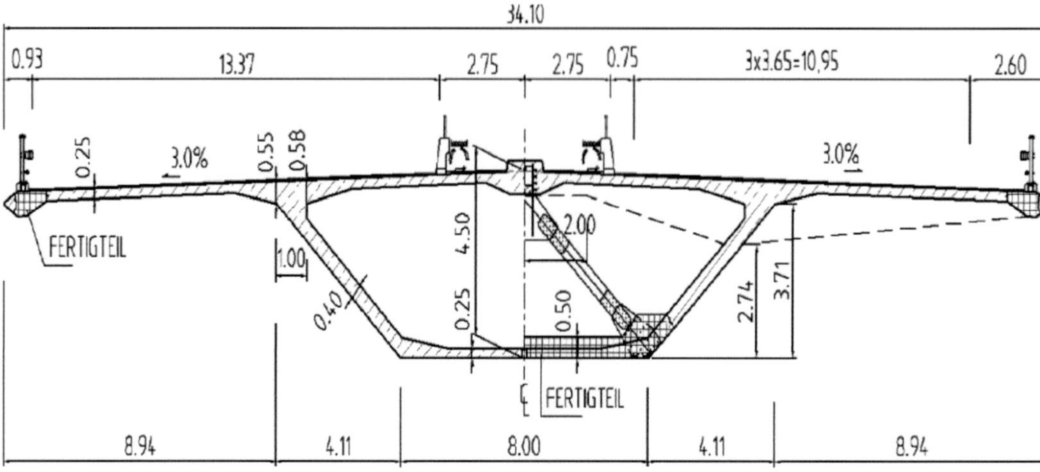

Fig. 9.14 Cross-section of the Centennial Bridge (Saul et al., 2005)

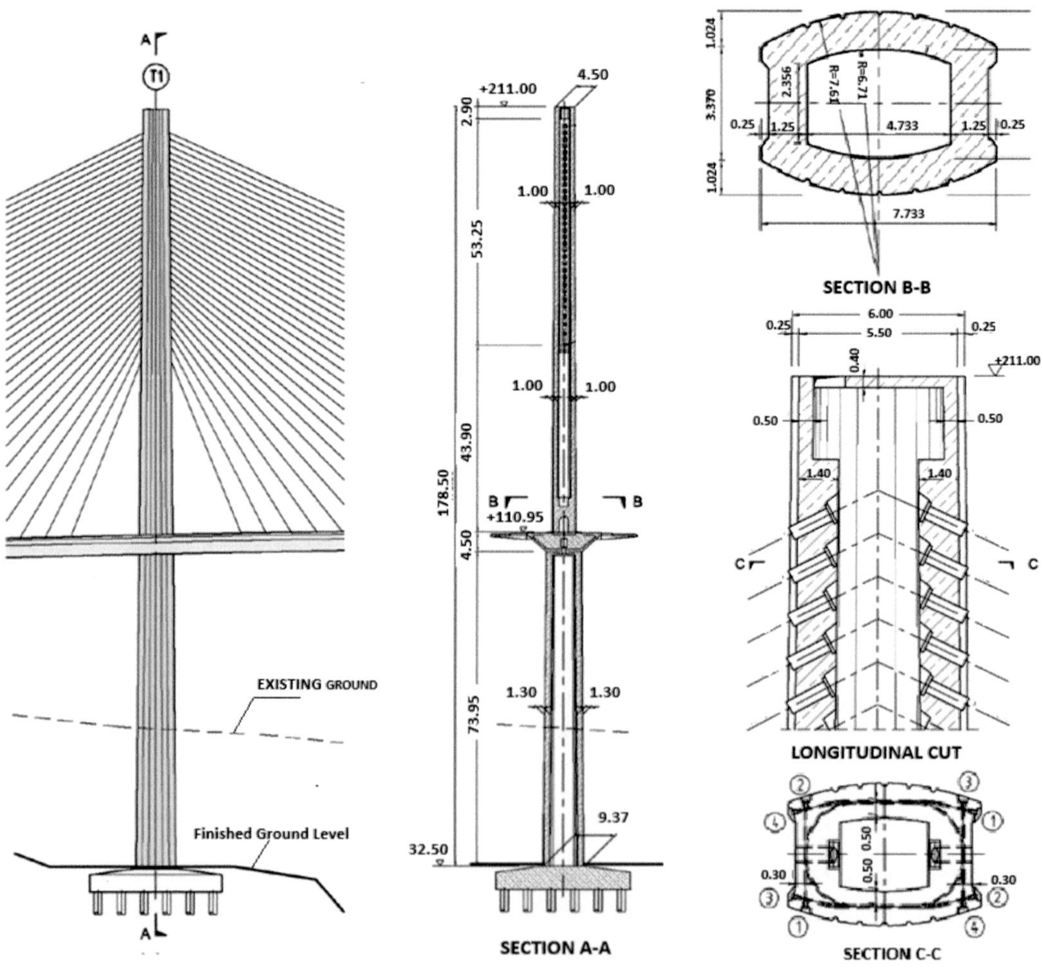

Fig. 9.15 Pylon of the Centennial Bridge (Saul et al., 2005)

Fig. 9.16 The Atlantic Bridge, Panama, 2019

The superstructure as shown in Figure 9.18 is an open cross-section that comprises two edge trapezoidal box girders with wall thicknesses ranging from 0.23 m to 0.28 m. The height of the deck is 2.6 m at the edges and 2.834 m at the center. The two edge girders are connected to each other by cross diaphragms that are spaced every 8.1 m. The pylons are diamond shaped with a hollow reinforced concrete section as shown in Figure 9.19.

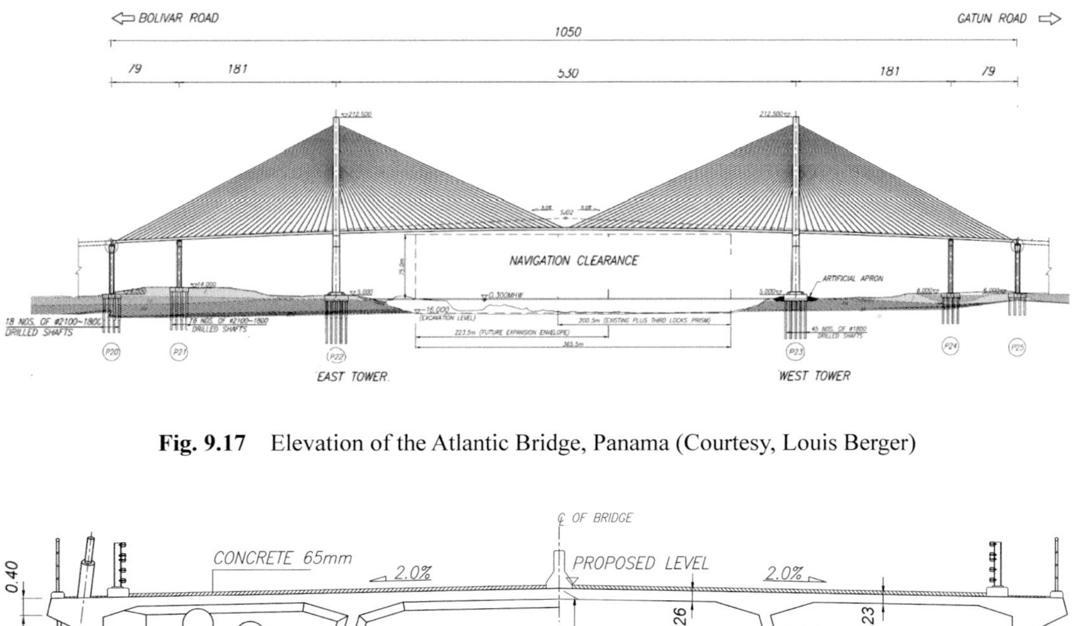

Fig. 9.17 Elevation of the Atlantic Bridge, Panama (Courtesy, Louis Berger)

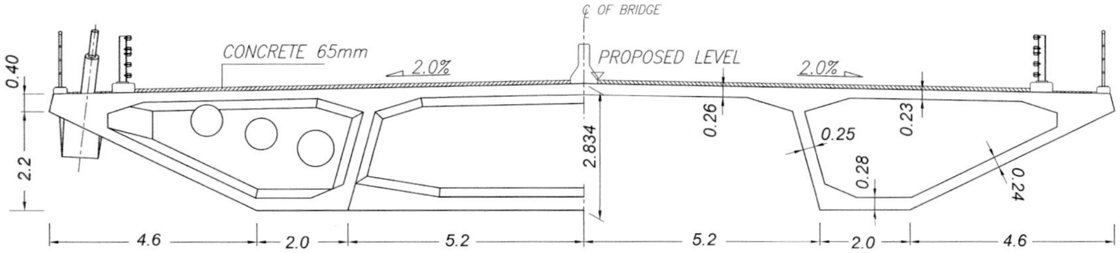

Fig. 9.18 Cross-section of the Atlantic Bridge, Panama (Courtesy, Louis Berger)

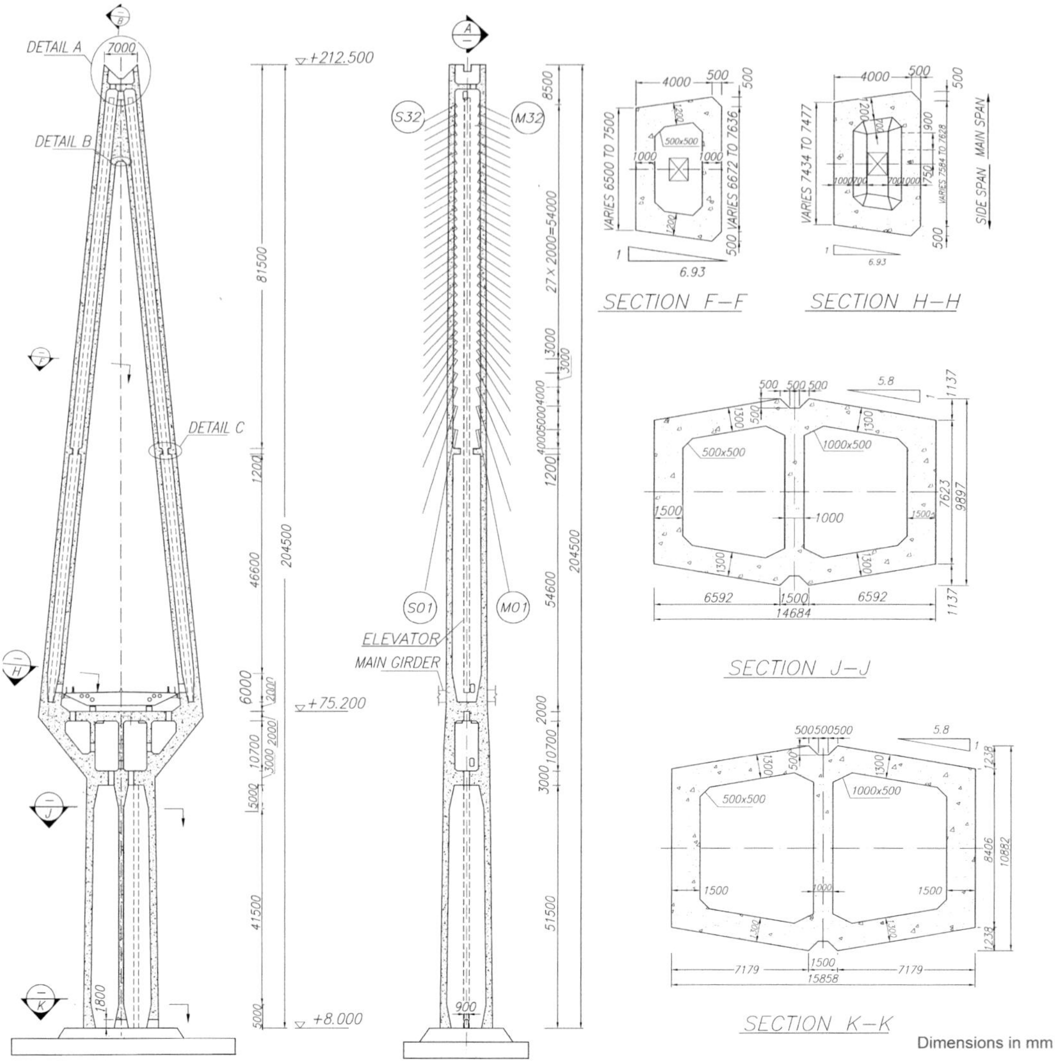

Fig. 9.19 Pylons of the Atlantic Bridge, Panama (Courtesy, Louis Berger)

The deck is linked to the pylon through vertical and lateral bearings. The vertical bearings are multi-rotational spherical steel bearings that permit limited translations in both the transverse and longitudinal directions. The lateral bearings are free sliding pot bearings that allow translations in the longitudinal direction. There are two viscous dampers attached between each pylon and the deck to dissipate seismic energy in the longitudinal direction. Restrain blocks are installed to prevent motion in the transverse direction. All other piers are provided with spherical bearings, two at each pier. Each plan of stay cables comprises 128 stays, i.e., 256 stay cables for the entire bridge. Cable diameters range from 58 mm to 99 mm depending on their location. Each cable includes a few parallel strands, each based on seven high-strength galvanized wires with a diameter of 5 mm. Petroleum wax and a tightly extruded High-Density Polyethylene (HDPE) sheathing were used to fill spaces between the wires.

9.1.4 Ecuador

Two cable-stayed bridges were opened in Ecuador in 2012, the Rio Napo Bridge and the Rio Aguarico Bridge. The Rio Napo Bridge (Figure 9.20) is located on highway Tiwino in the Coca-Auca district. It has a full length of 740 m and a main span of 312 m. The width of the deck is 16.40 m and accommodates two traffic lanes, pedestrian sidewalks, and a bicycle path. It provides a 15 m vertical clearance above the water level. The pylon is 85 m tall and with two legs connected with transverse ties underneath the deck and at an intermediate level. There are two planes of stay cables arranged in a semi fan configuration. The superstructure is a composite cross-section. The bridge was designed to withstand high intensity earthquakes according to AASHTO LRFD Specifications.

Fig. 9.20 The Rio Napo Bridge, Ecuador, 2012

The Rio Aguarico Bridge also known as Puente Monsenor Gonzalez Lopez (Figure 9.21) has a total length of 440 m and a main span of 270 m. It has a composite superstructure and a reinforced concrete pylon identical to that of the Rio Napo Bridge.

Fig. 9.21 The Rio Aguarico Bridge, Ecuador, 2012

9.1.5 Columbia

9.1.5.1 The César Gaviria Trujillo Viaduct

Two significant Cable-stayed bridges were built in Columbia as part of its viaducts network, the César Gaviria Trujillo Viaduct, and the Provincial Viaduct. The César Gaviria Trujillo Viaduct (Figure 9.22) is a cable-stayed bridge over the Otun River connecting the cities of Pereira and Dosquebradas. It was opened to traffic in 1997. The bridge has a total length of 440 m including a central span of 211 m. It carries four traffic lanes and two sidewalks, giving it a 26 m total deck width.

Fig. 9.22 The César Gaviria Trujillo Viaduct, Columbia, 1997

The superstructure is a composite cross-section with two edge plate girders and tansverse diaphragms spaced every 4.5 m. The two pylons are diamond shaped reinforced concrete structures having a maximum height of 55 m above the Otún River. A total of 76 cables transfer the load of the superstructure to the pylons arranged in a semi fan layout in two planes.

9.1.5.2 The Provincial Viaduct

The Provincial Viaduct (Figure 9.23) is in Bucaramanga, northeastern Colombia. It connects the sectors of Calle 45 with Carrera 9a and the Ciudadela Real de Minas. It was opened to traffic in 2015. The 550.80 m cable-stayed bridge comprises a main span of 292 m and two side spans of 129 m.

Fig. 9.23 Provincial Viaduct, Columbia, 2015

The bridge has a 30 m wide deck that includes three 3.5 m lanes of traffic in each direction, two 2.22 m sidewalks and a 3.5 m median. The main bridge deck is made from reinforced and post-tensioned concrete. The deck features a single central plane of cables and a continuous single cell concrete box girder 2.8 m deep with transverse post-tensioned ribs at the locations of the stay anchors (see Figure 9.24). The pylons are monolithic with the bridge deck whereas the two piers are supported on bearings that are free in the longitudinal direction and restrained in the transverse direction. The two single leg pylons are made of reinforced concrete having a maximum height of 133 m above the foundations. The bridge is supported by a single plane of cables along the centerline of the bridge. The cables are arranged in a semi fan configuration. A total of 80 cables transfer the load of the superstructure to the pylons.

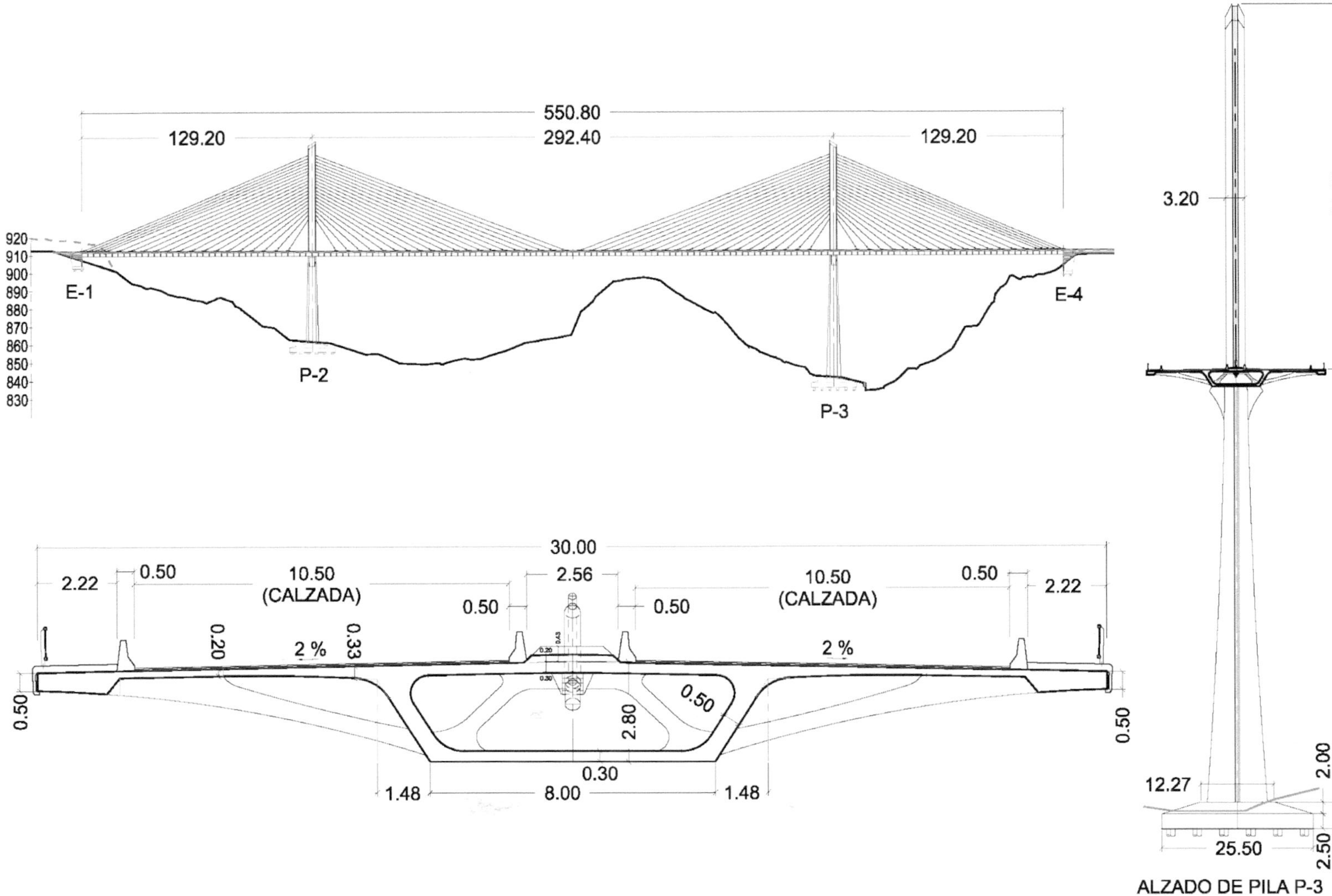

Fig. 9.24 General arrangement of the Provincial Viaduct, Columbia (Iglesias et al., 2012)

9.1.6 Peru

9.1.6.1 The bridge over the Nanay River

The bridge over the Nanay River (Figure 9.25) is a 2,157 m long viaduct that was opened in 2022 to traffic. It connects the districts of Bellavista on the right bank of the Nanay River with the district of Santo Tomas on the left bank of the river. The main span over the Nanay River is a Cable Stayed Bridge with a total length of 423.5 m. The approach structures are 1,184 m long on the right bank of the river, and 319.9 m long on the left bank of the river. The main structure over the navigational channel is a symmetric cable stayed bridge with a main span of 241.50 m and a total length of 423.5 m (see Figure. 9.26a). The bridge carries two lanes of traffic and two pedestrian walkways. The roadway has a total width of 12.00 m and pedestrian walkways, each 1.20 m wide.

Fig. 9.25 The bridge over the Nanay River, Peru, 2022

The superstructure is a steel-concrete composite cross-section with two longitudinal edge girders 15.7 m apart and floor beams spaced at 3.5 m intervals. The typical floor beam separation is 3.50 m. The concrete slab is 200 mm to 300 mm thick over the longitudinal edge girders. The edge girders are plate girders with a constant depth of 1.50m; the typical floor beams are variable depth plate girders, 1.00 m deep along the centerline of the bridge. A central longitudinal stringer is provided to reduce the width of the pre-cast slab panels (see Figure. 9.26b). The cable-stayed bridge has two planes of cables located at the edges of the super-structure.

The bridge has a total of 44 cable stays distributed in a semi-harp arrangement. The stays are connected to anchorage boxes bolted to the webs of the longitudinal girders. The typical anchorage spacing along the superstructure is 10.50 m. A closer anchorage spacing (3.50 m) is used for the cable stays over the anchorage piers. The stays use ASTM A882 epoxy-coated 15 mm – 7 wire strands with an ultimate strength of 1860 MPa. The number of strands per stay varies from 12 to 37 depending on the location of the stay. The back anchor stays and the stays adjacent to the mid-span have the largest number of strands per stay (31 and 27 strands per stay respectively).

The 2 legs pylons (see Figure 9.26c) have a total height of 80 m measured from the bottom of the pile cap. There are two cross beams connecting the two legs of each pylon at 18.50 m and 47.50 m above the pile cap level. The pylon columns are rectangular hollow reinforced concrete elements. Below the deck level the columns have variable dimensions: 3.00 m × 4.00 m over the pile cap, and 2.50 m × 4.00 m below the deck level. Above the deck level the columns have a constant 2.50 m × 4.00 m cross section. The cross beams are hollow reinforced concrete elements with a rectangular 4.00 m × 3.00 m cross-section.

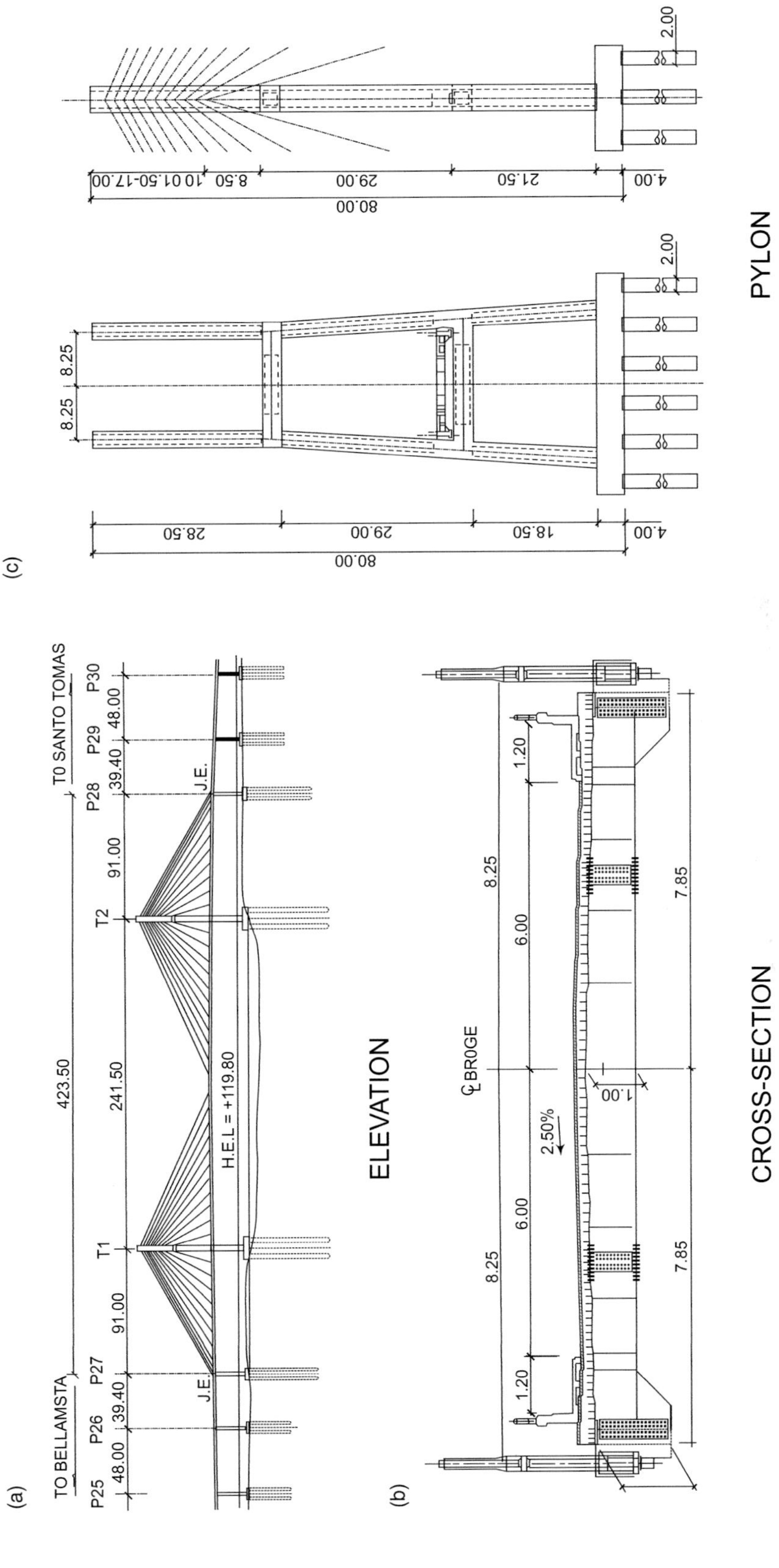

Fig. 9.26 Lopez-General Configuration of the bridge over the Nanay River (Acuna and Jara, 2014)

The super-structure is vertically and transversely restrained at the pylon and free to move in the longitudinal directions under service load conditions. In the event of an extreme seismic occurrence shear keys located at the level of the lower cross beam provide longitudinal restraint and transmit the forces from the super-structure to the pylons.

9.1.7 Venezuela

9.1.7.1 General Rafael Urdaneta Bridge

Venezuela is one of the first countries that implemented the cable-stayed bridge concept as the General Rafael Urdaneta Bridge (Figure 9.27), inaugurated in 1962. This bridge was designed by the renowned engineer Riccardo Morandi. It is located at the Tablazo Strait outlet of Lake Maracaibo, in western Venezuela. The bridge connects Maracaibo with the rest of the country.

Fig. 9.27 The General Rafael Urdaneta Bridge, Venezuela, 1962

The total length of the bridge is about 8.85 km. Its 17.4 m wide deck carries four lanes of traffic with a 1.22 m central median and two 0.90 m sidewalks. It includes five 236 m main cable-supported spans. Each span consists of 2 95 m cast-in-place decks at both ends and a 46.6 m suspended prefabricated deck.

In order to prevent damage from uneven foundation settlement or seismic forces, statically determinate central spans were required. As a result, the main span has a suspended center section that is simply supported and divided into cantilever sections. While the cable stays are supported on two A frames with a portal member at the top, the cantilever span is supported on X frames. The X and A frames are not connected in any way (see Figure 9.28).

The spans are supported by inclined ropes suspended from the top of 92.5 m high four-legged piers of the two inclined A-frames linked at the top by a transverse girder as shown in Figure 9.28. The three-cell box girder that makes up the continuous cantilever girder is 5 m deep and 14.22 m wide. The horizontal component of the cable force induces an axial prestress force into the girder. Therefore, conventional reinforcement was sufficient for the most part. For negative moments above the X frame support and the transverse cable-stay anchorage beams, more prestressing tendons were needed. Each of the six strands in the cables has a diameter of 73 mm. They have a bitumen covering to prevent corrosion.

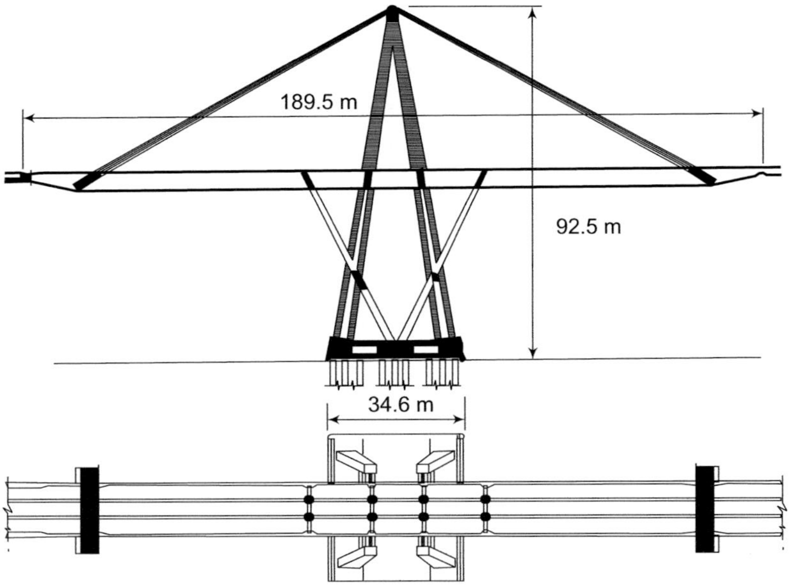

Fig. 9.28 Main Tower and X frames of the General Rafael Urdaneta Bridge (Podolny, 1973)

9.1.7.2 Orinoquia Bridge

Orinoquia Bridge (Figure 9.29) was inaugurated on November 13, 2006. It is a road/railroad bridge that crosses the Orinoco River close to Ciudad Guayana, Venezuela. Four lanes of traffic are carried across the roughly 4.5 km long crossing, which is divided in the middle by a single railroad track. It is made up of two 300-meter cable-stayed navigation spans, one for the north approach measuring 636 m and ten measuring 1320 m for the south, consisting of twenty-two 60-meter spans. The two navigation spans are supported by four 120-meter-tall H-shaped pylons via two planes of cable stays having a clearance of 41 meters above sea level.

Fig. 9.29 The Orinoquia Bridge, Venezuela, 2006

The main bridge is a double stay cable bridge with spans from $3 \times 60 - 300 - 4 \times 60 - 300 - 3 \times 60 = 1200$ m, which is the spanned area 1080 m long, Figure. There is a total of $2 \times 8 \times 11 = 176$ cables arranged in the shape of a fan. They are located at 12 m intervals on the superstructure and 24 m intervals on the pylons (see Figure 9.30). The superstructure consists of a 5.50 m high and 5.7 m wide box girder. Crossbeams, braces and struts are placed at intervals of 3.0 m at the cable anchorage locations. The road plate is 25 cm and 36 cm thick in the and 36 cm road and railway regions.

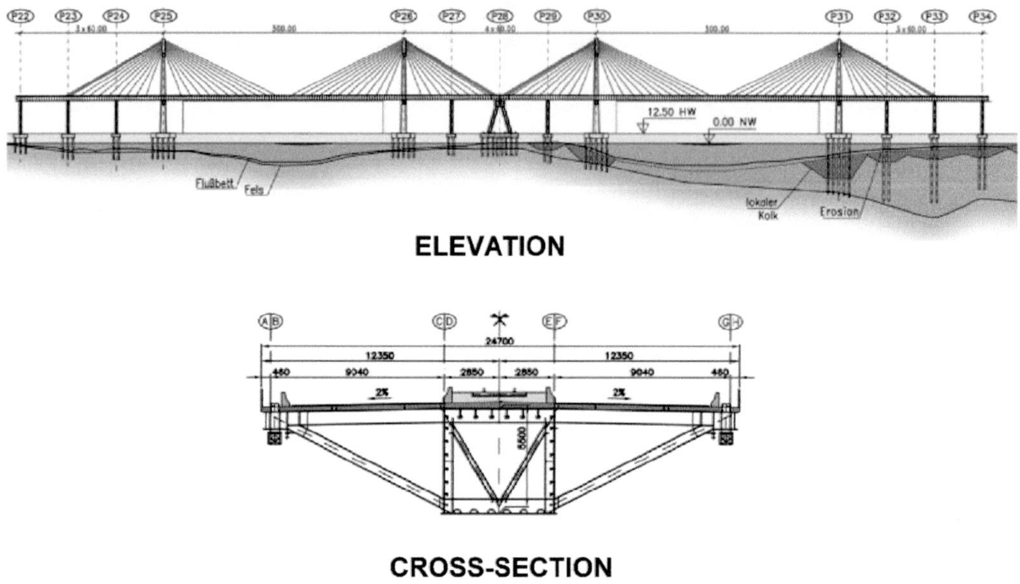

Fig. 9.30 Elevation of the Orinoquia Bridge (Saul et al., 2006)

9.1.8 Argentina

9.1.8.1 San Roque González de Santa Cruz Bridge

The San Roque González de Santa Cruz Bridge (Figure 9.31) is one of the two large cable-stayed bridges that were completed during the past three decades in Argentina. The bridge which was opened to traffic in 1990 crosses the Paraná River between the cities of Posadas, capital of Misiones Province, Argentina and Encarnación, capital of Itapúa, Paraguay. The total length of the crossing is 2550 m including a main bridge that is 560 m long. The approach structures at the Argentina side are 29 spans 55 m long, totalling 1595 m. The approach structures at the Paraguay side have 7 55 m long spans, totalling 385 m. The cable-stayed bridge is 560 m long with a 330 m main span and two 115 m side spans. The bridge provides a minimum vertical navigation clearance of 18 m.

Fig. 9.31 The San Roque González de Santa Cruz Bridge, Argentina-Paraguay, 1990

The deck is 18.9 m wide, and 2.94 m deep. Its section is a three-cell box, in which the lateral inclined slabs and the lower slab are tensile prestressed elements. A box girder was selected due to the eccentric railway track (see Figure 9.32). The suspended spans are supported by 128 stay cables arranged in a semi-fan configuration. The pylons are A-shaped with hollow legs. They have a height of more than 100 m above the riverbed. Since they are in the deepest part of the river, they had to rest on pairs of large concrete caissons that are lowered to sound rock levels by excavation under pressure with compressed air (Figure 9.33).

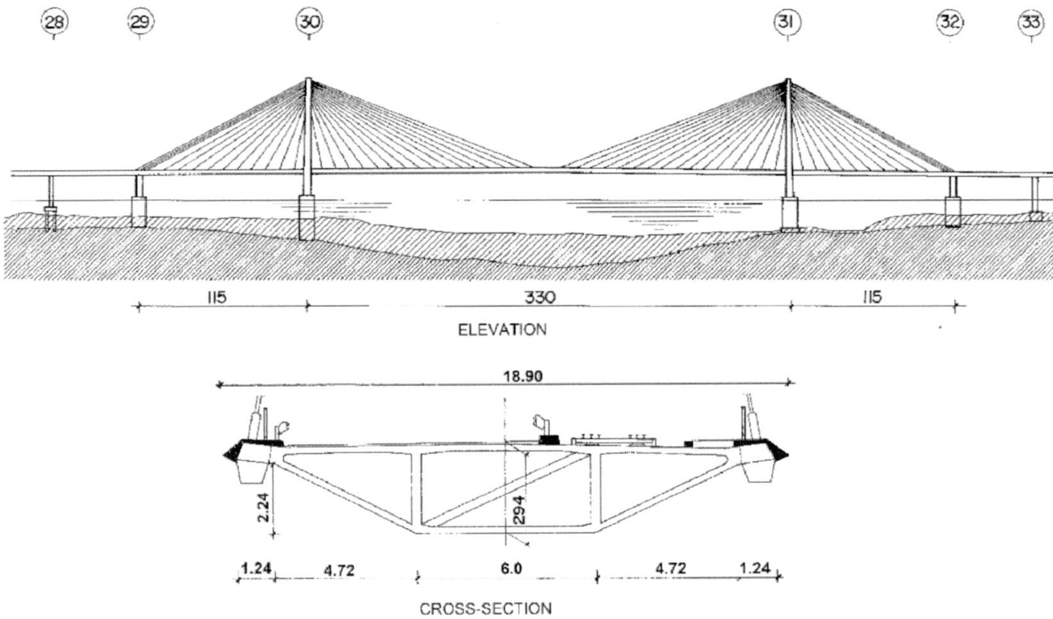

Fig. 9.32 General configuration of the San Roque González de Santa Cruz Bridge

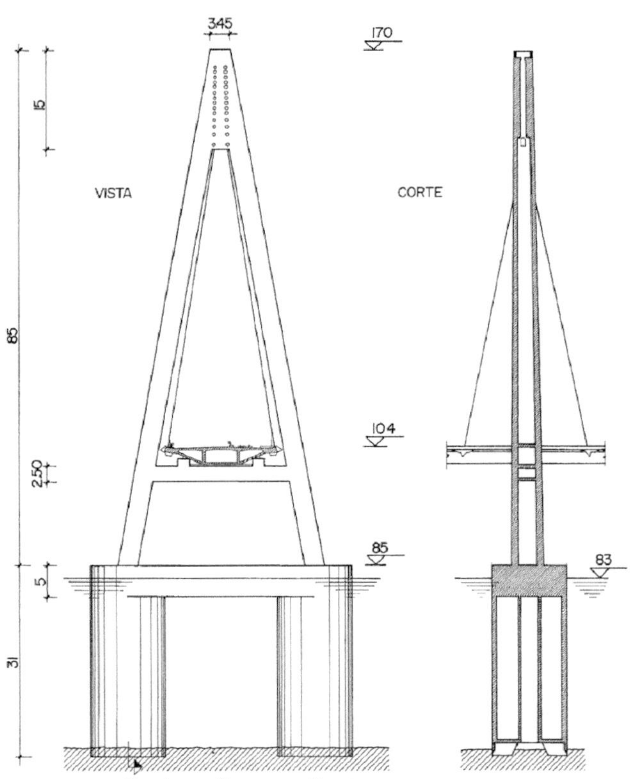

Fig. 9.33 Pylon of the San Roque González de Santa Cruz Bridge

9.1.8.2 Rosario-Victoria Bridge

The Rosario-Victoria Bridge (Figure 9.34), which spans the Paraná River, was completed in 2003. It consists of a 350 m central span cable-stayed pre-stressed concrete deck bridge that crosses the river's navigation channel, 3.5 km of access viaducts leading to the main bridge, 8.2 km of cast-in-place pre-stressed concrete bridges that span 60 m in the Paraná river delta, and about 44 km of hydraulic fill ramparts that finish the 56 km road. A vertical navigation clearance of 53 meters is offered by the bridge.

Fig. 9.34 The Rosario-Victoria Bridge, Argentina-Paraguay, 2003

The bridge provides a connection between the Argentinian cities of Rosario (province of Santa Fe) and Victoria (province of Entre Ríos). The main bridge is a cable-stayed bridge with spans of 120 – 350 – 120 m and 9m cantilevers on both sides to support the approach bridges (Figure 9.35 a). The

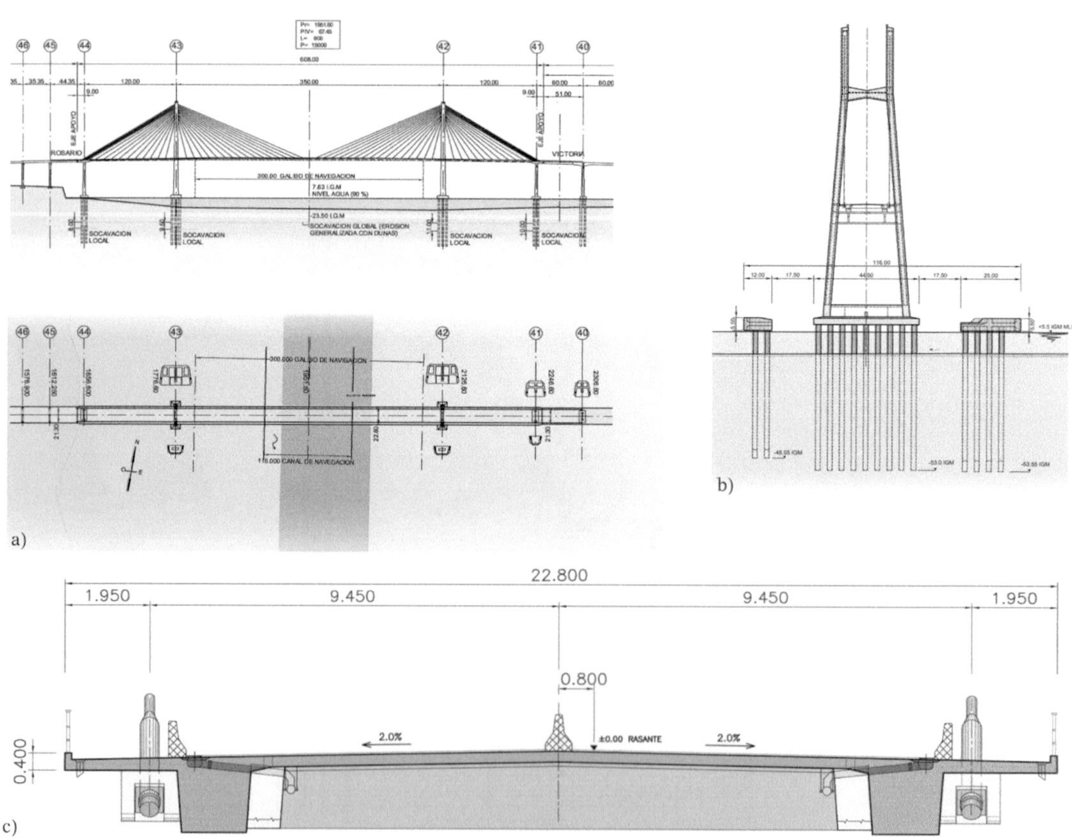

Fig. 9.35 General configuration of the Rosario-Victoria Bridge (Saul et al., 2003)

bridge carries four lanes of traffic with a total width of 22.8 m. The superstructure is an open concrete cross-section characterized by two edge prestressed concrete beams and floor beams spaced at the locations of the cable anchorages. The suspended spans are supported by 128 stay cables arranged in two planes in a semi-fan configuration. The cables are made of parallel strands wires. The pylons are reinforced concrete H type secured by crossbeams above the pile cap under the superstructure at the level of the the lowest cable stay.

9.1.9 Chile

9.1.9.1 Yelcho Bridge

The Yelcho Bridge (Figure 9.36) is the first cable-stayed bridge designed and built in Chile during the 1980s. The bridge is in the Los Lagos Region about 1,000 km from Santiago. The bridge has a total length of 250 m with a main span of 150 m. The deck is a reinforced concrete solid slab cross-section. It is characterized by two concrete frame pylons and cable anchorages at the pylon. A unique design of this bridge is that it has all its cables anchored in a horizontal plane at the pylon top. During the last few years, due to a lack of a maintenance program, there were some concerns about deficiencies and the structural safety conditions. Hence, the owner initiated a comprehensive program that includes in-depth inspection, structural analysis, and rehabilitation measures.

Fig. 9.36 The Yelcho Bridge, Chile, 1983

9.1.9.2 Treng-Treng and Kay-Kay Bridge

The Treng-Treng and Kay-Kay Bridge (Figure 9.37) was inaugurated on April 2, 2021. The bridge which crosses the Cautín River, solves connection problems between the cities of Temuco and Padre Las Casas. The bridge will improve the commute for people who use it daily to cross from one city to another.

The structure, that has a total length of 240m is made up of five spans including the main span of 140 m, with the resulting distribution being 23 + 27 + 140 + 27 + 23 m. The architectural design of the 70 m high pylon has an unusual twist in its form. So, it had to be prestressed to counteract the bending moment caused by the forces of the suspenders. The bridge accommodates two traffic lanes

Fig. 9.37 The Treng-Treng and Kay-Kay Bridge, Chile, 2021 (courtesy, Louis Berger)

in each direction,a median and sidewalks on each side. The superstructure is reinforced concrete consisting of a two-cell box 2 m deep and 16.5 m wide and two lateral cantilevered slabs 5.25 m long. Prestressed concrete transverse diaphragms are added at the main span at the cable anchorages' locations. The main span is supported by 2 planes of 12 cables, distributed according to a semi-fan arrangement. The rear suspension system consists of two pairs of four parallel cables anchored to the abutment.

9.1.10 Uruguay

9.1.10.1 Bridge of the Americas

The Bridge of the Americas (Figure 9.38) was inaugurated in 2005 over Giannattasio Avenue in Ciudad de la Costa in the Barra de Carrasco neighborhood, department of Canelones on the suburbs of the capital Montevideo. This bridge has become one of the most iconic works of reference in the country. It is in an area of constant growth, real estate, and commercial development since the city of Montevideo is constantly growing towards the east.

Fig. 9.38 The Bridge of the Americas, Uruguay, 2005 (courtesy, MC2)

The bridge is 488 meters long, with a suspended span of 140 meters, supported by 30 steel stays. The deck (Figure 9.39) which carries two traffic lanes is 11.25 m wide and consists of a solid slab with a constant section along the bridge, but with a slight variation in cross direction thickness i.e from a maximum of 42 cm at the center to a minimum of 30 cm at the edges. There are two longitudinal ribs, in which the anchorages of the cables allow a perfect distribution and transfer of the forces. The bridge has a single pylon initially made up of a single vertical shaft with a constant width of 2.00 m and a gently variable length between 2.90 m at the foundation level, and a minimum of 2.25 m at a height of 4.50 m, above the deck level. From this level, the length gradually increases to a maximum of 2.80 m, at 16.89 m above the foundation; at this point the pylon splits into a V-shape of two elements or arms slightly curved in elevation, with a constant length of 1.40 m and width of 2.00 m. These arms are tied horizontally at 37.41 m above the foundation, forming a kind of inverted delta. The left arm extends up for another 7.30 m for anchoring the side cables.

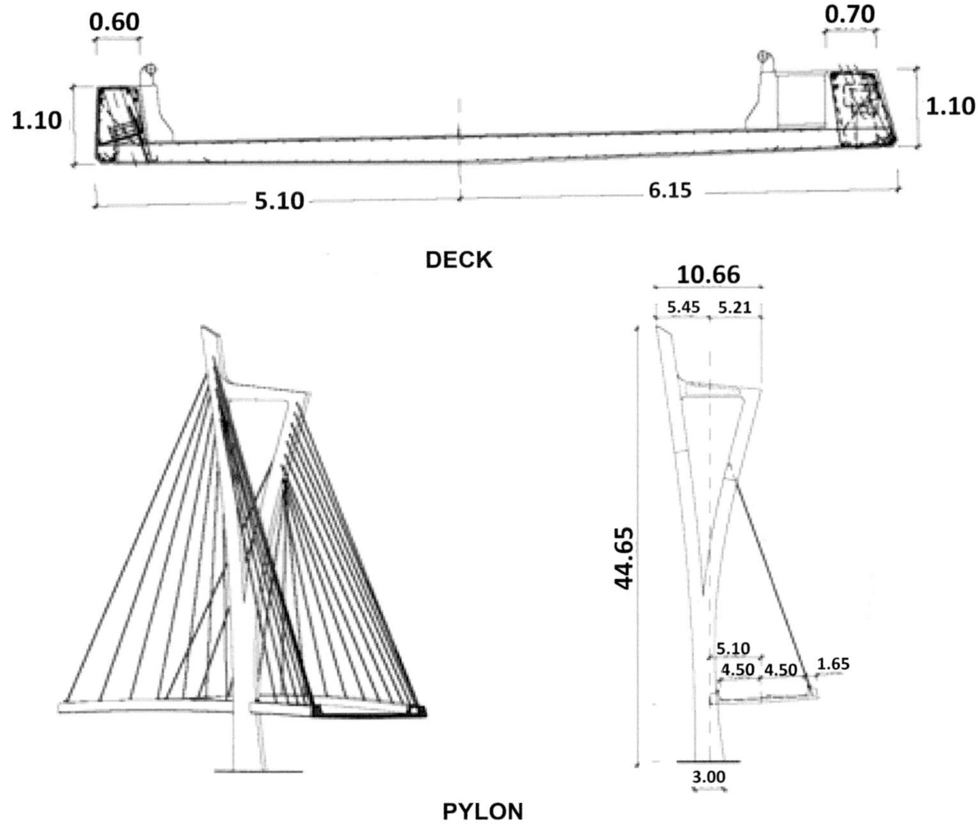

Fig. 9.39 Cross-section of the Bridge of the Americas (Calzon, 2005)

9.1.11 Brazil

Highways in Brazil have developed significantly in the last two decades. The urban extension of large cities required considerable extension of its roadway infrastructure and construction of new structures at significant intersections. In this context numerous cable-stayed bridges have been built across the country for spanning large rivers across the Amazon and elsewhere. Cable-stayed bridges were also built and employed as overpasses in large cities such as Sao Paulo and Rio De Janeiro as part of their growth and expansion. This section will provide brief outlines of some of these bridges.

9.1.11.1 Rio Guamà Bridge

Oened in 2002, the Guamá River Bridge (Figure 9.40) is located close to Belém in the Brazilian State of Para. The bridge is a part of a brand-new route that links Vila da Conde's harbor with Belem, the capital, spanning the Amazon forest. With a central bridge and two approach viaducts, it has an overall length of nearly two kilometers. The latter is a 320-meter-center-span, triple span cable-stayed bridge.

Fig. 9.40 Rio Guamà Bridge, Brazil, 2002 (Courtesy, Studio de Miranda Associati)

The bridge is 1930 meters long overall, with two 719- and 629-meter approach viaducts and a 584-meter cable-stayed bridge. Two traffic lanes, an emergency route, and two walkways for pedestrians are all accommodated on the deck. The central span of the cable-stayed bridge is 320 meters long, while the two side spans are each 132 meters long. The bridge is entirely made of concrete.

The superstructure is an opened reinforced concrete cross-section with a width of 14.20 m and two solid edge beams that are 1.45 m deep, connected by slab and floor beams. It is constructed from precast sections joined by high-performance concrete joints that are cast in place. Both longitudinally and transversely, the deck is prestressed. The deck is supported by stay cables spaced 7.60 meters apart, which is equal to the length of the segment. They are constructed from individual parallel galvanized strands that are covered in a sheath of high-density polyethylene (HDPE) and wax. Rain-induced vibrations are avoided through helicoidal ribs on the external HDPE duct. Each Pylon is made up of two vertical concrete masts with rectangular hollow sections and rounded corners, and each one is 98 meters above river level. Two transverse beams, one located directly below and the other 50 meters above the deck, connect them (Miranda, 2003).

9.1.11.2 Aracaju-Barra dos Coqueiros Bridge

The Aracaju-Barra dos Coqueiros Bridge (Figure 9.41) over the Sergipe River was opened in 2006. It connects Aracaju to the municipality of Barra dos Coqueiros, cities on the coast of Sergipe.

Fig. 9.41 The Aracaju-Barra dos Coqueiros Bridge, Brazil, 2006

The bridge has a total length of 1,350 m, a central span of 200 meters. It has a span distribution of 40 m – 80 m – 200 m – 80 m – 40 m. The 21.30 m deck accommodates four lanes (two in each direction) for vehicles, in addition to a bicycle path and an exclusive pedestrian path. The deck is continuous and integral with piers and pylons along the entire length of 440 m. The main piers, pylons and cable stays lie on two parallel planes. The pylons are free cantilevered above the bridge deck.

9.1.11.3 Newton Navarro Bridge

The Newton Navarro Bridge (Figure 9.42) is another significant river crossing that was opened in 2007. The bridge crosses over the Potengi River just before it reaches the Atlantic Ocean. It connects the city of Natal to other cities on the north coast.

Fig. 9.42 The Newton Navarro Bridge, Brazil, 2007

The crossing has a total length of 1,781.6 m including approach viaducts. The cable-stayed structure comprises a 212 m main span and two 94 m side spans. The 21 m deck accommodates four lanes for vehicles, two in each direction. It provides 56 m vertical navigation clearance. The main cable-stayed bridge is a prestressed concrete structure with a double plane of stay cables continuous between the expansion joints.

The superstructure is an opened reinforced concrete cross-section composed of two edge beams connected by the roadway slab and floor beams. There is a total of 144 stay cables supporting the deck arranged in a semi-harp configuration and anchored at each precast deck segment. They are made of parallel galvanized strands individually protected by wax and High-Density Polyethylene (HDPE) sheaths. The pylon risees 103.45 m above the foundations. The piers and pylons are box concrete sections. All constructions are concreted in situ. The approach viaducts have double column piers while the deck is made of parallel simply supported precast girders with 40-meter spans (Miranda, 2010).

9.1.11.4 Guarulhos City Viaduct

The Guarulhos City Viaduct (Figure 9.43) is a cable-stayed bridge that was opened to traffic in the Sao Paulo metropolitan area in 2008. It is an example of employing the cable-stayed bridge concept to resolve traffic issues at one of the very congested points of the infrastructure network in this region of the country. The city of Guarulhos, where Dutra Interstate, one of Brazil's main highways runs through, is part of the metropolitan area of Sao Paulo. The cable-stayed bridge was built to facilitate access to the Dutra Interstate. It connects Paulo Faccini Street to the Dutra highway easing traffic congestion in the city.

The bridge is 170 m long and has a main span of 96 m and a side span of 74 m. The only pylon of the structure is 61 m high. The reinforced concrete deck is 24 m wide and accommodates two lanes in each direction as well as two pedestrian and cycle paths. The bridge has a single plane of 28 stay cables arranged in a semi-harp configuration. The reinforced concrete pylon rises 61 m above roadway.

Fig. 9.43 The Guarulhos City Viaduct, Brazil, 2007

9.1.11.5 Octávio Frias de Oliveira Bridge

The Octávio Frias de Oliveira Bridge (Figure 9.44) was opened in May 2008 over the Pinheiros River in São Paulo. It connects Marginal Pinheiros to Journalist Roberto Marinho Avenue in the southern part of the city. It is the only bridge in the world that has two curved roadways, located at different elevations, supported by a single concrete mast. Both bridges have spans of 140 m on the Roberto Marinho Avenue side and 150 m on the Marginal Pinheiros side respectively.

Fig. 9.44 The Octávio Frias de Oliveira Bridge, Brazil, 2008 (Courtesy, Denise Paiva Rodrigues)

The bridge deck is 16 m wide, with two traffic and two pedestrian lanes 85 cm long. The superstructure is a reinforced concrete solid cross-section with two inverted edge girders 1.42 m in depth, and a reinforced concrete slab 48 cm thick as shown in Figure 9.45.

Each span has been suspended by a pair of "plans" composed of 18 stay cables each. The pairs of stay cables that are suspend the spans on the Roberto Marinho Avenue side connect to the pylon without crossing, while the "plans" on the Marginal Avenue side, over the Pinheiros River, cross each other at the top, near the pylon. The pylon is 138 m high. It consists of two legs with box sections of varying dimensions. The legs are connected at the bridge deck levels (12 m and 24 m high) by beams that support the initial bridge segments, and by two more beams at 81 m and 114 m high. The foundations are composed of 112 drilled shafts with diameters 1.30 m and 10 m inclined at a diameter of 0.41 m. The superstructure is attached to the pylon monolithically.

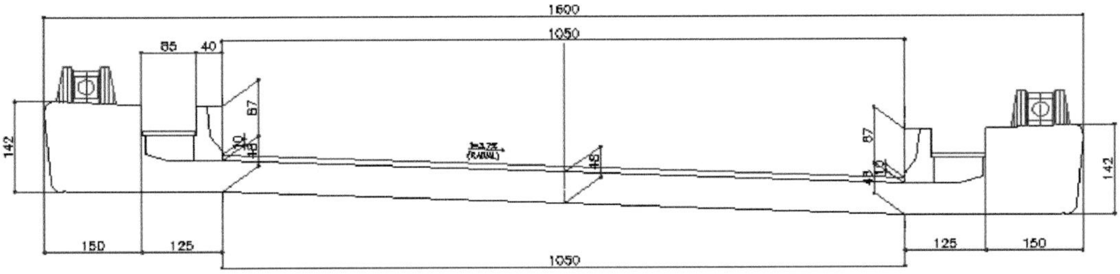

Fig. 9.45 Cross-section of the Octávio Frias de Oliveira Bridge (Ribeiro et al., 2008)

9.1.11.6 Ponte do Saber (Bridge of Knowledge)

On February 17, 2012, Rio de Janeiro commemorated the opening of its first cable-stayed bridge. The Ponte do Saber bridge is located in Guanabara Bay. The Federal University of Rio de Janeiro (UFRJ)'s University City on Fundão Island is connected to the mainland by means of the bridge. The two lanes on the 934-meter-long bridge are each 4.5 meters wide. The bridge's main span is 179.40 meters, and its tallest pylon is 94.00 meters. The concrete used in the design of the bridge's pylon and deck has a typical compressive strength of 50 MPa. 6 A view of the bridge is shown in Figure 9.46.

Fig. 9.46 The Ponte do Saber, Brazil, 2012

The deck is a cast-in-place prestressed concrete box girder. Since the bridge has one central cable plan, two post-tensioned diagonals are supplied to transfer the cable forces to the webs and bottom flange. The cross-section's dimensions and geometric details are displayed in Figure 9.47. The bridge's stay cables are made up of several 15.7 mm-diameter, 7-wire prestressing strands. The bridge's structure is supported by 21 stay-cables; 15 are located in the main span and three pairs are located on the back span.

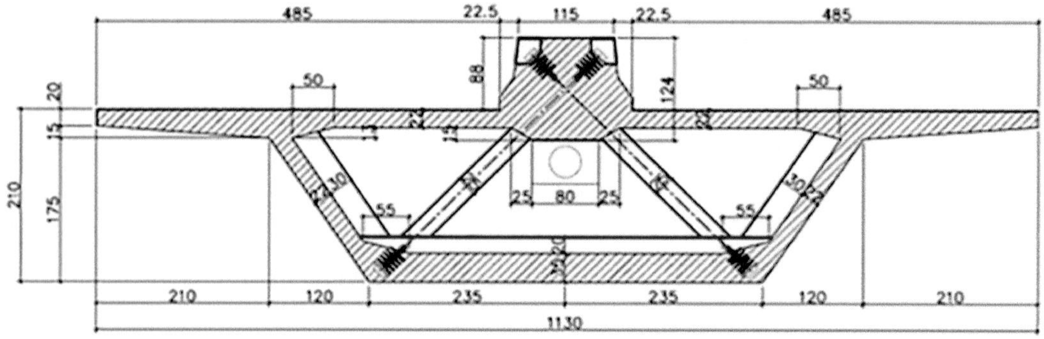

Fig. 9.47 Cross-section of the Ponte do Saber (Hoffman et al. 2022)

The cross-section of the pylon is hollow, with dimensions that vary with height. As shown in Figures 9.48, which depicts the general geometric properties of the pylon, there are two stiffening slabs inside the pylon where the girder deck's flanges connect, along with a number of additional stiffeners evenly spaced.

9.1.11.7 Avenida Ayrton Senna Bridge

Avenida Ayrton Senna Bridge (Figure 9.49) was opened to traffic in 2014 over the Jacarepaguà creek in Rio De Janeiro. The bridge is part of the Transcarioca, a high-speed road link between the Galeão International Airport, in the North area, and the Olympic City, located in the South-West. The cable-stayed bridge has three spans, a main span of 130 m and two side spans each, 39 m long. The superstructure is an opened concrete cross-section formed by a pair of longitudinal beams 1.90 m deep with a trapezoidal, cross-section. There are 64 cable stays that transfer the superstructure load to the pylons. They are arranged in a semi-fan configuration in two parallel vertical planes and made with parallel strands and protected with a high-density polyethylene sheath. The two-leg pylons have triangular cross-sections and a tapered profile (Miranda and Miranda, 2015).

9.1.11.8 Anita Garibaldi Bridge

The Anita Garibaldi Bridge (Figure 9.50) was completed in 2014 over Canal das Laranjeiras. It is located in Laguna, Santa Catarina State of Brazil. The bridge is a part of the Brazilian Highway BR-101 which connects the Northern and Southern parts of Brazil and connects Brazil with other Southern American countries. The 2,830 m long bridge comprises a 400 m long three-span cable-stayed concrete bridge with a semi-fan system, a 1,640m long East Viaduct and a 790 m long West Viaduct. The deck cross-section is a single box with a width of 8.5 m.

The cable-stayed bridge has a 200 m main span and two 100 m side spans. The total width of the deck is 25 m, which carries 2 lanes in each direction and two shoulders, one on each side. The deck cross section is a single box with 8.5 m width (9 m at the bottom slab level). The total width of 25 m

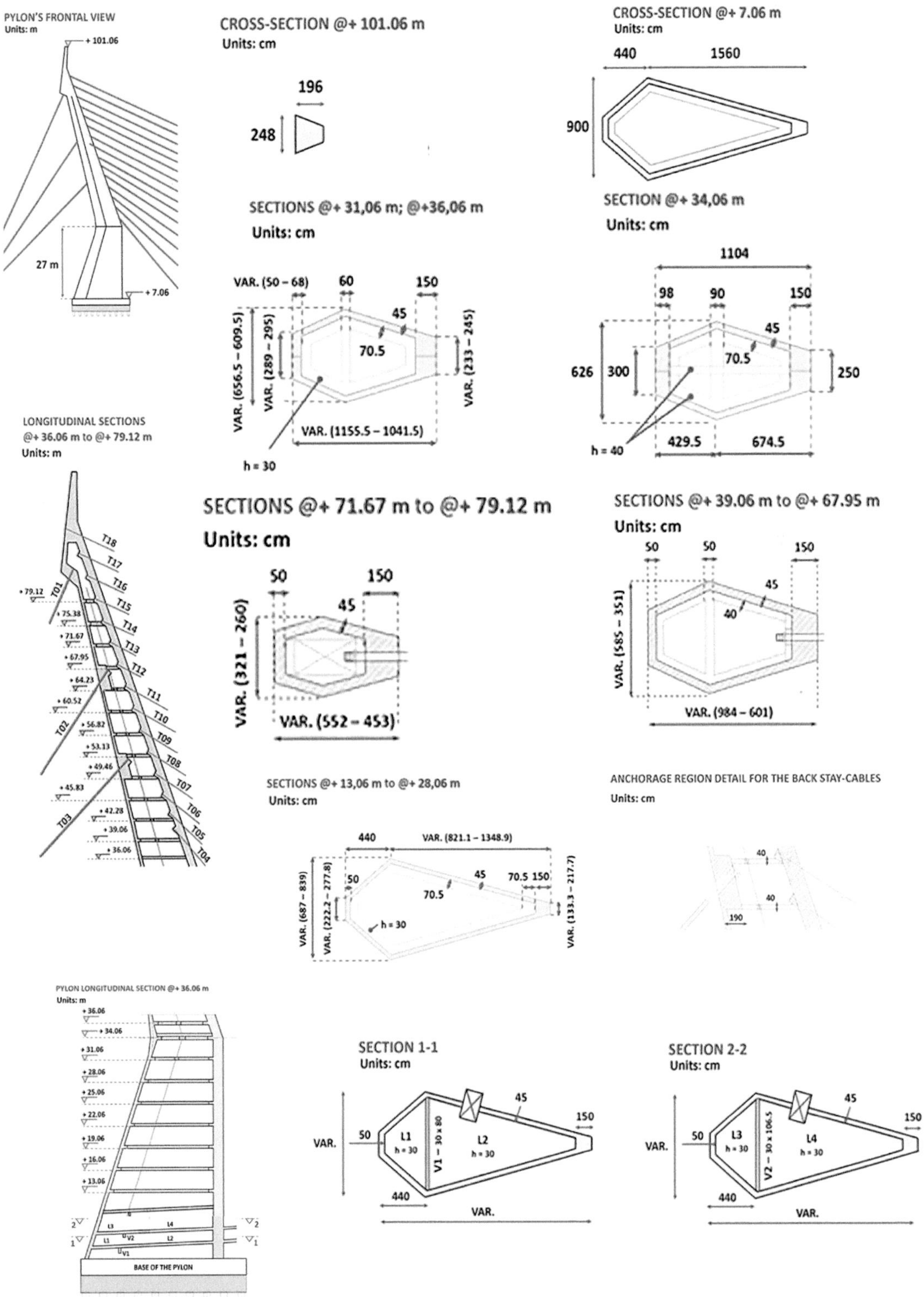

Fig. 9.48 Pylon of the Ponte do Saber (Hoffman et al. 2022)

Fig. 9.49 The Avenida Ayrton Senna Bridge, Brazil, 2014

Fig. 9.50 The Anita Garibaldi Bridge, Brazil, 2014

is attained by continuous precast slabs supported by precast concrete outriggers and finished with in situ compression slabs (see Figure 9.51). The bridge has a single plane of 56 stay cables arranged in a semi-harp configuration. The pylons have reinforced concrete rectangular hollow sections.

9.1.11.9 Oyapock River Bridge

The Oyapock River Bridge (Figure 9.52) spans the Oyapock River, linking the cities of Oiapoque in Amapá, Brazil and Saint-Georges-de-l'Oyapock in French Guiana, France. The construction of the bridge began in 2008 and the work completed three years later, in 2011, but the bridge was only inaugurated in 2017, after a long series of problems and controversies. The bridge is consists of entirely reinforced and prestressed concrete and has three spans of 66.50, 245.00 and 66.50 m. Its total length is therefore 378 m and carries two traffic lanes.

The deck which has a typical width of 15.60 m is a reinforced concrete open-cross-section structure, prestressed longitudinally and transversely. The cross-section (Figure 9.53) is characterized by two longitudinal edge girders 1.40 m deep attached transversely by floor beams that are spaced at equal intervals corresponding to the distance between the cables' anchorages at the deck. The roadway slab is supported by the floor beams, which transfer their loads to the longitudinal girders.

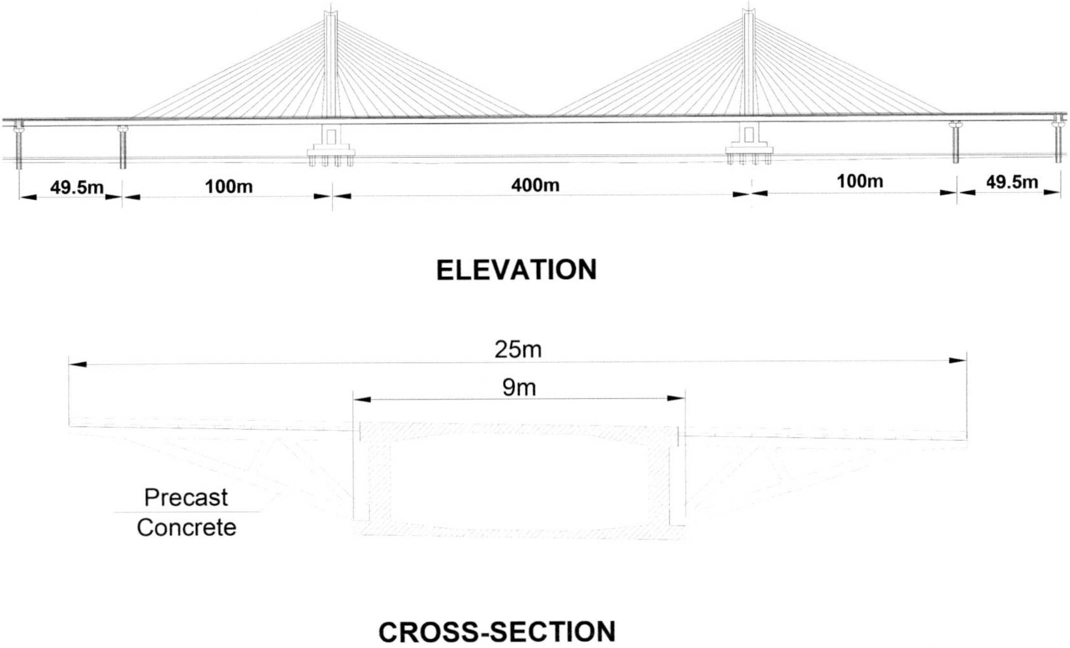

Fig. 9.51 General configuration of the Anita Garibaldi Bridge (Pacheco et al., 2021)

Fig. 9.52 The Oyapock River Bridge, Brazil-French Guiana, 2017

There are 96 cable stays (56 for the main span and 40 for the side spans) transferring the superstructure load to the pylons. They are arranged in a semi-fan configuration in two parallel vertical planes and made of parallel wire strands protected with a high-density polyethylene sheath. The stays are anchored to the deck at equal distances of 7.50 m.

The pylons have a hollow rectangular section, with rounded edges and variable thickness. The foundations of the pylons are laid on drilled shafts of 2.20 m diameters that are socketed to rock. The deck is monolithically attached to the pylons. At each end, the deck is integrated into mooring blocks, which counteract the vertical components of the stay tension forces, supported by elastomeric bearings capable of providing both vertical reactions and transverse and longitudinal elastic constraints.

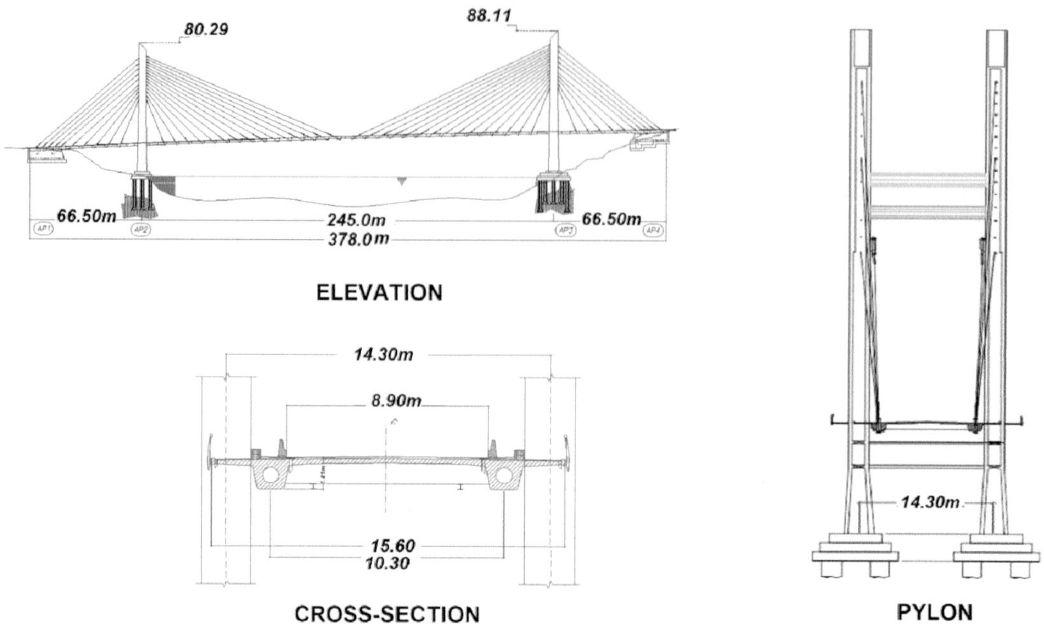

Fig. 9.53 General Configuration of the Oyapock River Bridge (Miranda, 2014)

References

Acuna, J.L. and Lopez-Jara, J. Design of the Nanay Bridge Iquitos – Peru, Proceedings of the International Symposium on Steel Bridges, Lima, Peru, 2014.

Calzon, J.M., Bridge of the Americas in Montevideo, Uruguay, *Third Congress of Ache-Bridges and Structures, the Structures of the 21st Century, Sustainability, Innovation and Future Challenges*, 2005.

Gomez, R., Pozos-Estrada, A., Sanchez-Garcia, R., Gomez-Johnson, R., Escobar, Sanchez, J. et al., Analysis of a Cable-Stayed Bridge: The Case of the Baluarte Bridge, *IABSE Symposium, Large Structures and Infrastructures for Environmentally Constrained and Urbanised Areas*, Venice, 2010.

Heckhausen, C.F., The San Roque González de Santa Cruz Bridge (in Spanish), *Concrete and Steel (Hormigon y Acero)*, Volume 42, No. 181, 133–154, 2012.

Hoffman, I., Lazzari, B.M., Campos, A., Lazzari, P.M. and Pacheo, A.L et al., Finite element numerical simulation of a cable-stayed bridge construction through the progressive cantilever method, *Structural Concrete Journal of the fib,* Volume 23, No. 2, pp 632–651, 2022.

Iglesias, C., Ayuso, G, Cano, A. and González, R, Simultaneous buckling of bridge piers: algorithm applied in push-over analysis for the seismic design of the pylons on Bucaramanga Bridge in Colombia (in Spanish), *Concrete and Steel (Hormigon y Acero)*, Volume 63, No. 263, 65–82, 2012.

Miranda, M., Cable-Stayed Bridge over the Guamà River, Brazil, *Structural Engineering International* Volume 13, No. 3, pp 171–173, 2003.

Miranda, M., New Developments in Brazilian Cable Stayed Bridges, *IABSE Symposium, Large Structures and Infrastructures for Environmentally Constrained and Urbanised Areas*, pp 516–17, Venice, 2010.

Miranda, M., The Oiapoque Bridge: a passage between Europe and Brazil, *Strade and Autostrade*, Volume 18, No., 106, pp. 27–31, 2014.

Miranda, M. and Miranda, M., In Brazil, the cable-stayed bridge on Avenida Ayrton Senna, *Strade and Autostrade*, Volume 19, No., 112, pp. 68–73, 2015.

Pacheco, P., Resende, A., Torres, A. and Coelho, H., Segmental Construction of the Anita Garibaldi Bridge Viaducts in Beazil using the LG50-S Launching Gantry, *e-mosty,* No. 4, 2021.

Podolny, W., Cable -Stayed Bridges of Prestressed Concrete, PCI Journal, Volume 18, No., 1 pp 68–79, 1973.

Ribeiro, C., Stucchi, F., Nogueira, H. and Hernando, C., Cable stayed bridge with two decks and a single tower, *17th IABSE Congress: Creating and Renewing Urban Structures – Tall Buildings, Bridges and Infrastructure,* pp 580–81, Chicago, 2008.

Saul, R., Hopf, S., Humpf, K., Patsch, A. and Bacher, A. et al., Innovative protection against ship impacts for the Rosario Bridge–Victoria across the Paraná (Argentina), *Stahlbau* 72, 7, pp 469–484, 2003.

Saul, R., Humpf, K., Hopf, S. and Patsch, A., The second bridge over the Panama Canal – a 420 m cable-stayed bridge Central opening and record construction time, (in German), Concrete and Reinforced Concrete Construction (*Beton-und Stahlbetonbau)* 100, 3, pp 225–235, 2005.

Saul, R., Humpf, K. and Lustgarten, M., The Orinoco Bridge in Ciudad Guayana/Venezuela Double cable-stayed bridge with composite superstructure for roads and railway freight transport *Stahlbau* 75, 2, pp 82–92, 2006

Zambrano Ramos, H., Armijo Mejia, M., Chauvin, Alain, The Tampico Bridge in Mexico, *IABSE Symposium: Mixed Structures, including New Materials*, Brussels, Belgium, 1990.

CHAPTER **10**

Analysis, Design and Construction Techniques of Cable-Stayed Bridges

10.1 INTRODUCTION

Cable-stayed bridges are considered highly static indeterminate structures that require some sort of computerized algorithms for their analysis and design. A comparatively small number of cables were used in early cable-stayed bridges. A greater number of cable stays were employed after computers and structural design software development which lessened the load on the deck girder and allowed for longer spans. An idealized cable-stayed bridge's deck would consist of several elastic supports with varied stiffnesses, arranged continuously on a beam. The diagonal cable stays support the vertical loads on the deck by transferring the loads to the pylons, which subsequently convey them to the foundations. At the pylons, the forces exerted by the cables at the main span's horizontal components balance those from the side spans. Likewise, the compression load components of the side spans and the cumulative horizontal compression load components from the main span are balanced. As a result, the deck and pylons' predominant compression forces maintain equilibrium throughout the bridge system. By using the classical first-order theory and ignoring the system's deformation when determining the equilibrium conditions, one can ascertain the deflections of the fundamental system under applied loads. The resulting equations for a statically determined basic system are linear in the loads and internal forces, and the internal forces resulting from various loads can be handled by linear superposition. Under the assumption that Hooke's law holds true, linear superposition also applies to the displacements. Therefore, the system's performance is linear, allowing manual design calculations if the analysis is generally predicated on the idea that a structure's elastic displacements are proportional to the applied load. However, this assumption has proven to be approximate and, for large cable-stayed bridges, may be inconvenient. Bending moments in the deck and pylons will increase, due to second-order effects, arising from the deflection of the structure (the P-δ effect) and their application will be non-linear.

10.2 STATIC ANALYSIS OF CABLE STAYED BRIDGES

The overall load-displacement relationships for a cable-stayed bridge structure are geometric nonlinear under normal loads, despite the material in the structure members exhibiting linear elastic behavior under the influence of service and environmental loads. This overall nonlinear behavior is a result of: the nonlinear axial force-elongation relationship for the inclined cable stays due to the sag caused by their own dead weight; the nonlinear axial force and bending force-deformation relationships for the pylons and longitudinal girder elements due to the interaction of large bending and axial deformations (P-delta effect); and the geometrical change caused by large displacements in this type of structure under normal as well as environmental design loads. All of these effects are due to geometrical changes which occur in the structure as it deforms under load. These effects are discussed in what follows.

10.2.1 Nonlinear Behavior of Cable

It is well known from engineering mechanics that a cable, supported at its ends and subjected to its own weight and an externally applied axial tensile force will sag into the shape of a catenary. The axial stiffness of the cable varies with changing sag, which in turn changes with displacements at its ends. The displacements of the cable ends, which occurs due to deformations in the structure subjected to the applied loads, have three different effects upon the cable material : variation in strain in the cable material; rearrangement of the individual wires in the cable cross-section under changing loads; and most importantly, the change in sag of the cable, exclusive of material deformation. The change in sag is governed by the length and weight of the cable, and its tensile force. It is this change in sag that causes the nonlinear force-deformation effects on the cable since the change in sag does not vary linearly with cable tension. As the cable sag increases its axial stiffness reduces. Additionally, as the cable's axial tension rises, the change in cable sag diminishes and material deformation becomes the primary cause of end movement. Because of this, the cable's apparent axial stiffness rises as its tensile stresses do.

10.2.2 P-Delta Effects

Assuming small deformations, the axial and flexural stiffnesses of bending members are typically regarded as uncoupled. However, when a concurrently applied axial force is applied to a laterally deflected, or bent, member, the additional bending moment developed in the member either increases or decreases the original bending moment in the member, a phenomenon known as the P-delta effect. This phenomenon occurs under large deformations. Therefore, under the combined influence of an axial force and bending moment, there is an interaction between axial and flexural deformations. The effective bending stiffness of the member decreases for a compressive axial force and increases for a tensile force as a result of this axial force-bending deformation interaction. Similarly, because the bending deformations appear to shorten the member, the presence of bending moments will impact the member's axial stiffness. Generally speaking, this coupling effect or interaction is negligible in conventional linear structures. However, as a flexible structure, a cable-stayed bridge is susceptible to large deformations, so any nonlinear analysis must take this interaction into account.

10.2.3 Geometry Change due to Large Displacement

It is assumed in linear structural analysis that the structure's joint displacements under applied loads are insignificant in relation to the original joint coordinates. The overall stiffness of the structure in the deformed shape can therefore be taken to be equal to the stiffness of the undeformed structure, disregarding the geometric changes in the structure. On the other hand, under typical design loads, large displacements are possible in cable-stayed bridges, which can lead to important modifications in the bridge geometry. In this situation, the bridge's stiffness in its distorted shape ought to be calculated using the structure's updated geometry.

10.3 NON-LINEAR ANALYSIS TECHNIQUES OF CABLE-STAYED BRIDGES

The set of linear simultaneous stiffness equations for a linear structural system can be solved with ease using matrix analysis to determine the static displacements (Weaver and Gere, 1980),

$$[K]\{\Delta\} = \{P\} \qquad ...(10.1)$$

In equation (10.1) $[K]$ is the global stiffness matrix of the structure, $\{\Delta\}$ is the vector of joint displacements, and $\{P\}$ is the vector of applied joint loads. The general assembly procedure can be used to build the global stiffness matrix $[K]$ from the stiffness matrices of the individual members of the structure. The constant terms in $[K]$ remain unaffected despite the linear structure's deformation. The stiffness coefficients for a nonlinear structural system vary as the structure deforms and the load changes. This somewhat complicates the analysis of such structures. In equation (10.1), the stiffness matrix $[K]$ is a function of the joint displacements $\{D\}$, which are not known at this stage. The set of equations for nonlinear stiffness is not east to solve. Therefore, in order to solve such nonlinear equations for the displacement vector $\{D\}$, numerical solution techniques are typically employed. They fall into three categories: (1) the Newton or incremental procedure; (2) the iterative or mixed procedure; and (3) the mixed procedure.

10.3.1 Incremental Procedure

The incremental procedure applies the load in small steps, with the assumption that the structure will respond linearly at each step. At the conclusion of the preceding load step, the structural stiffness is recalculated using the structural geometry and actions of end members. This process doesn't require iterations, but unless very small steps are used, errors will probably start accumulating after a few steps (Figure 10.1).

10.3.2 Newton-Raphson Iterative Methods

The iterative process involves the formation and decomposition of the stiffness matrix of the structure either once (modified Newton-Raphson iteration) or at the start of each iteration cycle (Newton-Raphson iteration) (Figure 10.2).

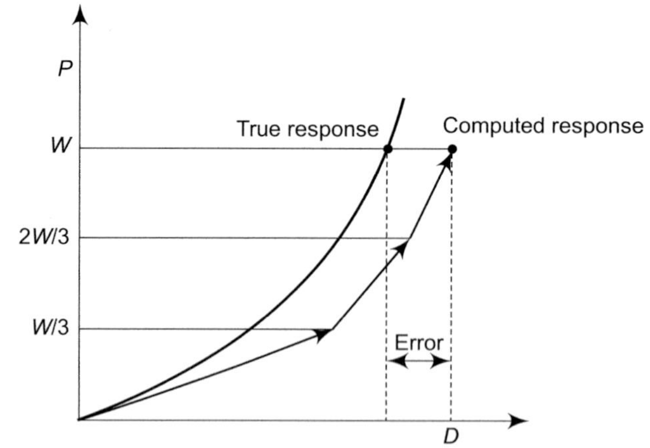

Fig. 10.1 Incremental procedure for cable-stayed bridges (Nazmy and Abdel-Ghaffar, 1987)

The modified Newton-Raphson technique has the advantage of saving computational time, but it converges more slowly than the Newton-Raphson iteration. As seen in Figure 10.2(c), the plot in the modified Newton-Raphson method may, however, never converge in some circumstances, especially if the load increment is sizable and the nonlinearity is strong.

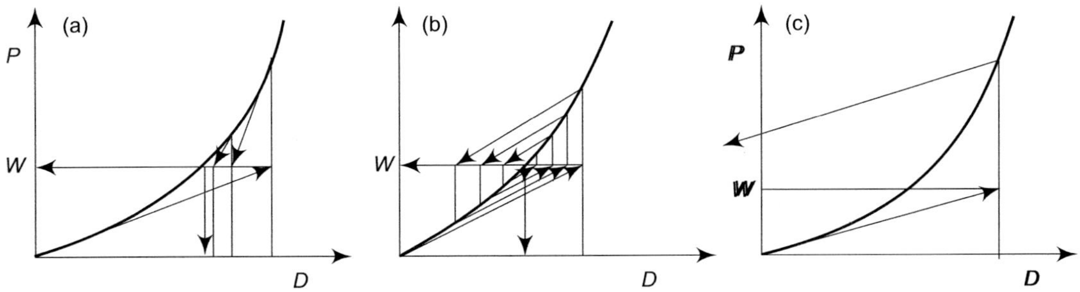

Fig. 10.2 Iterative procedures for cable-stayed bridges. (a) Newton-Raphson iterations. (b) Modified Newton-Raphson iterations (converging case). (c) Modified Newton-Raphson iterations (diverging case) (Nazmy and Abdel-Ghaffar, 1987).

10.3.3 Mixed Procedure of Non-linear Analysis

The mixed procedure implies a combination of the incremental and iterative methods (Fleming, 1979). The load is applied incrementally, and iterations are performed using either the Newton-Raphson or modified Newton-Raphson method. The unbalanced loads are applied progressively during each iteration cycle, assuming linear behavior of the structure during the application of each load increment. The unbalanced loads are obtained by summing the external applied loads and internal member forces at each joint in the structure. Following each load increment, the stiffness coefficients for the structure are recalculated using the actual deformed shape of the structure. Iterations are carried out until equilibrium is achieved at every joint within a specified tolerance. Although it requires more computational work, this process yields high accuracy. In the analysis of cable-stayed bridges, it is preferable to the Newton-Raphson iterative method, though, if the out-of-balance force in the first cycle is very large. Figure 10.3 illustrates this technique graphically with three equal load increments in the first cycle.

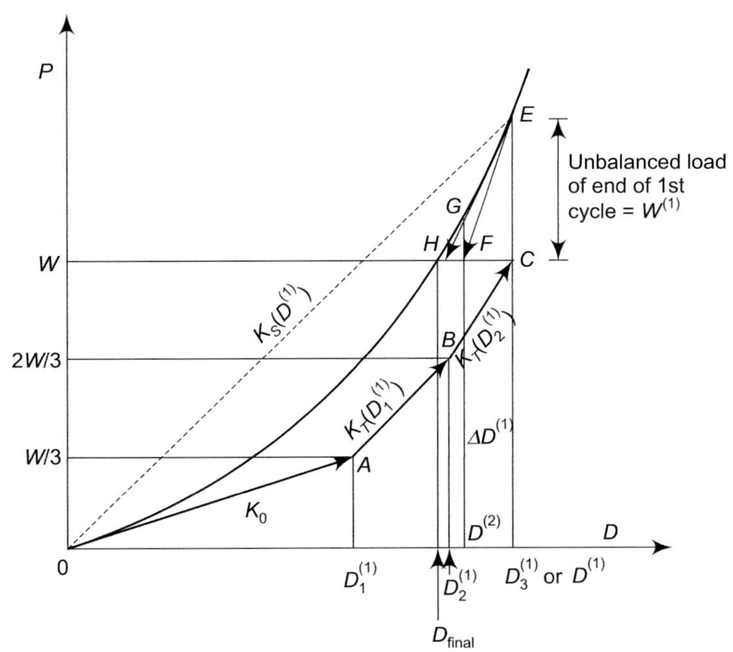

Fig. 10.3 Mixed procedure used for the nonlinear static analysis (Nazmy and Abdel-Ghaffar, 1987)

The tangent stiffness matrix of the undeformed structure is used to apply the first load increment. $[K_0]$, and the joint displacements $\left\{D_1^{(1)}\right\}$ are then calculated using equation (10.1) with $\{P\} = \frac{1}{3}\{W\}$. The tangent stiffness matrix $[K_T]^{(D_1^1)}$ that corresponds to the displaced shape of the structure is then evaluated, and used to compute the incremental joint displacements due to the second load increment. These incremental displacements are then added to the previously calculated joint displacements $\left\{D_1^{(1)}\right\}$ to obtain $\left\{D_2^{(1)}\right\}$, which corresponds to point B in the Figure. For the final load increment, the procedure is repeated until point C is reached, which represents the end of the first cycle. The computed displacements at the end of the cycle $\left\{D_3^{(1)}\right\}$ (or simply $\{D^{(1)}\}$) truly corresponds to loads at point E on the true load displacement curve. The unbalanced loads at the end of the first cycle, $\{W^{(1)}\}$, represented by line CE, are then computed and applied as a new set of joint loads during the second iteration cycle, using the tangent stiffness $[K_T](D^1)$, represented by line EF. The incremental displacements are then computed from:

$$\left[[K_T]^{(D^{(i)})}\right]\{\Delta D\} = \{W^{(i)}\} \qquad ...(10.2)$$

where $\{W^{(i)}\}$ is the out-of-balance force vector computed from:

$$\{W^{(i)}\} = \{W\} - [K_s]^{(D(i))}\,\{D^{(i)}\} \qquad ...(10.3)$$

in which $[K_s]^{(D(i))}$ is the secant stiffness matrix of the structure when the joint displacements are $\{D^{(i)}\}$. The new displacements are then computed from:

$$\{D^{(i+1)}\} = \{D^{(i)}\} + \{\Delta D^{(i)}\} \qquad ...(10.4)$$

The iteration process persists, with the computation of the unbalanced loads at the conclusion of each cycle. Another iteration cycle is necessary if the unbalanced loads, measured in some vector norm (usually the maximum norm), are not less than a predetermined acceptable tolerance. This process keeps going until point H, the convergence to the proper displacements, is reached.

10.3.4 Computation of Structural Stiffness

In equations (10.2) and (10.3) the global tangent stiffness or secant stiffness of the structure is obtained by the standard assembly procedure (Weaver and Gere, 1990), from the individual element stiffness matrices.

10.3.4.1 Inclined Cable Stays

The nonlinearity in the inclined cable stay is taken into consideration using the notion of equivalent modulus of elasticity that was previously discussed in Chapter 2. The axial stiffness of the equivalent member for any given combination of cable tension and sag is equal to the axial stiffness of the actual cable because the effects of geometric and material deformations are combined in the equivalent modulus of elasticity. If there is little variation in the cable's tension during a load increase, the axial stiffness of the cable won't change much (see also section 2.6). Equation (10.5) provides the equivalent modulus of elasticity for the cable as:

$$E_{eq} = \frac{E}{1 + \dfrac{[(wL)^2\ AE]}{12T^3}} \qquad ...(10.5)$$

In equation (10.5) E_{eq} = Equivalent modulus of the cable with sag, E = Modulus of elasticity of a straight cable, w = weight per unit length of the cable, L = projected horizontal length of the cable, T = tensile stress in the cable, and A = cable cross-sectional area. Equation (10.5) gives the tangential (or instantaneous) value of the equivalent modulus when the tension in the cable equals T (or when the cable tensile stress equals σ as shown in Fig. 10.4). If the tension in the cable changes from T, to T_f, during the application of a certain load increment (which is equivalent to a change in cable stress from σ_i to σ_f as shown in Figure 10.4), then the secant value of the equivalent modulus of elasticity over the load increment is given by (Fleming and Egeseli, 1980):

$$E_{eq} = \frac{\quad}{\left[\frac{(WL)\ (T_i\quad T_f)\,AE}{24\,T^2\,T^2}\right]} \qquad \text{...(10.6)}$$

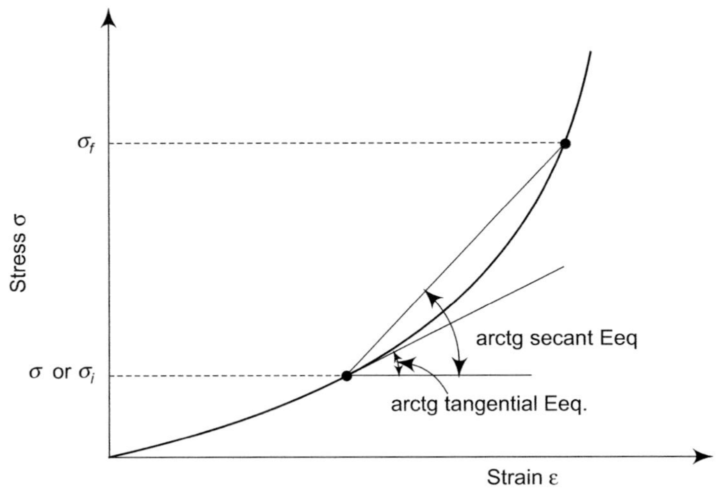

Fig. 10.4 Non-linear stress-strain relationship for a cable stay (Nazmy and Abdel-Ghaffar, 1987)

By using the concept of an equivalent modulus of elasticity the individual member secant (elastic) stiffness matrix for any inclined cable stays of chord length L_c, for the local coordinate system can be written in the standard form as:

$$K_{E_c} = \frac{AE_{eq}}{L_c}\begin{bmatrix} 1 & -1 \\ -1 & 1 \end{bmatrix} \qquad \text{...(10.7)}$$

In equation 10.7 stands for cable. The tangent stiffness matrix of the cable stay can be evaluated using the large deflection theory (Martin, 1965) as:

$$K_{T_c} = K_{E_c} + K_{G_c} \qquad \text{...(10.8)}$$

where, K_{T_c} is the element tangent stiffness matrix in its local coordinates, K_{E_c} is the elastic stiffness matrix, as given by equations (10.6) and (10.7), and K_{G_c} is the geometric stiffness matrix given by:

$$K_{G_c} = \begin{bmatrix} G_c & -G_c \\ -G_c & G_c \end{bmatrix}_{6x6} \qquad \text{...(10.9)}$$

where, G_c is a submatrix given by:

$$G_c = \frac{T}{L_c}\begin{bmatrix} 0 & 0 & 0 \\ 0 & 1 & 0 \\ 0 & 0 & 1 \end{bmatrix} \qquad ...(10.10)$$

10.3.4.2 Flexural Elements

By introducing the concept of stability functions, it is possible to consider the nonlinear behavior of other members of the structure, such as the pylons and bridge deck elements, under the combined effect of large bending moments and high axial forces (Weaver and Gere, 1990; Harrison, 1973; and Livesley and Chandler, 1956). These functions are multiplication factors that are used to adjust the member's axial and bending stiffnesses. The resulting 12 × 12 element secant (elastic) stiffness matrix, in the local coordinate system, the three-dimensional beam-column element shown in Figure 10.5, with 6 degrees of freedom at each end, is given by:

$$\begin{bmatrix} K(1,1) & K(1,2) & \cdots & \cdots & K(1,12) \\ K(2,1) & K(2,2) & \cdots & \cdots & K(2,12) \\ \cdots & \cdots & \cdots & \cdots & \cdots \\ \cdots & \cdots & \cdots & \cdots & \cdots \\ K(12,1) & K(12,2) & \cdots & \cdots & K(12,12) \end{bmatrix} \qquad ...(10.11)$$

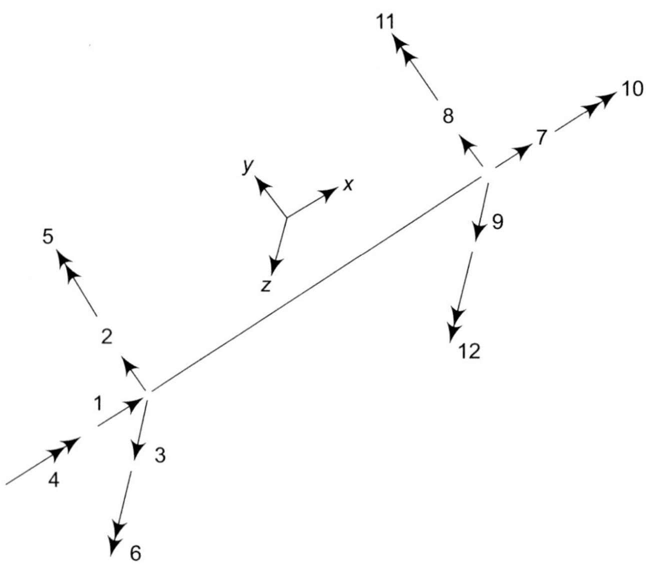

Fig. 10.5 Degrees of freedom of a flexural element in local coordinates

where,

$$K(1,1) = K(7,7) = -K(1,7) = -K(7,1) = \left(\frac{EA}{L}\right) S_0 \qquad ...(10.12a)$$

$$K(2,2) = K(8,8) = -K(2,8) = -K(8,2) = \left(12\frac{EI_z}{L^3}\right) SA_z \qquad ...(10.12b)$$

$$K(3,3) = K(9,9) = -K(3,9) = -K(9,3) = \left(12\frac{EI_y}{L^3}\right) SA_y \qquad ...(10.12c)$$

$$K(2,6) = K(6,2) = K(2,12) = K(12,2) = -K(6,8) = -K(8,6) = -K(8,12)$$
$$= -K(12,8) = \left(6\frac{EI_z}{L^2}\right)SB_z \qquad ...(10.12d)$$

$$K(3,5) = K(5,3) = K(3,11) = K(11,3) = -K(5,9) = -K(9,5)$$
$$= -K(9,11) = -K(11,9) = \left(6\frac{EI_y}{L^2}\right)SB_y \qquad ...(10.12e)$$

$$K(4,4) = K(10,10) = -K(4,10) = -K(10,4) = \frac{GI_x}{L} \qquad ...(10.12f)$$

$$K(5,5) = K(11,11) = \left(4\frac{EI_y}{L}\right)SC_y \qquad ...(10.12g)$$

$$K(6,6) = K(12,12) = \left(4\frac{EI_z}{L}\right)SC_z \qquad ...(10.12h)$$

$$K(5,11) = K(11,5) = \left(2\frac{EI_y}{L}\right)SD_y \qquad ...(10.12i)$$

$$K(6,12) = K(12,6) = \left(2\frac{EI_z}{L}\right)SC_z \qquad ...(10.12j)$$

where, E is the member material modulus of elasticity, A is the cross-sectional area, L is the member length, Z, and I, are the moments of inertia of the cross-section about the local principal y and z axes, respectively, as shown in Figure (10.5), Z, is the torsional moment of inertia of the cross-section, G is the member material shear modulus, and the S's are the stability functions. SA_y through SD_y, modify the bending stiffness of the member about the local y axis, while SA_z, through SA_z, modify the bending stiffness about the local z axis, and S_0 modifies the axial stiffness. If the axial force in the bending member is zero, all the stability functions become unity. The stability functions can be expressed in terms of the member axial force P, and the member end moments $M1$ and $M2$, at both ends about the member local y and z axes, as defined in Figure 10.6.

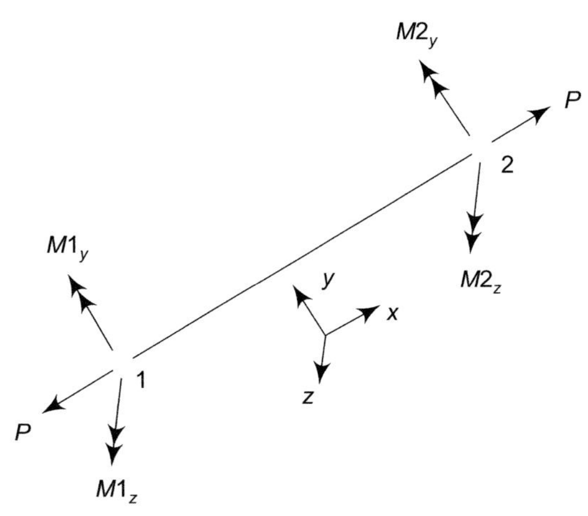

Fig. 10.6 Axial forces and end moments for a cable-stayed bridge flexural element

Stability functions for tension members SA_z through SD_z are as follows (Fleming and Egeseli, 1980):

$$SA_z = (\chi L)^3 \frac{\sinh \sinh(\chi L)}{12\Lambda_T} \quad ...(10.13)$$

$$SB_z = (\chi L)^2 \frac{[\cosh \cosh(\chi L) - 1]}{6\Lambda_T} \quad ...(10.14)$$

$$SC_z = (\chi L) \frac{[(\chi L)\cosh(\chi L) - \sinh(\chi L)]}{4\Lambda_T} \quad ...(10.15)$$

$$SD_z = (\chi L) \frac{[\sinh(\chi L) - (\chi L)]}{2\Lambda_T} \quad ...(10.16)$$

where,

$$\chi = \sqrt{\frac{P}{EI_z}} \quad ...(10.17)$$

and

$$\Lambda_T = 2 - 2\cosh(\chi L) - \omega \sinh(\chi L) \quad ...(10.18)$$

For compression members, the stability functions are determined as follows:

$$SA_z = (\chi L)^3 \frac{\sin(\chi L)}{12\Lambda_C} \quad ...(10.19)$$

$$SB_z = (\chi L)^2 \frac{[(\chi L)]}{6\Lambda_C} \quad ...(10.20)$$

$$SC_z = (\chi L) \frac{[\sin(\chi L) - (\chi L)\cos(\chi L)]}{4\Lambda_C} \quad ...(10.21)$$

$$SD_z = (\chi L) \frac{[(\chi L) - \sin(\chi L)]}{2\Lambda_C} \quad ...(10.22)$$

where, χ is as described in equation (10.17) and:

$$\Lambda_C = 2 - 2\cos(\chi L) - (\chi L)\sin(\chi L) \quad ...(10.23)$$

The stability functions SA_y through SD_y can be determined in the same way by replacing I_z by I_y in equations (10.13) through (10.23).

The stability function S_0 can be obtained for tension members as follows:

$$S_0 = \frac{1}{\left[1 - \frac{EA(\Lambda_{tmy} + \Lambda_{tmz})}{4P^3L^2}\right]} \quad ...(10.24)$$

where,

$$\Lambda_{tmy} = (\chi_y L)(M1_y^2 + M2_y^2)[\coth \coth(\chi_y L) + (\chi_y L)\operatorname{cosech}^2(\chi_y L)]$$
$$-2(M1_y + M2_y)^2 + (M1_y M2_y)[1 + (\chi_y L)\coth(\chi_y L)]$$
$$[2(\chi_y L)\ \operatorname{cosech}(\chi_y L)] \quad ...(10.25)$$

and

$$\Lambda_{tmz} = (\chi_z L)(M1_z^2 + M2_z^2)[\coth \coth(\chi_z L) + (\chi_z L)\operatorname{cosech}^2(\chi_z L)]$$
$$-2(M1_z + M2_z)^2 + (M1_z M2_z)[1 + (\chi_z L)\coth(\chi_z L)]$$
$$[2(\chi_z L)\ \operatorname{cosech}(\chi_z L)] \quad ...(10.26)$$

In equation 10.26,

$$\chi_y = \sqrt{\frac{P}{EI_y}} \text{ and } \chi_z = \sqrt{\frac{P}{EI_z}} \qquad ...(10.27)$$

The stability function S_0 can be obtained for compression members as follows:

$$S_0 = \frac{1}{\left[1 + \frac{EA(\Lambda_{cmy} + \Lambda_{cmz})}{4P^3L^2}\right]} \qquad ...(10.28)$$

where,

$$\Lambda_{cmy} = (\chi_y L)(M1_y^2 + M2_y^2)[\coth\coth(\chi_y L) + (\chi_y L)\,\text{cosech}^2(\chi_y L)] \\ -2(M1_y + M2_y)^2 + (M1_y M2_y)[1 + (\chi_y L)\coth(\chi_y L)] \\ [2(\chi_y L)\,\text{cosech}(\chi_y L)] \qquad ...(10.29)$$

and

$$\Lambda_{cmz} = (\chi_z L)(M1_z^2 + M2_z^2)[\coth\coth(\chi_z L) + (\chi_z L)\,\text{cosech}^2(\chi_z L)] \\ -2(M1_z + M2_z)^2 + (M1_z M2_z)[1 + (\chi_z L)\coth(\chi_z L)] \\ [2(\chi_z L)\,\text{cosech}(\chi_z L)] \qquad ...(10.30)$$

where, χ_y and χ_z are as defined in equation (10.27).

By using nonlinear strain-displacement relationships and the large deflection theory, the tangent stiffness matrix for a bending member can be found as follows:

$$K_{T_b} = K_{E_b} + K_{G_b} \qquad ...(10.31)$$

where K_{T_b} is the beam element tangent stiffness matrix in its local coordinates, K_{E_b} is the elastic stiffness matrix of the bending element as given by equation (10.11), and K_{G_b} is the geometric stiffness matrix give as (Yamamura, 1983):

$$\frac{P}{L}\left|\begin{matrix}
0 \\
0 & \frac{6}{5} \\
0 & 0 & \frac{6}{5} \\
0 & 0 & 0 & 0 \\
0 & 0 & \frac{L}{10} & 0 & \frac{2}{15}L^2 \\
0 & \frac{L}{10} & 0 & 0 & 0 & \frac{2}{15}L^2 \\
0 & 0 & 0 & 0 & 0 & 0 & 0 \\
0 & -\frac{6}{5} & 0 & 0 & 0 & -\frac{L}{10} & 0 & \frac{6}{5} \\
0 & 0 & -\frac{6}{5} & 0 & \frac{L}{10} & 0 & 0 & 0 & \frac{6}{5} \\
0 & 0 & 0 & 0 & 0 & 0 & 0 & 0 & 0 & 0 \\
0 & 0 & -\frac{L}{10} & 0 & -\frac{L^2}{30} & 0 & 0 & 0 & \frac{L}{10} & 0 & \frac{2}{15}L^2 \\
0 & \frac{L}{10} & 0 & 0 & 0 & -\frac{L^2}{10} & 0 & \frac{-L}{10} & 0 & 0 & 0 & \frac{2}{15}L^2
\end{matrix}\right. \qquad ...(10.32)$$

In 10.32 P is the axial force in the element, and L is its length. Prior to assembling the global stiffness matrix, the element stiffness matrix is transformed using the standard transformation formula from local to global coordinates:

$$[K] = [T_m]^T [K_m][T]$$

in which $[K]$ is the member stiffness matrix in global coordinates, $[K_m]$ is the member's stiffness matrix in local coordinates, and $[T_m]$ is its transformation matrix given by:

$$[T_m] = \begin{bmatrix} [R] & 0 & 0 & 0 \\ 0 & [R] & 0 & 0 \\ 0 & 0 & [R] & 0 \\ 0 & 0 & 0 & [R] \end{bmatrix} \qquad \text{...(10.33)}$$

The direction cosines of the member local axes with respect to the global axes are given by $[R]$, an order three submatrix [7]. The joint coordinates used in computing these direction cosines are based on the geometry of the bridge in its deformed state after applying the external loads, due to the large displacements that occur in cable-stayed bridges under normal design loads.

10.4 DYNAMIC ANALYSIS OF CABLE-STAYED BRIDGES

Cable-stayed bridges will respond to dynamic loads according to their dynamic characteristics i.e., natural frequencies and the modes of vibrations corresponding to these natural frequencies. Magnification of the response is expected if the exciting frequency is close or equal to the natural frequency of a certain mode. For a multi-degree of freedom system such as cable-stayed bridges, evaluation of the natural frequencies and mode shapes for the bridge is conducted for a free vibration system through the solution of the Eigen Value Problem independent of damping. To simplify the problem, the structural masses are assumed to be lumped at the nodes. Therefore, for a lumped mass formulation, the equations of motion for the bridge in a free vibration system are written in matrix form as:

$$[M] (\{\ddot{U}(t)\}) + [K]\{U(t)\} = \{0\} \qquad \text{...(10.34)}$$

In equation 10.34, $[M]$ is a diagonal matrix for lumped masses and $[K]$ is the structural stiffness matrix. Equation 10.32 represents a set of second order simultaneous, ordinary differential equations. For free undamped vibrations, the particular solutions to these equations must describe harmonic motions hence the displacement U and acceleration $\ddot{U}$ for node i at time t are written as:

$$U_i(t) = U_i \sin \omega t \qquad \text{...(10.35a)}$$

$$\ddot{U}(t) = -U_i\, \omega^2 \sin \omega t \qquad \text{...(10.35b)}$$

Substituting eq. (10.35) into eq. (10.34) yields:

$$([K] - \omega^2 [M])\{U\} = 0 \qquad \text{...(10.36)}$$

Equation (10.36) is a set of homogeneous algebraic equations for amplitude U_i in which frequency ⏋ is unknown. The problem of finding the unknown frequencies from eq. (10.36) is mathematically an Eigen Value Problem with a characteristic determinant:

$$|[K] - \omega^2 [M]| = 0 \qquad \text{...(10.37)}$$

The homogeneous algebraic equations can be used to determine the vibration mode shapes also known as eigen vectors, once the eigen values (natural frequencies) have been determined. For each frequency substituted in eq. (10.36), one mode is obtained. This approach is usually defined in most structural analysis software as modal analysis. An example of the first longitudinal, transverse, and vertical modes for a cable-stayed bridge is illustrated in Figure 10.7.

In reality, the bridge will not vibrate forever, but the vibration will cease due to damping. Damping is discussed in Chapters 11 and 12. Therefore, the equation of dynamic equilibrium of the bridge at time t_i is given, in matrix form, by:

$$[M]\,(\{\ddot{U}(t_i)\}) + [C]\{\dot{U}(t_i)\} + [K]\{U(t_i)\} = \{P(t_i)\} \qquad \text{...(10.38)}$$

In equation 10.38 [C] is the damping matrix of the system, and $\{P(t_i)\}$ is the vector of externally applied dynamic nodal forces at time t_i.

The method of step-by-step integration is the most effective way to solve the system of dynamic equations represented by 10.38 (Bathe, 1996). This method obtains the response at successive time steps Δt, which are typically assumed to be equal in length for computational ease. Every time interval starts with an evaluation of the dynamic equilibrium state. Next, we derive the response for an incremental time step Δt, assuming that the structural properties don't change over this time.

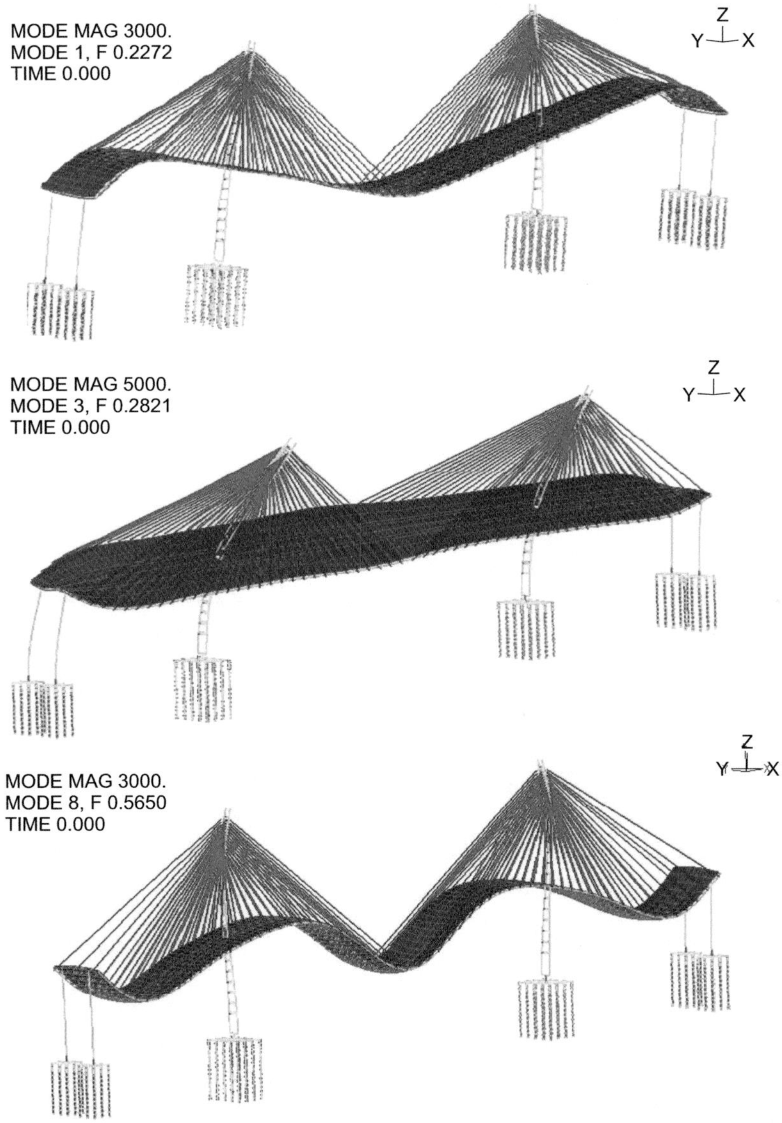

Fig. 10.7 Example of the results of eigen-value (modal) analysis for a cable-stayed bridge (only three modes are shown

By reevaluating the structural properties at the end of the time step and taking the tangential stiffness matrix into consideration, the non-linear nature of the system can be taken into consideration. The state of dynamic equilibrium is reached at the conclusion of this time step using an iterative method, and the computed accelerations, velocities, and displacements are then used as initial conditions for the following time interval. Therefore, from the point at which loading begins to any desired time, the solution is continued step by step. Seismic time history analysis finds the method to be highly convenient. A variety of techniques are derived from the step-by-step integration; the Newmark-β and Wilson-θ methods are the most widely used.

10.5 SIGNIFICANT CONSIDERATIONS IN THE DESIGN OF CABLE-STAYED BRIDGES

This section summarizes significant points that were previously discussed in Chapters 2, 3, and 4. These points, if taken into consideration at the very beginning of the design process will result in an optimal and economical design of the bridge in terms of the required amount of cable steel, beam, pylon and hold-downs at the bridge ends. These considerations are discussed herein.

1. Compared to other stay cables' longitudinal arrangements, the semi-fan arrangement has proven very efficient as it provides continuous elastic support and ensures a uniform distribution of the axial force through the deck and offers a significantly reduced concentrated force at each anchor point. This arrangement also results in lighter sections, simpler construction, and better appearance of the bridge. The cable anchor points are usually spaced at 1.5 m–2.5 m vertical intervals in order to provide enough space for anchoring.
2. The single plane arrangement of cables while advantageous due to its aesthetic nature and very convenient for divided highways where a central wide meridian strip can be used for locating the pylons in the center, requires a box girder with high torsional rigidity to accommodate unsymmetrical loading, which may require an extra amount of material for the deck structure. On the other hand, the double plane system has been most often used because no torsional rigidity is necessary for this system since the cables give a stiff support along each edge and the deflection is small, so that unsymmetrical loading gives only a minimal transverse inclination of the deck. Also, for aerodynamic safety this system does not require torsional rigidity.
3. The concrete cross section deck is applicable for main spans up to about 550 m. The all-steel bridge with an orthotropic plate deck becomes mandatory for large spans greater than 600 m to reduce the dead loads. The composite deck falls in between the ranges of the other two types.
4. If the width of the bridge is up to around 17 m, a massive concrete slab with or without longitudinal edge beams is necessary for bridges with wider decks. Cross girders arranged with a spacing in the range of 3 to 5 m are also necessary in such bridges.
5. Live load, if existing in the main span, will increase the stresses in the back stays whereas live loads in the side span decreases them. This creates a change in fatigue stress of these cables, which is governed by the ratio of the back span to main span. Therefore, a good ratio between the side span and main span is very important for an economic design of cable-stayed bridges and to keep fatigue stress amplitudes below the fatigue strength of the cables. It is important to keep in mind that long main spans produce high tensile forces in the back stays due to permanent loads. On the other hand, long side spans result in high stress change for fatigue due to live loads. Therefore, it is important that the ratio between the back span and the main span be less than 0.5 while 0.3 and 0.4 are optimal ratios for railroad and highway bridges.
6. Stay cable inclination influences the required stay size. Theoretically the most efficient stay angle is 45°. Practically, a stay inclination in the range of 25° (at the outer stay connecting the deck panel adjacent to the center of the main span to the top of the pylon) to 65° (located nearest the pylon) is very reasonable.

7. The ratio of pylon height above deck to the main span also influences the required amount of cable steel. The optimum pylon height comes to about 20% to 25% of the main span, taking into account the additional costs for higher pylons.
8. The spacing of the stay anchors along the deck must be limited to a reasonable range to maintain the capacity of the longitudinal girders under critical loading levels due to the sudden loss of a stay. Heavier concrete constructions require 5 m to 10 m stay spacing while 10 m to 15 m stay spacing is more suitable for steel composite constructions.
9. Determination of the superstructure depth must be selected precisely to ensure sufficient system stiffness and capacity. For bridges with two planes of cables, a 1.8 m deep girder is sufficient if the cables are spaced 12 m to 14 m apart. Box girders supported by a single cable plane would be about 3 m to 3.5 m deep, and the cable spacing may be about 6 m.

10.6 MODELING AND ANALYSIS FOR DESIGN

Designing a cable-stayed bridge is an iterative process. First, a preliminary set of sectional properties is estimated for each member of the system. Approximate manual calculation methods can be used for this purpose, wherein dead and live loads are the primary loadings at this stage. Next, the sectional properties estimated by approximate methods are analyzed using rigorous nonlinear statistical methods of analysis. Stresses and displacements under the given loads on the system are determined and compared with those allowed by the specifications. Depending on the outcomes of this step, a new set of sectional properties is chosen to satisfy the specification requirements and another non-linear analysis is repeated until the code and specification requirements are satisfied. Finally, the structure is checked for dynamic loads using non-linear methods of dynamic analysis.

It is very clear that approximate manual calculations are very significant because the level of accuracy in the estimation of the structural properties at this step will reduce the number of iterations and the desgn cost. Hence, preliminary design by manual calculation, should be considered as the first stage in the design process to provide a basis for a more rigorous analysis.

10.6.1 Approximate Design of Cables

Figure 10.8 illustrates the concept of an equivalent statically determinate system where hinges are placed in the longitudinal girder at every cable anchorage point assuming that the bending stiffness of the girders and pylons are very minimal compared to the stiffness of the cable stays and are neglected at his stage. According to this configuration, the cable force F_i due to dead and service loads P_i applied to node i is quantified as:

$$\frac{P_i}{\sin(\theta_i)} \quad \text{and} \quad P_i = W_i \, \psi \qquad ...(10.39)$$

In 10.39, θ_i is the inclination of cable i, and ψ is the spacing between cables and W_i is the tributary distributed load to cable i. The horizontal stay component H_i due to P_i is quantified as:

$$H_i = \sum_{i=1}^{n} \frac{P_i}{\tan(\theta_i)} \qquad ...(10.40)$$

Assessment of equations 10.39 and 10.40 is illustrated through manually calculating the cable forces and the horizontal compressive forces on the deck of the cable-stayed bridge illustrated in Figure 10.9. The displayed bridge has one plane of fan configured cables with a main span of 240 m and side spans of 114 m each. The superstructure is a reinforced concrete box girder of a 295.5 kN/m factored distributed uniform dead load. The cables are spaced at 12 m in the main span and the side spans except for the panels close to the pylon, which are spaced at 18 m. The backstay is attached to the anchor pier by hold-down devices. The pylon height above the deck is 50 m.

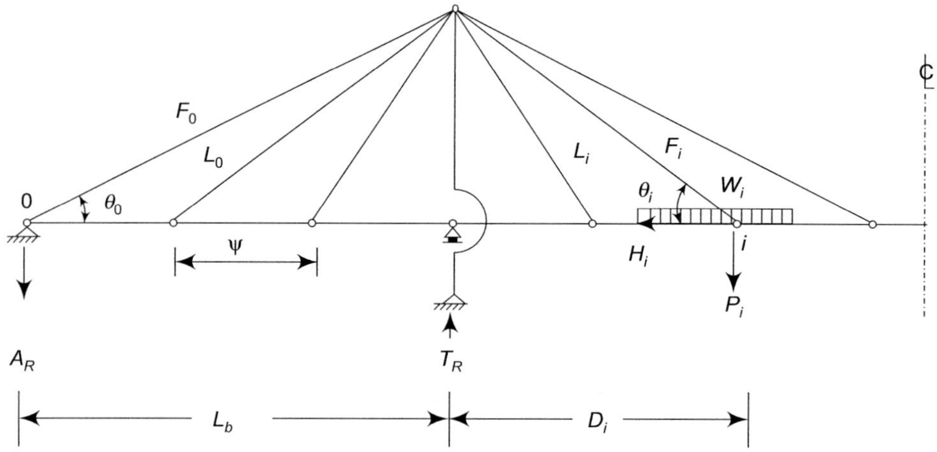

Fig. 10.8 Determination of normal forces in an equivalent truss system

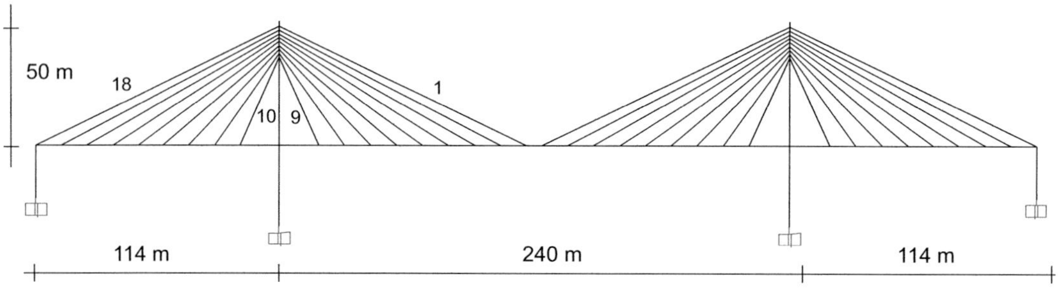

Fig. 10.9 Example of a cable-stayed bridge configuration

Only half of the structure is considered as it is symmetrical. As a first step, the cable forces are calculated using equation 10.39. Cable areas are calculated in a subsequent step and the number of strands are estimated. Table 10.1 summarizes the calculations as well as results by finite element modeling of the bridge. Distribution of the compressive forces in the deck is illustrated in Figure (10.10)

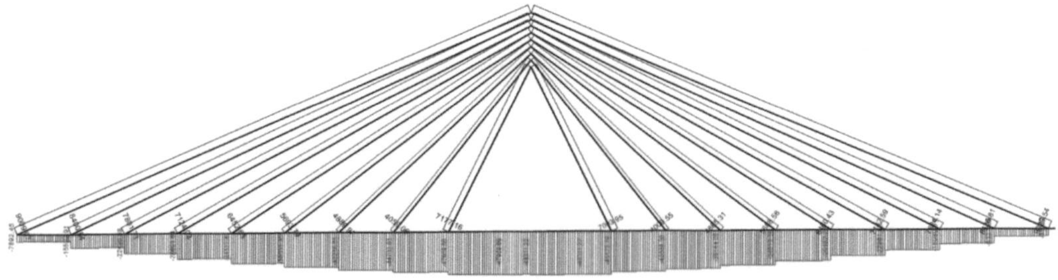

Fig. 10.10 Distribution of axial forces in cables and deck for the example of a cable-stayed bridge in Fig 10.9

As shown from the results of both the manually calculated forces, and the forces computed through finite element modeling, a favorable comparison exists. Differences range from 0.29% to 11.24% for the forces in the cables and from 0.07% to 4.34% for the horizontal forces in deck. The force in the middle panel of the deck was set to zero in the computer model to match the assumption of

Table 10.1 Summary of the calculations of the distributed axial forces in cables and deck

Cable number	*Distance from Pylon (m)*	*Cable Inclination (q)*	*Cable Force (eqn. 10.39 kN)*	*Required are (m^2)*	*Used Number of strands*	*Cable Force FEA (kN)*	*Absolute Error%*	*Horizontal Force (equ. 10.40 kN)*	*Horizontal Force FEA (kN)*	*Absolute Error%*
1	114	23.05	9055	0.011	73	9029	0.29%	8,332	8,327	0.07%
2	102	24.74	8471	0.010	68	8427	0.52	16,026	15,980	0.29%
3	90	26.82	7857	0.009	63	7824	0.42%	23,038	22,961	0.33%
4	78	29.43	7215	0.009	58	7143	1.00%	29,322	29,182	0.48%
5	66	32.78	6548	0.008	53	6460	1.34%	34,827	34,614	0.61%
6	54	37.21	5862	0.007	47	5688	2.97%	39,496	39,144	0.89%
7	42	43.24	5174	0.006	42	4865	5.98%	43,365	42,688	1.33%
8	30	51.71	4517	0.005	36	4009	11.24%	46,064	45,173	1.93%
9	18	63.75	7905	0.009	63	7093	10.27%	49,560	48,311	2.52%
10	18	63.75	7905	0.009	63	7177	9.21%	49,560	47,910	3.33%
11	30	51.71	4517	0.005	36	4059	10.13%	46,762	44,733	4.34%
12	42	43.24	5174	0.006	42	4890	5.50%	42,295	42,216	0.19%
13	54	37.21	5862	0.007	47	5694	2.87%	38,596	38,653	0.15%
14	66	32.78	6548	0.008	53	6453	1.45%	33,991	34,118	0.37%
15	78	29.43	7215	0.009	58	7124	1.26%	28,543	28,692	0.52%
16	90	26.82	7857	0.009	63	7802	0.70%	22,310	22,486	0.79%
17	102	24.74	8471	0.010	68	8401	0.82%	15,344	15,523	1.16%
18	114	23.05	9055	0.011	73	9009	0.51%	7,693	7,892	2.58%

the manual approach. This simple example illustrates that the designer can use a simple spreadsheet for a preliminary design of the main elements of the bridge such as the cables and the deck due to a permanent dead load. At the conclusion of this step, advanced finite element modeling is essential to check the structure due to the remaining loads i.e. transient, thermal, and extreme loads.

10.6.2 Aspects Associated with Modeling and Analysis of Cable-Stayed Bridges

At this stage, the designer has already manually sized the main components of the bridge such as the deck and cables due to manual calculations of permanent loads. The next step is to use rigorous finite element modeling and analysis to complete the design due to other types of loads. It is important to note that cable-stayed bridges are long span structures that are characterized by geometrical nonlinearity. Under the action of dead loads the cables will exhibit large strains that will result in large deflections of the structure as illustrated in Figure (10.11). Nevertheless, these large deflections need to be eliminated by applying initial actions to the cables to achieve the final configuration of the bridge at the conclusion of construction. This can be achieved through different ways according to the FE program used for the analysis. Some software packages require a load case where initial strains are applied to the cables to induce cable shortening first, followed by application of the dead load. Other softwares can include the two load cases together and run superposition internally. Also, some programs are provided with an optimizer to adjust joint displacements or member forces to desired levels set by the analyst.

10.6.2.1 Composite Cable-stayed Bridges

The procedure of eliminating large deflections due to permanent loads is illustrated in Figure (10.12) for composite cable-stayed bridges. Figure (10.12a) displays the moment action on the two edge girders of the composite deck for the cable-stayed bridge of Figure (10.11) due to elastic action of the permanent load. Figure (10.12b) displays the moment actions due to the application of internal strains. A realistic moment actions on the girders is displayed in Figure (10.12c) after superposition of the above two load cases. The initial strains are those induced on the cables due to the system's geometric nonlinearity, but with a negative sign, so as to produce shortening of cables. Superimposing results in a structure with minimal deflections to be employed in subsequent steps for live loads and other load combinations.

Introducing initial cable strains to adjust large deflections has a significant effect on the longitudinal moments of the pylon legs. Large deflections of the main spans due to permanent loads also cause the two pylons to lean towards each other inducing large longitudinal moments at the pylon legs, a scenario that will never occur for the final configuration of the deck. At the end of the deck construction the stays are re-stressed to give the required deck profile and the distribution of moments along the length of the deck after installation and stressing the cables on site. Figure (10.13) demonstrates the significance of introducing cable shortening through initial strains to adjust the deck deflection on the longitudinal moments of the pylon legs. This can be observed through comparisons of the longitudinal moments before and after introducing cable shortening in Figure 10.14.

As the live load gets applied to the structure, the deck exhibits considerable deformations and moments due to the non-linear behavior of the cables. Slenderness of the deck, particularly the two edge girders, play a significant role in accommodating these moments. The slenderness is expressed as the ratio of the structure depth to the main span length.

The designer needs at this stage to adjust the design to compromise between increasing the depth of the deck to accommodate the live load demands and the additional load to be taken by the cables and the pylons. It may take a few iterations to arrive at the optimal slender ratio that will satisfy the live load demands. Figure (10.15) displays typical moment envelopes for a cable-stayed

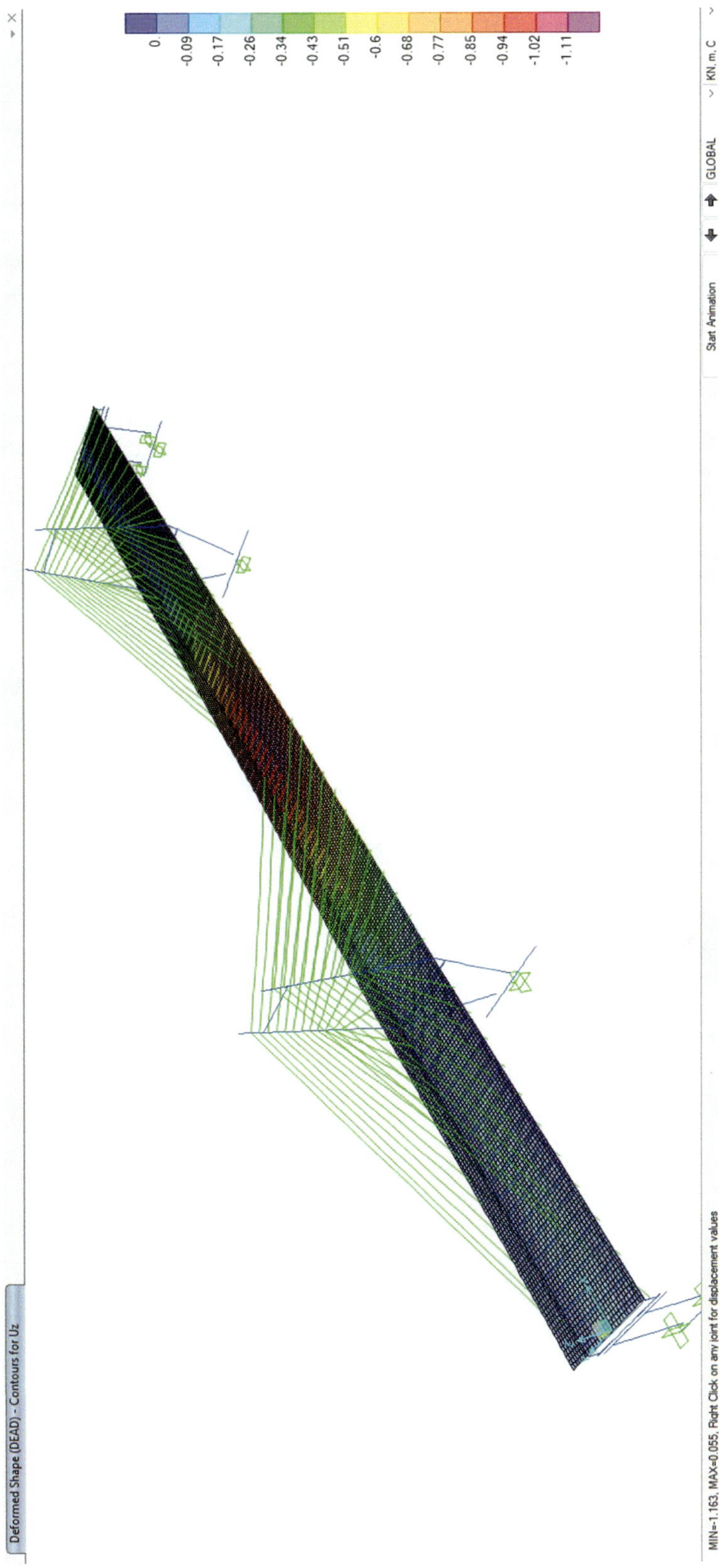

Fig. 10.11 Large deflections of a cable-stayed bridge due to permanent loads

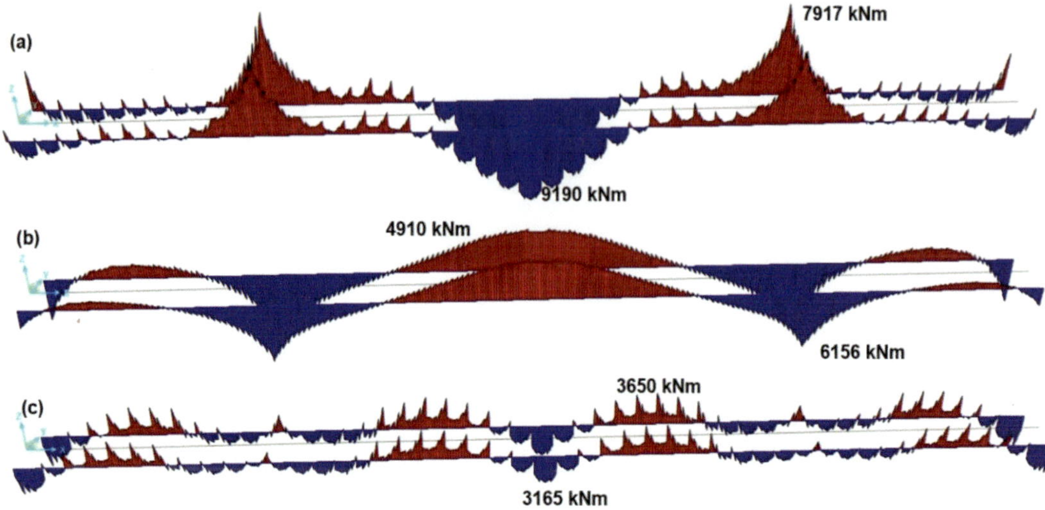

Fig. 10.12 Moment actions on deck due to permanent loads

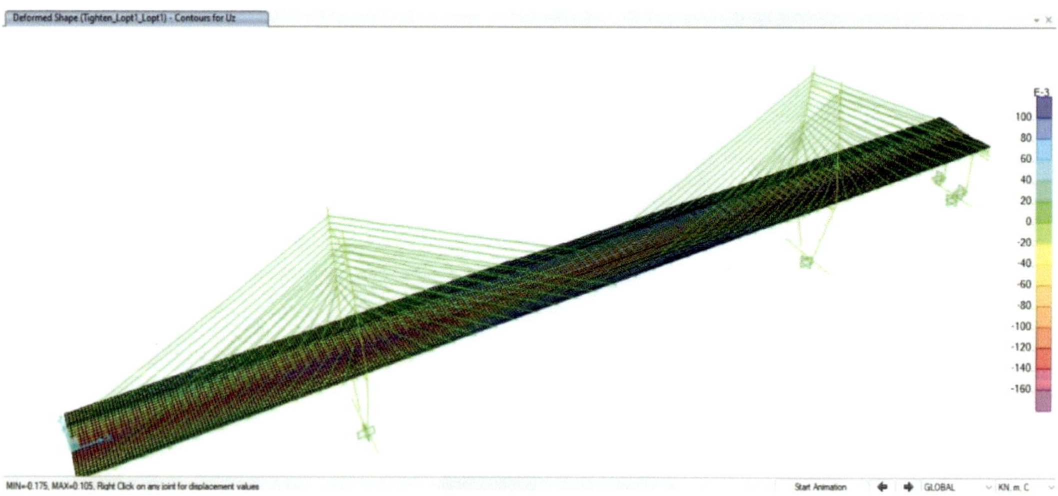

Fig. 10.13 Adjusted deflections for cable-stayed bridge

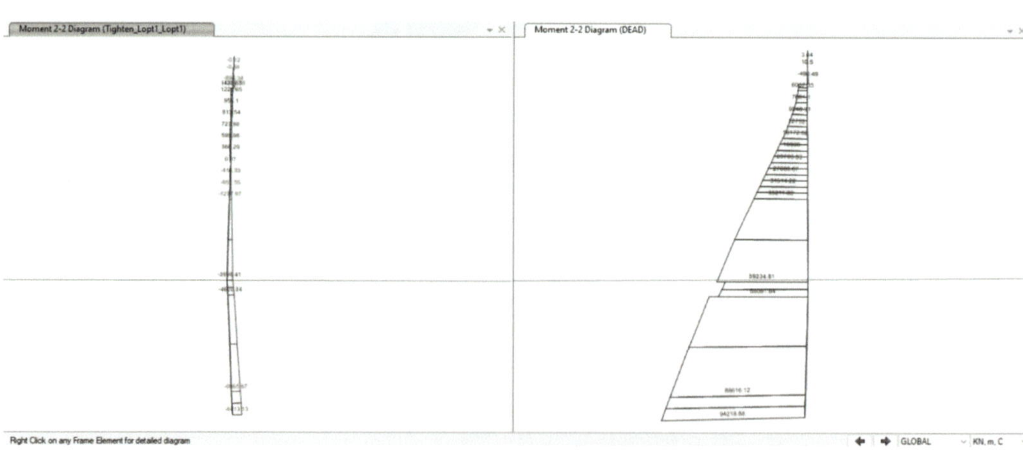

Fig. 10.14 Comparisons of Pylons longitudinal moments before and after cable shortening

bridge from live loads. The ratio of the positive to negative moment at the mid span is in the range of 2.5 to 1 and about 1.2 to 1.0 at the location of maximum moments at the side spans.

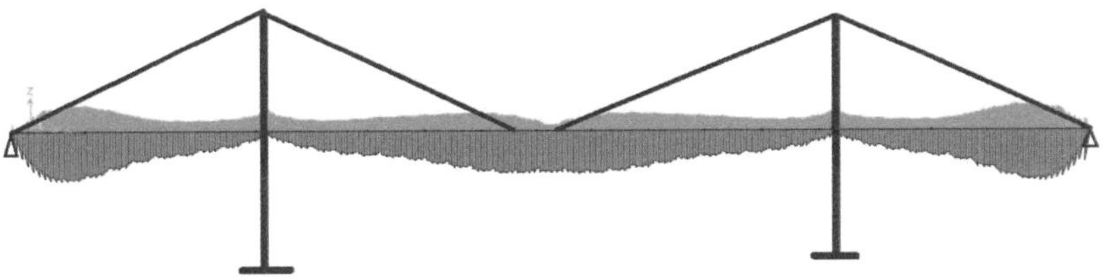

Fig. 10.15 Typical moment envelopes for a cable-stayed bridge due to live loads

10.6.2.2 Reinforced Concrete Segmental Cable-stayed Bridges

The analysis of cable-stayed bridges must incorporate the stage-by-stage construction of the deck particularly for reinforced concrete segmental cable-stayed bridges. The construction staging analysis must be conducted using software that supports this feature, wherein each stage of construction is included in the construction schedule. Large deflections in the pylon and deck cause p-δ effects to be significant, hence the structural analysis software used must handle geometric nonlinear

Wiz-Sched2 Options

	Task Description	Durat	Predeces	Finish Day	Operation	Object	Details	End Stage	End Case	Output Label
87	Cast Segment P02-SG01U	5		62	Add Structure	Segment P02-SG0	Age=1; SW=DEAD; SF=1	☐	☐	02-SG01-
88	Advance Traveler BTRAV1	1	87	63	Set Traveler for Segment	Traveler BTRAV1	to Cast Segment 'P02-SG02D'	☐	☐	
89	Advance Traveler BTRAV2	1	87	63	Set Traveler for Segment	Traveler BTRAV2	to Cast Segment 'P02-SG02U'	☐	☐	02-SG02-
90	Cast Segment P02-SG02D	2	89	65	Add Structure	Segment P02-SG0	Age=1; SW=DEAD; SF=1	☐	☐	
91	Cast Segment P02-SG02U	2	89	65	Add Structure	Segment P02-SG0	Age=1; SW=DEAD; SF=1	☐	☐	02-SG02-
92	Advance Traveler BTRAV1	1	91	66	Set Traveler for Segment	Traveler BTRAV1	to Cast Segment 'P02-SG03D'	☐	☐	
93	Advance Traveler BTRAV2	1	91	66	Set Traveler for Segment	Traveler BTRAV2	to Cast Segment 'P02-SG03U'	☐	☐	02-SG03-
94	Cast Segment P02-SG03D	2	93	68	Add Structure	Segment P02-SG0	Age=1; SW=DEAD; SF=1	☐	☐	
95	Cast Segment P02-SG03U	2	93	68	Add Structure	Segment P02-SG0	Age=1; SW=DEAD; SF=1	☐	☐	02-SG03-
96	Advance Traveler BTRAV1	1	95	69	Set Traveler for Segment	Traveler BTRAV1	to Cast Segment 'P02-SG04D'	☐	☐	
97	Advance Traveler BTRAV2	1	95	69	Set Traveler for Segment	Traveler BTRAV2	to Cast Segment 'P02-SG04U'	☐	☐	02-SG04-
98	Cast Segment P02-SG04D	2	97	71	Add Structure	Segment P02-SG0	Age=1; SW=DEAD; SF=1	☐	☐	
99	Cast Segment P02-SG04U	2	97	71	Add Structure	Segment P02-SG0	Age=1; SW=DEAD; SF=1	☐	☐	02-SG04-

Fig. 10.16 Example of a portion of a construction staging schedule for a segmental CSB.

behavior. An example of a portion of the construction staging schedule for a CIP segmental cable-stayed bridge is illustrated in Figure (10.16). Figure (10.17) illustrates a cable-stayed bridge at different stages during construction. Figure (10.17a) illustrates a stage where the pylons and piers are constructed, and construction of the concrete deck segments is in progress as both pylons with the travelers are shown for a better understanding of the process. Figures (10.17b) and (10.17c) illustrate the progress of construction at an intermediate stage and finally at the completion of the construction along with the installed and applied prestressing.

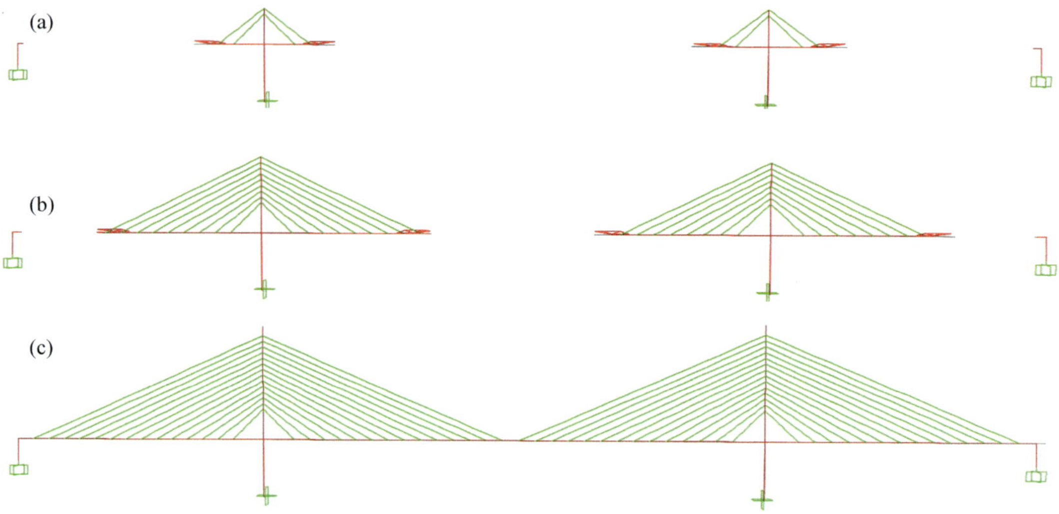

Fig. 10.17 Illustration of construction staging at a certain stage

It is very significant that the construction staging analysis captures all the details of construction since the distribution of internal forces in the structure varies according to each stage of the analysis and deck design is often controlled by these stresses and forces. Each stage must be checked for strength and serviceability requirements according to section 5.12.5.3 of AASHTO LRFD. At each erection stage the partially constructed structure is modeled with all the construction loads represented to simulate the actual moments, forces, stresses, and deflections. Figure 10.18 illustrates the distribution of moments in the deck and pylons during construction.

Fig. 10.18 Illustrating variations of bending moments on the deck and pylon during construction

Reinforced concrete segmental cable-stayed bridges are post-tensioned in the center of the bridge and near the anchor piers, where the normal forces from cables are small and the bending moment is large. The longitudinal compression induced by the stays adjacent to the pylon into the deck is usually enough to counteract any tensile stresses from the moments due to permanent loads. However, the little compression from the stays at midspan and near the ends of the deck are not sufficient to counteract tensions at these regions and the presence prestressing tendons is necessary to limit the stresses to acceptable values.

Figure (10.19) displays a typical axial force profile along the length of a reinforced concrete segmental cable-stayed bridge that is provided with prestressing tendons installed over the mid-span and at the ends of the deck. The profiles of these tendons are usually straight and can be installed either in the top slab or the bottom slab or in both and are anchored on blisters inside the box girder, with their numbers determined according to the stresses generated due to different types of loads.

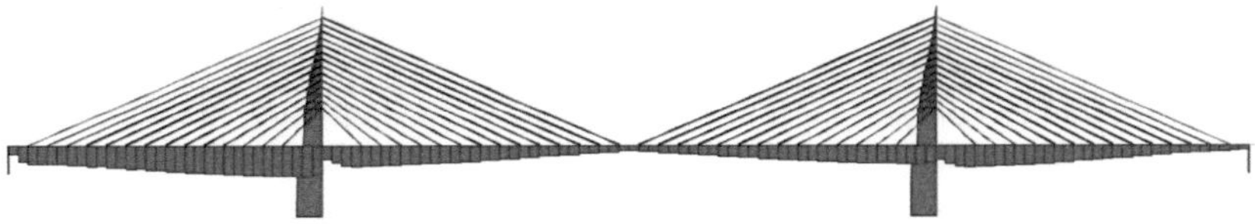

Fig. 10.19 Typical Axial forces distribution in a post-tensioned CSB deck

10.7 ASPECTS ASSOCIATED WITH THE DESIGN OF THE SUPERSTRUCTURE

10.7.1 Composite Superstructure

The composite superstructure is characterized by its ease of construction. Transverse floor beams at an average of 4.5 meters apart and two edge steel longitudinal plate girders make up the structure. The process of achieving integral composite action involves pouring concrete over the floor beams, edge girders, and lapping rebar and shear connectors in the strips between the precast panels. A layered view of a typical composite deck structure is shown in Figure 10.20. The structure is made up of floor beams, two edge steel girders, and high-strength precast deck slabs that are covered in a concrete wearing surface.

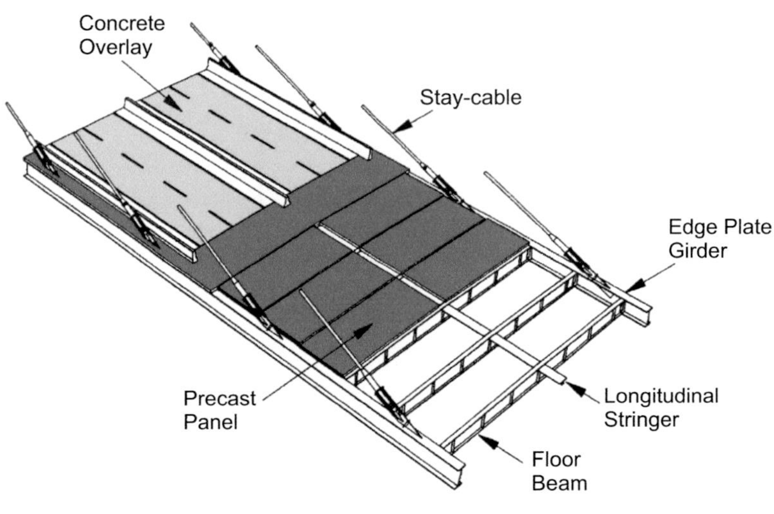

Fig. 10.20 Composite deck layered view

The floor beams and girders have a top flange provided by the composite concrete deck. The primary functions of the top steel flanges are to support the cast-in-place infill strips, shear connect, and stabilize the structure during erection. Together, the poured-in-place infill strips and precast deck panels create a broad, stiff deck diaphragm that helps bend floor beams and girders and distributes lateral and axial loads effectively. After the initial erection phase, deck cracking is not a major problem because the horizontal component of the stay-cable loads effectively compresses the deck in most areas. The following are important considerations for this kind of composite superstructure design (Taylor, 1987)

1. The critical aerodynamic stability is largely dependent on a shallower girder, which is produced by closely spaced cables, which also reduce girder bending moments.
2. Keep the backstay layout along the deck away from tie-down piers to prevent needless bending moment peaks from overloading the thin girder there.
3. Welding the cable gusset to the top flange is a simple arrangement that avoids creating biaxial stress states or secondary moments in that top flange zone.
4. It is preferable to add a central longitudinal stringer to control torsion in the floor beams during the erection of precast slabs.
5. The effect of initial girder imperfections like out-of-flat web panels requires consideration in both design and fabrication because axial girder stresses are high, especially in the areas near the towers.
6. The composite deck structural steel interface (Figure 10.21) requires careful detailing to allow for the required overlap of protruding rebars from the deck between the shear studs on the steel flange.
7. If good agreement with design geometry is to be achieved, creep effects in the deck must be taken into account in the bridge erection geometry calculations. Long-term tests of real job mixes should serve as the foundation for creep calculations. It is best to begin these tests as soon as possible.
8. If the deck panels are stored for sixty days before erection, the effects of shrinkage could be reduced.

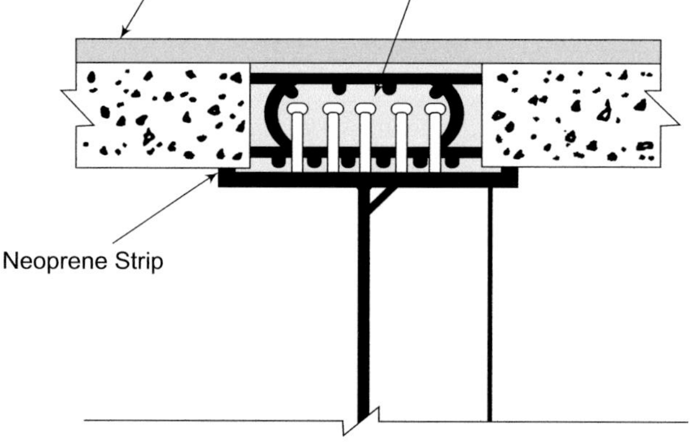

Fig. 10.21 Cast in place concrete closure detail

10.7.2 Orthotropic Deck

The orthotropic steel deck (OSD) system is typically comprised of a thin, flat steel plate that is stiffened by a row of longitudinal ribs that are closely spaced and positioned either orthogonally or at right angles to the floor beams, as shown in Figure 10.22. The deck is regarded as structurally anisotropic because of its distinct elastic properties in the longitudinal and transverse directions. The system was named orthogonal-anisotropic, or, in short, orthotropic, because the floor beams and ribs are both orthogonal and anisotropic (Troitsky, 1987). Because of their high compressive strength, OSDs are especially beneficial for cable-stayed bridges. Due to their ductile behavior and light weight, which lessen seismic inertial forces on piers and foundations, they are also a popular option for bridges in seismic zones. The main girders, FBs, deck plating, and ribs of the OSD bridge are all integrated into a single structural unit. In addition to distributing wheel loads, the deck panel serves as the top flange of the main girders and the FBs at the same time. Although the OSD uses materials very efficiently because it does all of these tasks, interactions still need to be considered.

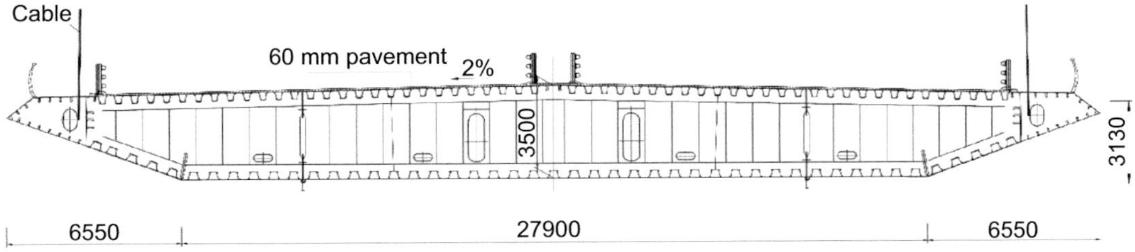

Fig. 10.22 Cross-section of a standard steel box girder panel

There are two types of OSD panels: torsionally stiff (closed) rib systems and torsionally soft (open) rib systems (Figure 10.23). To distribute wheel loads to floor beams, the ribs in either system are oriented in the bridge's longitudinal direction. The closed ribs' high torsional rigidity helps to better distribute concentrated transverse loads, which lowers the stresses in the deck plating. It is therefore favored over torsionally soft open-ribs. As a result, closed ribs are usually used to stiffen OSDs that are subjected to direct wheel loads, but open or closed ribs may be used to stiffen other structural elements. The closed ribs are trapezoidal-shaped, U-shaped, or V-shaped.

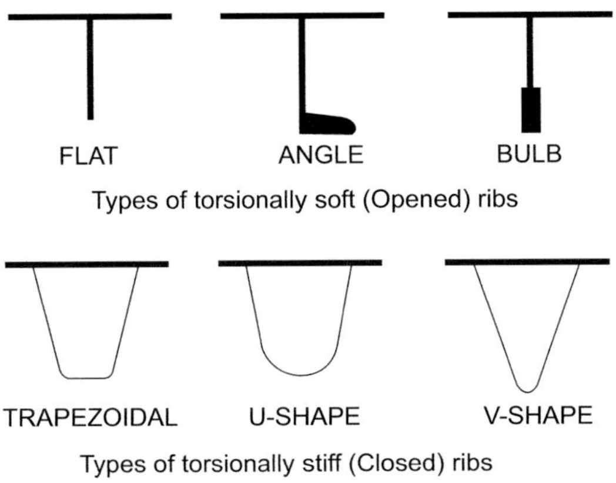

Fig. 10.23 Types of longitudinal ribs for orthotropic deck

10.7.2.1 Proportioning and Sizing

The deck consists of one continuous steel plate reinforced by a system of longitudinal ribs and transverse floor-beams. The thickness of the deck plate usually varies from 14 mm to 25 mm and depends on the rib spacing and loading requirements. Throughout the bridge's cross-section, open ribs range in size from 9 mm by 203 mm to 25 mm by 300 mm. They are formed of flat bars, bulb shapes, inverted T-sections, or angles. Most often, closed trapezoidal ribs are employed. Typically, floor beams are formed by rolling a section of steel in the shape of an inverted T-section or welding steel plates together; the deck plate forms the top flange. FBs are often separated by 3.05 to 6.1 m, based on the rib system being used. Selection of the rib spacing, span, stiffness, and thickness of the deck plate must be made during preliminary design, that is, before any analysis or testing is carried out. To improve the wearing surface's longevity and lessen stresses in the rib-to-deck weld, rib walls must be spaced apart such that localized bending of the deck plate and differential displacements between ribs from wheel loads are restricted. Side panels that are not subject to traffic loading can have their ribs positioned farther apart. Testing and/or analysis must be used to determine the final detailing dimensions, including the FB size and cut-out geometry (if used). Table 10.2 outlines the recommended limits for different elements of the OSD system (FHWA, 2012).

Table 10.2 Recommended limits for an Orthotropic system's elements

Detaling Dimension	*Limit*
Deck Plate Thickness	$t_d > 14$ mm
Rib Thickness	6 mm $< t_r <$ 12 mm
Rib Spacing-Direct Wheel Load	600 mm $< S <$ 762 mm
Rib Spacing-No Direct Wheel Load	600 mm $< S <$ 1000 mm
Floor-beam Spacing	$L <$ 6000 mm
Ratio pf Rib-to-Floor-beam Depth	$h_{rib}/H_{fb} < 0.4$
Floor-beam Web Thickness	10 mm $< T_{fb} <$ 20 mm
Ratio of Cut-out to Rib Depth	$h_{cuyout}/h_{rib} > 0.33$

10.7.2.2 Fatigue Issues of Orthotropic Decks

Because of its torsional and flexural rigidity, the closed rib system is most frequently utilized for orthotropic decks; however, the one-side partial penetration weld that connects the ribs to the deck plate (the RD connection) presents challenges. Furthermore, at the floor-beam intersection (RF connection), closed ribs are vulnerable to fatigue due to local secondary deformations and stresses. In particular, a lot of research has been done on this connection. Figure 10.24 shows details with and without cut-outs, internal bulkheads with and without, and different types of welds and plate sizes. Furthermore, a great deal of research has been done on the geometry of the cut-out itself in cases where they have been employed.

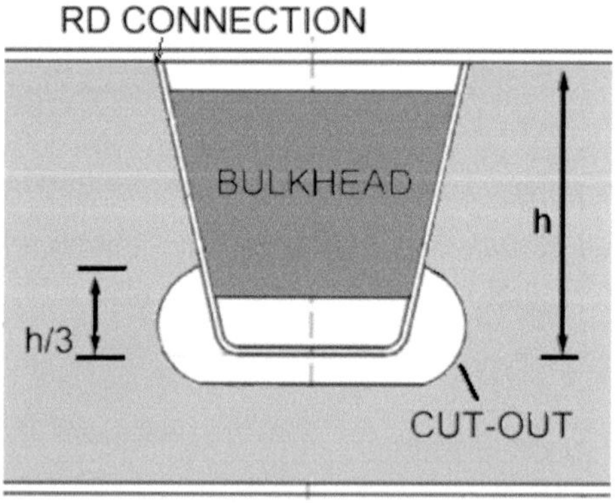

Fig. 10.24 Rib-to-floor-beam connection

A calculated, computed, or measured stress range is compared to a permissible stress range in order to evaluate fatigue. This comparison process, which makes use of rigorous finite element methods, ranges from the very simple to the very complex. It is crucial to remember that, even though they are "easier" to use, simplistic approaches can lead to extremely conservative and expensive designs. More importantly, unconservative designs can result from oversimplified analysis.

10.7.2.3 Design Levels

FHWA developed the idea of different design levels (FHWA, 2012). The principles are in place for a well-designed component, complete with all of its details, to potentially become a standardized modular part that can be applied to other similar applications down the road. Some bridges, however, have special qualities that call for a more thorough investigation. As a result, the Design Level is established based on the application type and any test data that the designer has access to. Three tiers of design exist. Level One Design is completed by choosing details that have been shown to have sufficient resistance in either new tests or by the availability of test data from earlier specimens with details and designs comparable to those suggested for a new project. Thus, it is not substantial.

Simplified 1-D or 2-D analysis techniques are used to analyze specific panel details as the bases for Level 2 Design. Calculations do not account for local concentrations; only nominal stresses are considered. For this purpose, the Vierendeel Model (Figure 10.25) can be applied (Haibach and Plasil, 1983). The upper chord in this model represents the real deck-plate, accounting for shear lag effects. After that, this chord is pinned to two-part vertical posts. The upper part has finite axial and flexural stiffness. It reflects the web area between the lower point of the cutout and the deck-plate. The stiffness characteristics are determined by considering the width of the undisturbed web at the halfway point of the rib's height. The lower part which is characterized by infinite stiffness reflects the undisturbed web area between the position of the lower chord of the model and the lowest point of the cutout. The lower chord comprising the lower flange of the floor-beam and the undisturbed web area is the last part. It forms an inverted T section in the substitute model.

Refined 3-D finite element analysis of the panel forms the basis of Level 3 Design, which precisely assesses the stresses on all parts and connections. As illustrated in Figure 10.26, computations take into account local stress concentrations at fatigue-vulnerable details. In order to model the bridge superstructure system, a global model of the panel's details must be nested inside a sub-model. In this case, the boundary conditions are defined by taking a small region from the larger model and utilizing displacements from the larger model.

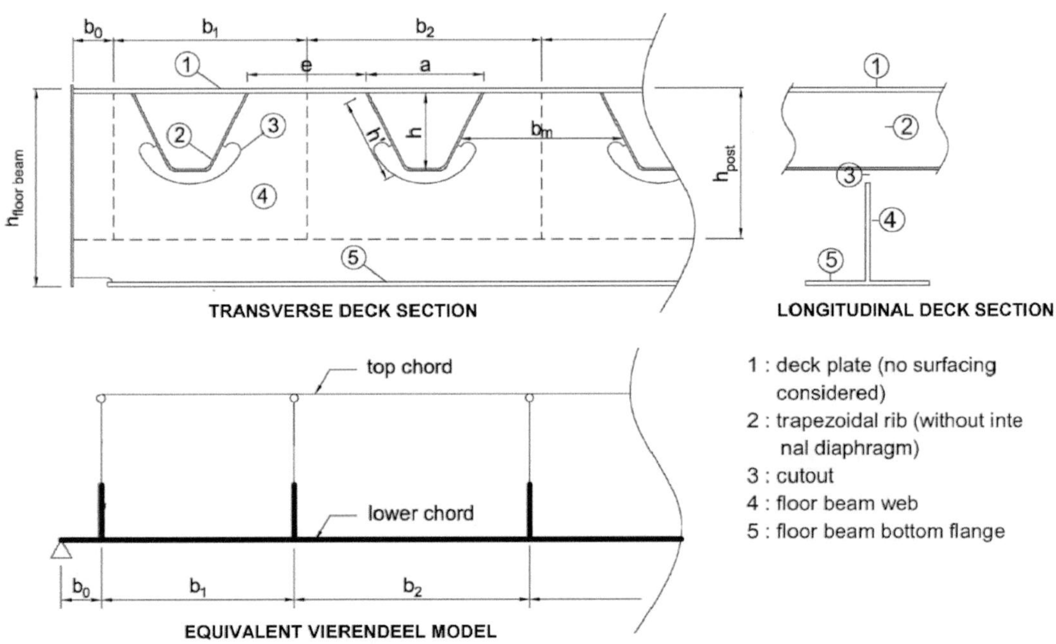

Fig. 10.25 Vierendeel model concept (DeCorte and Van Bogaert, 2007)

This permits the mesh inside the region to be refined to improve the accuracy over that obtained in the larger model. The displacements must be applied with a great degree of numeric precision.

Design Levels 1 and 3 are required for cases where no previous test data is available for a panel, unless it can be proved that floor-beam and rib distortions will not lead to fatigue cracking. Strength, service, and constructability only require a Level 2 design in general.

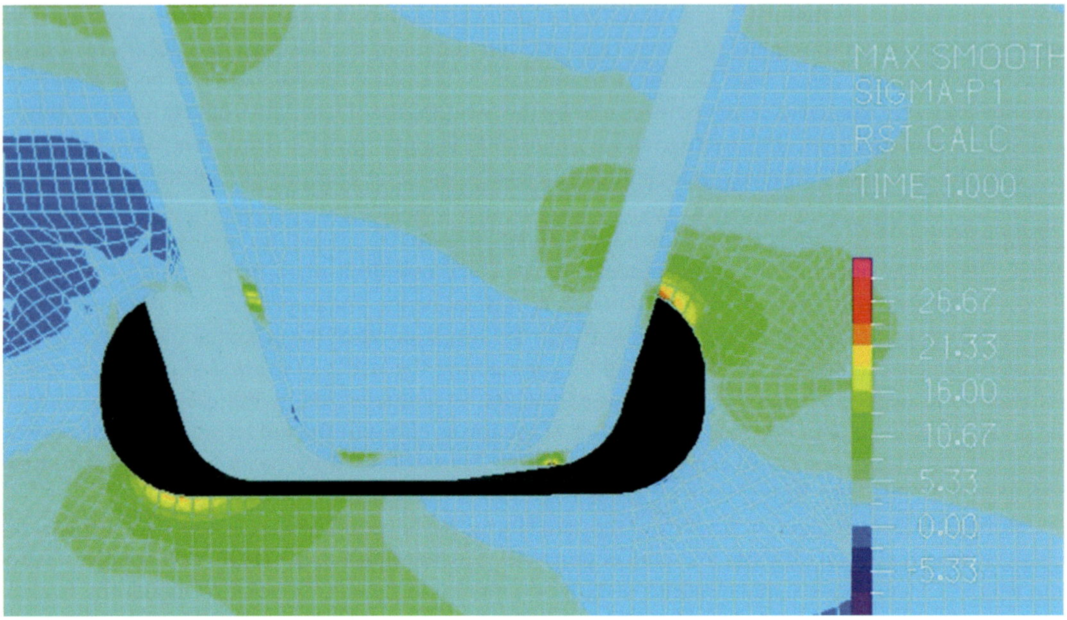

Fig. 10.26 Fatigue evaluation of Rib-to-floor-beam connection using FEM

10.7.2.4 Design Limit States

The design of orthotropic panels is generally governed by all strength limit states that contain both live load and dead load as the primary loads in the combinations. According to AASHTO LRFD (AASHTO, 2020), these are the Strength I and Strength II limit states. The states of these strength limits are necessary for yielding as well as buckling. While the strength II load combination is used with owner-specified permit loads, the strength I load combination is applied in conjunction with the HL-93 notional live-load model representing random traffic. The service I limit state must be satisfied for overall deflection limits for the deck plate (span/300) and the ribs (span/1000) and relative deflection of adjacent ribs (2 mm). The purpose of these deflection limits is to stop the wearing surface from deteriorating too soon. The design of FB and rib splices should take into account for the Service II limit state. For orthotropic deck systems, which are controlled by wheel loads and undergo millions of repetitive cycles of wheel loads, Fatigue I for infinite-life design must be applied.

10.7.3 Precast Box Girders

If the bridge is suspended with cables in the median only, the trapezoidal box sections provide good aerodynamic stability and the precast box-shaped section provides good torsional stability for carrying unsymmetrical loads acting transversely along the top flange. The required amount of torsional rigidity depends on the width of the deck and the length of the main span. The section is usually supplied with two diagonals to which the stay cable is anchored as illustrated in Figure (10.27). The diagonals are post-tensioned down to the webs and the vertical component from the stay cable is transferred into the web via the two diagonals. The tendons run through the top and bottom slabs and are anchored on blisters inside the box as illustrated in Figure 10.28.

In order to withstand bending stresses, the material for box girders is concentrated at the extreme fibers. Additionally, under ultimate loads, the large flange areas allow for the full utilization of post tensioning tendons without causing loss in the lever arm. Furthermore, for a given volume of concrete, the box section provides the least amount of post tensioning steel needed.

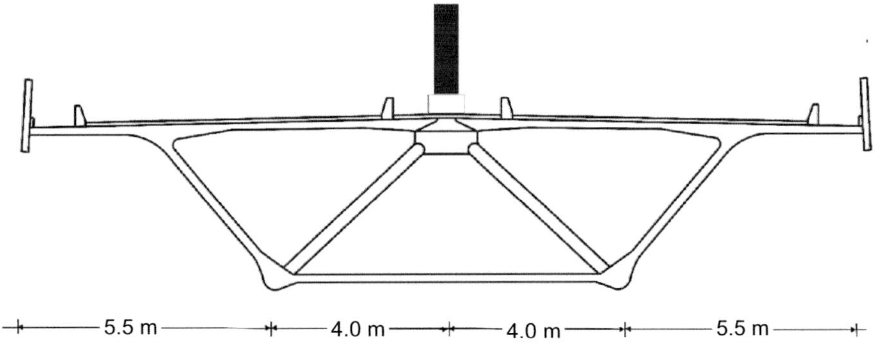

Fig. 10.27 Typical box girder for a single plan CSB

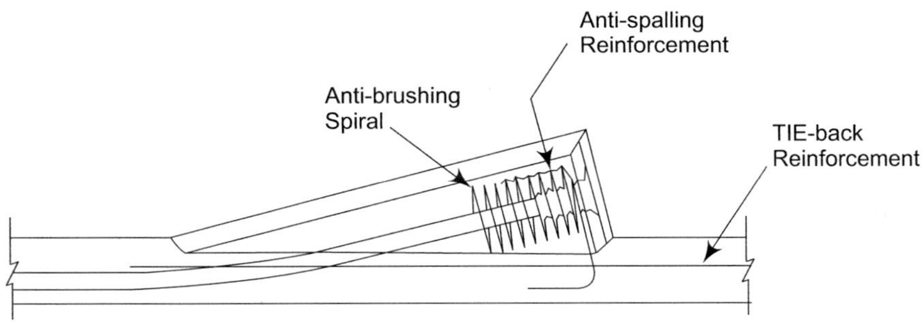

Fig. 10.28 Pos-tension blisters

The box sections efficiency was defined by Podolny and Muller (1982) as:

$$\rho = \frac{r^2}{c_1 c_2} \qquad ...(10.41)$$

where,

r = Radius of gyration, $\sqrt{\frac{I}{A}}$

c_1 = Distance from neutral axis to the extreme top fiber

c_2 = Distance from neutral axis to the extreme bottom fiber

I and A are the moments of inertia of the section and its cross-section area respectively.

10.7.3.1 Cross Section Proportioning

The following criteria can be used for sizing precast concrete box girder superstructures:

- The ratio of girder depth D to bridge width B for the cable layout of two planes can be taken as (Huang and Hu, 2020):

$$\frac{D}{B} \geq \frac{1}{10} \qquad ...(10.42)$$

- The ratio of girder depth to center span length generally ranges as (Huang and Hu, 2020):

$$\frac{D}{L} = \frac{1}{200} \text{ to } \frac{1}{100} \text{ (for two planes of cables)} \qquad ...(10.43a)$$

$$\frac{D}{L} = \frac{1}{100} \text{ to } \frac{1}{50} \text{ (for single plane of cables)} \quad ...(10.43b)$$

- The following range of the clear span to deck thickness ratio may be used

$$\frac{L_{\text{clear}}}{17} \leq t_{\text{slab}} \leq \frac{L_{\text{clear}}}{14} \quad ...(10.44)$$

- Linear haunches may be utilized at the web to save concrete and reduce weight. This solution can be pursued in lieu of thickening the entire slab. The length and amount of thickening for the haunches shall be determined by analysis and detailing considerations (AASHTO LRFD, 2020).
- Top and bottom flange thickness must not be less than 1/30 of the clear span between webs or 1/30 of the span between the thinner ends of the haunches where haunches are present in the flange span as illustrated in Figure (10.29). In general, top flange thickness must not be less than 200 mm (AASHTO LRFD, 2020).
- The minimum thickness of webs with only longitudinal or vertical post-tension tendons is 300 mm, or 380 mm for webs with both longitudinal and vertical tendons.

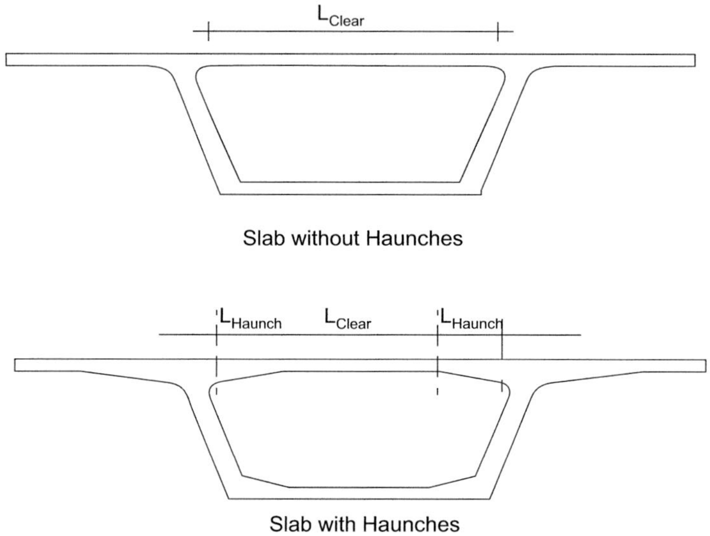

Fig. 10.29 Clear Span for slabs with and without haunches

- Where the clear span between the faces of webs is 4.5 m or larger, transverse prestressing of the top deck must be utilized.
- The cantilever length of the top flange measured from the centerline must not exceed 0.45 of the interior span of the top flange measured between the centerline of the webs.
- Cantilever length tip thickness ranges from 0.25 m to 0.30 m depending on the level of crash testing associated with the barrier being used. The thickness of the root at the intersection of the cantilever and the web varies with cantilever length. For cantilever length less than 1.5 m, the thickness of the root can be calculated as:

$$t_c = \frac{L_c}{1.5}(300) \geq t_{\text{tip}} \quad ...(10.45a)$$

and for overhang lengths larger than 1.5 m

$$t_c = 300 + (83.3L_c - 127) \quad ...(10.45b)$$

where, L_c = length of the cantilever wing (m)
t_{tip} = thickness of the overhang tip (mm)
t_c = thickness of the overhang root (mm)

An example is given to illustrate the use of the above equations. Consider a cross-section that handles 4 traffic lanes 3.6 m wide, a median to separate traffic traveling in opposing directions, and a 1.5 m shoulder to facilitate guard rails and bicycle traffic. The center suspended span is 150 m long and a single plane of stay cables is utilized. Therefore, by running the mathematics, a 2.5m central median to facilitate pylon and guard rails plus 4 lanes @ 3.6 m = 14.0 m and 2 shoulders @ 1.5 m = 3 m. Using these dimensions as guidelines, a width of 19.8 m is selected for the upper slab. The depth of the section will be taken as 3m, which is the upper bound of equation 10.43b to take advantage of the increased moment of inertia and lengthened lever arm for prestressing design. The cantilever length will be taken as 3.6 m, about 0.29 of the interior span of the top flange measured between the centerline of the webs and less than the maximum limit of 0.45. Equation 10.45b is used to calculate the thickness at the root of the cantilever. For a cantilever length of 3.6 m the thickness at the root is calculated as, 300 + (83.3*3.6 – 127) = 473 mm. The thickness at the tip is selected as 0.30 m. A value of 300mm is selected for the thickness of the web. By applying equation 10.44, with a clear length of 6m there is an interior brace acting as a support and thickening the top slab at the center, a value of 0.4 m since was used for the thicknesses of both top and bottom slabs. Finally, the width of the bottom slab is calculated by assuming a web batter of 1:4. Knowing that the cantilever length is 3.6 m, the width of the bottom slab is calculated as 11.4 m. In view of the minimum widths and thicknesses previously determined, the following dimensions, as shown in Figure 10.30, were chosen for the design.

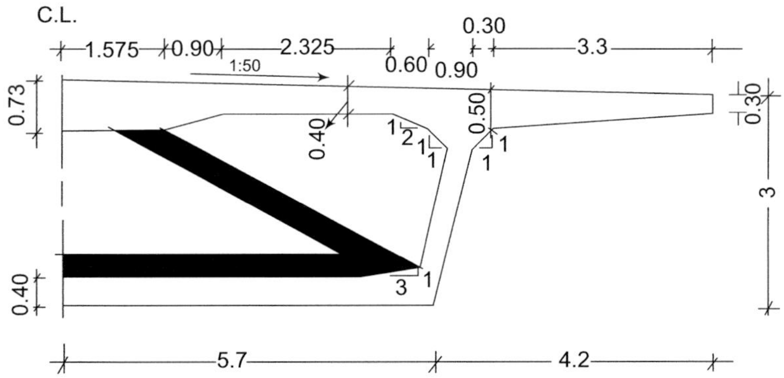

Fig. 10.30 Box section considered as example

10.7.3.2 Transverse Analysis

Due to both permanent and live loads, the cross section of a precast segmental box girder bridge exhibits bending moments acting transverse to the longitudinal direction of the bridge span. Depending on how the box girder deflects, the size and distribution of these moments vary along the length of the suspended and side spans. When general longitudinal deflections are distributed over a larger area of the span, they lower the maximum transverse moments within the span. On the other hand, localized bending moments are more concentrated close to piers and pylons, where deflections are restricted.

Quantifying transverse bending moments resulting from permanent loads, such as self-weight, barrier rails, sidewalks, wearing surfaces, and utilities attached to the box-girder superstructure, is the aim of the transverse analysis. Bending moments due to live loads, such as the HL-93 Design Truck and Design Tandem (AASHTO LRFD, 2020), must also be considered. A simplified, two-dimensional model that sufficiently accounts for longitudinal load distribution is considered an accepted approach for the transverse analysis. The typical cross section is modeled using beam elements. The transition from a typical cross section, to idealized beam members, to node and element layout is displayed in Figure 10.31.

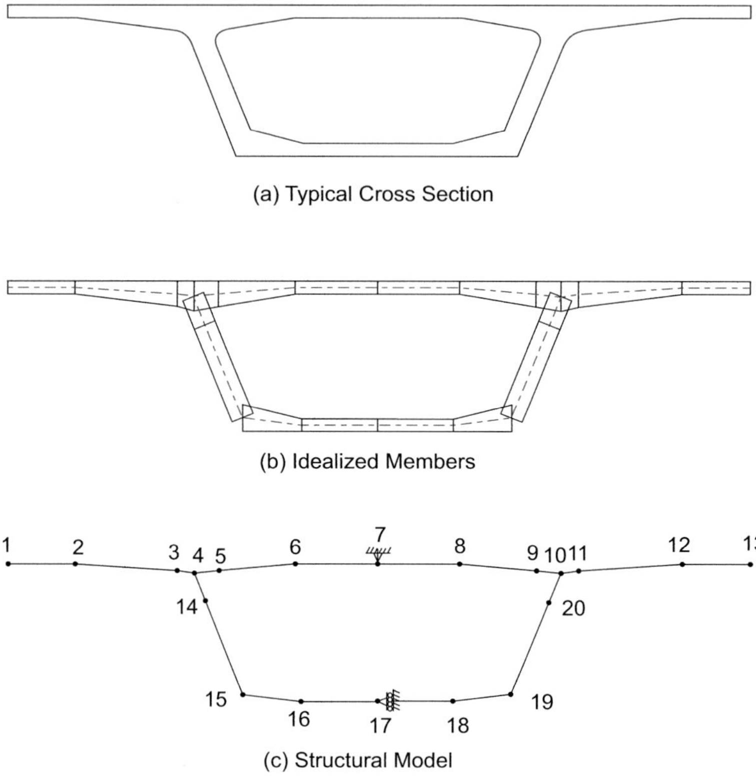

Fig. 10.31 Development of a two-dimensional transverse model (PCI, 2014)

A vertical support is placed at the location where a single plane of cables is attached to the structure as shown in Figure 10.31c. A second guided support that allows vertical movement and rotation can be added for model stability and to restrain the side-sway of the two-dimensional model. It also accounts for the torsional stiffness of the box section in an actual three-dimensional representation. Nevertheless, this additional horizontal support, however, can be problematic for two-dimensional transverse post-tensioning modeling. As shown, flanges are analyzed as variable depth sections, to capture the distribution of stresses realistically. The model shown in Figure 10.31b shows node-to-node connections by beam elements. It is left to the discretion of the engineer whether to use this approach or model a portion of these intersecting members as rigid elements.

Wheel loads must be arranged to provide maximum moments at critical sections, and elastic analysis shall be used to determine the effective longitudinal distribution of wheel loads for each load location. There are at least four-wheel load scenarios that need to be considered: case of wheel loads on the cantilever length; case of wheel load to produce critical negative load moments in the center of top flange; case of wheel loads to produce maximum negative moment in the top slab at the inside face of web; and case of wheel loads to produce positive live load moments at the centerline of the top flange. Figure 10.32 depicts these load cases. The number and placement of trucks and tandems inside the lanes are arranged in compliance with the AASHTO LRFD specifications in order to generate the greatest transverse bending moments at crucial points. The number of design lanes that should be applied for a particular critical section should be determined by considering an appropriate multi-presence factor.

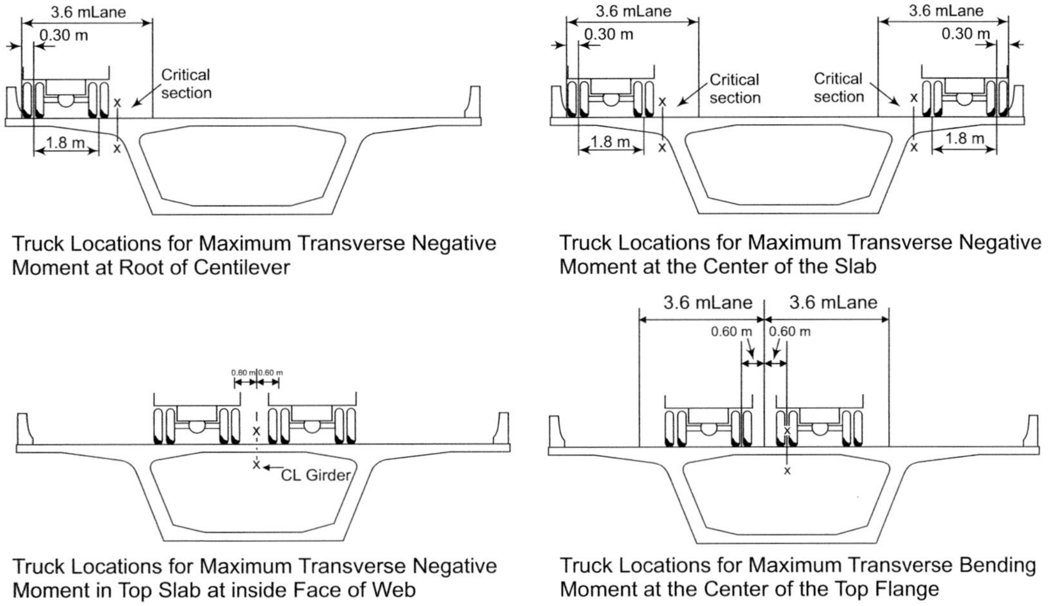

Fig. 10.32 Truck load Scenarios for maximum transverse bending moments (PCI, 2014)

The three-dimensional loading effects need to be considered in evaluating the concentrated wheel load for each case. This can be achieved through two different approaches. The first approach is by using theory based influence lines from the most popular sets of charts such as those by Baron (1941), or Homberg (1968). These charts are based on classical plate theory assuming small deflections and no shear deformations. Homberg charts display moments at a certain location due to unit loads positioned on the top slab. Several Homberg charts are available for various plate geometries, plate thickness, and support conditions. A Homberg chart provides the moment at a particular location such as the root of the cantilever (overhang), therefore, to determine the moment, the value on the chart at the load's location must be multiplied by its magnitude. Figures (10.33) and (10.34) display two Holmer charts that are usually employed for calculating transverse moments in the top slab of a precast, prestressed concrete segmental box girder bridge due to moving loads. Figure (10.33) would be used to analyze the bending moment at the root of the cantilever portion of the top slab. The top edge of the chart (the y axis) represents the fixed edge that is shown as point "3" in the figure of the cantilever below the chart. Figure (10.34) is for a doubly fixed plate, which can be used to analyze the negative fixed end moments at the top slab. Figure (10.35) is for a doubly fixed plate, which can be used to analyze the positive fixed moment at the center line of the top slab. In a subsequent step, the fixed end moment obtained from the charts is applied as external forces to the 2D frame to obtain the distribution of moments on other members. As shown, the charts are developed for a specific ratio for the support depth to the midspan, which limits its applicability to other deck configurations. Hence, this method is approximate and can be used only for preliminary design.

Three-dimensional finite element analysis is a more accurate approach for estimating the transverse moments. Nowadays, the availability of three-dimensional finite element software packages has enabled the longitudinal and transverse effects to be determined directly from a single analysis model. Most of the structural engineering programs such as CSiBridge, MIDAS and LARSA can develop influence lines for beam and frame elements and influence surfaces for shell elements. Also, several types of vehicles that cover most of the codes worldwide are available. Therefore, from three-dimensional analysis influence lines can be generated at any transverse section. Three-dimensional finite element analysis can further provide a direct envelope of maximum transverse moments at any section of interest. It is recommended that this method be used for final design. The results of the transverse analysis are used to design the reinforcement in the cantilever wings, top flange, webs, and bottom flange and post-tensioning in the top flange. Service Limit State I should be used for checking tension related to transverse analysis.

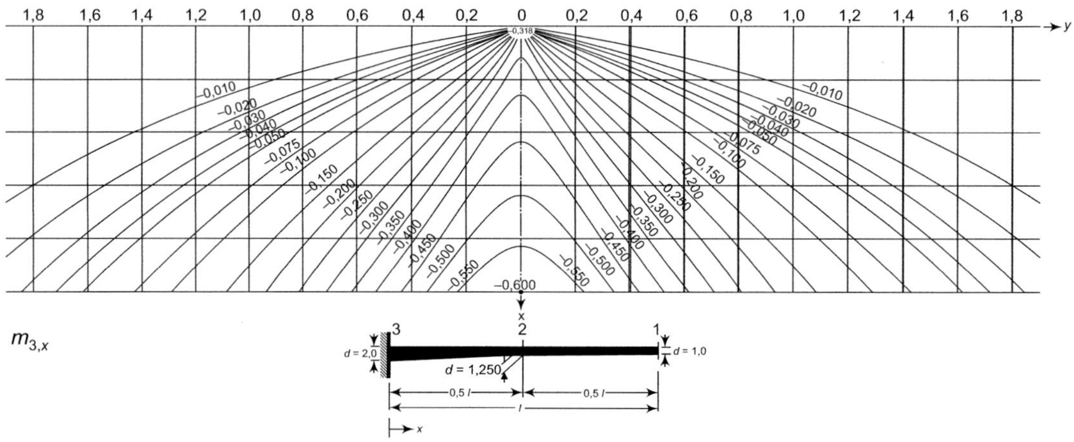

Fig. 10.33 Homberg Influence surface at location 3 (m_{3x}) for the cantilever flange

10.6.3.3 Longitudinal Analysis

Longitudinal analysis of segmental cable-stayed bridges must account for construction staging and the method of construction as well as the time-dependent effects of concrete shrinkage, creep, and prestress losses. AASHTO LRFD Section 5.12.5.3 provides construction load combinations, stresses, and stability considerations that need to be considered in an erection analysis. Both uniform temperature and temperature gradient must be included in service limit state load combinations. AASHTO LRFD requires a load factor of 1.0 for uniform temperature when checking stresses, and 1.2 for structural deformations. For a temperature gradient, a load factor of 1.0 is required at the service limit state when live load is ignored A load factor of 0.50 is required at the service limit state when live load is considered. The service limit state design of the superstructure requires a stress check for Service Limit State I, Service Limit State III, and an additional load case for segmental bridges that is defined by AASHTO LRFD Equation 3.4.1-2. This load combination has no live load; hence 100% of the temperature gradient needs to be included.

The service limit state stress calculations must account for the secondary moments due to prestressing. The serviceability stress check controls the prestress requirement. Determining the stresses on the section resulting from applied loads and construction effects, and then estimating the prestress required to keep these stresses within the allowable limits, yields the number of prestress tendons required at critical sections, such as mid-span, pylons, and piers. Since the prestress secondary moments are unknown, it is prudent to calculate the actual secondary moments after determining the necessary prestress layout and make sure the results match the estimated values in order to provide the final design. This process may require multiple iterations. Following the

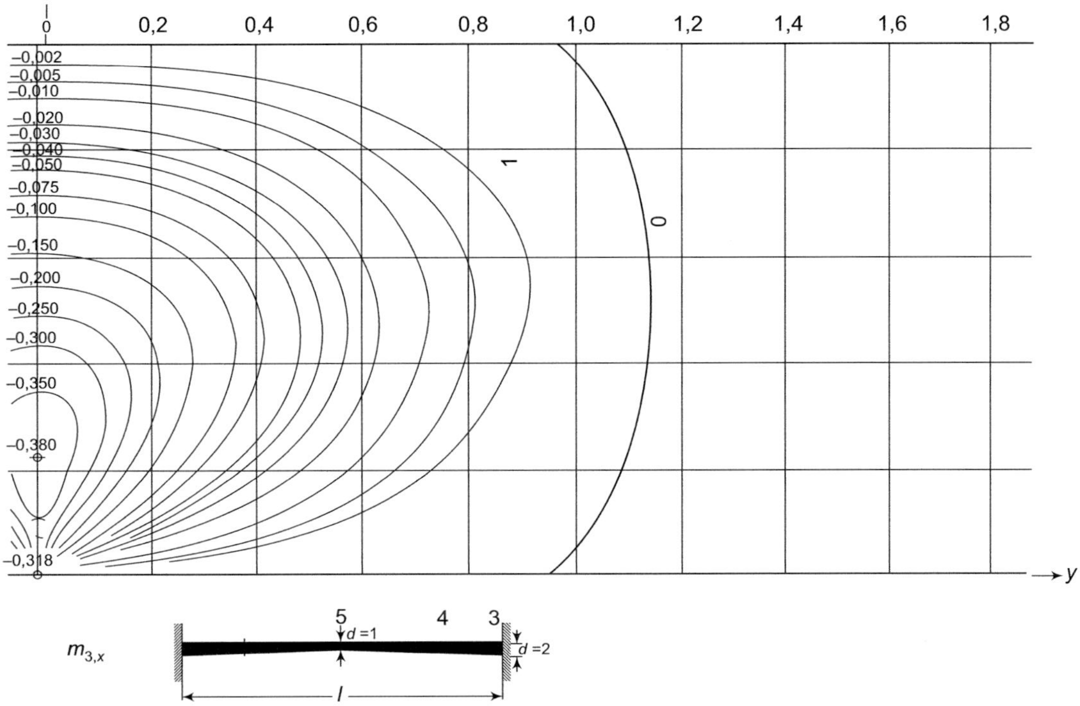

Fig. 10.34 Homberg Influence surface at location 3 (m_{3x}) for the negative moment at top slab

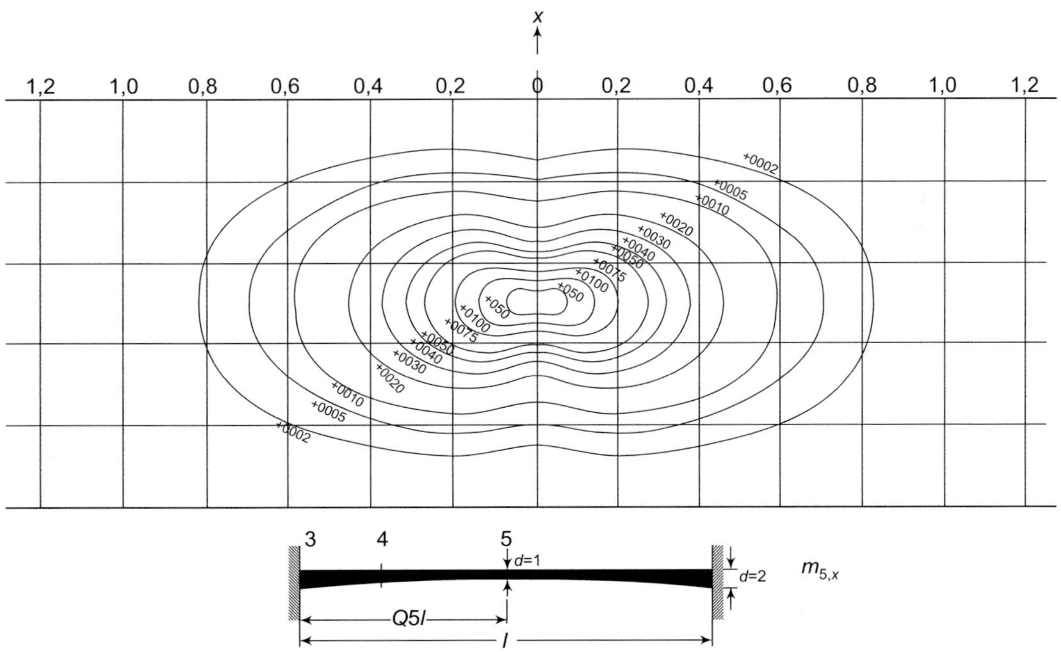

Fig. 10.35 Homberg Influence surface at location 5 (m_{5x}) for the negative moment at top slab

calculation of this step, the ultimate moment is examined in light of the strength limit states at the critical sections. If more tendons are required to guarantee the achievement of sufficient resistance, they may be added. Should more tendons be needed, it is necessary to reevaluate the serviceability stresses to make sure they remain within acceptable limits.

At the strength limit state, the nominal values of the secondary force effects caused by prestressing must be added algebraically to the other relevant factored loads. Redistribution of construction-stage force effects owing to internal deformations and modifications in support and restraint conditions, including cumulative locked-in force effects arising from the construction process, will be examined in the final structural system analysis.

10.7.3.4 Design of Segmental Box Girders

Principal tensile stresses in webs need to be checked. Diagonal tension cracks could occur if the concrete's permissible tensile strength is exceeded. Maximum principal tensile stress is typically restricted between $0.249\sqrt{f_c'}$ to $0.33\sqrt{f_c'}$ (N/mm^2). The combination of axial and shear stress that results in the largest principal tensile stress will be used to calculate the principal tensile stresses. If present, vertical compressive or tensile stresses must also be taken into account. Vertical post tensioning bars may be used to control the principal tension stress if analysis at the critical section indicates that the maximum principal tensile stress is greater than what is permitted. Altering the cross section (web thickness) or increasing the longitudinal compressive stress (additional strands) are two more ways to address the overstress. The principal tensile stress is checked by first determining the shear stress in the vertical web. The vertical shear stress is calculated as (AASHTO LRFD, 2020):

$$\tau = \frac{VQ}{Ib} \qquad ...(10.46)$$

where,

V = Vertical shear force
Q = First moment of an area with respect to the CG of the section
I = Moment of inertia about the CG of the section
b = total web thickness at the top of the web

The principal stress can now be quantified based on the analysis using Mohr's Circle as shown in Figure 10.36 as:

$$f_{\min} = \frac{1}{2}\left((f_{pcx} + f_{pcy}) - \sqrt{(f_{pcx} - f_{pcy})^2 + (2r)^2}\right) \qquad ...(10.47)$$

where,

$f_{\min}$ = minimum principal stress in the web, negative tension
f_{pcx} = horizontal stress in the web
f_{pcy} = vertical stress in the web

Checking the limit state of the flexural strength is prudent and must be checked. In general, Strength Limit State I governs. However, for longer spans where the ratio of the dead load to live load is large, Strength Limit State IV needs also to be checked for cable-stayed bridges. The top and bottom flanges need to conform to flexural strength requirements. Article 5.6.3.2 of AASHTO LRFD requires that

$$M_u \le \phi_f M_n \qquad ...(10.48)$$

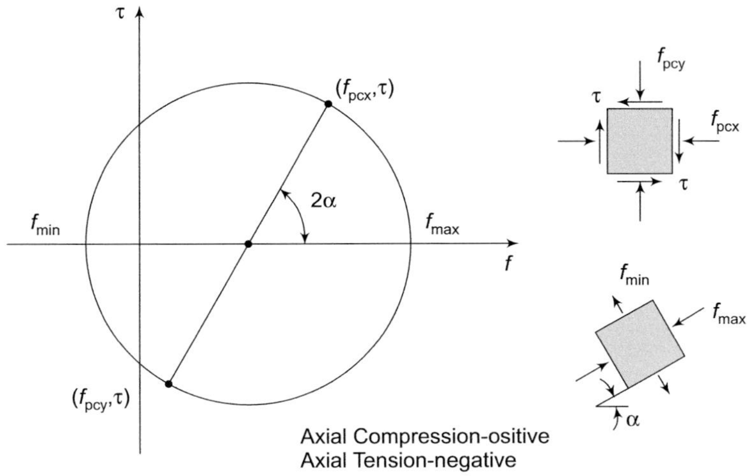

Fig. 10.36 Mohr's Circle of Principal Stresses in Box Girder's Web (AASHTO LRFD, 2020)

In inequality 10.58 M_u = factored applied moment; M_n = nominal flexural resistance (N-mm); and ϕ_f = resistance factor for flexure taken according to Article 5.5.4.2 of AASHTO LRFD. The nominal flexural resistance can be calculated as:

$$M_n = A_{ps} f_{ps}\left(d_p - \frac{a}{2}\right) + A_s f_s\left(d_s - \frac{a}{2}\right) \quad \text{...(LRFD 5.6.3.2.2-1 10.49)}$$

where, A_{ps} = are of prestressing steel and A_s = area of non-prestressed steel (mm^2); f_s = stress in the non-prestressed compression reinforcement at nominal flexural resistance (MPa), as specified in Article 5.6.2.1 of AASHTO LRFD; d_p and d_s = distances from extreme compression fiber to the centroid of prestressed and non-prestressed tensile reinforcement respectively; a = depth of the equivalent stress block (mm) evaluated in terms of the distance from the extreme compression fiber to the neutral axis c, and the stress block factor specified in Article 5.6.2.2 of AASHTO LRFD as, $a = c\beta_1$; and f_{ps} = average stress in prestressing steel at nominal bending resistance (MPa) evaluated for sections with bonded tendons as:

$$f_{ps} = f_{pu}\left(1 - k\frac{c}{d_p}\right) \quad \text{(LRFD 5.6.3.1.1-1 10.50)}$$

$k = 0.28$ for low relaxation strands; f_{pu} = specified tensile strength of prestressing steel. The distance from the extreme compression fiber to the neutral axis can be calculated for sections with bonded tendons as:

$$c = \frac{A_{ps} f_{pu} + A_s f_s}{\alpha_1 f_c' \beta_1 b + k A_{ps} \dfrac{f_{pu}}{d_p}} \quad \text{(LRFD 5.6.3.1.1-4 10.51)}$$

where, f_c' = compressive strength of concrete (MPa); α_1 = stress block factor specified in Article 5.6.2.2.

AASHTO LRFD adopts the modified compression field theory for shear and torsion design of concrete sections. For sections subjected to shear only, the following basic relationship must be satisfied at each section:

$$V_u \leq \phi_s V_n \quad ...(10.52)$$

where, V_u = factored shear force; V_n = nominal shear resistance of the box girder section; and ϕ_s = resistance factor for shear taken according to Article 5.5.4.2 of AASHTO LRFD. Following AASHTO LRFD-Article 5.7.3.3, V_n is determined as the lesser of

$$V_n = V_c + V_s + V_p, \quad \text{(LRFD 5.7.3.3-1 10.53a)}$$

$$V_n = 0.25 f'_c\, b_v\, d_v + V_p \quad \text{(LRFD 5.7.3.3-2 10.53b)}$$

where,

$$V_c = 0.083\beta \sqrt{f'_c}\; b_v\, d_v \quad \text{(LRFD 5.7.3.3-3 10.54)}$$

$$V_s = \frac{A_v f_y d_v (\cot\theta + \cot\alpha)\sin\alpha}{s} \lambda_{duct} \quad \text{(LRFD 5.7.3.3-4 10.55)}$$

$$\lambda_{duct} = 1 - \delta\left(\frac{\lambda_{duct}}{b_w}\right)^2 \quad \text{(LRFD 5.7.3.3-5 10.56)}$$

in which V_c = nominal shear resistance of the concrete (N); V_s = shear resistance provided by transverse reinforcement (N); V_p = component of prestressing force in the direction of the shear; positive if resisting the applied shear; b_v = effective web width taken as the minimum web width within the depth d_v (mm); d_v = effective shear depth as defined in Article 5.7.2.8 of AASHTO LRFD (mm); θ = angle of inclination of diagonal compressive stresses which can be determined as $\theta = 29 + 3500\,\varepsilon_s$; β = factor indicating ability of diagonally cracked concrete to transmit tension and shear evaluated as:

$$\beta = \frac{4.8}{(1 + 750\,\varepsilon_s)} \quad \text{(LRFD 5.7.3.4.2-1 10.57)}$$

ε_s is the net longitudinal tensile strain in the section at the centroid of the tension reinforcement.

Tensile strain ε_s may be determined by the following equation:

$$\varepsilon_s = \frac{\left(\dfrac{|M_u|}{d_v} + 0.5\,N_u + |V_u - V_p| - A_{ps} f_{po}\right)}{E_s A_s + E_p A_{ps}} \quad \text{(LRFD 5.7.3.4.2-4 10.58)}$$

where M_u = absolute value of the factored moment at the section, not taken less than $|V_u - V_p| d_v$ (N mm); N_u = factored axial force, taken as positive if tensile and negative if compressive (N); A_{ps} = area of prestressing steel; A_s = area of non-prestressed tension steel; $f_{p0} = 0.7 f_{pu}$ the specified tensile stress for prestressing steel; E_s and E_p = moduli of elasticity of steel reinforcement and prestressed steel.

The modified compression field theory acknowledges that shear causes tension in longitudinal steel. For sections not subjected to torsion, the capacity of the longitudinal reinforcement is expressed as follows:

$$A_{ps} f_{ps} + A_S f_y \geq \frac{|M_u|}{d_v \phi_f} + 0.5\frac{N_u}{\phi_c} + \left(\left|\frac{V_u}{\phi_v} - V_p\right| - 0.5\,V_s\right)\cot\theta \quad \text{(LRFD 5.7.3.5-1 10.59)}$$

ϕ_f, ϕ_v, and ϕ_c are resistance factors for moment, shear, and flexure taken from Article 5.5.4.2 of AASHTO-LRFD.

Torsional effects shall be investigated where:

$$T_u > 0.25\,\phi T_{cr} \quad \text{(LRFD 5.7.2.1-3 10.60)}$$

where, T_u = applied factored torsional moment (N mm); ϕ = applicable resistance factor; and T_{cr} = torsional cracking moment (N mm) which can be calculated as:

$$T_{cr} = 0.328\, K_0 \sqrt{f_c'}\, 2A_0 b_e \qquad \text{(LRFD 5.7.2.1-5 10.61)}$$

where, A_0 = Area enclosed by shear flow path (mm^2); and b_e = Minimum effective shear flow web or flange width to resist torsional stresses (mm); and K_0 is evaluated as:

$$K_0 = \sqrt{1 + \frac{f_{pc}}{0.328\sqrt{f_c'}}} \le 2.0 \qquad \text{(LRFD 5.7.2.1-6 10.62)}$$

The shear and torsion design approach for segmental box girder bridges is outlined in Article 5.12.5.3.8c of AASHTO-LRFD. The nominal shear resistance, V_n, shall be determined as the lesser of the following:

$$V_n = V_c + V_s \qquad \text{(LRFD5.12.5.3.8c-1-10.63)}$$

$$V_n = \sqrt{f_c'}\, b_v d \qquad \text{(LRFD 5.12.5.3.8c-2 10.64)}$$

The shear resisted by the concrete section is calculated as:

$$V_c = 0.166 K \sqrt{f_c'}\, b_v d \qquad \text{(LRFD 5.12.5.3.8c-3 10.65)}$$

The shear provided by transverse reinforcement is determined as follows:

$$V_s = \frac{A_v f_y\, d}{s} \qquad \text{(LRFD 5.12.5.3.8c-4 10.66)}$$

where, A_v = Area of transverse reinforcement in all webs in the cross-section within a distance s (mm^2); s = spacing of stirrups (mm); d = 0.8 h and K = stress variable parameter evaluated as:

$$K = \sqrt{1 + \frac{f_{pc}}{0.166\sqrt{f_c'}}} \le 2.0 \qquad \text{(LRFD 5.12.5.3.8c-5 10.67)}$$

When torsional effects are considered, the cross-sectional dimensions shall satisfy:

$$\left(\frac{V_u}{b_v\, d_v}\right) + \left(\frac{T_u}{2A_0 b_e}\right) = 125 \qquad \text{(LRFD 5.12.5.3.8c-6 10.68)}$$

Also, the longitudinal and transverse reinforcement must satisfy the following conditions:

$$T_u \le \phi T_n \qquad \text{(LRFD 5.12.5.3.8d-1 10.69)}$$

where, T_n = the nominal torsional resistance from transverse reinforcement A_t computed as:

$$T_n = \frac{2A_0 A_t f_y}{s} \qquad \text{(LRFD 5.12.5.3.8d-2 10.70)}$$

The minimum additional longitudinal reinforcement A_l should satisfy:

$$T_n = \frac{T_n p_h}{2A_0 f_y} \qquad \text{(LRFD 5.12.5.3.8d-3 10.71)}$$

p_h = perimeter of the outermost centerline continuous closed stirrups (mm); A_l = Total additional longitudinal reinforcement required for torsion (mm^2); A_t = Total area of transverse torsion reinforcement in the exterior web and flange.

10.8 DESIGN OF PYLONS

The pylons are a dominant element of the cable-stayed bridge due to their prominence in the landscape and their ability to deliver a cost-effective design concept. Since pylons are usually subjected to high compressive loads due to the dead weight of the superstructure, their self-weight and live loads, advances in the design and construction of reinforced concrete have made the use of concrete increasingly favorable for pylon construction. Therefore, only the design of concrete pylons is considered, a choice motivated by the fact that steel alternatives are generally less economical than concrete alternatives.

In the preliminary design, several configurations should be investigated, and comparisons are drawn to come up with a design that would technically meet economical and schedule requirements. Also, a design that provides a highly constructible pylon.

Recognizing the benefits of vertical pylon legs on constructability, a H frame pylon with a single cross beam above deck would be the first choice. The cross beam stiffens the pylon in the transverse direction thus reducing bending demand in the pylon legs. Also, the vertical legs would ease the construction with minimal requirements for geometry control since usually a jump form is used. On the other hand, an A-frame is more aesthetically appealing than the H-frame option and provides a more aerodynamic stable system. Nevertheless, this option requires a wide footprint and may require a longer construction schedule that renders it an exorbitantly expensive option. The diamond Shape pylon may fill this gap but will require a significant increase in the stiffness of the lower section below the deck to overcome wind and seismic loads.

10.8.1 Longitudinal Design

In general, the deflection of the pylon head is due to the change in length of the backstay cable, resulting in pylon moments with the maximum moment at the bottom leg. It is important to note that the pylons are usually stiffened by the stays in the longitudinal direction and the moments developed along their lengths due to non-linear effects are subject to significant change if the vertical components of the cable loads are taken into account. The non-linear effects reduce with increasing stiffness of the pylon. Longitudinal design of the pylons is intended towards checking the stresses and estimating the deflections These checks should cover not only forces due to live loads, but also effects of temperature changes and time-dependent effects on the concrete. It is worth noting that creep effects will considerably get reduced since the longitudinal bending moments under permanent loads are minimized by an adequate arrangement of the stays. Distribution of the normal forces as well as bending moments along the pylon will provide bases for checking the load-bearing capacity and completing the design. Pylon legs and cross beams are usually designed as hollow sections.

10.8.2 Transverse Design

For pylons with a single leg or with legs that are not transversely linked, the cross-wind is generally the governing force. If the height of the pylons above the deck is very large, special investigation of the wind effects has to be conducted on a wind tunnel on a scale model, taking into consideration the physical location of the bridge. This investigation provides an insight on the coefficient of dynamic drag to be used in the estimation of transverse wind loads. For pylons with legs linked transversely, the governing load is the transverse bending caused by any deviation of the stays with respect to the centerline of the legs at the fixity. Transverse bending of the pylons under permanent loads is normally a predominant part of the total bending, and the influence of creep can be important.

The transverse moments due to wind are relatively small for straight A-pylons with respect to pylons with vertical legs because the loads are resolved into normal and tangential components to the leg. For diamond shaped pylons, the legs are pulled together underneath the main transverse beam in a diamond shape. In this case significant moments are created in the lower pylon legs which require an increase in their transverse width to increase their flexural strengths.

10.9 DESIGN OF FOUNDATIONS

The foundations of the pylons and piers are usually selected as a group of drilled shafts joined by a rigid reinforced concrete pile cap. The length of each shaft is separated into two distinct sections. The lower portion of the shaft is usually reinforced concrete, and the upper portion is reinforced concrete with an integral steel casing. The purpose of the casing in the upper portion is to serve as a form for the concrete when placed in open water or soft soils as well as a load carrying element to enhance the lateral capacity of the shaft. The lateral design of the shafts is usually controlled by seismic loads or vessel impact. Therefore, it is preferable to consider designing a pedestal on top of the pile cap to prevent vessel impact on individual piles and to protect the pylon leg sections above.

To achieve the required factored vertical resistance, the shafts have to be extended deep into the dense soil layers and get socketed into the rock A conventional drilled shaft of this type that would primarily utilize skin friction within the socketed length, which stems from the shear stresses at the interface of the shaft and the surrounding rock. Skin friction is usually mobilized under relatively low levels of vertical loads as compared to the end bearing resistance at the tip. Therefore, it is conservative to count only on the skin friction in this case. If the site is not underlaid by rock or rock is practically very deep, the skin friction resistance would be utilized or the designer may compromise between the skin and end bearing resistances according to the nature of soil stratifications in the site. Chapter 12 provides more details on foundations and soil-structure interaction.

10.10 CONSTRUCTION OF CABLE-STAYED BRIDGES

Cable-stayed bridges are in particular easily constructible because their intermediate systems are built using the balanced cantilever method from one pylon towards both sides. This section discusses construction systems for superstructure, pylons, and foundations

10.10.1 Construction of Superstructure

10.10.1.1 Precast Concrete Segmental Construction

Cable-stayed structures are usually built by the balanced cantilever method. The superstructure segments are often heavy (130-300 t), which require easy access to the site. Since the main span of a cable-stayed bridge usually crosses a navigation channel, then a barge can be used to transport segments to site. Erection of precast segments can progress for low-level bridges, using a floating crane. On the other hand, high-level bridges may use a derrick on the deck to lift the segments off barges. Segments can also be delivered from the top if their weight is not excessive.

As shown in Figure 10.37, a typical cycle of erection for a concrete segmental cable-stayed bridge entails alternating between pouring or erecting segments and stressing the stay-cables. A picture of a normal precast box segment erection in a cantilever is shown in Figure (10.38).

Critical stresses such as maximum negative bending in the deck, or maximum tension in the last erected stay-cable are usually attained at the time of segment lifting or pouring. When pouring or lifting segments, segmental box girders are regarded as a rigid superstructure that can distribute the loads among the previously erected stay cables, minimizing the negative moment in the deck. In order to facilitate re-stressing of the stay-cable in between lifts and further minimizing moments in the deck, stay-cable spacing is typically set up at two or three segment lengths. It is significant to note that, as Figure (10.37) illustrates, when heavy construction equipment, like a crane, is on deck, stay-cable forces and deck bending moments during construction become more crucial.

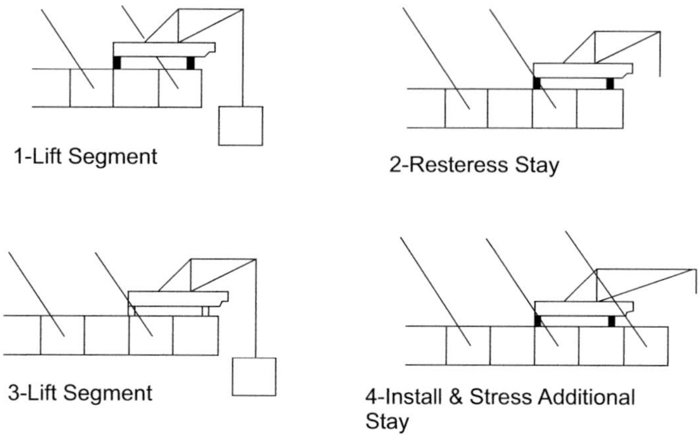

Fig. 10.37 Erection phases for box girder

Fig. 10.38 Typical erection of a box girder segment in cantilever (Bonzon, 2008)

During construction, the cantilevers can reach exceptional lengths, which may create asymmetrical dead and wind loading conditions on either side of the pylon that can result in critical moments in pylons and foundations. Therefore, it is significant to check the structural stability of the partially completed bridge during construction. Load combinations for checking stability during construction can be found in AASHTO LRFD Articles 3.4.2 and 5.12.5.3.2. Analytical investigations and wind tunnel tests must be conducted to assess the performance of the structure during construction. Wind tunnel testing and analytical wind studies must incorporate the dynamic properties of the partially completed bridge. Sometimes, temporary fairings during construction may be a practical solution even if not required in the final condition.

There are several ways to resist unbalanced loads during construction. One way is to design the lower part of the pylon to resist the unbalanced moments. The twin wall pier of the Sunshine Skyway Bridge in Florida is an example as shown in the photograph of Figure (10.39). Another method that

Fig. 10.39 Twin walls of the Sunshine Skyway Bridge

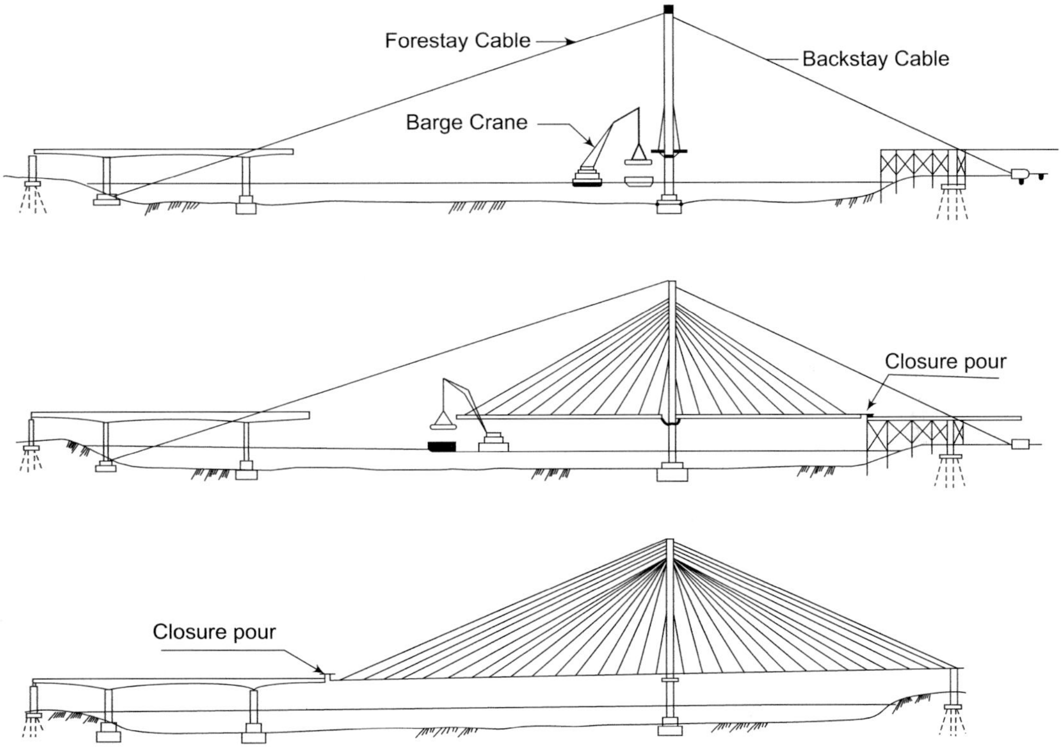

Fig. 10.40 Stabilizing Temporary cable during construction (Man-Chung Tang, 1987)

was employed during the construction of the East Huntington Bridge is to use temporary cables that can be connected to the top of the pylon and outside anchors, hence minimizing unbalanced moments in both the pylon and foundation (Figure 10.40). Temporary anchor cables, sometimes referred to as buffeting cables, in the pylon foundation to reduce moments in the pylon section at deck level are also a means of resisting unbalanced loads during construction (see Figure 10.41).

Moreover, temporary cables can be connected to an outside anchor as shown in Figure (10.42) to reduce moments in the pylon foundation and control buffeting.

Horizontal distribution of the stay-cable forces sometimes referred to as shear lag is another phenomenon that needs to be investigated during construction. It is very significant to ensure that the spread of the horizontal component of the stay-cable force into the section, with the vertical component effectively applied at the stay-cable anchorage, can be accommodated by the section during all phases of construction. This may be resolved by adding post-tensioning to the areas of the section furthest from the anchorage.

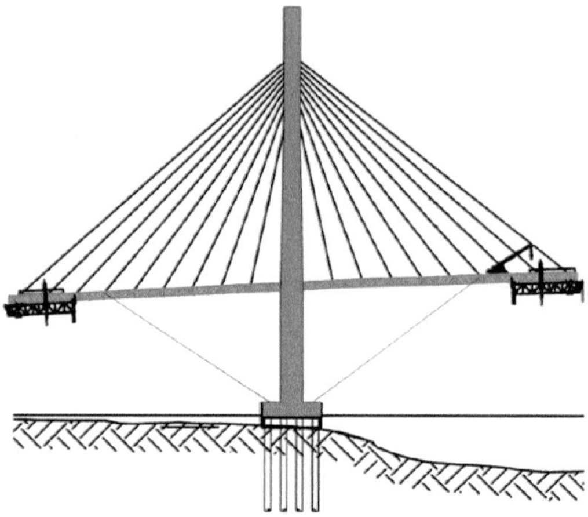

Fig. 10.41 Stabilizing temporary buffeting cable during construction

As discussed in Chapter 3, some cable-stayed bridges employ cable saddles for anchorage at the pylon. Unbalanced loads may be critical for stay-cable saddles. Therefore, the designer needs to ensure that sufficient friction is provided between stay-cables and saddle pipes to counteract unbalanced loads with an acceptable margin of factor of safety as per PTI recommendations.

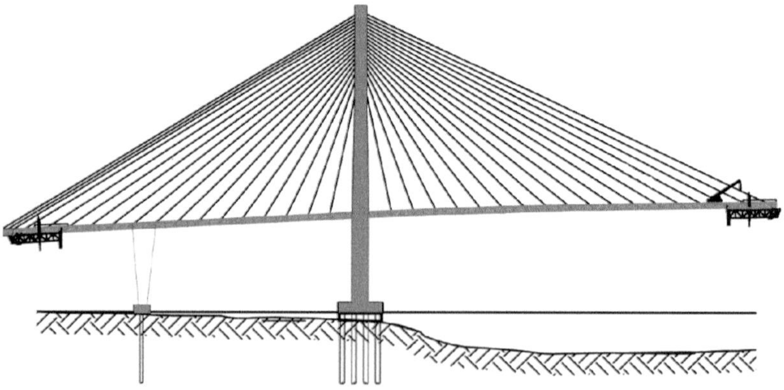

Fig. 10.42 Stabilizing temporary buffeting cable during construction

10.10.1.2 Composite Decks

Composite decks can be constructed using the balanced cantilever method. Construction generally starts with building the approach spans, usually on land, using traditional scaffolding. Construction is usually performed using derrick cranes located at each cantilever tip. Construction progresses from one pylon pier towards mid span at one end and towards side span from the other end. Temporary bents located at the side span, as shown in Figure (10.43), can be employed to stabilize the partially completed bridge for the construction wind and out-of-balance dead load forces. Construction materials for the main span can be fed to the derrick crane through a trestle that has to run parallel to the span or through a barge from the channel. Material for a typical deck erection cycle includes, floor beams, edge girders and deck panels all constituting one unit or grid that is typically in the range of 15 m in length. A typical erection cycle for superstructure construction involves the following steps: (i) erect structural steel grid and bolt, (ii) install and initially stress stay cables, (iii) release steel grid from the derrick, (iv) place precast panels, (v) adjust cable lengths, (vi) pour joints and block-outs (vi) finally provide tension in the cables. For the Audubon Bridge, the sequence was arranged to permit significant overlap between pylon construction and superstructure construction where deck construction could commence after completion of the pylon's lower cross beam under the deck and pylon construction did not need to be completed until each cantilever contained five deck segments. Pylon construction was performed at a rate where one 4 m segment was typically completed every 5 days (Schemmann et al., 2008).

10.10.1.3 Geometry Control

For cable-stayed bridges, as with other bridges, profile and alignment control are required. Nevertheless, geometry control must be used to verify that stresses in the structure and stay-cable forces meet design requirements. Geometry control of cable-stayed bridges can be established through development of casting curves. In this context, all segments must be cast following a pre-established casting curve to ensure the final profile matches the desired profile once all short- and long-term deflections have occurred.

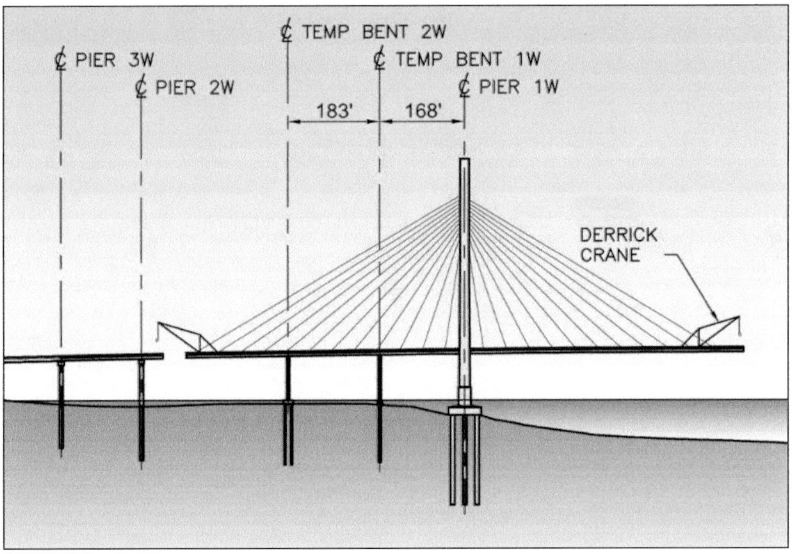

Fig. 10.43 Composite deck balanced cantilever construction (Schemmann et al., 2008)

There are several predetermined parameters that need to be established as they influence the casting curve. These parameters include: (i) the loads applied to the structure such as dead loads, construction loads, and stay cable forces; (ii) construction schedule and erection sequence; (iii) material characteristics such as concrete creep and shrinkage, elastic modulus of the concrete, and mechanical properties of the stay-cables including nonlinear effects due to sag. Establishment of an accurate casting curve is not a trivial process. It requires a software package that can handle the time-dependent effects in concrete as well as significant effects of geometric nonlinearity. An analysis that accounts for several effects such as concrete creep and shrinkage, and steel relaxation is required. It is vital to have a fully engineered construction analysis performed by a construction engineer (who may or may not be the designer). The analysis should include each stage of construction, with all loads applied properly.

The Incheon cable-stayed bridge constructed in South Korea in 2009 employed a very advanced Geometry Control procedure during design. The bridge as described in Chapter 8.1.1.2 constitutes an inverted Y-shaped reinforced concrete pylon 238 m in height supported by 3.0 m diameter drilled shafts, 24 in total for each pylon with precast pile caps. The bridge's 208 parallel wire strand cables, whose diameters range from 109 to 301.7 mm, support the continuous orthotropic steel deck. With an 800 m center span, the bridge's overall length is 1480 m. A floating crane with a capacity of 30,000 kN was used to erect the side span, which was constructed as four sizable blocks, on temporary bents for the superstructure erection. Together with the stay cables, the center span was constructed as a 15 m long block and assembled using a derrick crane. A hydraulic jack capable of generating 8000 kN was used to install and tension the cables. To get the design value, four cables were stressed at once, and the girder geometry and cable tension were controlled. For the geometry control system, the permissible tolerance of the girder level was set at ±200 mm at the center of the main span. A photo of the erection of the orthotropic deck is displayed in Figure (10.44).

Fig. 10.44 Erection of the orthotropic deck of Incheon Bridge (Yang et al., 2012)

In order to accomplish geometry control and cable adjustment during bridge construction, an integrated geometry control system was created for the long-span steel deck cable-stayed structure. The integrated data management system, the structural analysis system, the error adjustment and survey management systems, and the structural analysis system make up the geometry control system. The theoretical analysis data was made graphical by the integrated data management system so that it could be compared and used to make the appropriate adjustments during an erection. Data from the structural analysis system was compiled by this system. A construction sequence analysis and software that takes geometrical nonlinearity into account were used in the analysis of the bridge. As a result, before the cable installation process began, information on the initial equilibrium state for determining the initial geometry and the step-by-step analysis during construction were completed. In real time, the survey measurement system gathers the data required for geometry control, including cable tension, temperature, and geometric survey data for bridge members. The outcomes of the geometry control that was used are shown in Figure 10.45. When the analysis and measured values for the girder's vertical level were compared, it was found that the maximum error was +85 mm, and the error at the center section was −78 mm. The target range of ±200 mm for geometry control was fully satisfied.

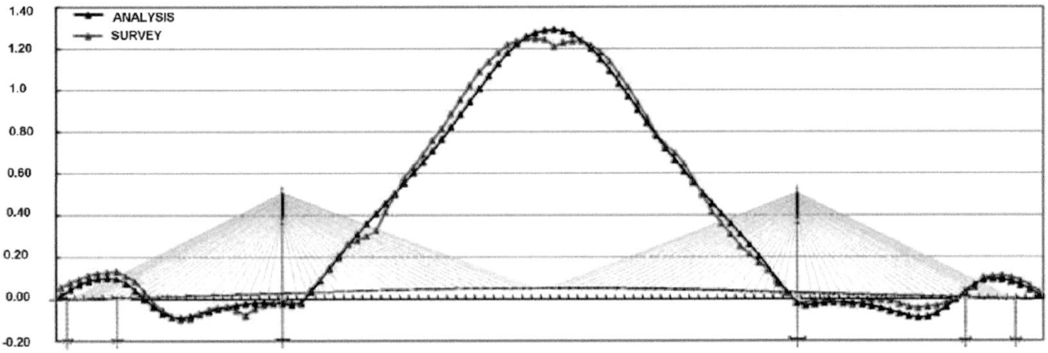

Fig. 10.45 Comparison of analysis measurement with geometry control measurements for Incheon Bridge (Yang et al., 2012)

10.10.2 Construction of Pylons

The pylons are usually constructed in 4 m lifts using an enclosed automatic jump form to accelerate the construction time and ensure safety. Temporary struts are used to reduce the bending moment of the pylon and control its displacement during construction. Typically, three struts are used for a pylon in the range of 125 m to 150 m with two cross beams. The sequence of construction is dependent on the shape and geometry of the pylon, in addition to the controlling wind loads affecting the pylon during construction. Construction may progress simultaneously in the two legs to a certain height above the deck level. In a subsequent step a temporary lower strut is installed a few meters above the deck level to control the forces and deflections in the pylon leg. It may also be employed for heavy lifting the lower precast crossbeam. Alternatively, the precast cross beam can be erected using a floating crane. Once the lower crossbeam is connected, construction of the deck can initiate in parallel with ongoing pylon construction, which necessitates pier table falsework that can be lifted using the installed temporary strut. Continuation of building the pylon legs and installing the second temporary intermediate strut can proceed. The intermediate strut is used to control the forces in the pylon legs. It is also used to support a catch platform for safety during the construction of the upper crossbeam. The lower strut can be removed at this point. Installation of the third temporary upper strut can progress in a subsequent step. The upper strut acts as both a brace for the pylon legs and as a support for the formwork of the upper crossbeam. Construction of the upper crossbeam is followed by removal of the intermediate and upper struts and completion of the pylon leg construction. Figure (10.46) displays photos of the typical construction of an H-shaped pylon. The upper portion of the legs is usually used to support the cables. A lot of contemporary cable-stayed bridges have stay anchorages built into the pylon. Nevertheless, this necessitates that designers use a lot of reinforcement and post-tensioning to offset the high tensile forces created between these anchors. For the same purpose, heavy, intricate, and costly steel anchorage boxes have been inserted into concrete pylons more recently. As an alternative, cable saddles serve the same function. The saddle system uses the natural capacity of concrete to effectively handle large compressive loads by transferring the load of the stay cables to the pylon concrete through direct radial compressive stresses rather than producing large tensile forces. The pylon concrete's cross-sectional area and reinforcement were reduced as a result of the more effective load transfer. Chapter 3.6 describes all these systems in detail.

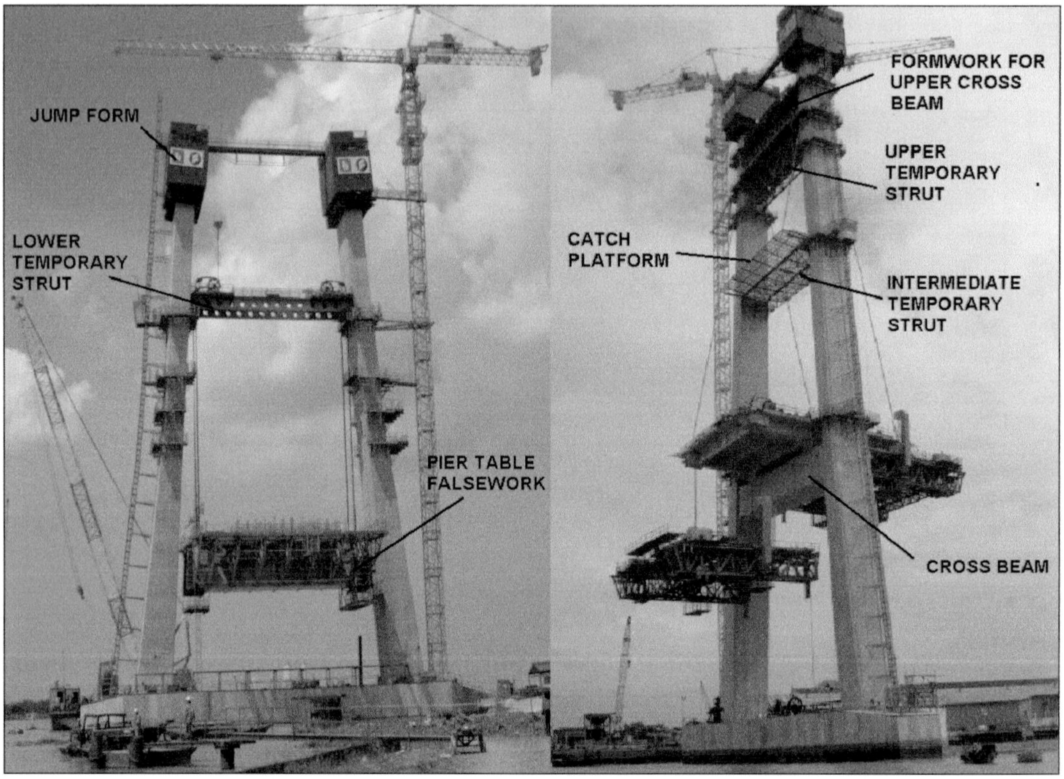

Fig. 10.46 Construction of pylons (Moir et al., 2010)

10.10.3 Construction of Foundations

Foundations for cable-stayed bridge pylons are either groups of drilled shafts or pneumatic caissons. Construction methods for drilled shafts may differ depending on the type of soil and ground water level. Therefore, the construction process can be divided into three main groups: (i) the dry method; (ii) the casing method; and (iii) the wet method (FHWA, 2010). The most advantageous conditions for the cost-effective use of drilled shafts are thought to exist in the dry method of construction. It can be used on rock and soil above the water table that won't cave in when the hole is drilled all the way down during the time needed to install the drilled shaft. Figure 10.47 shows the steps involved in building a dry hole: first, the shaft is excavated using augers, which are probably equipped with teeth to break up the soil. After that, the base is cleaned with a bucket or flat-bottom instrument to get rid of any loose debris and maybe a tiny bit of water. A full-length reinforcing cage is then positioned. Lastly, a drop chute or centering device is used to place the concrete.

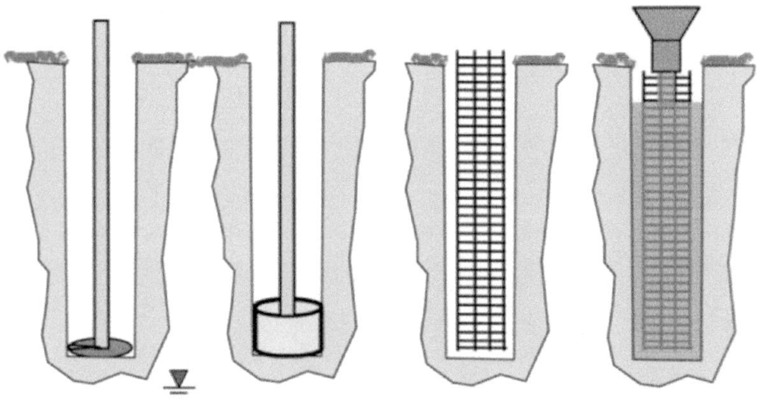

Fig. 10.47 Dry method of construction of drilled shafts (FHWA, 2010)

The casing method is applicable to sites where soil caving or excessive soil or rock deformation can occur during shaft excavation. The casing can also be used to extend the shaft excavation through water or permeable strata to reach a dry, stable formation. One way to accomplish the installation, as shown in Figure (10.48) is to excavate an oversized hole through the shallow permeable strata using a drilling fluid, then place and advance the casing into the bearing stratum. After the casing is sealed into the underlying more stable stratum, the drilling fluid can be removed from inside the casing and the hole advanced to the final tip elevation in the dry.

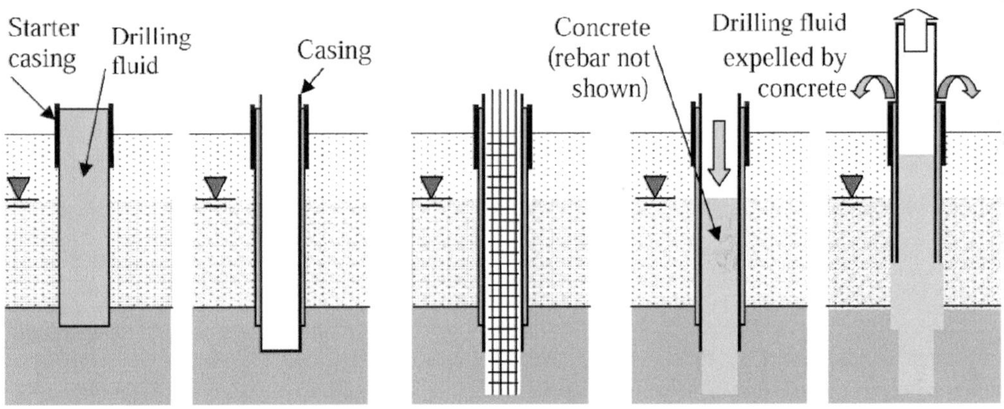

Fig. 10.48 Construction of drilled shafts using casing through drilling fluid (FHWA, 2010)

Alternatively, the casing can be advanced through the shallow permeable layers and into the bearing stratum before excavating the shaft as illustrated in Figure (10.49), and then excavating within the casing in the dry soil. The casing may be driven using impact or vibratory hammers or using a casing oscillator with sufficient torque and downward force to drive the casing through the soil ahead of the excavation. A larger upward force may be required to pull the casing during concrete placement.

Wet construction techniques are an option. when the state of the soil prevents dewatering. Using this method, the hole must be kept full of a fluid while the hole is being drilled, during the reinforcing, and the concrete is being placed. If the hole is stable and won't collapse, the drilling fluid can be either water or a prepared slurry that will keep the hole stable. A starter casing that can

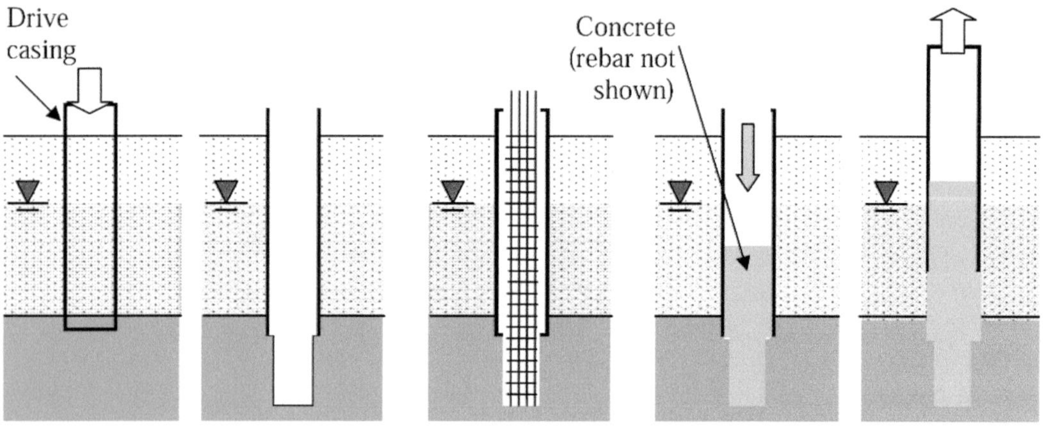

Fig. 10.49 Construction of drilled shafts using casing driven ahead of excavation (FHWA,2010)

extend above ground to raise the surface level of the slurry in the hole and as deep as needed to prevent surface failure would be part of a typical construction, as depicted in Figure 10.50. Knowing the elevation of the groundwater in advance is crucial to maintaining the slurry head at least five feet above the water's surface. The reinforcing cage is positioned once the excavation and base cleaning are finished, and a tremie is used to advance the concrete placement.

Caissons in general are large structures that can be constructed above ground or water level and then sunk as a single unit to the required depth to support bridge towers for suspension bridges and pylons of cable-stayed bridges. Their dimensions are large and can reach 69 m by 39 m such as the Verrazano Narrows Bridge in New York City. Sunken caissons have been used for their ability to distribute large loads from the towers or pylons to the deep bearing soils. The overlying soils are soft and make the use of driven piling which is impractical due to the lack of lateral support in the upper regions. However, sunken caissons have inherent risks that include control of geometry, lack of flexibility, and working at depths with pressurized air such as in the case of pneumatic caissons.

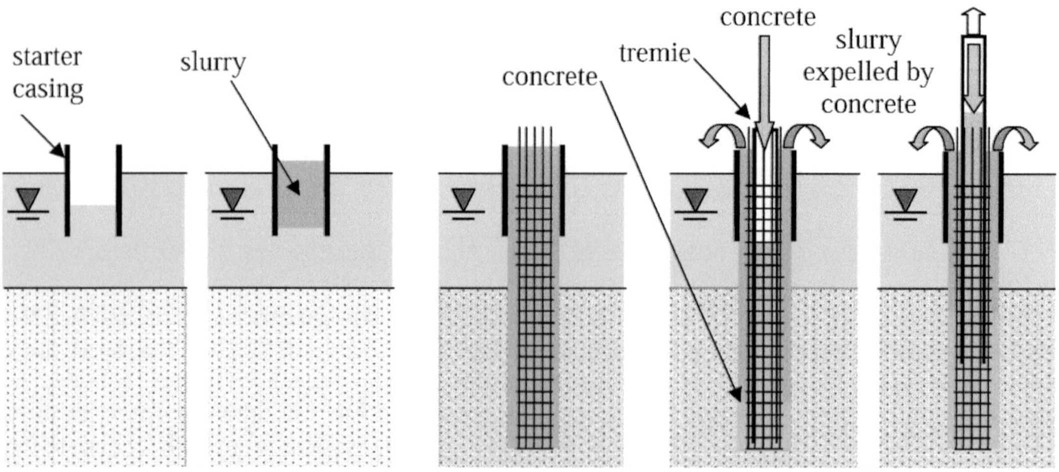

Fig. 10.50 Construction of drilled shafts using wet method (FHWA,2010)

The pneumatic caisson is provided with a working chamber inside its lower part. Figure (10.51) displays the traditional method of constructing this type of caisson during the old days. Nowadays

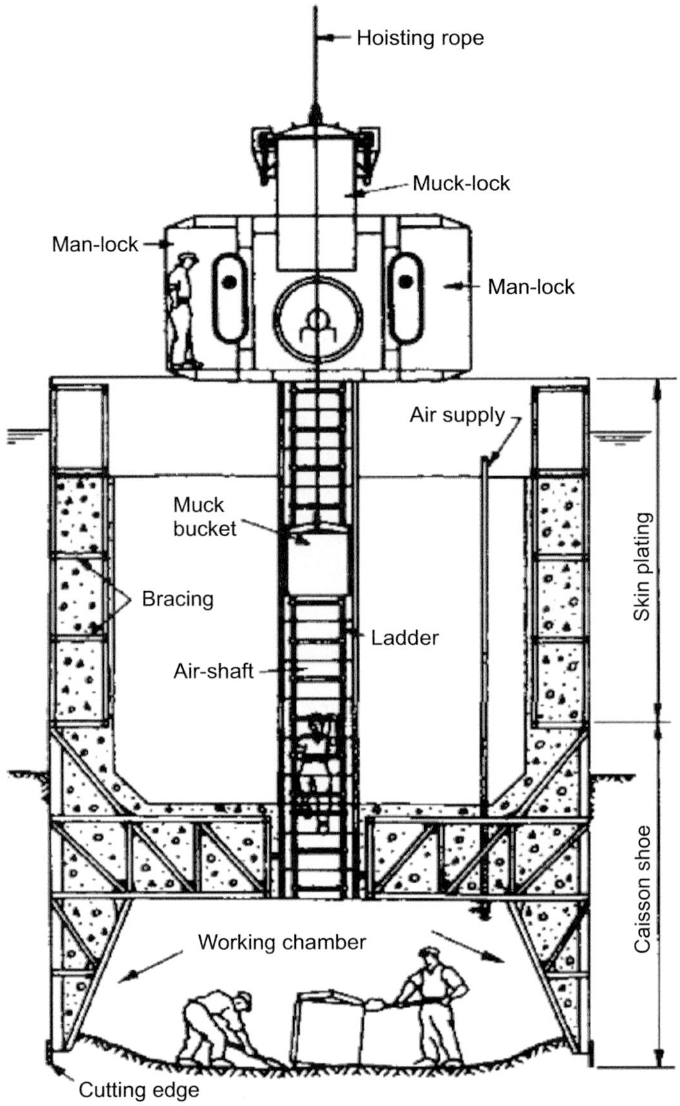

Fig. 10.51 Traditional method of Construction of pneumatic caissons (Tomlinson, 2001)

compressed air is introduced into the working chamber to expel the water while excavation progresses below the sea-bed using mechanical shovels operated by remote control from the surface. The Bai Chay Bridge in Vietnam employed pneumatic caissons where compressed air pressurizes the inside of the working chamber to compensate for the water pressure encountered during excavation as shown in Figure (10.52). The maximum excavation depth was 27.7 m at one of the pylons Piers where the air pressure was recorded at 0.245 MPa. Furthermore, the two main towers and the two anchorages of the Rainbow Bridge in Japan were designed and built using pneumatic caisson foundations (Nakamura et al. 2007). Robotic excavation was used for the pneumatic caissons. A video camera was used to remotely operate a computer-controlled caisson shovel. The materials that had been excavated were loaded onto an automated belt conveyor for removal. This technique guaranteed worker safety and construction efficiency in the caisson's chamber, even at 3.5 bar of atmospheric pressure. Additionally, the hard mudstone was excavated using specialized machinery.

Fig. 10.52 Pneumatic caisson construction for Bai Chay Bridge (Nakamura et al., 2007)

References

American Association of State Highway and Transportation Officials, *AASHTO LRFD Bridge Design Specifications*, 9th Edition, Washington D.C., 2020.

Bathe, K.J, *Finite Element Procedures*, Prentice Hall, New Jersey, 1996.

Baron, F.M., Influence Surfaces for Stresses in Slabs, ASME Journal of Applied Mechanics, pp. A1-A13, 1941.

Bonzon, W.S., The I-280 Veterans' Glass City Skyway: New Landmark Cable-Stayed Bridge, Ohio, *Structural Engineering International*, Volume 18, Issue 1, 43–48, 2008.

DeCorte, W., Van Bogaert, P. (2007) "Improvements for the Analysis of Floorbeams with Additional Webs Cutouts for Orthotropic Plated Decks with Closed Continuous Ribs." *J. Steel Composite Struct.*, 7, 1–18, 2007.

Federal Highway Administration, *Manual for Design, Construction, and Maintenance of Orthotropic Steel Deck Bridges,* Publication Number: FHWA-IF-12-027, Washington, D.C., 2012.

Federal Highway Administration, *Drilled Shafts: Construction Procedures and LRFD Design Methods,* Publication Number: FHWA-NHI-10-016, Washington, D.C., 2010.

Fleming, J.F., Nonlinear Static Analysis of Cable-Stayed Bridge Structures, *Computers and Structures*, 10, 621–635, 1979.

Fleming, J.F. and E. A. Egeseli, E.A., Dynamic Behavior of a Cable-Stayed Bridge, *International Journal of Earthquake Engineering and Structural Dynamics*. Vol. 8, pp. 1–16, 1980.

Haibach, E., Plasil, I., (1983) "Investigations into the operational stability of light steel carriageways with trapezoidal hollow stiffeners in railway bridge construction." (in German) *Der Stahlbau*, 9, 269–274.

Harrison, H.B., *Computer Methods in Structural Analysis*. Prentice-Hall, Englewood Cliffs, New Jersey, 1973.

Homberg, H., *Road Slabs with Variable Thickness* (in German), Springer-Verlag Berlin, 1968.

Huang, D. and Hu, Bo., Concrete Segmental Bridges Theory, Design, and Construction to AASHTO LRFD Specifications, CRC Press, 2020.

Martin, H.C., On the Derivation of Stiffness Matrices for the Analysis of Large Deflection and Stability Problems, *Proc. of the Conference on the Matrix Methods in Structural Mechanics*, Wright Patterson Air Force Base, Ohio, pp. 697–16 (1965).

Moir, G., Edmonds, C., Walser, P. and Romberg, M., Construction Engineering of Phu My Cable-Stayed Bridge, Vietnam, *Structural Engineering International*, Volume 20, No. 3, pp 331–337, 2010.

Nakamura, T., Tsuchida, K., Ohno, H. and Nagamoto, N., B., Bai Chay Bridge, Vietnam, *Structural Engineering International*, Volume 17- No. 3, pp 210–213, 2007.

Nazmy, A.S. and Abdel-Ghaffar, A., *Seismic Response Analysis of Cable-Stayed Bridges Subjected to Uniform and Multiple-Support Excitation*, Report No. 87-SM-1, Department of Civil Engineering, PrincEton University, 1987.

Newmark, N.M., A Method of Computation for Structural Dynamics, *ASCE Journal of Engineering Mechanics Division*, Vol. 85, pp. 67–94, 1959.

Podolny, W. and Muller, J., *Construction and Design of Prestressed Concrete Segmental Bridges, John Wiley and Sons,* 1982.

Precast/Prestressed Concrete Institute, *Bridge Design Manual*, 3rd Edition, 2014.

Schemmann, A.G., Bergman, D.W. and Shafer, G., John James Audubon Bridge-Design-Build Delivery of the Longest Span Cable-stayed Bridge in the United States, *Proceedings of the International Bridge Conference*, IBC-90-61, Pittsburgh, 2008.

Tang, M.C., Construction of East Huntington Bridge, *PCI Journal*, Vol. 32, Issue 6, pp 32–48, 1987.

Taylor, P.R., Composite Cable-stayed Bridges the Concept with the competitive Edge, *Engineering Journal*, American Institute of Steel Construction, Vol. 24, pp. 157–163, 1987.

Tomlinson, M.J., Foundation Design and Construction, 7th Edition, Addison-Wesley Longman Ltd, 2001.

Troitsky, M.S. (1987) *Orthotropic Bridges Theory and Design*. 2nd Ed. The James F. Lincoln Arc Welding Foundation. Cleveland OH.

Weaver, W. and Gere, J., *Matrix Analysis of Framed Structures.* 2nd Edn. D. Van Nostrand. New York, 1980.

Wilson, E.L., Farhoomand, I. and Bathe, K.J., Nonlinear Dynamic Analysis of Complex Structures, *International Journal of Earthquake Engineering and Structural Dynamics*, Vol. 1, pp. 241–252, 1973.

Yamamura, N., A *Seismic Analysis Program for. 3-Dimensional Framed Structures*, Internal publications of the Hitachi Zosen Company for Steel Structures, (in Japanese), 1983.

Yang, J.H., S., Im, D.K., Kim, C.H., Ahn, S.S. and Lee et al., Construction and Geometry Control of Incheon Cable-Stayed Bridge, *Structural Engineering International*, Volume 22, No. 1, pp 49–52, 2012.

CHAPTER 11

Wind Effects and Aerodynamic Stability

11.1 INTRODUCTION

There has been a great evolution during the past two decades in the design of cable-stayed bridges. This interval, which marked a period that was characterized by an ever-increasing length of the main span, has brought a great awareness among the engineering professionals about the significance of the aerodynamics of such bridges. Cable-stayed bridges are flexible structures, their stability against wind dynamic forces is extremely significant. Aerodynamics evaluation includes site specific wind studies, analytical modeling, and wind tunnel testing. These studies and investigations establish stability, loading criteria and validate design decisions. In general, the configuration and number of stays along with the slenderness of the deck play an important role in the aerodynamic stability of cable-stayed bridges.

11.2 PERFORMANCE OF EARLY BRIDGES

In the 19th and early 20th century, a significant number of medium-span cable-supported bridges were built. A considerable number of these bridges exhibited instability and collapsed under windstorms. The causes, as well as physical backgrounds, were not disclosed in the historical records. Further development of cable-supported bridge structures to cross longer spans was achieved by using parallel wires for cables. The introduction of a design procedure based on the elastic theory was also quite essential. The development of the deflection theory and its use in the design of the first Tacoma Narrows Bridge led to installation of a thin plate–girder deck, which resulted in the collapse of the bridge. This section outlines several bridges that exhibited problems due to wind and provides an in-depth discussion on the failure of the first New Tacoma Narrows Bridge.

11.2.1 Wind Failures of Nineteenth Century Cable Supported Bridges

Many years back, prior to the invention of cable-stayed bridges, between 1818 and 1889, several cable-supported bridges exhibited major damage or collapsed in windstorms. These bridges were located as displayed in Table 11.1 in Great Britain, United States, France, and Germany (Farquharson, 1950).

Table 11.1 Bridges Severely Damaged or Destroyed by Wind

Bridge	*Location*	*Designer*	*Span (m)*	*Failure Date*
Dryburgh Abbey	Scotland	John and William Smith	80	1818
Union	England	Sir Samuel Browns	137	1821
Nassau	Germany	Lossen and Wolf	75	1834
Brighton Chain Pier	England	Sir Samuel Browns	78	1836
Montrose	Scotland	Sir Samuel Browns	132	1838
Menai Straits	Wales	Thomas Telford	177	1839
Roche-Bernard	France	Le Blanc	195	1852
Wheeling	USA	Charles Ellet	308	1854
Queenston-Lewiston	USA	Edward Serrell	318	1864
Niagara-Clifton	USA	Samuel Keefer	384	1889

The Dryburgh Abbey Bridge in Scotland had a span of 80 m and was only 1.20 m wide. It was observed that the bridge had a very profound vibration when crossed by any person. In 1818, six months after the completion of the bridge, it collapsed during a violent windstorm (Drewry, 1832). The Union Bridge across the Tweed near Berwick was the first suspension bridge in England to be designed for the passage of loaded carriages. It was also the first eye-bar chain bridge in Great Britain. It was blown down during a windstorm six months after its completion (Jakkula, 1941). The Brighton Chain Pier Bridge on the English Channel consisted of four spans of 78 m and was stiffened by an iron railing. It was partially damaged during a windstorm in 1883 and completely damaged during another violent windstorm in 1838. The Montrose Bridge over the South Esk in Scotland had a span of 132 m and was 8 ft wide. The bridge collapsed in 1830 due to overloading during a boat race. The bridge was reconstructed but was destroyed by wind in 1838.

The Menai Straits Bridge over Menai Straits Wales is considered one of the famous bridges in Great Britain. The bridge had a suspended span of 176.5 m with two side pans of 79 m tied down to a series of arches, 8.5 m wide. The bridge had two traffic lanes each 3.75 m and one 1.2 m sidewalk. The bridge exhibited successive damage in the windstorms of 1826 and 1836 and was completely destroyed by a storm in 1839. It was restored and altered following the storm of 1839 and rebuilt with the exception of towers in 1939 (Jakkula, 1941). The Nassau Bridge over the Lahn River, Germany had a main span of 75 m and two side spans of 12 m and a width of 6.7 m. The bridge had 16 eye-bar chains and no stiffening. It was opened in June 1830. Later, it was damaged by wind during the winter of 1833-1834, 12 chains being broken. The bridge was replaced by another suspension bridge completed in 1927 (Jakkula, 1941). The Roche-Bernard Bridge over the Vilaine River, France had a span of 195 m and used 4 wire cables. It was damaged by a hurricane on October 26, 1852, and counter cables were added in the restoration. The Wheeling Bridge was built in West Virginia, wheeling, over the Ohio River. It has a main span of 308 m and width of 7.25 m. The deck was built without stiffening and supported by 12 cables, each composed of 550 wires. In 1854 a violent tornado turned over the floor and broke all but 2 of the cables. It was rebuilt again in 1862 (Finch, 1941).

The Queenston–Lewiston Suspension Bridge was completed in March 1851. The bridge had a span of 257 m and was 6 m wide. On February 1864, a wind storm caused partial collapse of the bridge. The cables and sections of the broken deck kept hanging over the river for almost 35 years, as shown in the photo of Figure 11.1, until the bridge was replaced with a new suspension bridge that was opened in 1899 (Finch, 1941).

The Niagara-Clifton Bridge was built 275 m north of the American Falls. The suspension bridge had a span of 386 m and a 3 m wide timber deck with a stiffening truss and timber towers supporting the cables at each end of the bridge. The bridge was completed and officially opened in

January 1869. In October of 1887, the bridge underwent an extensive renovation, which allowed the replacement of all the wooden components with steel. Also, the bridge deck was widened to 5m. On the night of January 9th, 1889, a ferocious storm lashed Niagara. On the morning of January 10th, 1889, at 3:20 a.m., the bridge broke loose and crashed into the river below.

Fig. 11.1 Remnants of the Queenston-Lewiston Suspension Bridge Over the Niagara River

11.2.2 Wind Problems of Early Twentieth-Century Cable Supported Bridges

The Fyksesund Bridge, Norway as listed in Table 11.2, was completed in 1937. The bridge was built without a stiffening truss. Hence its original deck span to depth ratio of 511:1 was much larger than other bridges of this era. The bridge started to exhibit oscillations in the main span under wind with amplitudes reported up to 0.8 m (Farquharson, 1950). The designer, Arne Selberg, studied this phenomenon and realized the role of aerodynamics in this behavior. He later designed retrofit measures in the form of girder ties below the main span. The additional stays were installed in 1945 and are still in use today as illustrated in Figure 11.2.

Table 11.2 Early Twentieth-century Bridges which have Oscillated in Wind

Bridge	*Year Built*	*Span (m)*	*Type of Stiffening*
Fyksesund (Norway)	1937	229	Rolled I-beam
Golden Gate	1937	1280	Truss
Thousand Islands	1938	244	Plate Girder
Deer Isle	1939	329	Plate Girder
Bronx-Whitestone	1939	701	Plate Girder

The Golden Gate Bridge is the only bridge listed in Table 11.2 that has a stiffening truss with a depth-to-span ratio of 1:164. Nevertheless, motions have been observed in this structure. The wind reached 100 km/h during a storm on February 11, 1941. Amplitudes up to 0.60 m were estimated (Anon, 1946). The Thousand Island Bridge is an international bridge system connecting the United States to Canada over the Saint Lawrence River in northern New York. The system contains two suspension bridges with spans of 244 m and 228 m on the American side and the Canadian side respectively. Plate girder depths of 2 m and 1.8 m for the two spans, respectively, produced depth-to-span ratios of 1:125. Mild vertical undulation had been observed in both spans during the late stages of construction. The undulation was especially pronounced in the larger American span. Motion was first observed during the installation of the formwork of the concrete floor and placement of the floor did not resolve the issue. Undulation at times involved the side spans as well, with one large half-wave raising the main span up 61 cm while lowering the side spans less so,Steinman the bridge designer sought to restrict movement by stiffening the main cables through installing a system of center cable ties and cable stays (Steinman, 1945).

The Deer Isle Bridge is a suspension bridge that connects Maine mainland to Deer Isle island. The bridge was designed by David Steinman and has a main span of 329 m. The 2 m depth of the plate girders made the suspended structure exceptionally shallow in relation to its length as it produced a depth-to-span ratio of 1:164. The bridge encountered wind stability problems prior to its completion in 1939.

Fig. 11.2 Fyksesund Bridge, Norway

The designer decided to stabilize the bridge against this troublesome wind-induced motion through the addition of diagonal stays running from the stiffening girders at the towers to the main cables in both side and main spans. After the completion of the bridge, during the winter of 1942-1943, the bridge exhibited significant motion during a violent windstorm, which resulted in damage of some of the stabilizing stays. The bridge was then retrofitted by installing a more extensive system of longitudinal and transverse diagonal stays as illustrated in Figure 11.3. Since then, there have been no further reports of wind aerodynamic problems.

The Bronx-Whitestone Bridge was designed in 1935 by Othmar Amann over the East River at New York. It has a 701 m span and a plate girder 3.3 m deep, which produced a depth-to-span ratio of 1:209. The bridge has shown a tendency for noticeable vertical motions that were felt by workers during construction. Vertical motions were accompanied by sway along the span's axis. Vertical waves with double amplitude of up to 90 cm were also observed most frequently when winds blew at an oblique angle to the bridge.

Several successive steps were taken to stabilize the structure. Midspan diagonal stays and friction dampers at the towers were first installed. These devices were not entirely adequate and in 1946 the suspended structure was effectively stiffened by the addition of truss members mounted above the original plate girders, the latter becoming the lower chords of the trusses. In 1988, after extensive aerodynamic evaluation, a tuned mass damper system was installed to replace the friction damper at the tower. In 2002, a deck replacement for the entire suspended span was designed to increase

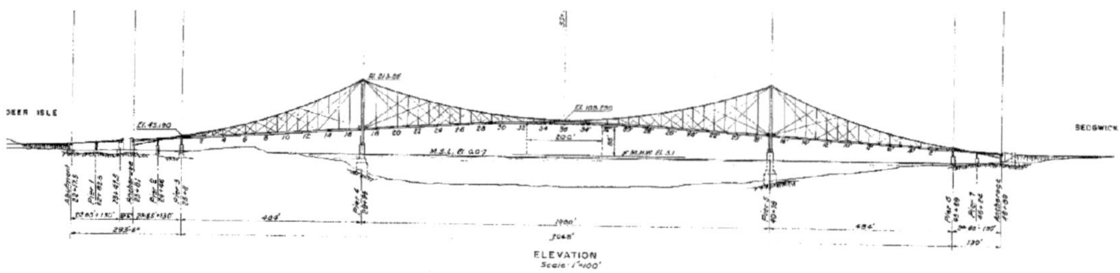

Fig. 11.3 Deer Isle Bridge-Current Configuration of Cables and Stays

the load-carrying capability of the main cables. The project involved the replacement of the original concrete-filled steel grid deck with an orthotropic deck, removal of the stiffening trusses, and a new lateral system. The aerodynamic stability was re-evaluated under the revised configuration, which led to designing wedge-shaped Fiber Reinforced Polymer (FRP) fairings on the bridge as shown in Figure 11.4. The fairings were designed to stabilize the bridge in winds up to 190 km/h.

11.3 LESSONS LEARNED FROM TACOMA NARROWS COLLAPSE

The First Tacoma Narrows suspension bridge collapsed four months and six days after its opening. The dramatic collapse of the bridge on November 7, 1940, due to wind effects has urged bridge engineers and scientists to study and investigate the main reasons behind this catastrophic incident. Four months and six days after the bridge's opening, its oscillations in a gale increased to destructive amplitudes until the main span broke up, ripping loose from the cables and crashing into the water of Puget Sound.

Fig. 11.4 Installation of FRP Fairings at Bronx-Whitestone Bridge

The bridge was by far the most flexible of all modern suspension bridges. Whereas authorities had formerly recommended for suspension bridges a minimum width of 1:30 of the span, the width center to center of cables of the Tacoma Narrows Bridge was only 11.9 m or 1:72 of the span. This lateral flexibility of the bridge was not the major factor in the failure. What proved critical, however, was the vertical slenderness of the span. The bridge had a depth to span ratio of 1:350, almost a tenth of the 1:40 of bridges designed using the traditional elastic theory with stiffening trusses such as the Williamsburg Bridge. The resulting extreme vertical flexibility was a major factor in the failure as it contributed to the main failure factor which was the newly discovered aerodynamic instability phenomenon.

Designed by Leon Moisseiff, the bridge had a main span of 853 m, with side spans of 335 m. It can be noticed from Table 11.1 that 51 years between the failure of the Niagara Clifton Bridge and the spectacular collapse of the Tacoma Narrows Bridge were characterized by built suspended spans of accrescent length and load-carrying ability. The First Tacoma Narrows Bridge was designed according to the Theory of Elastic Distribution (Moisseiff and Lienhard 1933), an updated version of the Deflection Theory in the lateral direction. They demonstrated that the stiffness in the main cables via the suspenders can absorb up to one half of the static wind pressure pushing a suspended structure laterally. This energy would then be transmitted to the towers and anchorages. According to the theory, the lower the span's sag ratio, the lower the horizontal displacement. This in turn meant that less lateral stiffness in the suspended structure was required. Also, the deck would be subject to less horizontal pressure. This theory had made it possible to introduce a new form of suspended structure, the plate girder in lieu of a rigid stiffening truss system. The New Tacoma Narrows Bridge had a sag ratio of 1:12, which in Moisseiff's judgment provided the necessary rigidity including lateral stiffness against static wind pressure. He also proposed a solid plate girder just 2.4 m deep, which resulted in reducing the weight of the superstructure by 50% than if it was designed using a traditional stiffening truss system (Scott, 2001).

It was noticed that the bridge started to exhibit wind-induced oscillations during the final stages of work. Professor Farquharson of the University of Washington was retained to develop a 1/100 dynamic model of the bridge that was able to model the span's motion. The vertical undulations continued shortly after the opening of the bridge on July 1, 1940, particularly at low wind speeds. Vertical waves of up to almost 81 cm double amplitude were observed. Mindful of stronger autumn winds, engineers added hold-down cables to restrain the side spans with wires anchored in wooden piles driven offshore. Meanwhile, wind tunnel tests were conducted at the University of Washington to examine the effect of galloping, a self-sustaining vertical motion phenomenon. Early tests suggested that lift forces were influenced by the deck cross-section and the angle of wind attack. Wind tunnel tests suggested two possible remedies. The first was deflecting the wind using fairings attached to the girder fascia. A second remedy, which was less favored due to its irreversible nature, involved punching holes in the plate girders to reduce the effect of wind pressures.

Around November 2, a simple 23 cm tall, curved fairing was designed to be supported 1.9 m from the girder face by two struts and centered about the deck level. Engineers prepared to order materials. On the night of November 6, an expected storm started to excite the bridge and by 8:00 A.M. on November 7, the wind had increased to 61 km/h. A series of observations and photographs taken around 10:00 a.m. showed that the bridge was moving with usual vertical motion with a frequency of 36 cycles per minute and suggested a wind of 68 km/h attacking the bridge obliquely. A violent change in motion, which appeared to take place without any intermediate stages were noted. The wave frequency had changed from 36 cycles per minute to 14, and the main cable started to vibrate out of phase while the entire bridge started to vibrate on a torsional mode as depicted in Figures 11.5 and 11.6. At this stage of the torsional motion progressive failure of the girders' lateral bracing took place. Meanwhile, a plate girder buckled over the west half of the main span while several suspenders snapped from their sockets. At 10:30 a.m., a floor panel broke apart and fell into the Narrows and around 11:00 a.m., a 183 m section of the superstructure peeled away near the west quarter point of the main span and crashed into the water of Puget Sound as depicted in Figure 11.7 (Ammann et al., 1941).

The disaster had revealed the engineering profession's ignorance of the effects of the dynamic effects of winds on bridges. For decades, engineers accounted for different kinds of static loads, even wind was dealt with from a static perspective. Failure of the First Tacoma Narrows Bridge revealed that the dynamic effect of wind combined with the extreme flexibility of the span were the two factors that caused the catastrophe. It was learned that these unknowable phenomena would affect suspension bridge performance.

Fig. 11.5 Torsional Oscillation, Tacoma Narrows Bridge (Courtesy of the University of Washington Library)

Fig. 11.6 Torsion of Floor Center Span, Tacoma Narrows Bridge (Courtesy of the University of Washington Library)

The Public Work Administration (PWA) established a Board of Engineers led by their Head, John Carmody. The Board later was known as the Carmody Board. It included Othmar Ammann who had retired from the Triborough Bridge Authority in NYC and opened his own firm; Glenn Woodruff, Engineer of Design for the Transbay; and Theodore von Karman, director of the Aeronautical Laboratory at California Institute of Technology at Pasadena.

Fig. 11.7 Collapse of the main span, Tacoma Narrows Bridge (Courtesy of the University of Washington Library)

The Carmody Board had performed wind tunnel tests on sectional and full models of the Tacoma Narrows Bridge at the California Institute of Technology Aeronautical Laboratory to investigate the effect of aerodynamic damping. Von Karman realized the role of vortices and discussed it in the report but failed to convince the other two Board members of any possible role of vortex shedding, a phenomenon related to low amplitude oscillations due to the shedding of vortices, in the Tacoma Narrows Bridge collapse. The Board members submitted their report in late March 1941 and did not point to a significant role of vortex shedding in the oscillations of the bridge. Moreover, the Board demoted the role of aerodynamic instability in the vertical oscillations and instead stated that these oscillations are attributed to the turbulent character of wind action. The Carmody Board also concluded that the amplitude was independent of wind speed and that a critical velocity at which torsional motion began depended on the frequency of vibration. The Board considered that the vital event in the collapse was the change from a vertical to a torsional mode of oscillation. They attributed the initiation of the torsional oscillations to the slipping of the cable band on the north side of the bridge to which the center ties were connected. They had blamed these torsional movements for the failures at various points of the superstructure and further structural damage followed almost immediately. The Board also felt fairings would have a minor effect on the vertical oscillations of the bridge and might even have an unfavorable influence on the torsional stability. The Board also admitted that further installations such as diagonal stay ropes from the top of tower to the floor would have increased the rigidity but would have been insufficient to compensate for the extreme flexibility of the structure. The Board also indicated that the designer relied on the mass of the suspended structure and the low cable sag to provide the vertical stiffness, which had a minor role in counteracting the aerodynamic action on the bridge. It had been overestimated and the light plate girder had practically no stiffening effect on the main cables.

Nowadays, after several decades of the First Tacoma Narrows Bridge failure, the engineering profession recognizes that the resistance of suspension bridges to dynamic vibrations due to wind and earthquakes is a function of the rigidity, inertia and structural damping of the superstructure. The First Tacoma Narrows Bridge was deficient in all three respects. Furthermore, the bridge had solid web longitudinal girders as well as solid floor beams which resulted in a cross-section, which is peculiarly sensitive to aerodynamics Both the cross-section and the lateral flexibility were main factors for the lack of resistance to the dynamic effect of the wind load. The main lesson learned from the Tacoma narrows failure is that while the bridge was perfectly safe for all the loads and forces for which it had been designed such as dead load, live load, temperature, and moreover the static effect of the wind, its failure was due to overlooking the dynamic effect of the wind load. This meant that the effect of a steady wind acting on the flexible superstructure was to produce a fluctuating resultant force, which synchronizes in timing and direction with the harmonic motions of the bridge so as to cause a progressive amplification of these motions to unsafe amplitudes (Steinman and Watson, 1945). This phenomenon, namely flutter, is explained in detail in the following sections.

11.4 STRUCTURAL DYNAMICS BASICS

Prior to introducing wind aerodynamics, basic principles of structural dynamics are outlined in this section. We start first with the dynamics of the single degree of freedom system (SDOF) which is illustrated in Figure 11.8 and consists of a mass m attached to a massless member AB, which is fixed at A. A dynamic harmonic load $F(t) = F_0 \sin \omega t$ is applied to the mass at B. The displacement $x(t)$ is opposed by the restoring force $Kx(t)$, where K is the stiffness of the SDOF system, and the damping force is $C\dot{x}(t)$, where C is the viscous damping coefficient of the system, a measure of the energy dissipated in a cycle of vibration. According to Newton's second law, the mass multiplied by its acceleration $M\ddot{x}(t)$ is equal to the total force applied on the mass, therefore the equation of motion of the system is:

$$M\ddot{x}(t) + C\dot{x}(t) + Kx(t) = F_0 \sin \omega t \quad ...(11.1)$$

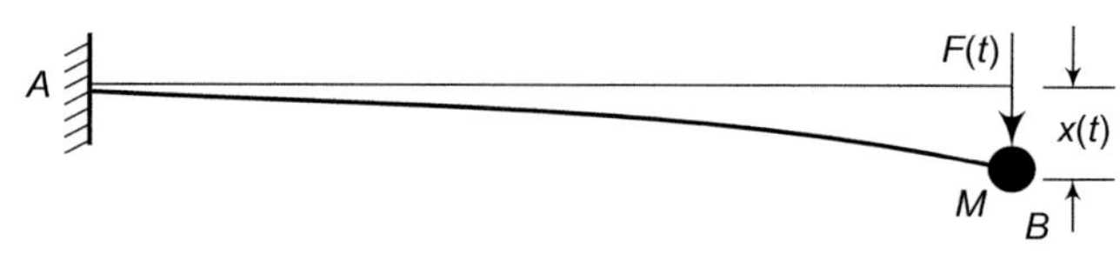

Fig. 11.8 Single Degree of Freedom System

where, ω is the circular frequency of excitation. If it is assumed that this SDOF system vibrates freely without damping it can be shown that the circular natural frequency of this system in rad/s is $\omega_n = \sqrt{\frac{K}{M}}$. The natural circular frequency is related to the natural frequency as in $\omega_n = 2\pi f_n$. where f_n is defined as the number of cycles completed in one second, cycles/s. The natural period, in seconds, of the system $T_n = \frac{1}{f_n}$ and defined as the time required to complete one cycle of motion. If damping is introduced and the system vibrates freely, it can be shown that the critical damping coefficient of the system $C_{cr} = 2\sqrt{KM}$ which is the minimum damping required to prevent the system from vibrating. The damping ratio can be defined as the ratio of the damping coefficient of the system to its critical damping coefficient $\xi = \frac{c}{2\sqrt{KM}}$. Solution of equation (11.1) is of the form:

$$x(t) = x_0 \sin(\omega t - \phi) \quad ...(11.2)$$

The velocity and acceleration can be obtained by differentiating equation (11.2) once and twice. Therefore:

$$\dot{x}(t) = \omega x_0 \sin\left(\omega t - \phi + \frac{\pi}{2}\right) \quad ...(11.2a)$$

$$(\ddot{x}(t) = \omega^2 x_0 \sin(\omega t - \phi + \pi) \quad ...(11.2b)$$

Substitution of equations (11.2) in equation (11.1) and rearranging will yield

where, $$D = \frac{1}{\sqrt{(1-r^2)+(2\zeta r)^2}} \qquad ...(11.3)$$

D is defined as the magnification factor, which is the ratio of the dynamic displacement to the static displacement and $r = \frac{\omega}{\omega_n}$ is the ratio of the circular frequency of excitation to the natural circular frequency of the system. ϕ is a phase angle dependent on ω, ω_n and ξ. Note that for the case of $\omega = \omega_n$ the amplitude of the response is largest and is inversely proportional to the damping ratio ξ. In this case the motion exhibits resonance.

A typical time history of a damped free vibration is illustrated in Figure 11.9. This system is vibrating with a damping ratio $\xi < 1$. As shown the system oscillates with a decaying amplitude and a damped natural frequency $\omega_d = \omega_n (1 - \xi^2)^{1/2}$ and the damped natural period in seconds is $T_d = \frac{2\pi}{\omega_d}$. It is shown that the successive peak amplitudes denoted U_n, U_{n+1}, U_{n+2} are occurring at a time interval, T_d. The ratio between successive peaks, U_n/U_{n+1} is the same for all values of n and it can be proved that this value can be expressed in terms of the damping ratio as $\frac{U_n}{U_{n+1}} = \exp(\xi\omega_n T_d)$. The natural logarithm of this ratio is called the logarithmic decrement, designated as:

$$\delta = \ln\frac{U_n}{U_{n+1}} = \xi\omega_n T_d = \frac{2\pi\xi}{(1-\xi^2)^{1/2}} \qquad ...(11.4)$$

Using 11.4 we can express the damping ratio in terms of the logarithmic decrement as:

$$\xi = \frac{\delta/2\pi}{\left[1+\left(\frac{\delta}{2\pi}\right)^2\right]^{1/2}} \text{ or if } \delta \text{ is small, } \xi \approx \delta/2\pi \qquad ...(11.5)$$

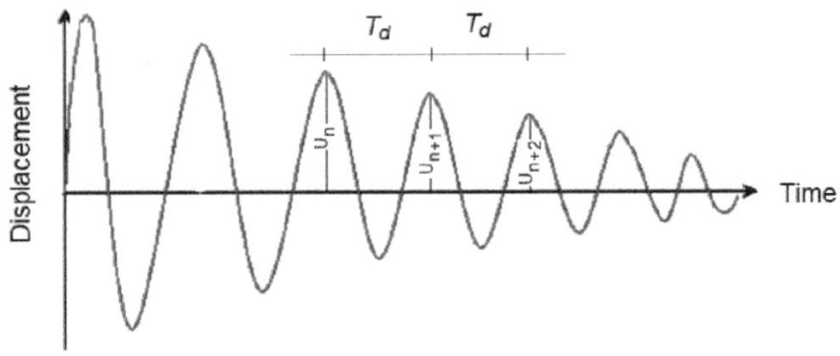

Fig. 11.9 Oscillation of a damped Single Degree of Freedom System

The SDOF system above has one natural frequency and one mode of vibration (translation in direction of x). Similarly, for a multi-degree of freedom system, there is a number of natural frequencies equal to the numbers of degrees of freedom of the system and each natural frequency has an associated mode shape. Evaluation of the natural frequencies and mode shapes of the system is defined as modal analysis and can be evaluated mathematically through eigenvalue analysis. As an example, Figure 11.10 displays the first vertical mode shape of a cable-stayed bridge associated with a natural period of 1.77 s.

Fig. 11.10 A vertical mode shape example for a Cable-Stayed Bridge

11.5 WIND ENGINEERING FUNDAMENTALS

11.5.1 General

Wind is defined as the movement of air due to the differential temperature of the surroundings. This motion can be categorized into vertical components named current and horizontal components named wind. Wind is further decomposed into two components known as along wind and crosswind depending on the orientation of the structure. Meteorological features of the wind flow include dependence of wind speed on averaging time; variation of wind speed with height, which depends on surface roughness and topography; and the atmospheric turbulence, which affects the wind speed and the aerodynamic loading (Simiu and Miyata, 2006). Frictional forces play an important role in the balance of forces on the moving air as the surface of the earth is approached. For thunderstorms, this zone is around 100 m but extends up to 500 to 100 m height for larger storms. The term boundary-layer means the zone of wind flow affected by friction on the earth's surface.

The variation of the mean wind speed with height above the ground can be represented by the power law (Holmez, 2003), which relates the wind speed at any height Z, with that at 10 m being,

$$U(Z) = U_{10}\left(\frac{Z}{10}\right)^{\alpha} \quad ...(11.6a)$$

In equation 11.6a the exponent α depends on the terrain characteristics. The other relationship used to approximate the wind profile is the logarithmic law expressed as:

$$U(Z) = \frac{1}{k}U_{*} \ln\left(\frac{Z}{Z_0}\right) \quad ...(11.6b)$$

In equation 11.6b, U_* is a friction velocity $= \sqrt{\frac{\tau_0}{\rho}}$ where τ_0 is the wind stress at the ground level and ρ is the air density =1.2256 kg/m^3 at sea level and 15°C; and k is von Karman constant = 0.4 approximately, based on wind tunnel experiments (Taly, 1998).

Usually, the wind load on structures is proportional to velocity pressure q defined in terms of the mean wind speed U as

$$q = \frac{1}{2}\rho V U^2 \quad ...(11.6c)$$

11.5.2 Laminar and Turbulent Wind Flows

Wind flow at very low speeds tends to be orderly and is defined as laminar. As wind speeds up, a transition occurs, and it crinkles up into complicated, random turbulent flow. It fluctuates in time and space' i.e., the wind speed is a random function of time at any one point in space. Turbulent wind flow is of interest in structural engineering because of several associated phenomena that can lead to serious consequences and will be discussed in detail in the following sections. Due to turbulent fluctuations of wind flow, the definition of wind speeds depends on averaging time.

11.5.3 Wind Speeds and Averaging Times

The hourly wind speed is of a great interest as it is commonly used as a reference wind speed in wind tunnel simulations. It is the speed averaged over 1 hour at 10 m above the ground. The peak 3-s gust speed is a storm's largest speed averaged over 3 s at 10 m above the ground. Also, turbulence can affect the wind flow around a structure and hence the distribution of wind forces. Most importantly, the flow fluctuations induce dynamic effects in flexible structures such as cable-stayed bridges. It is for this reason that wind tunnel tests attempt to simulate basic features of turbulent wind flow.

11.5.4 Reynolds Number

Reynolds number (R_e) defined as the ratio of inertial to viscous forces in the fluid is a very significant parameter for turbulence representation in both wind tunnel tests and Computational Fluid Dynamics (CFD). The Reynolds number is function of the velocity of flow, viscosity of air, and a characteristic dimension L of the flow domain, usually taken as the height of the flow domain therefore:

$$R_e = \frac{UL}{\upsilon} \quad ...(11.7a)$$

For air flow at atmospheric pressure conditions equation 11.7a reduces to (Simiu and Miyata, 2006):

$$R_e = 67000UL \quad ...(11.7b)$$

In 11.7b U and L are expressed in m/s and m respectively. The transition from laminar to turbulent flow starts around $R_e = 10^4$ and the flow is considered turbulent beyond 10^5 and may reach high values to the order of 10^8 to 10^9 for large problems such as flow of clouds in space (Warhaft, 1997).

11.5.5 Strouhal Number

Figure 11.11 depicts a cross-section of a rectangular cylinder immersed in uniform, smooth wind flow. It is shown that as the wind blows across the cylinder, vortices are shed alternately from one side to the other. Wake alternating vortices shed at a dominant frequency f_v given by the relationship:

$$f_v = S_t \frac{U}{D} \quad \text{or} \quad S_t = \frac{f_v D}{U} \quad ...(11.8)$$

Where the Strouhal number St depends on the cross section of the body, U is the velocity of oncoming flow, which is assumed linear, and D is the characteristic dimension of the body projected on a plane normal to the wind flow. The fluctuating flow on the cylinder's wake will generate pressure zones that induce on the cylinder a transverse fluctuating load normal to the oncoming flow, namely lift force.

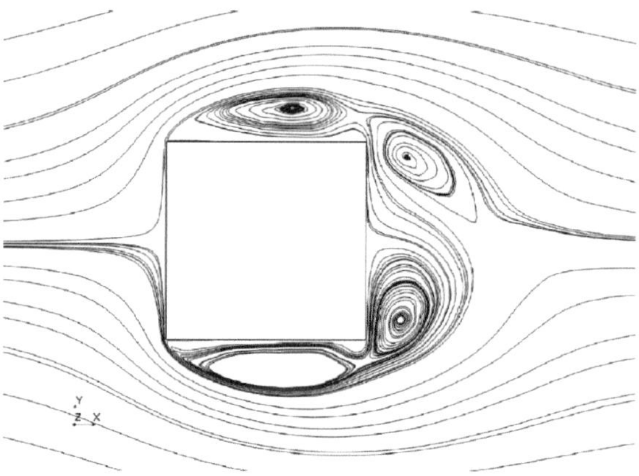

Fig. 11.11 Flow around a rectangular Cylinder (Peters and Uddin, 2019)

11.5.6 Scruton Number

The Scruton number is an important parameter when considering mechanisms of dynamic wind excitation. It is defined as:

$$S_c = \frac{m\xi}{\rho D^2} \qquad ...(11.9)$$

where,

m is the mass per unit length (kg/m); ξ is the damping as a ratio of critical damping; ρ is the air density (kg/m^3); and D is a characteristic dimension (m) that can be taken for example as the stay cable diameter when investigating cable vibrations.

11.5.7 Wind Turbulence

Turbulence or 'gustiness' in the wind speed, can be measured by its standard deviation, or root-mean-square. The ratio of the standard deviation to the mean value is known as the turbulence intensity. The wind spectrum is a means of describing how slowly or quickly the wind speed varies with time, in other words, it describes the distribution of turbulence with frequency. It is defined so that the contribution to the variance (σ_u^2, or square of the standard deviation), in the range of frequencies from ω to $\omega + d\omega$, is given by $S_u(\omega)\, d\omega$, where $S_u(\omega)$ is the spectral density function for the velocity as a function of time $u(t)$. Integrating over all frequencies, the variance can be defined as the area enveloped by the spectrum curve $\sigma_u^2 = \int_0^\infty S_u(\omega) d\omega$. The most common and widely used spectrum adopted for wind engineering for the longitudinal velocity component (parallel to the mean wind direction) is the von Karman spectrum (Harris, 1968)

11.6 AERODYNAMIC FORCES ON CABLE-STAYED BRIDGES

Structural motions on cable-stayed bridges are classified into two categories: self-excited or buffeting. Structural motions that increase the aerodynamic action of the wind flow on the structure are called self-excited and the behavior associated with them is termed aeroelastic. On the other hand, buffeting is a random type of vibration caused by turbulent wind flow. It may cause strength or fatigue problems for the structure.

11.6.1 Effects of Self Excited Structural Motions

11.6.1.1 Vortex-Induced Oscillations

Vortex-induced oscillations originate from the alternate and regular shedding from both sides of a bluff body such as a bridge deck or tower. This shedding of vortices in the wake of the body gives rise to fluctuating lift forces. If the vortex-shedding frequency f_v approaches one of the natural frequencies of the system, oscillations will occur due to resonance effects. Once resonance occurs, the vortex shedding frequency no longer follows equation (11.8) but becomes locked to the body's natural frequency. The behavior of the bridge in this case is an aeroelastic phenomenon because the wind flow affects the body motion, and the properties of the body in turns affect the flow to the extent that the vortex-shedding frequency synchronizes with the natural frequency of the moving body.

Vortex shedding usually occurs in smooth wind flow and tends to decrease as the turbulence of the wind increases. Peak vortex-induced vibrations may occur for wind speeds of the order of 10 m/s for typical bridge superstructures with deep or shallow box decks with or without overhangs. Figure 11.12 depicts results of wind tunnel investigations on the Long Creek Bridge's box deck to mitigate the effect of vortex-induced oscillations. A soffit plate and fairings with various dimensions were added to the original section for this purpose (Simiu and DogHun, 2019).

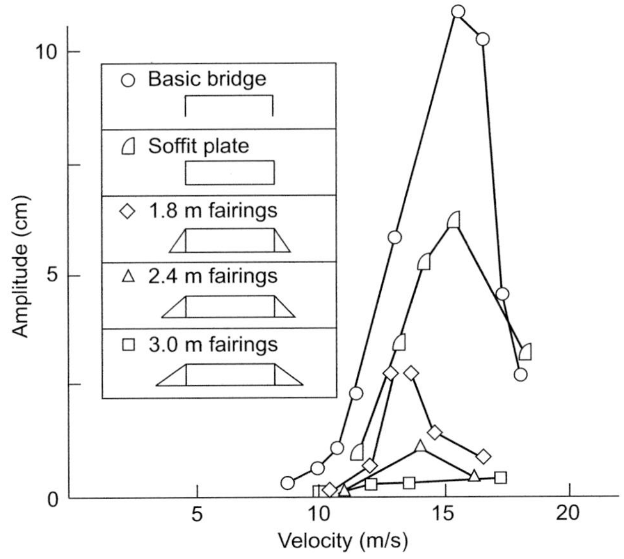

Fig. 11.12 Vertical amplitude of vortex-induced oscillations for several options of bridge deck sections proposed for Long Creek Bridge (Simiu and DogHun, 2019)

Generally, vortex-induced oscillations do not exert serious problems to the bridge deck as long as the amplitude of oscillations is not excessive. Nevertheless, it may result in problems in the long term due to fatigue. Also, in the case of footbridges, if the deck weight is light with low damping, the oscillations may reach a level of amplitude that may cause discomfort to bridge users.

11.6.1.2 Galloping

Galloping occurs at frequencies much lower than the vortex shedding frequencies for the same sections and is usually associated with the natural vertical mode of the deck. It occurs when the angle of wind attack changes as the body in motion deviates from its equilibrium position and consequently the relative velocity of the attacking wind changes. The changed relative velocity will develop an asymmetrical pressure distribution that enlarges the body's motion rather than decreasing it. The oscillations take place in a plane perpendicular to the oncoming wind flow velocity. Galloping tends to occur on narrow bridges with slenderness ratios (width/depth) less than 5.

11.6.1.3 Flutter

Flutter is a self-excited phenomenon and involves oscillation amplitudes that grow in time and can cause disastrous structural consequences. It initiates due to forces caused by the relative motion between the bridge deck and the approaching wind. Following a small perturbation, the deck will revert to its position of static equilibrium, owing to stabilizing self-excited forces associated with the perturbation. If the wind speed increases, the aerodynamic forces acting on the deck change until the wind speed reaches a critical value, at which, the oscillation amplitudes amplify in time. The self-excited forces that cause these increasing oscillations can be viewed as producing a negative damping effect. Therefore, flutter induced forces may result in the wind either suppressing the bridge motion under the stable phase of oscillation or promoting the bridge motion under the unstable phase of oscillation.

Flutter may involve torsional motion only (torsional flutter) or coupled torsional and vertical motion (classic coupled flutter). Pure torsional flutter is exhibited by bridge decks that are narrow, i.e., small width-to-depth ratio. The collapse of the first Tacoma Narrows Bridge on November 7, 1940, was due to the mechanism of torsional flutter, which led to the collapse. This was caused by dynamic vortex separation from the windward edges of the stiffening girder, which descended along the floor web, always synchronized with torsional motion, as shown in Figure 11.13 (Miyata, 2003).

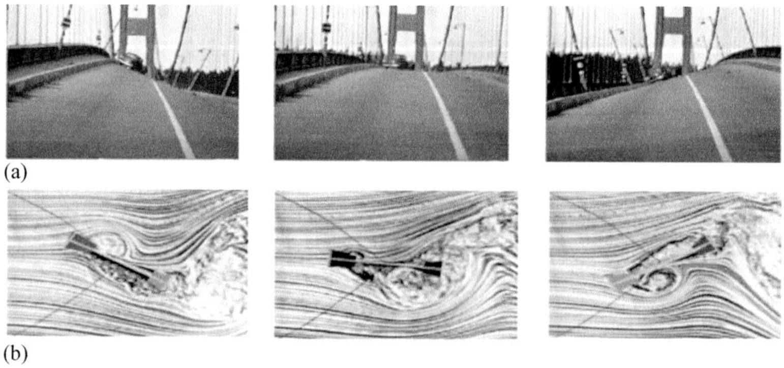

Fig. 11.13 The collapse of the Tacoma Narrows Bridge: (a) torsional motion led to collapse; and (b) wind tunnel separated vortex descending by wind flow visualization (Miyata, 2003).

The wind speed, at which the onset of torsional flutter occurs, can be viewed as the speed at which aerodynamic forces induce the torsional damping with negative aerodynamic and structural components. If this condition is reached, oscillations can grow to levels that may cause failure of the bridge. Wide decks with streamlined cross-sections will tend to exhibit classic (coupled) flutter. If the deflection of a vertical mode of vibration is different significantly from that of a torsional mode, then it is unlikely that the two modes will combine in flutter instability. Coupled motions will only occur for similar torsional and vertical modes. The critical wind speed for flutter tends to be less for decks with low torsional stiffness. Details at the edge of the deck may play a significant role in this type of instability.

11.6.1.4 Analytics of the Flutter Problem

Consider the bridge section as depicted in Figure 11.14. Let h represent the vertical deflection of the local c.g. of the section, and α the rotation coordinate (angle) about the c.g. m represents the mass per unit span and I, the mass moment of inertia about the c.g. per unit span. Then (neglecting lateral motions as unimportant to flutter) the two sectional equations of motion are (Scanlan, 1981):

$$m[\ddot{h} + 2\tau_h \omega_h \dot{h} + \omega_h^2 h] = L_h \quad \text{...(11.10a)}$$

$$I[\ddot{\alpha} + 2\tau_\alpha \omega_\alpha \dot{\alpha} + \omega_\alpha^2 \alpha] = M_\alpha \quad \text{...(11.10b)}$$

In equations 11.10a and 11.10b τ_h are τ_α are the critical damping ratios and ω_h, ω_α are the natural circular frequencies, respectively in h and α motions, and L_h, M_α are the aerodynamic force and moment per unit span acting on the section.

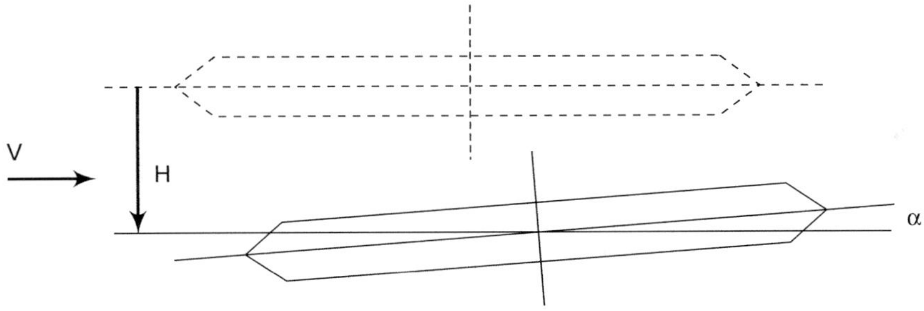

Fig. 11.14 Flutter analytics notations

The aerodynamic force and moment at the c.g. are of the linear self-excited type and are given by:

$$L_h = H_1 \dot{h} + H_2 \dot{\alpha} + H_3 \alpha \quad \text{...(11.12a)}$$

$$M_\alpha = A_1 \dot{h} + A_2 \dot{\alpha} + A_3 \alpha \quad \text{...(11.12b)}$$

In equation 11.12b the coefficients H_i, A_i ($i = 1, 2, 3$) are aerodynamic in origin and must be evaluated from experiments for the shape of the considered deck. The coefficients H_1, pertaining to $\dot{h}$, and A_2 and A_3, pertaining to $\dot{\alpha}$ and α, are the direct coefficients, and the others (H_2, H_3, and A_1) are the coupling coefficients.

A nondimensional form of the coefficients H_i and A_i is necessary so that their values evaluated in scaled model tests, can be transferred for use in full scale. This is accomplished by writing equations (11.12) in the form of lift and moment as:

$$L_h = \frac{1}{2}\rho U^2 B\left[KH_1^* \frac{\dot{h}}{U} + KH_2^* \frac{B\dot{\alpha}}{U} + K^2 H_3^* \alpha\right] \quad \text{...(11.13a)}$$

$$M_a = \frac{1}{2}\rho U^2 B\left[KH_1^* \frac{\dot{h}}{U} + KH_2^* \frac{B\dot{\alpha}}{U} + K^2 H_3^* \alpha\right] \quad \text{...(11.13b)}$$

where,

ρ = air density =1.225 kg/m^3,
U = cross wind velocity,

$K = B\omega/U$,
B = deck width,
ω = circular frequency of flutter oscillation,

The non dimensional aerodynamic coefficients H_i^* and A_i^* are functions of K and assume the following relations to H_i and A_i:

$$H_1^* = \frac{mH_1}{\rho B^2 \omega} \; ; \; A_1^* = \frac{IA_1}{\rho B^3 \omega} \quad \text{...(11.14a)}$$

$$H_2^* = \frac{mH_2}{\rho B^3 \omega} \; ; \; A_2^* = \frac{IA_2}{\rho B^4 \omega} \quad \text{...(11.14b)}$$

$$H_3^* = \frac{mH_3}{\rho B^2 \omega^2} \; ; \; A_3^* = \frac{IA_3}{\rho B^4 \omega^2} \quad \text{...(11.14c)}$$

Examples of the values experimentally obtained for H_i^* and A_1^* are illustrated in Figure 11.15 for decks whose sectional forms are sketched on the figures. These graphs are given as functions of $\frac{U}{NB} = \frac{2\pi}{K}$, where N = frequency of flutter oscillation = $\omega / 2\pi$.

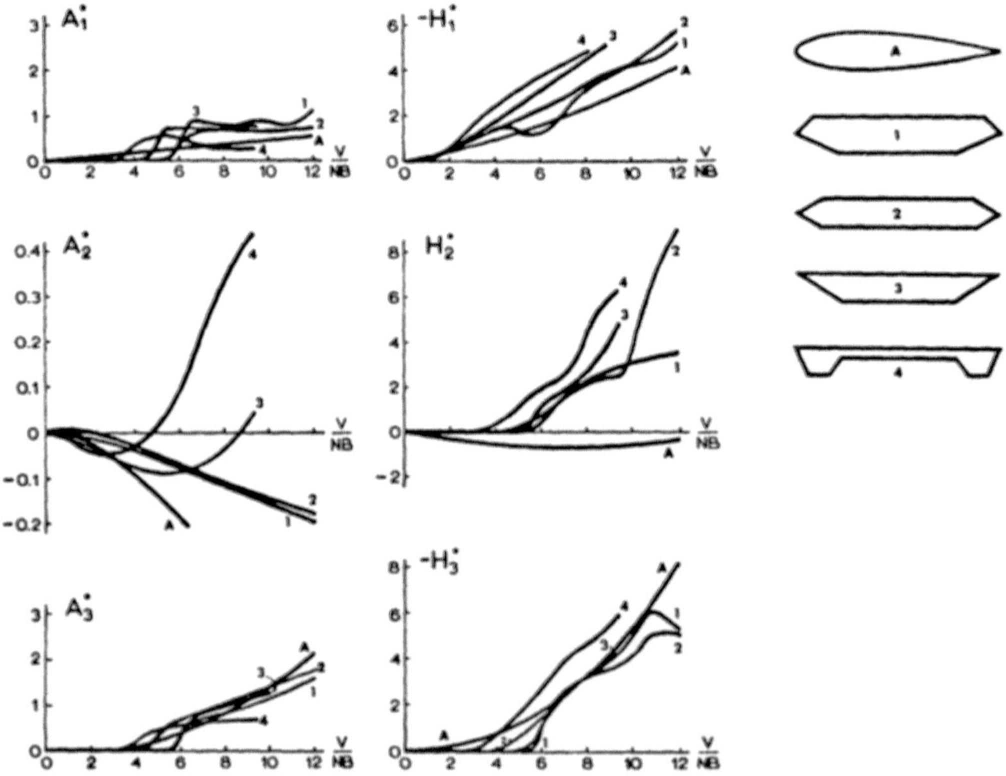

Fig. 11.15 Flutter analytics aerodynamic coefficients

It is important to note that the products KH_1^* and KA_1^*, play the roles of flutter derivatives. This can be shown by considering the term $H_1\dot{h}$ as an example. It has the dimensions of a force per unit span length. If written in classic aerodynamic lift force form, it would translate to:

$$\frac{1}{2}\rho U^2 BKH_1^* \frac{\dot{h}}{U} = H_1\dot{h} = \frac{1}{2}\rho U^2 BC_L \approx \frac{1}{2}\rho U^2 B\frac{dC_L}{d\alpha}\frac{\dot{h}}{U} \qquad ...(11.15)$$

where, $\alpha = \frac{\dot{h}}{U}$ is an effective angle of attack. Thus, $KH_1^* = \frac{dC_L}{d\alpha}$ where, $\frac{dC_L}{d\alpha}$ is the derivative of a lift coefficient C_L with respect to the attack angle.

It's important to note that all aerodynamic derivative values must be determined experimentally using wind tunnel testing, and that they change as functions of decreasing velocity. U/NB. Examples of these flutter derivatives established for the Luling Bridge are displayed in Figure 11.16. The manner in which the coefficient A_2^* evolves with U/NB is of particular interest as it is proportional to torsional aerodynamic damping, and it plays a central role in many cases of bridge flutter susceptibility, since it often changes sign from stable to unstable with increasing U/NB values in some cases.

If the bending and torsional modes interact, a self-excited motion may occur, and the oscillations can diverge to a destructive level on the bridge. Therefore, it is imperative to keep the design flutter wind speed always significantly higher than the expected wind speeds at the site. Simplified criteria and wind tunnel section model tests are typically adequate to estimate the critical wind speeds for flutter and galloping. Nevertheless, simplified criteria may be used at early phases of design and cannot substitute standard section model tests. Selberg (1963) proposed an empirical equation for critical flutter speed, U_F, which, in its simplest form, can be written as:

$$U_F = 0.44B\sqrt{(\omega_T^2 - \omega_V^2)\frac{\sqrt{\upsilon}}{\eta}} \qquad ...(11.16)$$

where, $\upsilon = 8\left(\frac{r}{B}\right)^2$ and $\eta = \frac{\pi\rho B^2}{4m}$; r is the radius of gyration of the cross section($I = mr^2$); ω_T and ω_V are the circular frequencies in the first torsional mode and first vertical bending modes respectively.

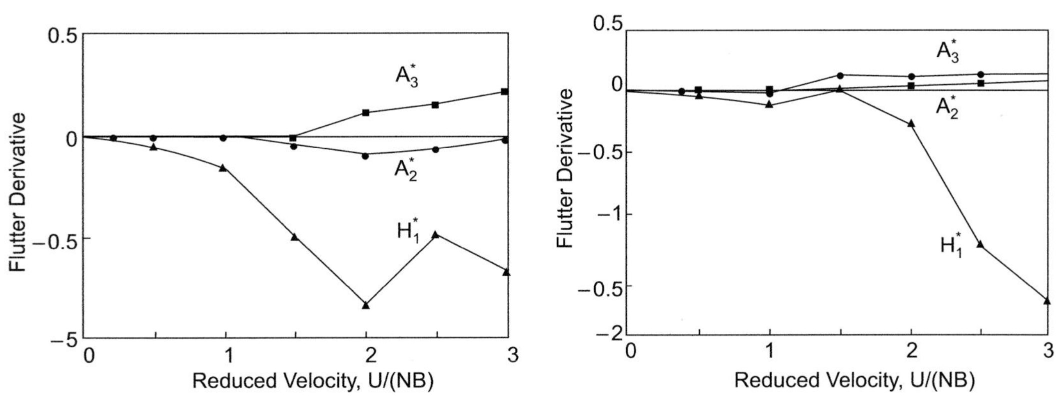

(a) 0-deg angle of attack

(b) 6-deg angle of attack

Fig. 11.16 Flutter derivatives for cross-section of Luling Bridge

11.6.2 Effects of Buffeting Motions

The nature of wind is turbulent. It strikes the bridge deck from various horizontal angles. Thus, there are variations in transient unsteady pressure distributions that are random in both space and time due to the fact that wind velocity vectors are not all simultaneously oriented in the mean horizontal direction. As wind speed increases, these velocity vectors can cause a variety of vibration modes.

The primary goals of assessing the buffeting response are to guarantee the bridge's structural integrity in the event of turbulent wind; determining whether this potential excitation will be detrimental to the bridge in the long run, either in terms of user fatigue or comfort; and determining which design parameters will mitigate the effects of buffeting. It is crucial to conduct wind tunnel tests in turbulent wind flow for a variety of wind directions in order to identify the most critical case of wind loading and wind-induced deflections because velocity vectors may excite the bridge from different directions.

Since cable-stayed bridges are significant structures, buffeting loads need to be evaluated for windstorms with a 100-year return period. Generally, theoretical buffeting analysis is the basis of the evaluation of the wind loads acting on the bridge. Since the load distribution for buffeting is assumed to be random in space and time, stochastic analysis in the frequency domain is usually conducted to evaluate such loads in which wind turbulence is represented by spectrum and cross-spectrum. This analysis considers the response of each of the bridge's modes of vibrations separately and superimposes the results. Input parameters include the static force and moment coefficients on the bridge deck as obtained from wind tunnel tests; structural properties of the deck such as mass, mass moment of inertia and deck dimensions; modal frequencies and shapes; structural damping; and turbulence properties. This method is not explored further in this book and the reader can refer to Davenport, 1962; Irwin, 1977, or Scanlan and Jones; 1990 for further explanations on the derivation of the approach.

Figure 11.17 shows the order of the dynamic response of a cable-stayed bridge deck under different types of aerodynamic excitation displayed in terms of wind speed.

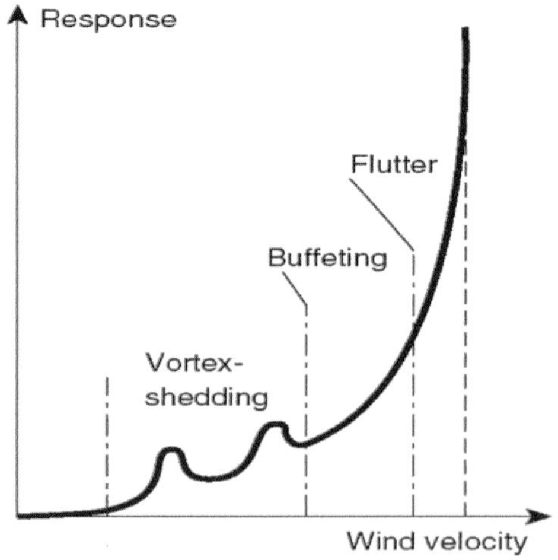

Fig. 11.17 Dynamic Response of a bridge deck under aerodynamic excitation (Gimsing and Georgakis, 2012)

11.7 AERODYNAMIC INVESTIGATION OF CABLE-STAYED BRIDGES

Aerodynamic stability for cable-stayed bridges is established through a combination of wind climate and site analysis studies; wind tunnel tests; and numerical analyses. Wind loads are evaluated based on the aerodynamic stability and should account for all the effects of the above investigations.

11.7.1 Wind Climate and Site Analysis

The wind statistics used to determine the design wind speeds at the bridge site should be based on onsite wind speed measurements but in most cases historical data from nearby wind recording stations are typically used. In this regard, records of up to 50 years should be used from local meteorological stations. Extremal Statistical models of wind speed and direction are usually fitted to the meteorological wind data to evaluate wind speed as a function of return period and also to evaluate the component of the wind velocity normal to the bridge span as a function of return period. Results are usually presented as mean-hourly (i.e.,1-hour mean) speeds, which are directly applicable for design, or as 10-minute mean speeds, which is typical for aerodynamic stability criterion. Figure 11.18 illustrates an example showing various mean-hourly wind speeds of 10 m/s for an open terrain as a function of return period for the location of the Olivier-Charbonneau Cable Stayed Bridge in Quebec, Canada. This relationship is compared to the mean hourly speeds from CAN/CSA-S6-06, for 10, 25, 50, and 100 years. The mean-hourly wind speed can be converted to the 10-minute mean using relationships of Figure C6-2 of the ASCE 7-05.

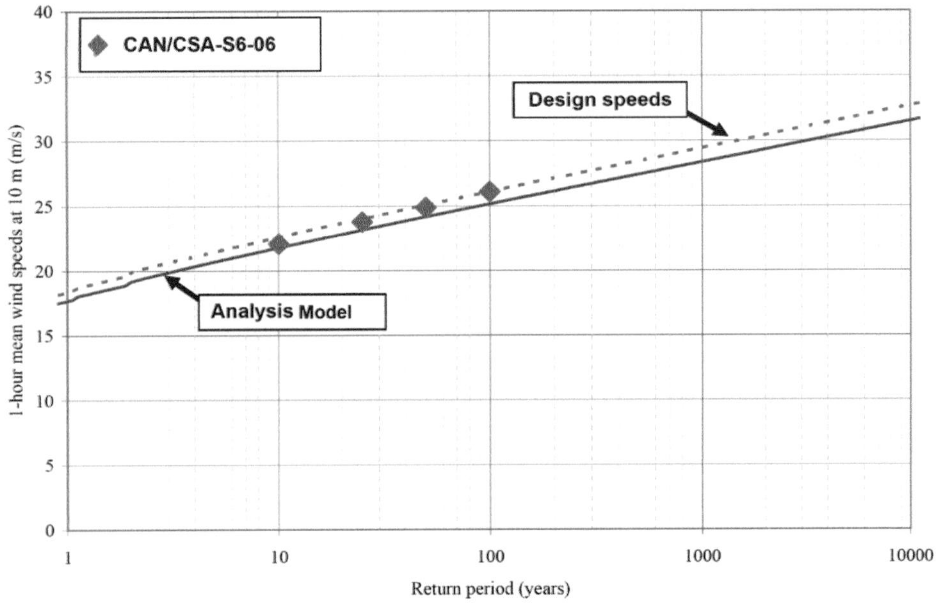

Fig. 11.18 Illustration of the wind return period versus 1-hour mean wind speed (Courtesy, Parsons Corporation)

For the structural design of cable-stayed bridges, a return period of 100 years is typically used. At sites of extreme winds such as hurricanes, wind speed predictions for very long return periods are required such as 100, 1000 or more return periods. In this case, additional numerical simulations that would extend the time length of data are deemed necessary. For flutter instability of the completed bridge, the recommended return period is 10,000 years which is based on a 10-min mean. The long return period is usually considered because, if flutter occurs, it is likely that the bridge collapses.

Wind directionality effects at the bridge site can be defined by fitting the wind data at the site with the Weibull probabilistic distribution function. Wind distribution results are usually displayed in terms of the percentage of time where wind speeds for specified return periods, blowing in any direction, would be exceeded with regard to the bridge alignment. An example of this distribution is illustrated in Figure 11.19. It is observed for the bridge alignment shown in the figure that the most probable directions for strong winds (e.g., once in 100 years) are from the northeast (20 degrees to about 50 degrees) and from the southwest (220 degrees to about 280 degrees). Both wind directions are close to the normal of the bridge span.

Topographical model studies using Weather Research and Forecasting Model (WRF) could provide additional information at the site regarding turbulence intensities, scaling factors, and the mean angle of wind attack. These additional resources can be used to enhance the data quality and predictions collected from other resources on the wind speed measurements. Local terrain studies should take into consideration the topographic effects of tall local buildings and/or large bridges.

Turbulence properties at the bridge site need to be evaluated. Turbulence intensities for different velocity components since they are significant for the buffeting response of cable-stayed bridges to strong winds.

Results from wind climate analysis shall be documented in a report with sufficient detail using tables and diagrams. The report should include but not be limited to description of the bridge site with a general classification such as open, suburban terrain; wind speeds for return periods from 10 to 10,000; wind directionality and varying attack angles; effect of complex topography (if present) on the mean wind and gust speed profiles; effect of tall buildings and large structures if present; and turbulence described in terms of turbulence intensity and coherence.

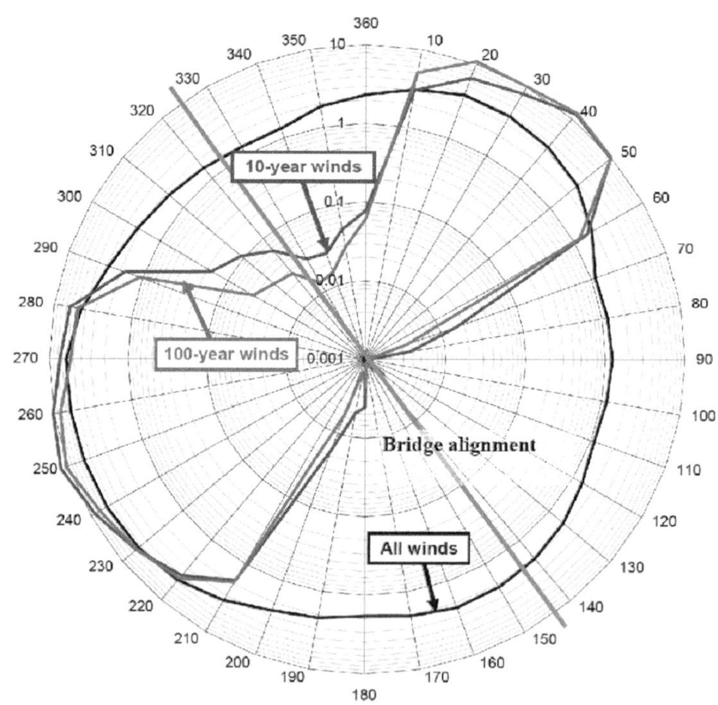

Fig. 11.19 Wind directionality distribution (Courtesy of Parsons Corporation)

11.7.2 Wind Tunnel Tests

Wind tunnel tests are used to reliably evaluate the wind-induced structural performance. The most widely used wind tunnel techniques include sectional model tests and aeroelastic model tests. The Section model test is a simple technique of assessing aerodynamic stability and occupies a central position in bridge aerodynamic simulations because of several factors: it can be the largest in scale and, therefore, most faithful as to geometric detail; it is the most useful model in determining the principal elements of geometric contours necessary for aerodynamic stability; extraction of accurate non dimensional aerodynamic coefficients from section model studies enables several subsequent full-bridge aerodynamic studies.

An aeroelastic model test may be used in the final design phase to further investigate the impact of some parameters that cannot be captured by the sectional model method, such as the effect of any complex terrain and surroundings; three dimensional effects; and wind directionality effects.

11.7.2.1 Sectional Model Tests

Typically, a section model represents a segment of the full-scale bridge deck together with its proper degrees of freedom, their frequencies, and effective rotation points so that all prototype features are simulated by the scaled model. The scaled length of the segment depends primarily on the dimensions of the structure and the size of the wind tunnel. Model scaling implies essentially a typical geometric scale ranging from 1/25 to 1/100, other parameters, like mass and time, will be properly scaled as well.

A sectional model is usually constructed from different materials including wood, aluminum, and steel. It is designed to simulate the scaled mass and mass moment of inertia about the center of gravity. It is very significant that all the details of the prototype structure cross-section such as parapets, and barriers, be modeled in the sectional model. Table 11.3 summarizes the relations required among the full-scale properties of a bridge and the corresponding model scale values used for the sectional model tests. Scaling requirements further include the equal reduced velocity condition defined as

$$\left(\frac{U}{NB}\right)_{\text{Model}} = \left(\frac{U}{NB}\right)_{\text{Prototype}} \qquad \text{...(11.17)}$$

Table 11.3 Relations required for sectional model scaling

Quantity	*Symbol*	*Scale*
Length	λ_L	$\frac{L_{\text{mo del}}}{L_{\text{prototype}}}$
Wind Velocity	λ_U	$\frac{U_{\text{mo del}}}{U_{\text{prototype}}}$
Density	λ_ρ	$\frac{\rho_{\text{mo del}}}{\rho_{\text{prototype}}}$
Frequency	λ_N	$\frac{\lambda_U}{\lambda_L}$
Time	λ_T	$\frac{\lambda_L}{\lambda_U}$
Damping ratio	λ_T	$\frac{\tau_{\text{mo del}}}{\tau_{\text{protototype}}} = 1$

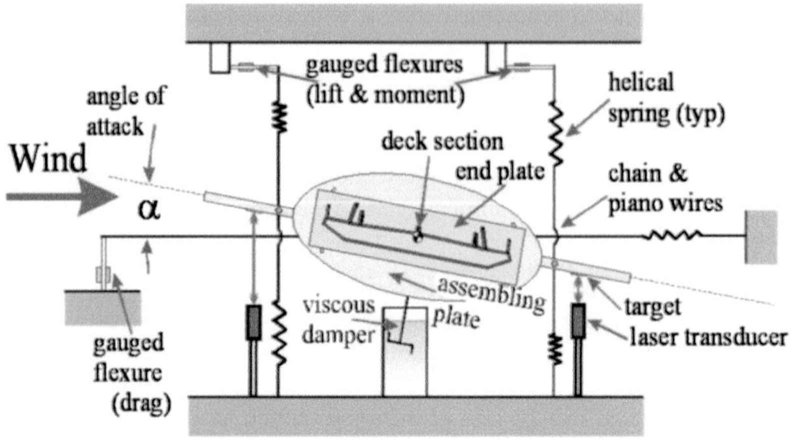

Fig. 11.20 Sectional model setup (Courtesy, Rowan Williams Davies & Irvine Inc.)

The model is mounted on a spring suspension system as depicted in Figure 11.20, which depicts a view from the wind tunnel outwards. The tunnel wall (not shown) is located between the end plate and the assembling plate. The suspension system with the springs are placed outside the tunnel walls and built directly into the side walls of the wind tunnel. The suspension system is adjusted to model effectively the stiffness of the deck both in the vertical direction and torsion. This suspension system allowed 2DOF vertical and torsional motions to be simulated. Selection of an appropriate stiffness and spacing of the springs permitted tuning of the model to the desired vertical and torsional frequencies for testing. The first symmetric vertical modal frequency and first symmetric torsional modal frequency of the completed bridge are usually selected as target values for the wind tunnel tests. These modes are selected because of their potential coupling during wind-induced response, which could result in coupled vertical-torsional flutter.

The model is driven sinusoidally through prescribed motions, where laser displacement transducers are used for the measurement of these vertical and torsional motions. The loading on the section is measured using strain-gauged flexures attached to the model's center of rotation and the end spring supports. Damping is added to the system by energy absorption devices located outside the tunnel. These devices allow the structural damping for vertical and torsional motions to be adjusted as desired.

In general, the structural damping level for sectional model tests is set to a value ranging from 0.5% to 1.0% of critical damping for cable-stayed bridges Simiu and Miyata (2006). The damping ratio is proportional to the amplitude of vibrations. Therefore, the model structural damping is usually set at an amplitude associated with the admissible vortex-shedding motions of the bridge deck. A typical sectional model in the wind tunnel is shown in Figure 11.21.

It has been customary in the past to conduct sectional model tests in conditions with smooth wind flow. The deck stability against flutter can be evaluated under these conditions. Nevertheless, smooth flow tests may give conservative results for a vortex-induced response. On the other hand, turbulent flow tests are more representative of the bridge performance in strong winds because the natural wind tends to be turbulent Therefore, it is important to induce low model turbulence, expected at full scale for the prototype as it is awkward to simulate the large-scale turbulence well on a sectional model. Testing procedures are implemented in the following steps:

- The wind tunnel tests to evaluate vortex-induced oscillations and flutter are conducted for a range of wind speeds and attack angles both in smooth, uniform flow and in low turbulence flow.
- The wind speed is increased gradually in small steps and the motion in both the vertical direction and in torsion are recorded for each attack angle.

Fig. 11.21 Sectional model in the wind tunnel (Courtesy, RWDI)

- To calculate the aerodynamic damping, vertical and torsional motions of the model are manually induced so that the decay and logarithmic decrement of the oscillations is recorded.
- The wind speed is increased gradually until the model exhibits flutter instability or until the maximum tunnel wind speed is reached.
- The critical wind speed for flutter is determined as the speed at which the total damping including structural and aerodynamic components become negative at a given peak amplitude in smooth flow.

Figure 11.22 displays example results of a sectional model test. For this bridge the mean hourly wind speed for a 100-year return period at the bridge deck was evaluated as 27 m/s. This is the

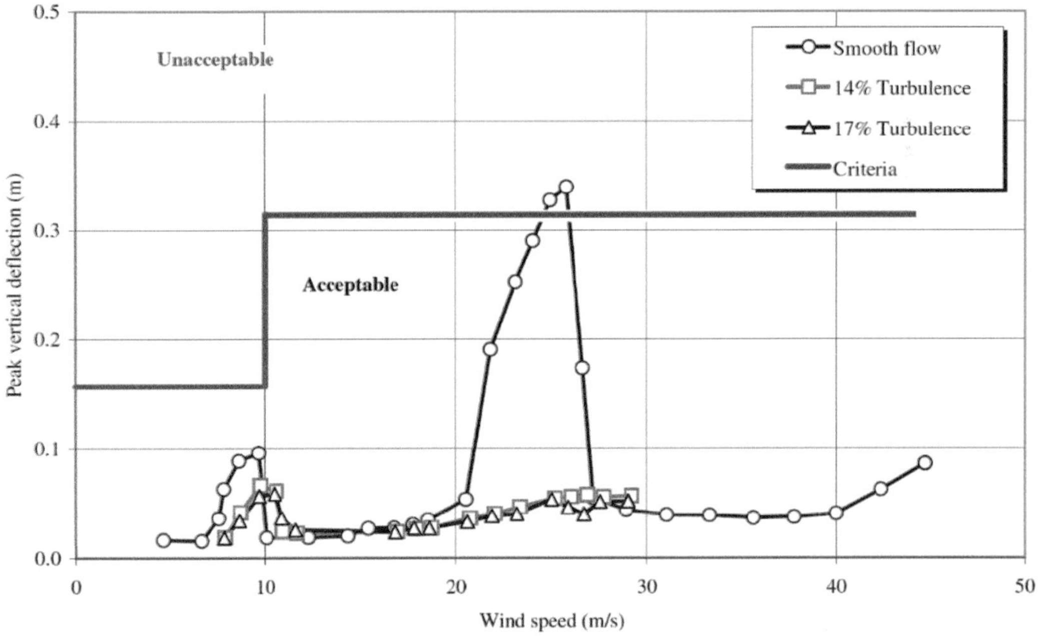

Fig. 11.22 Example illustrating the performance of a cable-stayed bridge deck under sectional model testing (courtesy, Parsons Corporation)

limiting speed for vortex shedding. The 10-min mean wind speed for a 1000-year return period was evaluated as 36 m/s. The sectional test was subjected to varying wind speeds at varying attack angles that happens to be zero degrees for the results displayed. It can be observed that flutter was not observed below the 36 m/s minimum flutter speed. The unacceptable vortex-induced vibrations are only observed for smooth flow in a very narrow speed range. This is not an issue as smooth flow testing is expected to yield exaggerated deflections with respect to turbulent flow tests. It is important to note that winds above 26 m/s become less significant as the main concern is wind buffeting.

The same suspension rig (but with much stiffer springs and added damping to minimize motions of the model) can be used to measure the static force and moment coefficients on the bridge deck. Force coefficients are measured for a range of wind speeds and attack angles. Under these conditions, static force and moment coefficients can be calculated by normalizing the forces and moments measured on the deck section as follows:

$$C_y = \frac{F_y}{\frac{1}{2}\rho U^2 D}, \quad C_z = \frac{F_z}{\frac{1}{2}\rho U^2 B}, \quad C_{mx} = \frac{M_x}{\frac{1}{2}\rho U^2 B^2} \qquad \text{...(11.18)}$$

In 11.18 F_y, F_z and M_x are lateral force (drag), vertical force, lift and moment per unit length; B is the deck width; and D is the representative depth of the deck. The coefficients are with respect to the deck axis system as depicted in Figure 11.23 where F_D and F_L are drag and lift in the wind axis system.

The turbulence lateral force coefficient C_y, vertical force coefficient C_z, pitching moment coefficient C_m, and their weighted slopes can be calculated as a weighted average of the values evaluated for the range of attack angles.

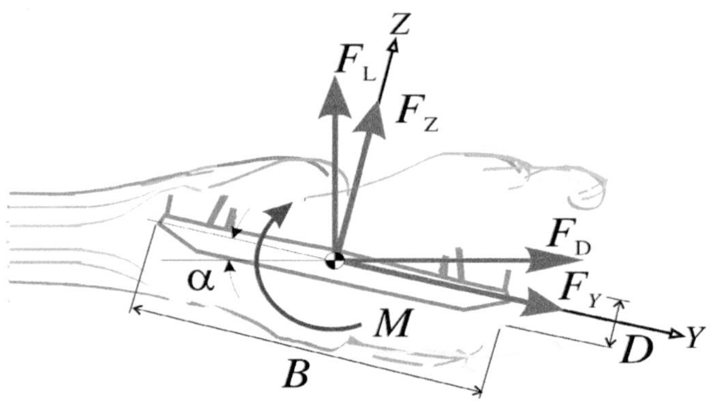

Fig. 11.23 Forces acting on a bridge deck due to wind flow (Courtesy, Rowan Williams Davies & Irvine Inc.)

For designing purposes wind loads need to be applied simultaneously as static in combination with other types of loads such as dead, live, friction, and temperature loads as per the design codes and guidelines. Theoretical buffeting analysis is usually carried out for the derivation of the wind loads acting on the bridge. Aerodynamic parameters such as drag, lift and moment coefficients as per equation (11.18) are obtained from sectional testing and become input to this analysis. Equivalent static load representation from the multimode buffeting response of bridges is formulated in terms of either a weighted combination of modal inertial load components, or the background and resonant load components. An equivalent static load representation in terms of background and resonant load distributions leads to a physically meaningful and realistic load description (Davenport, 1983; Holmes, 1994). The background component of the wind load can be treated as a quasi-static load and can be determined based on the load-response-correlation (LRC) approach (Kasperski and Niemann, 1992). The resonant load component follows the distribution of the inertial load and can be expressed in terms of modal inertial loads. The total response is then calculated by combining the background and resonant responses utilizing the square root of the sum of squares (SRSS) combination approach or the complete quadratic combination (CQC) approach. The application of this approach in combining the section model tests for the equivalent wind loads on bridges has been presented by Davenport and King (1984).

11.7.2.2 Aeroelastic Model Tests

The principal advantage of the aeroelastic test is the ability to include surrounding terrain in the model. The aeroelastic model as illustrated in Figure 11.24 is a reduced scale geometric duplicate of the entire prototype bridge that includes all structural elements, the towers, the cable stays, the road deck and the road deck parapets and barriers.

For dynamic studies, it is necessary, as well, to model the mass, the mass distribution and the elastic characteristics of the prototype according to well-established scaling principles. The scale ratio for a long span bridge may be very small; for example, the model of a 900 m long bridge would have to be constructed at a scale ratio of about 1 : 400 if it were to be tested in a 2.50 m wide wind tunnel test section.

Fig. 11.24 Aeroelastic model of the Kosciuszko Bridge in the Wind Tunnel (Courtesy, New York State Department of Transportation)

The aeroelastic model is built following a set of similarity principles for structural dynamics and aerodynamics. The stiffness of various structural components, such as the deck and the towers are simulated by internal metal spines. The aerodynamic geometry of the bridge is achieved by mounting very light and stiff outer shells on the spines. The outer shells are usually built using the stereolithography modeling technique. The models are instrumented with accelerometers, strain gauges, and displacement laser transducers at selected critical locations, such as the top and the base of the pylon.

In general, an aeroelastic model furnishes comprehensive information of turbulence effects at high speeds; three dimensional influences; and construction stages. Furthermore, it allows more precise simulations of the natural turbulence and as a result, permits better full-scale predictions to be achieved. It has been demonstrated from aeroelastic model tests that the natural turbulences have a stabilizing effect on instabilities and that these are difficult to obtain from sectional model tests. This is particularly true for the stability estimations against vortex shedding and flutter.

11.8 NUMERICAL ANALYSIS

Wind flow in the atmosphere can be theoretically idealized as a flowing fluid, wherein basic principles of fluid dynamics can be applied. Within the context of fluid dynamics, the fluid flow must satisfy conservation of mass, momentum, and energy. Thus, the flow is expressed in terms of a set of partial differential equations (PDE) namely the Navier-Stokes (N-S) equations.

Theoretically, solving a set of PDE of equations is very complicated and impractical for most problems. Nevertheless, the theoretical and experimental approaches were the two basic tools to investigate fluid motion before the invention of computer hardware and software. Today with the great evolution in the modern digital computer industry and with the substantial reduction in computing costs, CFD became the dominant tool for fluid flow analyses rather than either theory or experiments.

By employing CFD, the fluid is idealized on the basis of the continuum hypothesis and the macroscopic properties associated with the bulk fluid can also be associated with a minute particle of the fluid. This allows the analyst to discretize the fluid medium to a number of points (nodes) and identify at each point, its own set of equations and then consider the entire volume of the fluid to be a continuous assemblage of these points.

Some of the major advantages of CFD compared to experimental fluid dynamics are:

- simulates very complicated flow problems, which are difficult to represent experimentally
- provides comprehensive information about the flow and its effect on the structure compared to the wind tunnel test, which provides information only at discrete locations
- well suited for design optimization, wherein several options or design parameters can be modified to end up with the optimal design of the structure
- it is less expensive than wind tunnel testing.

The finite element (FE) method is widely used in structural analysis, wherein the structure is discretized to a number of nodes with appropriate boundary conditions and using the computer, a set of equilibrium equations for the structure, that are governed by the constitutive laws of the material are solved together to provide strains, deformations, stresses, and forces at every location in the structure. This algorithm is also applicable to CFD problems. Keeping in mind that fluid deforms continuously when subjected to shear stress and knowing that the wind flow is Newtonian, the flow will be governed by a linear constitutive relationship between the rate of deformation and the shear stress, i.e., the fluid's viscosity is a key parameter for the solution. Knowing the constitutive relationship of the flow and employing the continuum approach in CFD, the FE method is used to discretize the fluid medium including the structure of interest to a number of nodes. By applying relevant boundary and initial conditions the solution of the N-S equations for the continuum is handled numerically by approximating the PDE through a system of algebraic equations solved by the computer. The numerical solution of the system's algebraic equations provides the pressure as well as the velocity distributions at every location of the structure. The CFD-FE approach is summarized in Figure (11.25).

11.9 WIND INDUCED VIBRATIONS OF STAY CABLES

11.9.1 Introduction

Wind vibrations may produce a low-amplitude, high-cycle fatigue condition on the main tension element, ancillary elements associated with the stay system, or connections with the bridge. Based on research, it seems unlikely that vortex shedding from the cables will cause a significant vibration issue for cable-stayed bridges. Effective suppression of vortex excitation will be achieved by introducing a small amount of damping (Simu and Miyata, 2006). However, large amplitude vibrations can also result from flow forces acting during other self-excited vibrations, like galloping. The foundation for characterizing stay cable vibrations and the environmental conditions that cause them has been provided by field observation programs. The term "rain-wind-induced vibrations" came about because it was determined that large oscillation incidents occurred when there was moderate rain and moderate wind. Typical stay cables have a relatively small inherent damping with a logarithmic decrement of the order of 0.01 for individually protected main tension elements (MTE) made of parallel strands or parallel wires and down to 0.001 for cement grouted stay cables. Hence, they cannot dissipate much of the dynamic excitation energy making them susceptible to fatigue. The designer must evaluate the significance of these vibrations on stay cable and bridge performance and incorporate mitigation measures such as external damping devices, stabilizing cables (cross ties), or special surface treatments used to minimize cable vibrations.

11.9.2 Vortex Excitation of Stay Cables

When the wind is roughly perpendicular to the cable axis, the alternating shedding of vortices from the two sides of the cable excites the vortex in that single cable. The wind velocity at which the vortex excitation frequency matches the natural frequency f_n is determined in terms of Strouhal number S_t and the cable diameter D as:

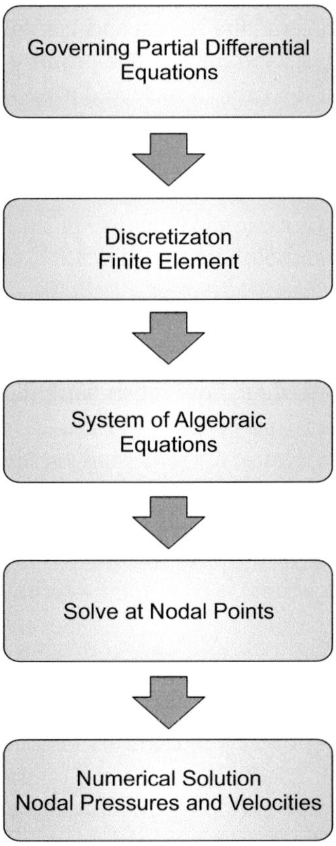

Fig. 11.25 Steps involve in computational fluid mechanics numerical methods

$$U_v = \frac{f_n D}{S_t} \qquad ...(11.19)$$

Inherent cable damping ratios are realistically estimated to be between 0.001 and 0.005. The upper end of this range is more typical of shorter cable stays with grouting and possibly some external damping, while the lower end is typical of very long cable stays without any grout infill. The cable oscillations' amplitude is inversely related to the Scruton number S_c. The Scruton number is increased, and oscillation amplitudes are thereby decreased by increasing the mass and damping of the cables. The Scruton number for stay cables typically ranges from 10 to 15, and the oscillation's amplitude is roughly 0.5% of the cable diameter. As a result, vortex shedding from the cables is not expected to cause significant vibration issues for cable-stayed bridges.

11.9.3 Rain-wind-induced Vibrations

Rain and moderate wind speeds can cause high-amplitude cable vibrations at low frequencies, usually less than 3 Hz, in the range of 0.25 to 1.0 m. Rain and moderate wind speeds (8–15 m/s) angled 20° to 60° to the cable plane, with the cable declining in the wind's direction, are the usual conditions that cause these vibrations. In certain situations, violent movements may cause neighboring cables to collide. After being initially noticed on the Meiko-Nishi cable-stayed bridge in Japan, observers have noted this phenomenon on a number of other cable-stayed bridges, such as the Fred Hartman Bridge in Texas, the Sidney Lanier Bridge in Georgia, the Talmadge Memorial Bridge in Georgia, the Faroe Bridge in Denmark, the Aratsu Bridge in Japan, the Tempohzan Bridge in Japan, the Erasmus Bridge in Holland, and the Nanpu and Yangpu Bridges in China (FHWA, 2007).

According to studies conducted in wind tunnels, the primary cause of this aeroelastic instability was water rivulets that flowed down the cable's surface during rainy conditions (Matsumoto et al., 1989). Water rivulets caused cyclical changes in the aerodynamic forces, which in turn caused the wind to feed energy into oscillations by changing the effective shape of the cable and moving as it oscillated.

The following tentative stability criterion in equation 11.20 for the Scruton number can be used for rain-wind-induced vibrations of smooth circular cables:

$$S_c = \frac{m\xi}{\rho D^2} > 10 \qquad ...(11.20)$$

This criterion can be used to determine how much damping needs to be added to the cable in order to reduce vibrations caused by wind or rain (PTI DC45.1-18, 2018). Given that the rivulets that form on the cable surface are what cause the rain/wind oscillations, it is likely that the instability is sensitive to surface roughness. To address the issue, a number of researchers have attempted to use tiny protrusions on the cable surface. With a negligible increase in drag coefficient, helical filets placed 1.5 mm high on the cables have shown to be effective. For newly constructed cable-stayed bridges, this kind of cable surface treatment is becoming a common design element (FHWA, 2007).

11.9.4 Galloping

The elliptical movement known as galloping is brought on by changes in drag and across-wind forces for cables trailing behind other objects like towers or other cables. This occurs at high wind speeds and results in large amplitude oscillations. The minimum wind velocity U_{CRIT} above which instability can be expected due to wake galloping effects can be evaluated approximately in terms of Scruton number as (FHWA, 2007):

$$U_{\text{crit}} = CfD\sqrt{S_c} \qquad ...(11.21)$$

where, C = constant, f = natural frequency of cable in Hz, D = cable diameter; and S_c = Scruton number. The constant C is dependent on the cable spacing. $C = 25$ for closely spaced cables (2 D to 6 D spacing). $C = 80$ for normally spaced cables (generally 10 D and higher). Since more damping results in an increase in the Scruton number by putting spacers or crossties along the cables to reduce the effective length of cable for the vibration mode of concern, the natural frequency can be raised.

The equations listed in this section are further explained with a numerical example. Assume that the mean wind speed is 90 km/h (25.0 m/s) and that the smooth stay cable has an exterior diameter of 0.25 m, mass per unit length of 185 kg/m, and an inherent damping ratio of 0.005. The cables are spaced at 1.5 m and the modal frequencies of the cable are computed to be $f_1 = 1.25$Hz, $f_2 = 1.75$ Hz, and $f_3 = 2.5$ Hz respectively. Equation 11.18 is used to check the stability criterion for rain-wind-induced vibrations of a smooth circular cable:

$S_c = \dfrac{185 * 0.005}{1.225 * 0.25^2}$ Therefore, no further measures are needed for rain-wind-induced vibrations.

Equation 11.19 is employed to check the galloping vulnerability. Since the cables are spaced at 1.5 m i.e., 5 D, therefore $C = 25$ is used. For the first natural frequency:

$$U_{\text{crit}} = CfD\sqrt{S_c} = 25 * 1.25 * 0.25 * \sqrt{12.08} = 27.15\text{m/s} > 25.0 \text{ m/s}$$

Similarly, for $f_2 = 1.75$ Hz $U_{\text{crit}} = 37.8$ m/s and for $f_3 = 2.5$ Hz $U_{\text{crit}} = 54.3$ m/s > 25.0 m/s (OK).

11.9.5 Mitigation Methods

Mitigation methods for stay cable vibrations include:

Raise damping: Suppressed aerodynamic instability can be achieved most effectively by increasing the damping. Significant increases in stability can be achieved by adding relatively small amounts of damping at or close to the cable ends. A number of methods are available, such as the use of visco-elastic material in the cable anchorage pipe, neoprene bushings at the cable anchorages, petroleum wax in-fill in the guide pipes, and external or internal viscous dampers (Figures 11.26 and 11.27).

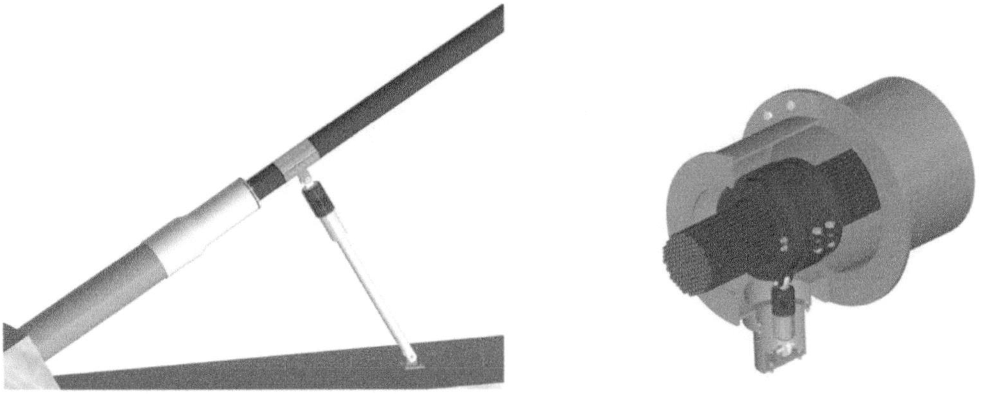

External Viscous Damper   Internal Viscous Damper

Fig. 11.26 Viscous dampers (Courtesy, BBR_HiAM)

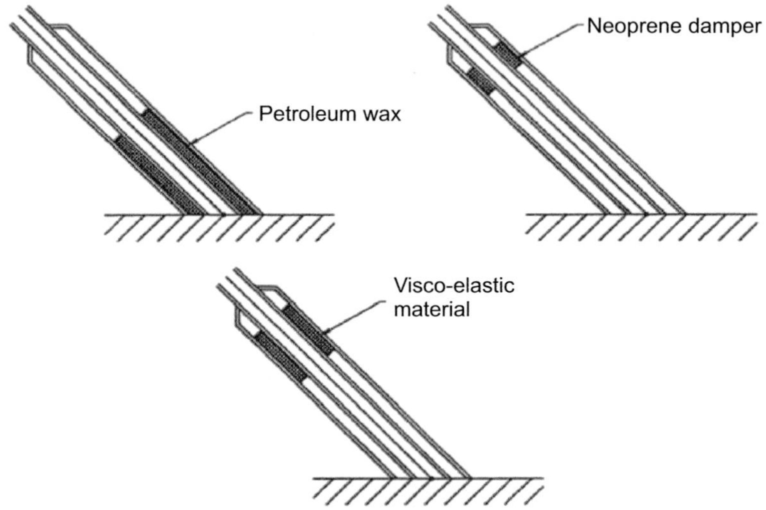

Fig. 11.27 Material dampers (FHWA, 2007)

Raise natural frequencies: Since the natural frequency is dependent on the cable's mass, tension, and length—all of which cannot be easily changed without affecting other design constraints—this is accomplished by adjusting the effective cable length. In cable arrays, the effective length can be adjusted by joining the cables transversely using secondary cable crossties (Figure 11.28). This procedure makes it easy to raise the natural frequency susceptible to aerodynamic excitation above the critical value. Aerodynamic instability begins at a higher wind velocity when the cables' natural frequencies are raised. This approach was applied to several bridges in the US including Dames Point Bridge (Florida), Greenville Bridge (Mississippi), and U.S. Grant Bridge (Ohio).

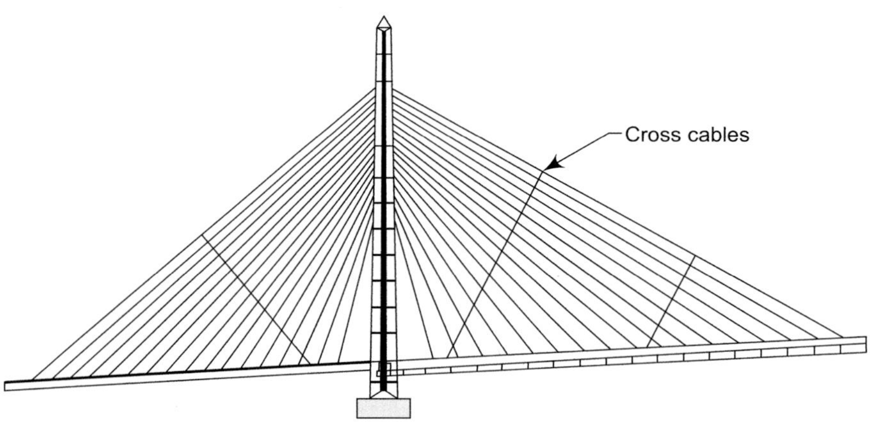

Fig. 11.28 Cable crossties (FHWA, 2007)

Cable surface treatment: Given that the rivulets that form on the cable surface are what cause the rain/wind oscillations, it is probable that the instability is sensitive to surface roughness. One solution to the rain/wind issue (Figure 11.29) is to add small protrusions that are coiled around the surface of the cable or run parallel to its axis. On new bridges, the most prevalent spiral bead formation is the double-helix. This kind of cable pipe is offered by all major cable suppliers to reduce vibrations caused by wind and rain.

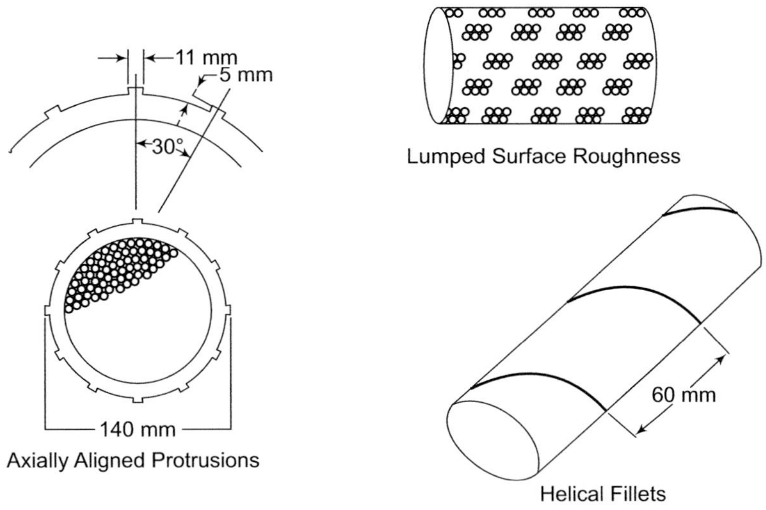

Fig. 11.29 Cable surface treatment (FHWA, 2007)

References

Ammann, O.H., Von Karman, T. and Woodruff, G.B., The Failure of the Tacoma Narrows Bridge, a Report to the Federal Works Agency, 1941.

Anon, Golden Gate Bridge Vibration Studies, Engineering News Record, August 8, 1946.

Davenport, A.G., Buffeting of a suspension bridge by storm winds. J. Struct. Div., ASCE, 88(3), 233–268, 1962.

Davenport, A.G. The Relationship of Reliability to Wind Loading. *Journal of Wind Engineering and Industrial Aerodynamics*, Volume 13, pp. 3–27, 1983.

Davenport, A.G. and King, J.P.C., Dynamic Wind Forces on Long Span Bridges using Equivalent Static Loads, IABSE, 12th Congress, Session VI, Vancouver, B.C., 1984.

Drewry, C.S., A Memoir on Suspension Bridges, Longman, Rees, Orme, Brown, Green and Longman, London, 1832, p. 24.

Farquharson, F.B., Aerodynamic Stability of Suspension Bridges with Special Reference to the Tacoma Narrows Bridge-Part 1: Investigation Prior to October 1941, Technical Report: The Structural Research Laboratory, University of Washington, 1950.

Finch, J.K., Wind Failure of Suspension Bridges or the Evolution and Decay of the Stiffening Truss, Engineering News Record, March 13, 1941.

FHWA, Wind-Induced Vibration of Stay Cables, Publication No. FHWA-HRT-05-083, 2007.

Gimsing, N.J., and Georgaki, C.T., Cable Supported Bridges Concept and Design, John Wiley, and Sons, 2012.

Harris, R.I., On the spectrum and auto-correlation function of gustiness in high winds, Electrical Research Association. Report 5273.1968.

Holmes, J.D., Along-wind response of lattice towers. Part I: derivation of expressions for gust response factors, Engineering Structures 16: 287–292, 1994.

Holmes, J.D., Wind Loading of Structures, Spon Press, London, England, 2003

Irwin, P.A., Wind tunnel and analytical investigations of the response of Lions' Gate Bridge to a turbulent wind, Report NAE-LTR-LA-210, National Research Council, Ottawa, Canada. 1977.

Jakkula, A.A., A History of Suspension bridges in Bibliographical Form, Texas A. & M. College of Engineering Experimental Station. Bulletin No., 57, 1941.

Kasperski, M. and Niemann, H.-J., The L.R.C. (Load-Response-Correlation) method: a general method of estimating unfavorable wind load distributions for linear and non-linear structural behavior, *Journal of Wind Engineering Industrial Aerodynamics*, Volume 43, pp 1753–1763, 1992.

Matsumoto, M., Shiraishi, N. and Shirato, H., Inclined-cable aerodynamics: Structural design, analysis & testing proceedings. *Proceedings of the ASCE Structures Congress*, San Francisco, CA, 1989.

Miyata, T., Historical view of long-span bridge aerodynamics, *Journal of Wind Engineering, and Industrial* Aerodynamics, Volume 91, pp 1393–1410, 2003.

Moisseiff, L.S. and Lienhard, F., Suspension Bridges Under the Action of Lateral Forces, *ASCE Transactions*, Volume 98, 1933.

Peters, B. and Uddin, M., Impact of Longitudinal Acceleration and Deceleration on Bluff Body Wakes, Fluids, Volume 4, No. 3, 2019.

Post-Tensioning Institute (PTI). Recommendations for Stay-Cable Design, Testing, and Installation. Farmington Hills: PTI Standard No. DC45. 1–18, 2018.

Scanlan, R. and Jones, N.P., Aeroelastic Analysis of Cable-Stayed Bridges, Journal of Structural Engineering, Vol. 116, No. 2, pp 279–297, 1990.

Scanlan, R., State-of-the-Art Methods for Calculating Flutter, Vortex-Induced, and Buffeting Response of Bridge Structures, Federal Highway Administration Report FHWA/RD-80/050, Washington DC, 1981

Scott, R., *In the Wake of Tacoma*, ASCE Press, 2001.

Selberg, A., Aerodynamic effects on suspension bridges, Proceedings, International Conference on Wind Effects on Buildings and Structures, Teddington, U.K. 26–28 June, 462–486, 1963

Simiu, E. and Miyata, T., Design of Buildings and Bridges for Wind, John Wiley and Sons, Hoboken, NJ, 2006.

Simiu, E. and Yeo, D, *Wind Effects on Structures Modern Structural Design for Wind*, Wiley-Blackwell, 2019.

Steinman, D.B., Rigidity and Aerodynamic Stability of Suspension Bridges, *ASCE Transactions*, Volume 110, 1945.

Taly, N., Design of Modern Highway Bridges, McGraw-Hill, New York, 1998

Warhaft, Z., *An Introduction to Thermal-Fluid Engineering*, Cambridge University Press, UK, 1997.

CHAPTER **12**

Performance-Based Design of Cable-Stayed Bridges

12.1 INTRODUCTION

Structural design requires that the designer faces multiple goals to achieve an optimal design. These goals include but are not limited to safety, economy, serviceability, sustainability, and robustness. Nevertheless, force-based oriented codes will constrain the designer to follow a prescriptive path that may not optimally satisfy some of these goals. As a result, stakeholders are getting designs that are targeted towards addressing structural conditions with ambiguous reliability because traditional design is not based on structural performance, rather it is based on structural strength. The majority of structural design in use today is not performance-based, despite the fact that almost all contemporary design specifications lean toward achieving some level of performance. Instead, typical design processes are based on strengths. Furthermore, the ability of the structure to achieve the expected performance is not explicitly verified by the designer using traditional design methods. The performance-based design, or PBD, makes it possible to create structures with predictable behavior under specified loading conditions. It reverses the design process by defining the end goal as the starting point because it is based on specifying performance objectives that the structure will meet under a specific loading condition. Specific performance levels for different members are established for the design. Finally, performance of the structure needs to be verified through analytical simulation. Defined performance objectives are essential since they establish the prospects for the design. They most often include statements of the likelihood that a damage level will be exceeded if a specified hazard occurs with a specific intensity on the structure. Nevertheless, establishing performance objectives and getting all the stakeholders concurring on them is not only challenging but also critical to the process of PBD.

Performance-based design offers several advantages over prescriptive design. Properly conducted PBD enables targeted performance to be accomplished with revealed confidence and reliability. Stakeholders can select through PBD the performance goals that are appropriate and satisfy economical, serviceability, and sustainability requirements for the project. PBD will entitle structural engineers to include creativity in their design and employ innovation by engaging basic principles of engineering and material mechanics, and plasticity. PBD will allow structural engineers to work with all stakeholders to identify the expected performance of the structure and demonstrate structural compliance to the criteria. Through PBD, structural engineers can address through their designs the needs for resilience, sustainability, and robustness.

It is noteworthy that a considerable number of engineers, along with other professionals and stakeholders in the construction industry, are still not familiar with PBD. It is necessary to transform PBD principles into useful guidelines that all parties involved can comprehend, support, and use in upcoming initiatives. The usual practice for structural engineers is to write strength-based prescriptive provisions. Instead, they should use PBD, a process that outlines the desired behavior of the structure and the procedure for verifying that performance requirements are met. It is recommended that structural engineers begin their education with a focus on the new competencies needed for designing to performance-based codes. They also need to develop efficient methodologies in order to attain high design quality. PBD ought to be a standard operating procedure for intricate, valuable, and mission-critical structures because the communities they serve stand to gain from the creativity and innovation that performance-based approaches encourage (SEI, 2018).

PBD has found application for countering three hazards: Seismic; Structural Fire Engineering; and Wind. While seismic performance-based design was applied for the design of several cable supported bridges and investigations of the performance of cable-stayed bridges due to vehicle fires have been recently conducted, PBD for wind was mainly applicable to buildings and hence will not be pursued further in this chapter. In the following sections, applications of PBD to seismic and fire hazards will be emphasized.

12.2 SEISMIC PERFORMANCE-BASED DESIGN OF CABLE-STAYED BRIDGES

The concept of seismic performance-based design (PBD) of cable supported bridges has been adopted in the design or evaluation of several projects such as the New Carquinez Bridge in California, the New Tacoma Narrows Bridge in Washington State, the Throgs Neck Bridge in New York City, the Gerald Desmond Bridge in California, the Olivier Charbonneau Bridge in Quebec, and more. The essence of this approach is that the overall performance goals of the bridge are tied to specific damage levels of its components.

Bridges with cable stays fall under the category of complex bridges. In 2009, AASHTO released the first edition of the Guide Specifications for LRFD Seismic Bridge Design, acknowledging that displacement-based design principles are more suitable in high seismicity areas. These specifications, though, are only applicable to traditional bridges. Complex bridges like suspension bridges, cable-stayed bridges, and arch bridges are expressly excluded from Section 3.1 of the Guide Specification, which also mandates that the owners of these bridges provide the necessary seismic design specifications. Depending on the specifics of each project, detailed seismic performance-based design criteria (PBDC) are developed for cable-stayed bridges. There are two main reasons why seismic PBDC is developed. First and foremost, PBDC is required to offer a measurable approach to derive a reasonable degree of confidence that a structure's performance across a range of seismic events, with varying degrees of hazard, satisfies desired expectations, such as post-earthquake functionality and repair cost. Second, PBDC maximizes the required structural investment while facilitating the achievement of the targeted performance levels. (Shama and Jones, 2020). The steps involved in seismic PBD are outlined in the following sections.

12.2.1 Seismic Hazard Levels

Throughout the course of the bridge's life, a bridge site may be exposed to a wide range of seismic hazard levels with variable probabilities of occurrence. The seismicity of the site and the response of the structure to this seismicity are the two independent variables that make up the core of the PBDC and must be assessed in order to achieve the desired performance across the continuum of seismic hazards. The current trend for major bridge projects is to specify two levels of seismic hazard: an upper-level seismic event with a very low probability of exceedance and a lower-level seismic event with a high probability of exceedance during the structure's life. However, any number of seismic hazard levels could be specified when developing performance-based design criteria. As a result, PBDC establishes the levels of seismic hazard and then the structure's target performance for each of these seismic hazards.

Several current seismic criteria and guidelines adopt two hazard levels, as a minimum, for the seismic performance-based design and evaluation of bridge structures. The lower-level seismic event is defined as the design earthquake and has a relatively high likelihood of occurrence within the life of the bridge. This earthquake ranges from 100-year return period (50% probability of exceedance in 50 years) to 475-year return period earthquake (10% probability of exceedance in 50 years). Examples of projects that used this range for the lower-level event include the design of the Carquinez Straits Bridge in California that used a 300-year return period; the design of the New Tacoma Narrows Bridge in Washington State that used a 100-year return period; and the seismic evaluation of the Throgs Neck Bridge in New York City that used a 475-year return period.

The upper-level seismic event is defined as the maximum considered earthquake and represents a large but unlikely event and has a relatively low probability of exceedance within the life of the bridge. This earthquake ranges from a 1000-year return period (5% probability of exceedance in 50 years) to a 2475-year return period (2% probability of exceedance in 50 years). Examples of projects that used this range for the upper-level event include the San Francisco to Oakland Bay Bridge (SFOBB) and Carquinez Straits Bridge in California; both used a 1500-year return period; Gerald Desmond cable stayed bridge that used a 1000-year return period; and the New Tacoma Narrows Bridge that used a 2500-year return period.

12.2.2 General Performance Requirements

In addition to defining the seismic hazards, PBDC needs to specify how the bridge will function under these conditions. Bridge service is used by the current PBDC to define an overall performance objective, and damage levels are allocated to the bridge for every level of seismic hazard. Stakeholders and bridge owners may choose these levels together. The bridge typically has an operational performance goal for the low-level seismic hazard. Immediate service, or having full service available for all vehicles within a few hours after a guided inspection of the bridge at pre-identified critical areas, is linked to the operational performance objective. In this instance, the damage is minimal. The bridge is often given a life safety performance objective for the upper-level earthquake. The repairable or significant damage is linked to this performance goal. Significant service interruption is anticipated if major damage is chosen. The bridge might need to be replaced even though it won't collapse.

Bridge component responses are assigned to allowable damage levels based on overall performance. These levels can vary from an elastic response with no expected damage to significant damage with no collapse to preserve life safety. The damage assignment takes into account the significance of the bridge, the owner's objectives for bridge accessibility following a seismic event, the allowable cost of repairs, and the expected service life. For instance, the AASHTO LRFD specifications (AASHTO, 2020) are based on a single-level (earthquake design) seismic design procedure in which a force-based approach and response modification factors are used to determine the required performance of a bridge based on its importance grade. When determining the performance objective for the seismic evaluation of bridges, the expected service life plays a big role. Damage levels are usually expressed in four categories as follows:

1. *Negligible damage*: Elastic behavior is expected from the bridge at this performance level. There may be signs of movement, but an examination of the bridge should reveal no significant damage and full serviceability. There would be no need for any member replacements or repairs. Other than expansion joints, damage to nonstructural elements would be permitted.

2. *Minimal damage*: Any repairs for this damage state could be completed in a non-emergency situation. The onset of secondary steel member yielding or a limited, narrow cracking in concrete are examples of the minor inelastic response that characterizes minimal damage. Avoiding irreversible deformations is the goal.
3. *Repairable damage:* The damage should be sufficiently contained to allow the structure to be rebuilt without needing to be closed off in its pre-earthquake state. It is typified by the occurrence of an inelastic response that causes minor yielding of structural steel, minor spalling of cover concrete, minor concrete cracking, and reinforcement yielding. Permanent offsets ought to be minimal and shouldn't affect the bridge's ability to function after repairs are completed.
4. *Significant damage:* Permanent offsets could happen even in the absence of collapse; concrete members may crack and yield at the reinforcement; steel members may buckle locally and extensively. In most cases, PBDC also offers restricted empirical specifications for the planning and construction of significant bridge elements like concrete towers. While strict empirical design specifications are more typical of prescriptive-based design than of PBDC, certain bounding limits are still preferable unless they are explicitly contested and supported on an individual basis These empirical bounding limits ensure that the structure is appropriately represented by the inventory of current laboratory test specimens used to establish material strain limits or drift limits. They are based on experience that has produced constructible structures.

12.2.3 Local Damage Levels

The assessment of local member damage levels is the primary method used to confirm that the bridge performance objectives are met. Local damage levels are attributed to different structural members of the bridge and are achievable in light of the global damage level. Generally speaking, two methods are used to evaluate these damage levels based on the dynamic analysis method and the member's location and functionality within the structure.

In most cases, the deformation-based approach is used for members that are predicted to show varying degrees of ductility during the seismic event. Therefore, during the analysis, the damage can be monitored. A member's ductility, which can be expressed in terms of a corresponding limiting value of strain, rotation, displacement, or curvature, is typically used to express the local damage to that member. These limiting values are determined using fiber element analysis of a structural section of this member, three-dimensional finite element modeling of the structural member detail under investigation, or experimental results. Figure 12.1 displays the inelastic behavior of a model for the New San Francisco-Oakland Bay Bridge tower shear links under lab testing. Typical moment-curvature relationships established at the base of Gerald Desmond bridge pylons (conceptual design) in the longitudinal direction are displayed in Figure 12.2.

The force-based approach is the traditional approach that is employed by most of the bridge design codes for the seismic design of bridges. Through this approach member capacities are compared to seismic structural demands (C/D ratio) and a factor of safety is established. Member ductility is usually expressed in terms of response modification factors and cannot be tracked explicitly during the analysis. In brief, the deformation-based approach is usually used for the portions of the bridge that are expected to exhibit nonlinear behavior during the seismic excitation in conjunction with the non-linear direct integration method of analysis for the main bridge. The force-based approach is usually employed for other members that are expected to behave linearly. As an example, Table 12.1 summarizes the criteria for different performance levels in terms of strain and residual drift limits included in the PBDC for the Gerald Desmond cable-stayed bridge in Long Beach, California (Shama and Jones, 2020).

Fig. 12.1 Inelastic local buckling behavior of a steel element under cyclic loading (Shama et al., 2001)

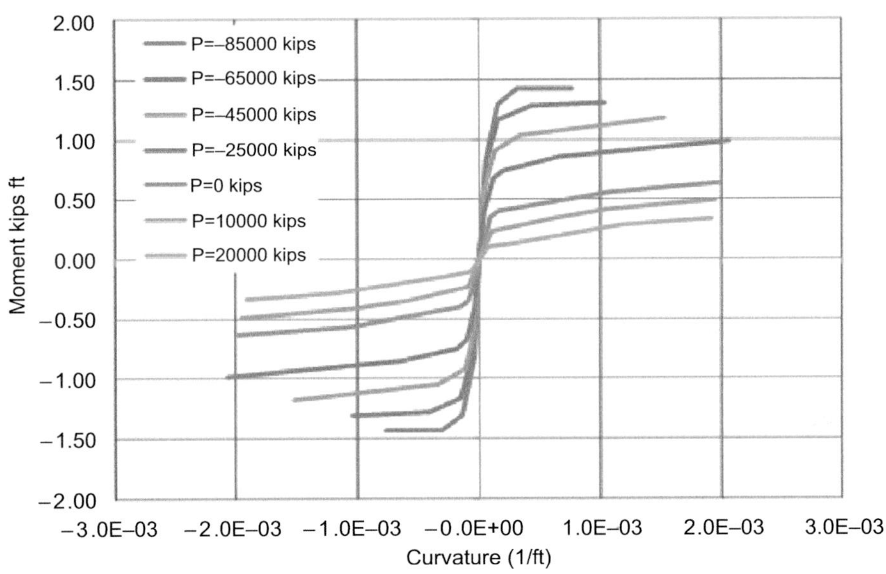

Fig. 12.2 Example of Longitudinal direction M-C curves for Gerald Desmond Pylon Base (Shama and Barbas, 2014)

Shake table tests were performed on 33 columns of standard ordinary bridges that ranged in geometric scale from 1/5 to 1/3 by Vosooghi and Saiidi (2010). The input earthquake motions varied from near-fault motions to spectrum matched motions. They released data relating the maximum longitudinal and transverse steel strains to Damage States (DS) at the plastic hinge zone based on these tests. For 17 of the tested columns, their data, as shown in Table 12.2, gives the mean value and standard deviation.

It is noteworthy that the Seismic Design Criteria (SDC) (Caltrans, 2019) for ASTM A706, grade 60 reinforcement, calls for a reduced ultimate tensile strain of 0.09 for #10 bars and smaller, and 0.06 for #11 bars and larger. SDC makes the assumption that strain hardening will range from 0.015 for #8 bars to 0.005 for #18 bars. Using a confined concrete model, SDC defines the ultimate concrete strain (ε_{cu}) as the point where strain energy equilibrium is reached between the confined concrete and the confinement reinforcement. Mander's stress strain model for confined concrete is commonly used to determine ε_{cu} (Mander et al., 1988). As shown SDC has implemented some degree of conservatism into the design process when using ultimate design values to determine displacement capacities since laboratory testing typically provides greater ultimate displacement capacities.

Table 12.1 Example of Performance-Based Criteria for a Cable-Stayed Bridge (Shama and Jones, 2020)

Component	*Seismic Event Level*					
	SEE			*FEE*		
	Maximum Material Strain		*Residual Drift*	*Maximum Material Strain*		*Residual Drift*
	Conc.	*Steel*		*Conc.*	*Steel*	
Drilled Shafts and Piles	[1] $0.40\ \varepsilon_{cu}$	[4] 0.015	Permitted within specified strain limits	[2] 0.003	(5)	None Permitted
Pylons & End Bents	[1] $0.40\ \varepsilon_{cu}$	0.015	Six inches in any direction at deck level relative to foundation	[2] 0.003	(5)	None Permitted
Spirals/Ties of Pylon & End Bents	–	0.05	–	–	(5)	None Permitted
Superstructure	(3)	(3)	–	–	(6)	None Permitted
Cable system	–	(6)	–	–	(6)	None Permitted
Expansion Joints	–	–	Significant Damage allowed	–	–	None Permitted

[1] Ultimate strain ε_{cu} determined by Mander's model for confined concrete (Mander et al. 1988).
[2] Strain at exterme face of concrete component and not the confined core.
[3] Strains not specified. Essentially elastic performance required by limiting elastic force Demand/Capacity ratios as determined by AASHTO LRFD to 1.25
[4] Strain shown is for concrete reinforcement. Steel shell strain limited to 0.02 tension and 0.01 compression.
[5] Strain limits as specified for AASHTO LRFD elastic design – No Damage performance.
[6] Not to exceed nominal elastic capacity as determined by AASHTO LRFD.

Table 12.2 Damage states for columns of ordinary standard bridges based on test results (Vosooghi and Saidi, 2010)

Damage State (DS)	*DS-1*	*DS-2*	*DS-3*	*DS-4*	*DS-5*
	Surface Crack	*First Spalling*	*Major Spalling*	*Exposed Reinforcement*	*Core Shedding*
Mean Longitudinal Steel Strain	0.0092	0.018	0.026	0.034	0.042
σ Longitudinal Steel Strain	0.007	0.0052	0.0069	0.012	0.013
Mean Transverse Steel Strain	0.00031	0.00069	0.0011	0.0017	0.0031
σ Transverse Steel Strain	0.00014	0.00028	0.00044	0.00055	0.0016

The ductility capacity assessment for the tower legs of a suspension bridge is an example of the deformation capacity approach. (Shama and Barbas, 2013). A finite element model was developed for a section where maximum moments are anticipated, to the point of inflection of moments. The shell elements that make up the model have six degrees of freedom, are applicable to a large displacement formulation, and employ Gauss numerical integration through the shell thickness. For the steel, the von Mises yield condition-based plastic multi-linear constitutive material model was used. With incremental drifts of 0.014%, drifts up to a maximum value of 1.4%, static pushover of the finite element model is carried out in displacement control mode in both the transverse and longitudinal directions. The deformed structure at the conclusion of the static pushover analyses is depicted in Figure 12.3, along with a section showing the contours of the accumulated plastic strain at the zone of local buckling. It is evident that local buckling occurred in a few of the cells along the loading direction. Figure 12.4 shows the relationship between moment and rotation. It is demonstrated that the tower legs have a notable degree of ductility. Furthermore, it is demonstrated that plate yielding on the tension side may occur before compression side buckling. It is evident that the behavior is too complex to be forecast using theoretical equations. The limiting rotational drifts were determined and incorporated into the seismic PBDC of the bridge based on these relationships. As demonstrated by the previous example, finite element analysis is a useful technique for establishing deformation-based criteria for various structural components by establishing limiting values for their deformations.

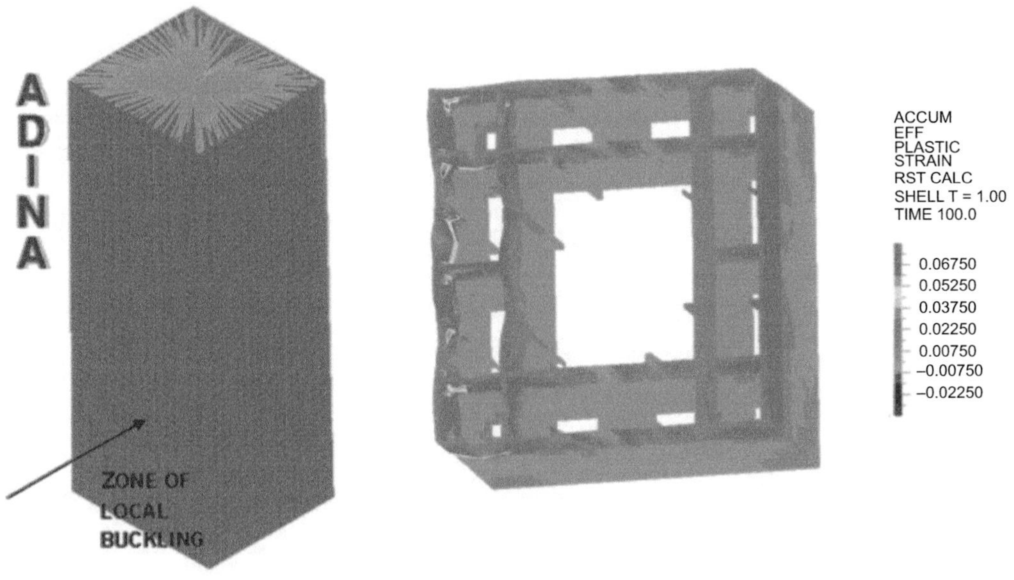

Fig. 12.3 Deformed Shape of FEM of a suspension bridge steel tower leg to establish its damage levels

Recent cable-supported bridges with PBDC have permitted Minimum Damage to both the superstructure and other bridge elements. This low damage level performance can be measured using a force-based method by capping these bridge elements' elastic force Demand/Capacity ratio at 1.25.

12.2.4 Modeling Techniques for Seismic PBD

Currently, linear response spectrum analysis (LRSA) is the standard procedure for seismically analyzing typical bridges in high seismic regions. While there are various approaches to putting an LRSA into practice, the general objective is to match the maximum displacements expected during the seismic design event with the structure's displacement capacity. In order to handle the

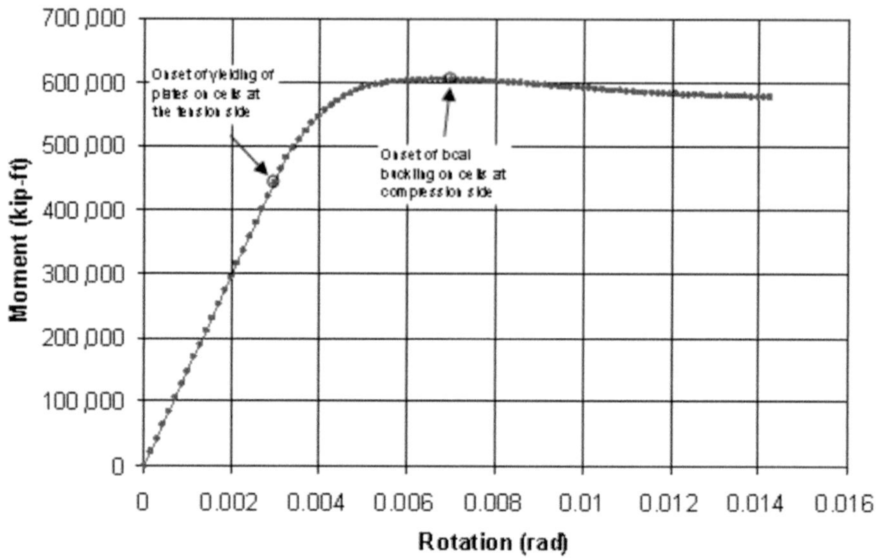

Fig. 12.4 Example of Moment rotation relationship to establish damage levels

highly non-linear response of the structure that is anticipated in a seismic event and the post-elastic material response of the structure and foundation elements, an indirect method is to estimate the displacement capacity of the structure when these materials reach a certain limit, compare that displacement with the structure, and then assume that the seismic event has resulted in a pure linear response. Unfortunately, this method is unable to accurately represent the material-level response of any bridge element to the actual expected seismic ground motions at a given site. On the other hand, PBD sets forth limiting damage levels and verifies these limits are not exceeded during the seismic event, which adds considerable complexity to the seismic analysis. Therefore, LRSA is no longer appropriate for Seismic PBD. When damage limits are specified by limiting concrete, reinforcing steel, or structural steel strains as well as residual drifts, the analysis must be sophisticated enough to capture these parameters during a seismic event. To attain this approach, PBDC typically requires employing non-linear time-history analysis (non-linear time integration methods), which provides the ability to track the behavior of critical elements and record their actual damage during the course of the seismic event in measurable terms. It will also capture the final displaced shape of the bridge if permanent residual displacements occur.

Critical elements are those that are anticipated to exhibit a non-linear response, which typically can include foundations, pylons, energy dissipating devices, and seismic bearing isolation. If the non-linear response is difficult to quantify with high levels of confidence for certain bridge members, PBDC will specify that these members must remain elastic. An example is the response of the main cables of suspension and stays of cable-stayed bridges. PBDC typically stipulates ‘No Damage’ for these elements, essentially requiring linear behavior of the cable system. The vertical response of the superstructure at cable-stayed bridges may cause stay-cable unloading. Owing to the unpredictability of stay-cable performance during unloading, PBDC has required a minimum force level, say 10% of the dead loading during seismic events, for earlier projects.

Low damage level performance for some members can be captured by modeling them as elastic elements and limiting their elastic force Demand/Capacity ratio in order to simplify the analytical model. Material strain limits are usually needed at higher damage levels, necessitating non-linear modeling to capture the resulting strains. In order to account for possible residual displacements, non-linear modeling of these components is also essential.

12.2.5 Damping

Structural elements exhibit two forms of damping through non-linear vibration under earthquake loads: linear viscous damping and material damping. Linear viscous damping constitutes several mechanisms for energy dissipation such as thermal effects of repeated elastic straining of the material, opening and closing of micro-cracks, and friction at steel connections. To simplify this complexity, an equivalent viscous damping is assumed for each mode of vibration of the structure that constitutes all sources of energy dissipation. Material damping is related to the energy dissipated through the non-linear hysteretic behavior of the material under seismic load. Examples of inelastic hysteretic loops for a steel sacrificial shear link element for the SFOBB are illustrated in Figure 12.5

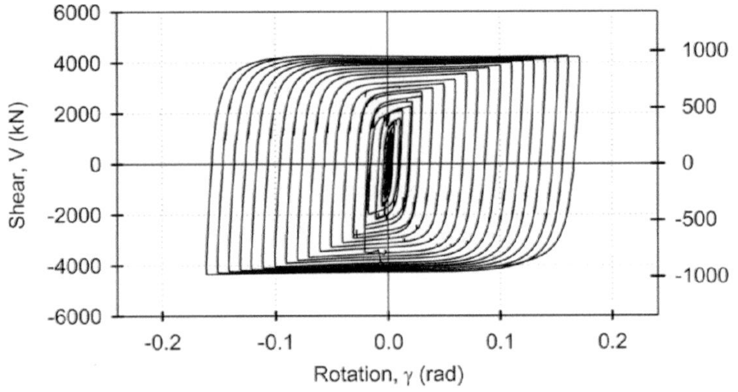

Fig. 12.5 Inelastic hysteretic loops for a steel sacrificial shear link element for the SFOBB (Buckle et al., 2005)

Rayleigh damping, which combines mass and stiffness proportional damping to provide equivalent viscous damping in dynamic time-history analysis, is a highly suitable representation for time integration techniques. Typically, the mass proportional term is chosen to match the fundamental period with the largest mass participation. In both horizontal and vertical directions, the stiffness proportional term should ideally be chosen to correspond to the highest frequency that captures at least 90% of the mass participation. During the dominant periods of the different bridge elements, the target maximum and minimum damping values are represented by the range of Rayleigh damping values. They are typically within the following ranges: reinforced concrete columns: 4% - 5%; reinforced concrete pylons: 4% - 5%; steel superstructure: 2% - 3%; concrete superstructure: 3% - 5%; foundations: 4% - 5% in addition to modeling radiation damping in soil. Examples of typical Rayleigh damping ratios established for the seismic analysis of the Gerald Desmond Bridge-conceptual design phase are illustrated in Figure 12.6.

Higher damping is expected at foundations where significant soil-structure interaction occurs; this should be taken into account when developing the analysis model. In global models that include soil-structure interaction elements as non-linear soil springs with hysteretic damping, total foundation damping should be investigated to avoid overestimating damping for these critical structural elements. How more damping is implemented will depend on how the soil-structure interaction is captured. The massive 80-by-130-foot supporting caissons for the new parallel Tacoma Narrows Suspension Bridge were modeled as discrete nonlinear Winkler soil springs in the global model, with nonlinear springs chosen to allow for residual displacements. Between five and ten percent of effective damping was recorded during the highest caisson rocking cycles. With a target value of 3.5 percent, between 2 and 5 percent equivalent viscous damping was permitted

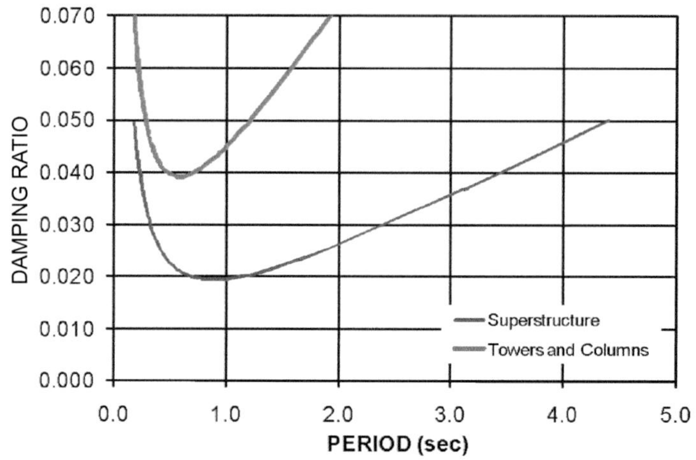

Fig. 12.6 Typical Rayleigh damping ratios established for the seismic analysis of the Gerald Desmond Bridge-conceptual design

for the analysis. Hysteretic damping connected to the non-linear Winkler supporting springs was responsible for the increased damping experienced during caisson rocking cycles. The amount of damping should be confirmed, regardless of the damping implementation strategy selected. To make sure the caissons were never over-damped during the analysis, the logarithmic decrement of free vibration oscillation for the Tacoma Narrows bridge was assessed (Jones et al., 2004).

12.2.6 Definition and Selection of Ground Motions

The techniques for identifying the ground motions for analysis must be specified in the seismic performance criteria. Typically, each set consists of two horizontal and one vertical ground motion component. Ground motions are typically based on a probabilistic seismic hazard analysis for new bridge design, which takes into consideration all seismogenic sources and produces a uniform hazard spectrum for ground motion development. It is typically required by criteria that the developed ground motions' spectral amplitudes match the uniform hazard spectrum. For this, wavelet analysis-based spectral matching is typically employed (Shama, 2012). The process of modifying an actual earthquake accelerogram to match a target design spectrum is known as spectral matching. In this method, a recorded accelerogram is scaled iteratively so that the resulting time history is compatible with the target spectrum by breaking it down into its wavelet components at different frequencies using a continuous wavelet transform algorithm that uses a mother wavelet. An example of spectral matching is illustrated in Figure 12.7

12.2.7 Foundations and Soil-Structure Interaction

Foundations of ordinary standard bridges in high seismic regions are usually designed as capacity protected members by forcing any anticipated yielding mechanism in the pier column supported by the foundation elements. This approach precludes the formation of plastic hinging within the pile if not designed as capacity protected as shown in Figure 12.8, which illustrates soil reactions and distribution of moments along the pile. There are two locations for potential plastic hinging, one at the pile to cap connection and another plastic hinge within the embedment of the pile. Its location is a function of several parameters related to both the pile and the soil (Shama et al., 2001). The philosophy of capacity protection of the foundations is emphasized in Caltrans Seismic Design Criteria (Caltrans, 2019), which requires in most cases foundation elements to resist the plastic

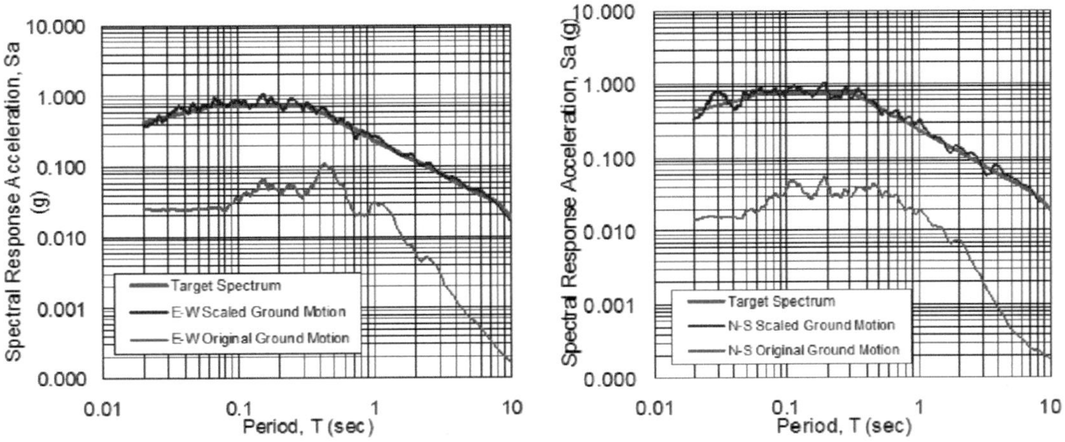

Fig. 12.7 Spectral matching technique for ground motion development

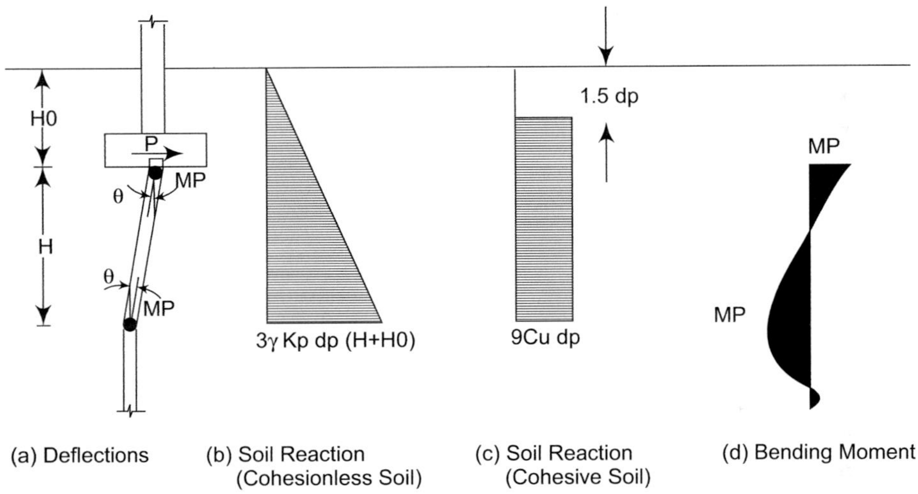

Fig. 12.8 Plastic hinges within the pile if not designed as capacity protected (Shama et al., 2001)

hinging over-strength of the supported column or pier wall. The main advantage of this approach is that it restricts damage to the column or pier wall and facilitates both the assessment of the damage and any required repair.

Since the pylons for cable-stayed bridges are typically large structures, applying the above philosophy—which is used for standard bridges—of designing the foundation for the over-strength capacity of the pylon may add a significant amount of cost to the bridge in order to achieve the life safety performance objective. SSI's involvement in the seismic analysis can shed more light on the anticipated seismic response of the bridge, given that the level of analysis for cable-stayed bridges is far more advanced than that of standard bridges. In such a case, the SPBDC can allow for some sort of nonlinear behavior of the soil, while keeping the piles behaving without any damage. The permanent deformations at the end of the seismic event shall be consistent with the main goal of the life safety performance. Therefore, the bridge can meet the seismic performance goal for this event with less foundation capacity.

The interaction between the foundations and the supporting soils needs to be taken into account in the dynamic analyses, which is why PBDC will define methods of including soil-structure interaction in global models for cable-stayed bridges, which feature large pylons and anchor piers. Beams supported on the non-linear Winker Model for pile foundations is the current trend used in the assessment of cable-stayed bridges in high seismic zones. This method can be implemented in the structural model without any difficulty as the number of large-diameter drilled shafts or piles for a typical bridge foundation is limited. This model as shown in Figure 12.9 is based on representing the soil by a series of nonlinear springs. It provides a complete representation of the soil–structure interaction. Furthermore, nonlinear behavior can be included as well as soil yielding. Nonlinear soil-structure interaction (SSI) elements in the near field can be used to represent material damping in the soil. A viscous dashpot connected in series with the near-field SSI element can serve as the far-field SSI element for modeling radiation damping through the soil. The displacement earthquake ground motion time histories are applied to the individual elements at each level. The ground motion shall be spatially variable in the vertical direction to account for the kinematic interaction in addition to the inertial effects. Springs to represent the skin friction and end bearing resistances can also be included as shown in Figure 12.9.

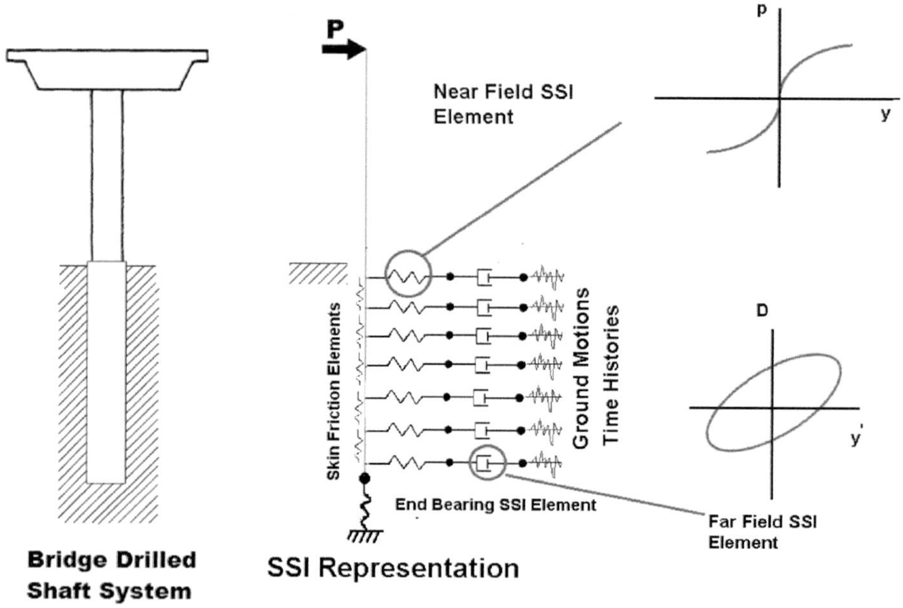

Fig. 12.9 Soil structure interaction representation for deep foundations

There are two analytical techniques for determining the nonlinear near-field springs. They are the strain wedge method and the *p-y* method. In the *p-y* model, *y* is the pile's corresponding deflection and *p* is the soil reaction per unit length at a specific point along the embedment length of the pile into the soil, merely describing the force displacement relationship of a soil spring. The fact that the *p-y* curves are semi-empirical relationships based on a small number of full-scale lateral load tests conducted on piles with diameters ranging from 12 to 24 inches represents a significant shortcoming of this approach. Therefore, for drilled shafts with larger diameters, this method might produce inaccurate results. A P-multiplier is also used by the *p-y* method to take group effects into account. Various codes and specifications still disagree on how to choose the values for this parameter.

All these problems are solved by the strain wedge method that Ashour and Norris (1998) developed. This approach yields non-linear springs that are based on modeling the three-dimensional interaction between the shaft and the surrounding soil. When creating the *p-y* curve, the method takes into account the alternative effects of the soil properties and the pile properties (stiffness and cross-section). Additionally, the strain wedge method does not require the use of P-multipliers because it takes into account the group effects by evaluating the overlap of the passive wedges that form as a result of pushing the shaft in a lateral direction. Depending on the spacing, diameter, and location of the shafts within the group, the strain wedges for various shafts will interfere with one another, reducing and softening the group's strength. When applied, the Masing hysteretic behavior of the soil automatically incorporates the energy dissipation related to material damping into the model. This is possible with the non-linear spring modeling technique.

12.3 PERFORMANCE-BASED FIRE SAFETY DESIGN

12.3.1 SFPE Framework for Performance-based Fire Safety Design

A framework for performance-based fire safety design is provided by the Society of Fire Protection Engineers' (SFPE) Engineering Guide to Performance-Based Fire Protection (SFPE, 2007). This process, which consists of multiple steps, is meant to be flexible enough to be customized to the specific needs of each performance-based design project.:

i. The first step is defining the project scope, which identifies the desired features for design, the characteristics of the structure, the codes or regulations that are applicable, and the project stakeholders.
ii. Once the scope is identified, the next step is to define the goals for the design project. Goals could be unique for different projects, based on the stakeholders' needs and desires. There are four fundamental goals for fire safety (SFPE, 2007): life safety, property protection, mission continuity, and environmental protection.
iii. The next step is to develop design objectives, which are further expressed in engineering terms. The purpose of identifying goals is to facilitate understanding and concurrence on how the structure is intended to perform in the event of fire.
iv. The fourth step in the design process is to develop performance criteria to be met by the design. These criteria are numerical values to which the expected performance of the trial designs can be compared. Criteria may include threshold values for temperatures of materials, gas temperatures, carboxyhemoglobin (COHB) levels, smoke obscuration, and thermal exposure levels.
v. The first part of design is to identify design fire scenarios to meet performance criteria. They are descriptions of possible fire events and consist of fire and structure characteristics. The fire scenarios identified are subsequently filtered (i.e., combined or eliminated) into a subset of design fire scenarios against which trial designs are evaluated.
vi. Once fire scenarios are established, preliminary or trial designs can be carried out to meet the performance-based criteria project requirements. The trial designs include proposed fire protection systems, construction features and operations that are provided in order for a design to meet the performance criteria when evaluated using the design fire scenarios.
vii. Trial designs are then evaluated using each design fire scenario to determine whether it meets the performance criteria. Trial designs that meet the performance criteria can be considered as a final design proposal. If multiple trial designs are evaluated, further analysis is needed to select a final design.
viii. Once the final design is identified, design documents need to be prepared. The documentation should include the fire protection engineering design brief, a performance design report, detailed specifications and drawings, and a structure operations and maintenance manual.

12.3.2 Fire Performance of Cable-stayed bridges in Past Incidences

Cable-stayed bridges are considered vital links in transportation networks, hence, their safety when exposed to fires must be assessed and ensured. Understanding the fire characteristics of this kind of bridges, such as fire growth, during accidental fires is essential to develop prevention strategies for potential damage. Fuel leaks from road or rail tankers can accidentally catch fire and result in fire incidents in cable-stayed bridges. Certain incidents also involve trucks carrying regular cargo. Stiffening girders, pylons, cable systems, and other structural elements could be negatively impacted in a deck fire. Since the flames are geometrically within the cables' reach and high temperatures cause a significant drop in the material performance of high-strength cable steel, the fire performance of cables is typically a major cause for concern. According to Liu et al. (2023), fires can also cause thermal bowing by raising the temperature of the pavement and girder deck. The temperature may seep into the area around a pylon if there is a fire, harming the pylon. Though they only affect a small number of members due to their localized nature, fires on cable-stayed bridges have the ability to alter the structural behavior of the entire structure (Liu et al., 2021).

Fire below the deck of a cable-stayed bridge due to operational facilities or ships in the navigation channel can impose a direct threat to the structure.

Compared to pylons and girders, cable systems are most likely affected by fires because instigator vehicles are usually located on the two sides of the deck, neighboring the cables. A cable is a complicated assembly consisting of high-strength steel wires, galvanized layer, filament tape, void, and more. The flammability of the HDPE sheath around the stay cables gives cable-stayed bridges unique fire characteristics when compared to other types of bridge structures. One cable failed and another was damaged in December 2004 due to a cable fire on Greece's Rion-Antirion Bridge, which was caused by lightning (Garlock et al., 2012). A tractor-trailer fire that started at the deck level of the Mezcala Bridge in Mexico in March 2007 caused one stay cable to fail. One stay cable on the Boston-based Zakim Bridge was partially burned in April 2014 due to a fire started by a car accident (Quiel et al., 2015). In October 2014, a fire caused by a rebar cutting operation occurred on an under-construction bridge, Red Stone Bridge, in Hunan China, which led to rupture of nine steel cables and collapsing of the bridge deck (Zhang et al., 2020). In 2015, a cable connected to a pylon of the Seohae Bridge broke during a fire caused by lightning. To replace the cable, the bridge was shut down for 20 days, and firefighting monitors and foam hydrants were installed on the bridge by the Korea Expressway Corporation to effectively manage fire accidents in the future (Kim et al., 2020).

12.3.3 Performance-based Fire Safety Design of Cable-Stayed Bridges

The SFPE Engineering Guide to Performance-Based Fire Protection is mostly applicable to buildings. Due to the loss of life and financial damages, bridge failures frequently have serious and disastrous effects on the community that surrounds the bridge. 52 bridges in 18 states across the nation failed due to fire in 2008, according to a survey conducted by the New York State Department of Transportation (NYSDOT). This figure is higher than the number of earthquake-related failures. In fact, a study of 1,746 bridge failure causes revealed that fire was the third most common cause of bridge failures (Garlock et al. 2012).

The most straightforward approach to fire-resistance analysis is, of course, fire experiments. Numerous extensive fire experiments have been conducted, including the Runehamar Tunnel fire test (Ingason and Lonnermark, 2015) and pool fire experiments (Skarsbo, 2011). These tests reveal details about the temperature distribution and the structural resilience under specific fire conditions. The analytic method and numerical simulation continue to be the preferred methods for performing performance analyses under bridge fires, despite the associated costs in terms of money, pollution, and possible safety issues.

Fire risk assessment for bridges became more popular during the past few years. It constitutes three steps: evaluation of the vulnerability of bridges during different fire scenarios, estimation of time to failure, and determination of the remaining strength of the structure after fire. Recent studies proposed a framework of performance-based design for cable-stayed bridges under vehicle induced fire accidents (Rujin et al., 2019 and Xu et al., 2021). Based on these studies, the framework of fire safety PBD for cable stayed bridges can be divided into the following steps:

12.3.3.1 Define Performance Objectives

A cable-stayed bridge may be affected by fire scenarios that involve vehicle fires of different strengths as the design load. Therefore, the maximum temperature that can be reached on the structure surface in the fire field and the time it takes to reach the maximum temperature are taken as the determining parameters for the damage level. Therefore, for vehicles that have a high frequency of occurrence, such as compact cars, buses and trucks, the objective is property protection, i.e., to enable the bridge to persist for a long time in the fire field caused by these vehicles with minimal damage. This global performance objective is transformed to local or component performance levels that are further expressed in engineering terms. As an example, for this performance objective, the performance level of the stay cable is that it cannot have strength reduction caused by elevated temperature and the sheath cannot be ignited. The maximum burning time can be taken as 90 minutes to be consistent with the requirement of the Post-tensioning Institute (PTI, 2012). For tanker trucks with a low frequency of occurrence and a high fire intensity the objective is continuity of operations, i.e., to enable the bridge to exhibit some damage without jeopardizing its structural integrity. For this performance objective, the performance level is failure prevention of the stay cables and the sheath can be ignited.

12.3.3.2 Identification of Bridge Fire Scenarios

The general procedure for the definition of bridge fire scenarios is detailed as follows (Rujin, et al., 2019):

1. ***Determine the potential combustibles and relative parameters, according to the bridge site and traffic forecast results:*** The combustion process is described using the t^2 fire model (J. DiNeno et al. 2002) as:

$$\{\dot{Q} = \alpha_f t^2 \quad 0 \leq t \leq t_g \; \dot{Q} = \dot{Q}_{\max} \quad t_g < t \leq t_d\} \qquad ...(12.1)$$

where, $\dot{Q}$ = fire heat release rate in kW (HRR); α_f = fire growth factor (kW/s^2); t_g = fire duration (s); t_d = fire end time (s). Since the cause of a bridge fire is similar to that of a tunnel fire, bridge fire definition can be derived from tunnel fire research and standards. It should be noted that there are some clear distinctions between a bridge fire and a tunnel fire (Alos-Moya et al. 2017). Due to the ample oxygen supply, bridge fires typically burn for shorter periods of time than tunnel fires. Additionally, a bridge fire accident may have a larger fire than a tunnel fire due to the possibility of fuel spreading over a large distance. This is due to a variety of environmental factors, including wind direction and speed, vertical alignment, transverse super elevations of the bridge deck, and road roughness. These factors may turn determination of the real size of a fire a great challenge. These limitations can be the subject of further research to determine the real fire size considering the spread effect. Table 12.3 displays recommended values of the fire scenario parameters based on recent studies on cable-stayed bridges (Rujin et al. 2019 and Xu et al. 2021).

Table 12.3 Parameter definitions of typical fire scenarios

Fire Scenario	*Fire Model*	*Fire Growth Factor (kW/s^2)*	*Fire Duration (s)*	*Maximum Heat Release Rate (MW)*	*Fire Size (m^2)*
Car Fire	t^2 Model	0.011	2700	1.5 ~ 10	1.5 × 4
Bus Fire	t^2 Model	0.123	5400	20 ~ 30	2 × 6
Truck Fire	t^2 Model	0.5	6000	30 ~ 200	4 × 6
Tanker Fire	t^2 Model	1	7200	100 ~ 300	(3.5 ~ 10.4) × 12

2. ***Determine the possible fire locations and vulnerable structural components, according to the bridge characteristics:*** Liu et al. (2023) indicated that cable-stayed bridges have up to four critical fire locations, as shown in Figure 12.10. Location 1, where the vehicle fire is close to cables, may lead to cable failure at high temperatures. An example of such a scenario is the On-deck truck fire caused by the traffic accident on the Mezcala Bridge in 2007, led to the failure of one stay cable and damage to an adjacent cable. Location 2 considers the fires inside the pylons. It may happen during construction such as the case of the Chishi Bridge in Hunan, China. Fires in Location 2 may also be triggered by lightning. Examples of this kind of accident include the Rion Antirion Bridge in Greece and the Seohae Bridge in South Korea.

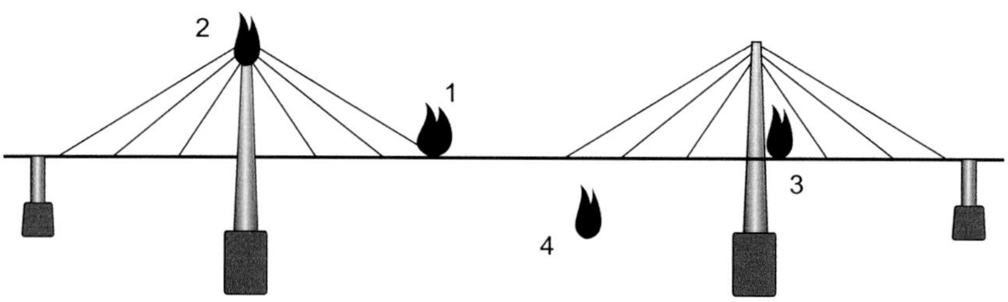

Fig. 12.10 Critical Fire Locations in Cable-stayed Bridges

Fires in Location 3 is the scenario in which fires occur close to the pylons. They threaten the stability of pylons, especially those consisting of steel portions carrying tremendous compression forces. Fires in Location 4 are caused in most cases by cargo tankers in the navigation channel or perhaps by vehicles in the under-crossing road and the bridge purpose is to act as an over-crossing (Kotsovinos et al. 2016).

3. ***Determine whether the wind effect should be considered, according to the environment of the bridge site:*** When predicting the behavior of the fire and the resulting thermal response of the bridge, winds play a crucial role. To ascertain the mean wind speed and the anticipated top speed in a year, meteorological data from the bridge will be examined. Every fire scenario needs to take into account these two wind speeds.
4. ***Determine the most representative fire scenarios based on the analyses results of the above steps:*** It is possible to create several design fire scenarios in every girder section and all the way across the bridge. For instance, six design fire characteristics were established in each section of the Minpu Bridge in China (Liu et al., 2023), and four fire features were taken into consideration in the longitudinal direction of the bridge. As a result, a total of 24 fire scenarios were investigated.

12.3.3.3 Computational Fluid Dynamics-based Fire Scenarios Analyses

Models of computational fluid dynamics (CFD) offer comprehensive data on turbulent fires. This analysis discretizes the fire domain into several meshing cells, which are then determined by mass and species conservation equations. The analysis can be handled by any general-purpose finite element program that has a CFD module. Nonetheless, the most widely used CFD-based fire engineering instrument is the NIST-developed Fire Dynamics Simulator (FDS). Based on fundamental concepts of fluid mechanics, FDS is an open-source computational fluid dynamics (CFD) program. The National Institute of Standards and Technology is in charge of its development. By solving the Navier-Stokes equations numerically, it creates a large-eddy model to replicate the fluid flow caused by a fire (McGrattan et al. 2022). Determining the mesh domain in which the simulation will run is necessary for FDS fire models. Discretized into multiple rectilinear cells, the domain is bounded on its edges. Along with the fire source and its combustion properties, obstructions that might potentially impact the fire-driven gas flow must be included. The structural geometry of the bridge and its surroundings, the heat release rate, and the fire source's intensity and footprint size are among the parameters that are input into each analysis model in a given fire scenario. Usually, the dimensions of the vehicle used in each fire scenario are used to define the combustion surface dimension. Although wind can be accurately modeled, coupled wind-fire analyses present significant challenges and must be handled carefully. According to Wegrzynski and Lipecki (2018), there is a lack of established interfaces for CWE and FSE coupled modeling, and there is also a dearth of information regarding the validation of these coupled problems. At the conclusion of this step, the simulation results provide the spatial-temporal temperature distribution over the structural boundary surfaces.

12.3.3.4 Evaluation of the Thermomechanical Response of the Bridge

Once, a CFD analysis for a specific fire scenario is completed, the next step is to implement the results into a finite-element (FE) model to calculate the thermomechanical response of the bridge to this specific scenario. A general-purpose finite element program such as ABAQUS, ANSYS, or ADINA needs to be employed for this modeling. When doing a subsequent heat transfer analysis, the FE model receives the predicted spatial temporal distribution of temperature over the exposed surface from the FDS as a boundary condition. The heat-transfer analysis is used in the second step to determine the structural model's internal temperature distribution. Next, nonlinear heat conduction is used to calculate the temperature profile inside the structure. The stress analysis must contain the temperature profile of the structural member that varies over time as a predefined field. The material model must take the material's temperature-dependent mechanical and thermal properties into account. Finally, nonlinear static analysis can be used to determine the structure's stress/strain field. Evaluation of different bridge components regarding their anticipated performance is undertaken and implementation of fire protection measures can be determined accordingly.

References

Alos-Moya, J., Paya-Zaforteza, I., Hospitaler, A. and Rinaudo, P., Valencia bridge fire tests: Experimental study of a composite bridge under fire, *Journal of Constructional Steel Research*, Volume 138, pp 538–554, 2017.

American Association of State Highway and Transportation Officials, *AASHTO LRFD Bridge Design Specifications*, Washington D.C., 2020.

Ashour, M. and Norris G., Lateral Loading of a Pile in Layered Soil Using the Strain Wedge Model, *Journal of Geotechnical and Geo-environmental Engineering*, Volume 124, Issue 4, 1998.

Buckle, I., Dusicka, P. and Itani, Ahmad, Development of Built-up Shear Links as Energy Dissipators for the Seismic Protection of Long-Span Bridges, *Bridge Structures*, Vol. 1, No. 1, pp 19–27, March 2005.

CALTRANS, *Seismic Design Criteria Version 2.0*, State of California Department of Transportation, 2019.

DiNeno, J., Drysdale, P., Beyler, D., Walton, C.L., John R. Hall, L.P.R et al., *SFPE Handbook of Fire Protection Engineering*, National Fire Protection Association, 2002.

Garlock, M., Paya-Zaforteza, I., Kodur, V. and Gu, Li, Fire hazard in bridges: Review, assessment and repair strategies, *Engineering Structures*, Volume 35, pp 890–898, 2012.

Ingason H, Li Y.Z and Lönnermark A., Runehamar tunnel fire tests. *Fire Safety Journal*, Volume 71, pp 134–149, 2015.

Jones, M., S. Treyger and P. Pence, Seismic analysis of the New Tacoma Narrows Suspension Bridge, *Proc., 13th World Conf. on Earthquake Engineering*, 1–15. Vancouver, BC, 2004.

Kim, M.O., K. Kim, J.H. Yun and M.K. Kim, Fire risk assessment of cable bridges for installation of firefighting facilities, *Fire Safety Journal*, Volume 115, pp 103–146, 2020.

Kotsovinos P., Flint Graeme and Walker Gary, Qualitative Assessment of the Fire Hazard Beneath Bridges, *Proceedings of 14th international Conference and Exhibition on Fire Science and Engineering INTERFLAM*, Royal Holloway College, UK. 2016.

Liu, Z., Silva, J., Huang, Q., Hasemi, Y., Huang, Y. et al., Coupled CFD–FEM Simulation Methodology for Fire-Exposed Bridges, *Journal of Bridge Engineering*, Volume 26 No., 10, 2021.

Liu, Z., Li, G., Paya-Zaforteza, Cai, C. and Huang, Q. et al., Fire Hazards in Bridges: State of the Art, Recent Progress, and Current Research Gaps, *Journal of Bridge Engineering*, Volume 28 No., 7, 2023.

Mander, J.B., M.J.N. Priestley, and Park, R., Theoretical Stress Strain Model for Confined Concrete, *Journal of Structural Engineering*, 114 (8): 1804–1826, 1988.

McGrattan,K., R. M., Simo Hostikka, R.M. and Floyd, J. Fire dynamics simulator. Gaithersburg, MD: NIST, 2022.

Quiel, S.E., T. Yokoyama, L.S., Bregman, K.A., Mueller and S.M., Marjanishvili et al., A streamlined framework for calculating the response of steel-supported bridges to open-air tanker truck fires, *Fire Safety Journal* Volume 73 pp 63–75, 2015.

Post-Tensioning Institute (PTI). Recommendations for Stay-Cable Design, Testing, and Installation. Farmington Hills: PTI 2018. Standard No. DC45. 1–18.

Rujin, M., Chuanjie, C., Mingeli, Ma, Airong, C., Performance-based design of bridge structures under vehicle-induced fire accidents: Basic framework and a case study, *Engineering Structures*, Volume 197, 2019.

SEI, *Advocating for Performance-Based Design*, Report to the Structural Engineering Institute Board of Governors Submitted by: Task Committee on Performance-Based Design, 2018.

Shama, A. and Jones, M., Seismic Performance-Based Design of Cable-Supported Bridges: State of Practice, *Journal of Bridge Engineering*, American Society of Civil Engineers, Volume 25, No. 2, 2020.

Shama, A., Mander, J., Blabac, B. and Chen, S., *Experimental Investigation and Retrofit of Steel Pile Foundations and Pile Bents Under Cyclic Lateral Loadings*, Technical Report Multidisciplinary Center of Earthquake Engineering MCEER-01-0006, Buffalo, NY, 2001.

Shama, A. and Barbas, J., Seismic Performance Based Criteria of Suspension Bridges, *Proceedings of the 8th International Cable Supported Bridge Operator's Conference*, Edinburgh, Scotland, 2013.

Shama. A., Spectrum-Compatible Earthquake Ground Motions by Morlet Wavelet, *20th Analysis and Computation Specialty Conference*, Chicago, Illinois, 2012.

Skarsbo, L.R., *An Experimental Study of Pool Fires and Validation of Different CFD Fire Models*, A thesis submitted in partial fulfilment of the requirements for the degree of Master of Science in the subject of Physics; Process Safety Technology, University of Bergen, Norway, 2011.

Society of Fire Protection Engineers, *SFPE Engineering Guide to Performance-Based Fire Protection*, 2007.

Vosooghi, A. and Saiidi M., *Post-Earthquake Evaluation and Emergency Repair of Damaged RC Bridge Columns using CFRP Materials*, CA Dept. of Transportation Research Rep. No. 59A0543. Reno, NV: Univ. of Nevada, 2010.

Wegrzynski, W. and Lipecki, Wind and Fire Coupled Modelling—Part I: Literature Review, Fire Technology, 54, 1405–1442, 2018

Xu, L.c., Huo, N.N. and Ruan X., Performance-based Vehicle Fires Protection Method for Cable-stayed Bridge, *Life-Cycle Civil Engineering: Innovation, Theory and Practice-Chen, Ruan and Frangpool (Eds)*, 2021.

Zhang, Y., Fang Z., Jiang R., Xiang Y., Long H. et al., Static performance of a long-span concrete cable-stayed bridge subjected to multiple-cable loss during construction, *Journal of Bridge Engineering*, Volume 25 (3): 159–171, 2020

CHAPTER 13

Structural Health Monitoring of Cable-Stayed Bridges

13.1 BACKGROUND AND MOTIVATION

Structural Health Monitoring (SHM) for cable-stayed bridges can collect large amounts of data about the damage state of various structural components at the global and local level. This information can serve as a significant supplement of knowledge on the performance history of the bridge and hence a valuable documentation source for decision makers. The SHM system for cable-stayed bridges is designed to monitor the structural response parameters set by the bridge engineer so that the behavior of the bridge can be evaluated periodically under different loading conditions.

SHM is defined as the approach of using nondestructive sensors to characterize in real-time the structure condition of the structure. This process includes analysis of the bridge system characteristics such as structural response for the purpose of detecting changes to material; geometric properties; boundary conditions; or structural connectivity, which may indicate damage or degradation. The damage identification process is generally categorized into different levels:

Level I: Determine whether damage is identified
Level II: Determine the location of damage if it has occurred
Level III: Identify damage, determine its location, and estimate its severity
Level IV: Identify damage, determine its location, estimate its severity, and estimate its impact on the structure or determine the remaining useful life of the bridge

The output of this process is updated information down the road regarding the ability of the bridge to function as intended despite the unavoidable degradation resulting from operational environments. SHM is also used after extreme events for rapid condition assessment of the structure.

As the profession is moving towards adopting performance-based design methods, advanced methods for structural performance detection such as SHM may be more adequate than traditional methods of analysis. Therefore, SHM is mainly employed tactically to evaluate structural performance that may change progressively over time, as well as to determine the structure's response to extreme events. SHM is currently gaining wide acceptance among bridge authorities for several reasons:

- Currently, cost and time of inspection of bridges in general is relatively high. Implementing a SHM can optimize inspection budgets.
- Real-time condition data avoids closures for routine and reduces unnecessary life-cycle costs.

- Current design methods for cable-stayed bridges aim to extend its design life. This can be achieved through SHM by extending the remaining life of the bridge through reduction of failures due to early detection.
- Main members of cable-stayed bridges such as the pylons and stays are susceptible to non-predictable performance under some loads such as wind and earthquakes.
- By implementing an SHM strategy for a bridge that efficiently monitors structural performance under operational and environmental variations, damage can be detected and repaired at early stages, which may result in substantial savings in maintenance expenses.
- Bridge operators can be alerted about potential defects and anomalies through an automated analysis and sequential alarm system that is available 24/7.
- Continuous monitoring of cable-stayed bridges will result in more homogenous statistics in terms of rating.

The main goals of SHM for cable-stayed bridges may, therefore, be summarized as:

- To provide data on the structural dynamics' characteristics to verify design assumptions.
- To gather data for the purpose of improving existing design guidelines or developing future design guidelines.
- To provide data to enable assessing structural damage of replaceable members such as bearings, sacrificial link elements, or lock-up devices.
- To shed some light on the behavior of non-replaceable elements such as pylons and to optimize the ideal behavior of these members.
- To provide real-time data to detect potential problems between intermittent inspections.

Monitoring typically includes tracking two parameters: load effects and structural response. Instrumentation used in SHM includes but is not limited to load cells; seismometers for earthquake excitation; anemometers for measuring wind speed and direction; triaxial accelerometers for structural dynamic response and system identification (St-Id); velocity gauges to track vibration response due to wind and earthquake; Global location and position (GPS) to measure displacements; displacement gauges; and thermometers. The following sections will cover the elements of SHM including a detailed description of each instrumentation used in this process.

13.2 INSTRUMENTATION AND TOOLS USED IN SHM

This section summarizes the main items that are used for cable-stayed bridge SHM and areas of application. Examples of applications of each item are discussed later under examples of implementation of SHM.

13.2.1 Displacement Transducers

A displacement transducer (DT) as illustrated in Figure 13.1, is an electromechanical device used to convert mechanical motion or vibrations, specifically rectilinear motion, into a variable electrical current, voltage or electric signal, and vice versa. The distance an object has traveled is measured in millimeters (mm) or inches (in) by the linear displacement sensor's output signal, which can have a positive or negative value. Conversion theories or output signal types are included in the categorization of electromechanical transducers. Position sensors also include linear potentiometers. They are applied to displacement measurements in all directions. Linear potentiometers measure displacement in a linear fashion.

Linear variable differential transformers (LVDT) are also used to measure displacement. They work on the more complex principle of variable inductance and require an alternating current supply. They are more expensive than potentiometers DTs, but more responsive because they do not require an additional sliding connector in their setup. An outer transformer with a primary coil

Fig. 13.1 Displacement Transducer

and two secondary coils wound in opposite directions, positioned on either side of the primary coil, houses the central core of an LVDT (Figure 13.2). After applying an alternating current (AC) excitation to the primary coil, a magnetic flux develops and is coupled to the secondary windings via the ferromagnetic core. When the core is placed in the magnetic center of the transformer, the two secondary coils cancel each other out because of their opposing windings, preventing any voltage readings at the output. The circuit's output voltage differential is caused by the unequal amount of induced flux that is coupled into the two secondary coils when the core deviates from this central position. which has a linear relationship with the core's displacement.

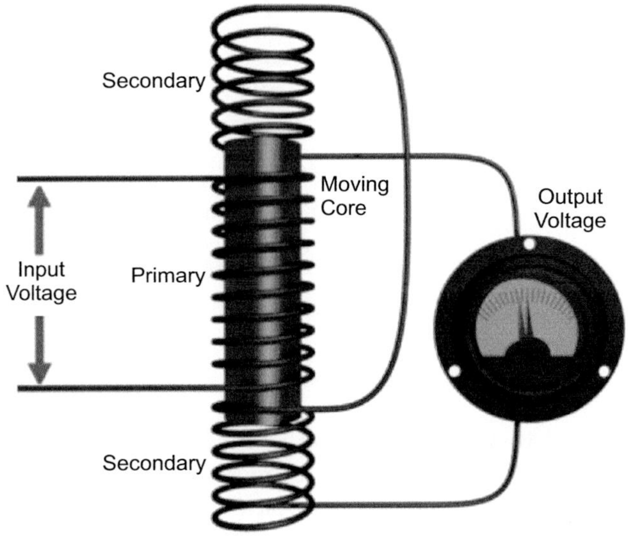

Fig. 13.2 Schematic of Linear variable differential transformers

The LVDT has a few distinct features due to its principles of operation and construction. When the core is properly supported, there is no friction between the sensing elements. This makes the LVDT a perfect choice for creep or low-friction type of testing. The mechanical life is controlled by the core support system so it can have an infinite fatigue life if properly designed.

13.2.2 Accelerometers

In its simplest form an accelerometer is a transducer that converts acceleration to a low impedance voltage signal. It is basically a mass-spring-damper system mounted inside a rigid casing (Figure 13.3) that is attached to the surface whose motion is to be measured. Piezoelectric accelerometers utilize the piezoelectric effect of quartz to convert acceleration to voltage. The Piezoelectric accelerometers are often used for test and measurement purposes. They're easy to install, provide a wide frequency response as well as high sensitivity thus, making them a preferred choice for bridge vibration testing. Piezoresistive accelerometers are different from piezoelectric accelerometers in their mechanism. They are made with a Piezoresistive material and will be deformed after a force is applied to it where the change in resistance is measured. Capacitive accelerometers measure the capacitance changes in the seismic mass under acceleration.

Because these accelerometers measure vibration in only one direction, multiple accelerometers need to be collocated at the measurement stations on the structure to permit vibration responses in multiple directions simultaneously. On the other hand, triaxial accelerometers are used to obtain vertical, horizontal (transverse and longitudinal) orthogonal accelerations. Triaxial accelerometers are also superior to other sensors in that they do not depend on the propagation of electromagnetic waves, and therefore avoid issues of signal refraction and line-of-sight problems with terrestrial or space objects and are unhindered by adverse weather conditions.

Fig. 13.3 A piezoelectric accelerometer on magnetic base

For many cable-stayed bridges the dominant structural frequency is usually less than 1 Hz. Traditional capacitive or resistive accelerometers are more suitable for such applications.

Accelerometers on cable-stayed bridges are typically placed on the stay cables, towers, and deck to address dynamic effects.

Fiber optic accelerometers (FOA) are opto-mechanical acceleration sensors. A FOA is provided with a micromechanical Silicon mirror (MEMS) which deflects a light beam proportional to the acceleration. The sensor head is available in a housing of either stainless steel or ceramic material (MACOR). The steel housing is hermetically sealed and has a laser welded collimator. The standard measurement direction is along the axial direction. These sensors can be placed in a severe environment insensitive to magnetic and electrical fields.

13.2.3 Strain Gauges

Strain gauges can be used for a multitude of functions on steel and concrete long-span bridges. They are used to measure local static strain, curvature, concrete creep and shrinkage. The most common kind of strain gauge consists of an insulating flexible backing which supports a metallic foil pattern. Mechanical strain gauges are often used to measure larger movements of foundations or crack growth.

Sensitivity of strain gauges will vary by the material used, temperature range, and the amount of strain measured. Sensitivity is proportional to approximately 1 + 2 times the Poisson's ratio, which is in the range of 0.25 to 0.35 for most metals encountered on bridges, so a normal sensitivity range would be 1.5 to 1.7.

Constantan alloy is the oldest and most widely used material in strain gauges since it has a relatively low temperature-induced strain in the range of −30 C to + 193 C (−20 F to +380 F). Hence, it is considered to have self-temperature compensation. It has almost constant sensitivity across a wide range of strain, which makes it particularly convenient for a broad range of applications. However, the resistance of Constantan drifts continuously at temperatures above +65 C (+150 F), which can become troublesome over a long period of time or at a high temperature. Constantan's sensitivity is higher than average, at 2.1.

Isoelastic alloy is widely used for dynamic strain measurements for vibration and impact. Its sensitivity is 3.6, higher than that of Constantan. It has better fatigue properties than many strain gauge materials. However, it is normally unsuitable with temperature fluctuations as it does not possess self-temperature compensation properties.

Karma alloy has similar overall properties to Constantan, including self-temperature compensation in the range of −73 C to +260 C (−100 F to +500°F), but it has a higher cyclic strain resistance than Constantan.

Strain gauges can be used on fatigue-sensitive details on existing or new long-span bridges to measure stress/strain for fatigue cycles, or out-of-plane bending. Strains greater than a pre-set maximum value can trigger an alarm, or flag for investigation or response, by the bridge owner. Strain gauge sensitivity factors are usually provided by the manufacturer, but bridge design engineers need to choose the right gauge wire for their application.

13.2.4 Fiber Optic Sensors

In field applications, strain gauges are notorious for their comparatively short lifespan. Fiber optic sensors (FOS), a potentially more robust technique for long-term structural health monitoring applications, have been used in more recent strain measurement applications. Optical fibers have better qualities. They can be arranged into any shape and are geometrically flexible. Optical fibers are immune to electromagnetic and electric interference and are small, usually with a diameter of 250 microns. Furthermore, because optical fibers can function as both the medium for transmitting signals and the sensing element, the electronic instrumentation can be placed far from the measurement site. For the structural health monitoring (SHM) of cable-stayed bridges, FOS are a good fit. Applications include strain, cable dynamics, deformations, accelerations, and the monitoring of crack growth and development in bridge substructures. Unlike traditional measuring devices, such as electrical sensors, fiber optic sensors are robust enough to endure harsh environments and are resistant to electromagnetic interference. The four main types of fiber-optic sensors are polarimetric, modalmetric, intensiometric, and interferometric. The kind of application, the ease of integration into the structure, the number of optical fibers required per sensor, the ability to localize the sensing region (i.e., using insensitive lead-in and lead-out optical fibers), and the possibility of a single-ended sensor are the main determinants of the fiber-optic sensor type that is chosen. A wide class of incredibly sensitive optical fiber sensors is represented by interferometric sensors. These work mainly by detecting the phase shift that occurs in light after it travels through a single-mode optical fiber. This group includes the Fabry–Perot fiber-optic sensor (FPFOS) and the fiber Bragg grating sensor (FBG). The fundamental working principle of FBG-based sensors is the examination of wavelength shifts in optical fibers caused by strain. In order to apply Bragg gratings into optical fibers, a powerful UV laser must pass through a mask to expose the fiber's core and achieve a desired wavelength of reflectivity. FBG sensors have a wide range of uses and are very sensitive. They can perform dynamic measurements as well.

A Fabry–Perot fiber-optic sensor, as shown in Figure (13.4), consists of a single optical fiber with a sensing region delineated by a cavity made up of two mirror surfaces that are perpendicular to the optical fiber's axis and parallel to each other. The interference pattern is produced by the interference of the reflected signals that the two partial mirrors combine to form in the cavity. With a clearly defined sensing region, FPFOS is an extremely sensitive sensor with excellent spatial resolution. In addition to their high sensitivity, the majority of these sensors can perform dynamic measurements.

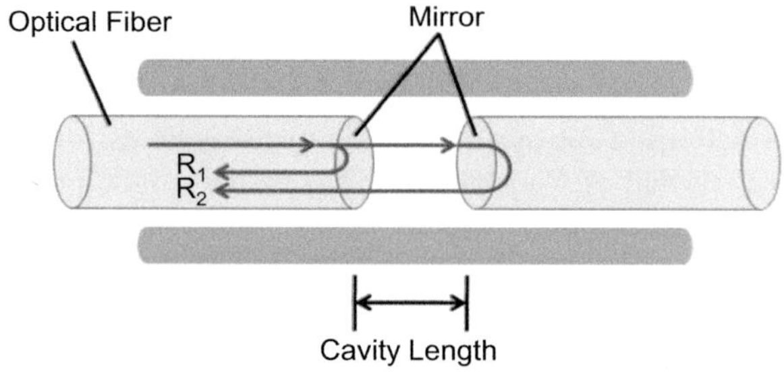

Fig. 13.4 Schematic of Fabry-Perot Interferometric sensor (courtesy, Elsevier)

Intensiometric sensors depend on a variation of the radiant power transmitted through a multimode optical fiber. This sensor's most basic version detects whether light is present or not. An optical fiber break, for instance, can function as the foundation for a damage-sensing system. This includes optical time domain reflectometry (OTDR) sensors, which use Fresnel reflection from fractures to locate breaks in an optical fiber using a time-of-flight measurement.

The ability of fiber optic sensors to distribute sensing and measurement in parallel or series can be a tempting feature as it can lead to extensive structural health monitoring of cable-stayed bridges. The ability of fiber optic sensors to measure strain and temperature while embedded in bridge cables is another benefit. Given the fragility of FOS structures, sensor package design must accurately sense the parameter of interest and safeguard the optical fiber from structural damage.

13.2.5 Global Positioning Systems (GPS)

GPS are becoming an alternative to Conventional sensors, which became impractical for long-span bridges under long-term monitoring of absolute displacement. Traditional displacement transducers are useful for relative displacement only, whereas laser transducers and total stations have been proven unreliable over the long term for monitoring purposes.

A GPS can be used to monitor vertical and horizontal displacement of pylon tops on cable-stayed bridges, or midspan deck vertical deflections. It can also be used for monitoring relative horizontal displacements across large expansion joints and verifying creep/shrinkage and movements due to temperature changes.

It can be used to monitor vertical and horizontal displacement of pylon tops on cable-stayed bridges, or midspan deck vertical deflections. It can also be used for monitoring relative horizontal displacements across large expansion joints and verifying creep/shrinkage and movements due to temperature changes.

GPS is traditionally referring to the North American global positioning system, or satellite positioning system. GPS is one of the 5 Global Navigation Satellite Systems (GNSS) used around the world. The 5 GNSS constellations include GPS (US), QZSS (Japan), BEIDOU (China), GALILEO (EU), and GLONASS (Russia). GPS instruments are used to measure displacements, especially the very low frequency displacements on cable-stayed bridges such as the deflection produced by wind loading. Such pseudo dynamic motions have very little accompanying acceleration and are often undetected by accelerometers. The basic idea behind relative position GPS motion monitoring is that with one GPS antenna located at a non-moving position near the structure, and with another on the moving structure, one can accurately measure the relative displacement. Because one is fixed, this relative motion is the absolute motion. The distance between the stationary reference GPS unit and the units located on the structure should be close within a few miles to ensure the satellite signal traveling through the atmosphere to the antennae is assumed to be the same.

It is important to note that the current GPS satellite distribution across the sky is uneven in the mid and high latitude regions, which affects the precision of position measurements. The effect of these issues can be further augmented with a supplement of additional GNSS to meet bridge structural health monitoring needs.

13.2.6 Elasto-Magnetic Stress Sensors

Measuring stress in stay cables during the service life of a cable-stayed bridge is quite challenging using traditional methods. Elasto-magnetic (EM) sensors, also sometimes referred to as magnetostrictive sensors, employ a low-cost technology for measuring the actual stress in ferromagnetic materials such as that of steel wires, strands, and steel bars. EM sensors are becoming a favorable tool because of their non-destructive and non-contact properties, which also include corrosion resistance and long service life.

The fundamental principle of EM sensors is the magneto-elastic phenomenon of the ferromagnetic material. The magnetic properties of a ferromagnetic material change under the application of stress and under the influence of temperature. Also, the permeability of steel to a magnetic field changes with the stress level in the steel. This means that, by measuring the change in a magnetic field, the magnitude of the stress in the steel element can be determined.

The sensor system as displayed in Figure 13.5 consists of a solenoid composed of a primary coil and a secondary coil that are insulated from each other by plastic or other polymers and work

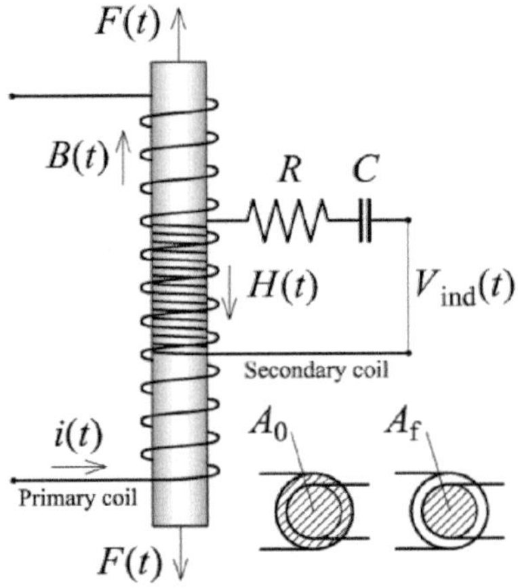

Fig. 13.5 EM measurement principle (Courtesy, MDPI)

together to formalize the elastic, magnetic characterization of the material. Pulsed current passes through the primary coil, and the secondary coil picks up the induced electromotive force that is directly proportional to the rate of change of the applied magnetic flux and the relative permeability of the member whose properties are to be measured. The secondary coil is linked to a power stress reading device, which measures the magnetic permeability of the member through the sensor and shows the force on the display.

Therefore, EM sensors can be employed for a wide number of applications for bridges and structures wherein steel elements that are subject to high axial forces are key elements, such as the cables for cable-stayed bridges, for attaining the scheduled performance of the structure. These sensors can substitute other traditional cumbersome methods for load calculation that require lift-off equipment or other cost-intensive techniques.

13.2.7 Weather Sensors

13.2.7.1 Temperature Sensors

Temperature sensors are very beneficial and necessary for a wide range of applications. They are used to measure ambient temperature, wet-bulb and dry-bulb temperature for determining relative humidity and determining the temperature of steel and concrete. They measure temperature readings via electrical signals. They contain two metals that generate an electrical voltage or resistance when a temperature change occurs. The sensors are categorized based on their connection to Contact Temperature Sensors (CTS) and Non-contact Temperature Sensors (NCTS). CTS such as thermostats measure the degree of hotness or coldness of an object via direct contact. They are generally used to detect a wide range of temperatures in different solids, liquids, or gases. On the other hand, NCTS are never in direct contact with an object or substance and measure how hot or cold something is via radiation emitted by a heat source.

Temperature sensors are mounted in the pavement on approaches to cable-stayed bridges as well as in the pavement (often at midspan), to help assess roadway freezing temperatures to optimize the use of roadway salt for deicing in Northern climates. Temperature sensors are also used to measure the air temperature above and below the girder levels and can be used to automatically notify the bridge owner's staff of possible issues with expansion joint movements, or possible kinks in roadways at pinned joints on cable-stayed bridges. They are also used in the field of geotechnical monitoring and renewable energy.

13.2.7.2 Anemometers

Wind pressure, which is correlated with speed, and wind speed can both be measured with an anemometer. For measuring wind speed, cup anemometers (Figure 13.6) are commonly utilized. The wind speed is measured by the cup anemometer using the principles of aerodynamics. When there is wind, the anemometer begins to rotate, allowing the number of rotations per second to be used to calculate wind speed.

The propeller anemometer, windmill (or Byram) anemometer, hot-wire anemometer, ultrasonic anemometer, bi-directional anemometer, acoustic resonance anemometer, and laser anemometer are the primary instruments that measure wind speed.

By design, anemometers that measure wind pressure function differently. Some work by measuring the force that the wind applies to a plate held in front of it, while others use a U-tube with a built-in pressure gauge to measure the wind pressure. Although its primary purpose is to measure wind pressure, the plate anemometer can also measure wind speed. Plate anemometers use the compressive force on a spring with a specified spring constant to measure the pressure that the wind exerts on a plate. Although plate anemometers can be imprecise when reading high winds and

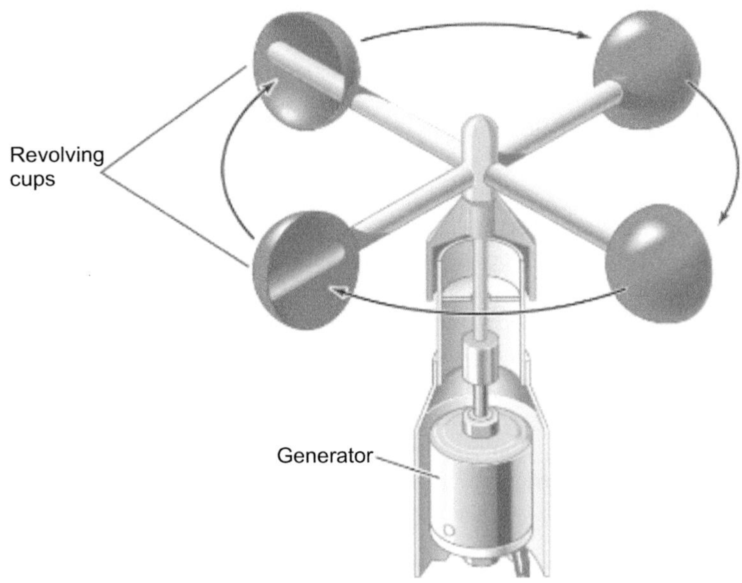

Fig. 13.6 Cup Anemometer (Courtesy, Britannica)

can be slow to react to varying wind speeds, they have been used for high wind alarms on bridges. They also do not respond well to light winds. Ball, tube, and Pitot tube anemometers are additional examples of this kind of anemometer.

Ultrasonic anemometers measure wind speed by detecting the difference in time taken for an ultrasonic pulse to travel in each direction between pairs of transducers caused by movement of the air. Since there is no mechanical inertia to overcome, this method permits accurate measurement even at low wind speeds. Precise time-dependent measurements are possible due to the high frequency of measurements made possible by the absence of mechanical inertia.

Anemometers have been placed on numerous bridges as part of weather station monitoring; they have also been integrated into the data collection algorithm on numerous cable-stayed bridges as will be discussed later in this chapter.

13.2.7.3 Barometers

A barometer is an instrument measuring atmospheric pressure and used specifically in forecasting the weather and determining altitude. Barometers produce measurements in atmospheres. One atmosphere is the average air pressure at sea level at a temperature of 15 degrees Celsius. This is equal to approximately 101325 pascals of pressure. Mechanical (aneroid) barometers have been used on land and sea continuously for several applications.

Electronic barometers are small pressure sensors read and controlled by microprocessors with a digital and sometimes graphic display. They have been used in science laboratories for many years. Affordable models intended for public use have become popular in the past few years. As with aneroid barometers, there is a wide range of quality in electronic barometers. Electronic barometers offer higher accuracy and a greater recording range than aneroid barometers and the ability to perform further data analysis on the captured data, including automated use of the data to forecast the weather, High-quality electronic barometers are sealed with an all stainless-steel case, which makes them ideally suitable for severe environments. Their high operating temperature, broad compensated range and excellent temperature compensation allow for stable readings in bridge applications.

13.2.7.4 Hygrometers

A hygrometer is an instrument used to measure the relative humidity of air, or the amount of invisible water vapor in each environment. The main categories of hygrometers include mechanical hygrometers, psychrometers, and electronic hygrometers. The mechanical hygrometer uses a strand of hair that is held by a spring to create slight tension. The spring is connected to a needle gauge, which provides the humidity level in the air based on the natural movement of the hair strand. The psychrometer features two thermometers side-by-side. One thermometer is a "dry" thermometer measuring temperature while the other thermometer is a "wet" thermometer immersed in a bulb of liquid. Low levels of atmospheric moisture result in evaporation and loss of heat from the wet thermometer. Eventually, the wet thermometer will read lower than the dry thermometer since it loses heat through evaporation and the dry thermometer remains constant. Calculating the difference between the thermometers will produce relative humidity levels. Electronic hygrometers provide instant digital humidity readings by monitoring either capacitance or resistance of an air sample. Capacitive sensors sense water by applying an AC signal between two plates and measuring the change in capacitance caused by the amount of water present. The resistive sensors use a polymer membrane which changes conductivity according to absorbed water.

13.2.7.5 Rain Gauge

A rain gauge is an instrument used to gather and measure the amount of liquid precipitation over a predefined area, over a period. The three major types of rain gauges are the standard gauge, tipping bucket gauge and weighing gauge.

The standard rain gauge (Figure 13.7) works by catching the falling rain in a funnel-shaped collector that is attached to a measuring tube. The area of the collector is 10 times that of the tube; thus, the rain gauge works by magnifying the liquid by a factor of 10.

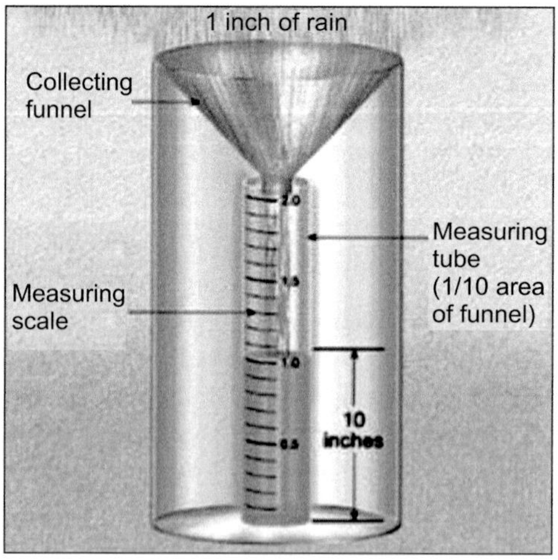

Fig. 13.7 Standard Rain Gauge (Courtesy of Maximum Weather Instruments)

Magnifying the rain in this way allows precise measurements down to a one-hundredth of an inch. Amounts that exceed the tube capacity are caught in the outer shell of the gauge, allowing the recorder to pour out the liquid in the tube and fill it back up if needed.

Electronic rain gauges typically operate on the "tipping bucket" principle and meet national Weather Service specifications for statistical accuracy. The tipping bucket rain gauge is provided with a receiving funnel which leads to one of two small buckets (Figure 13.8). Filling of one bucket occurs at one-hundredth of an inch. The result is a "tipping" of the liquid into the outer shell of the gauge, triggering the second bucket to take its place. The process then repeats itself, allowing for precise measurement of rainfall intensity and amount. This gauge has become standard for wireless weather stations.

Fig. 13.8 Tipping Bucket Rain Gauge (Courtesy, Maximum Weather Instruments)

The weighing Rain Gauge is precise in measuring rainfall intensity as the weighing mechanism at the bottom of the collector can be used to measure depth and time simultaneously. Recording is carried out much in the same way as the older versions of the tipping bucket gauges.

13.2.8 Bearing Temperature Sensors

Bearing sensors are designed to monitor the temperature of mechanical or rotating bridge bearings. For bearings that rotate or slide on a frequent basis, a reliable indicator of bearing conditions may be the localized temperature of the bearing metal beneath the shoe. Bearing sensors are suitable for areas where there are room limitations. They are miniature low mass sensors which respond fast.

13.2.9 Corrosion Sensors (Cells)

Metallic materials oxidize during the electrochemical process of corrosion, which results in mass loss and structural degradation. An anode, a cathode, and an electrolyte make up the basic components of an electrochemical system. The steel corrodes at the anode, and the reduction reaction takes place at the cathode. Ionic conductivity connects the anode and cathode through the electrolyte, which is typically an aqueous solution containing dissolved salts (like NaCl) and corrosive species. This allows the electron transfer between the two electrodes to be balanced. The corrosion process can be accelerated in anaerobic subsurface wellbores by the ubiquitous acidic gases CO2 and H2S, which can dissolve into the electrolyte, lower pH, and encourage cathodic reactions.

Monitoring corrosion rates facilitates the assessment of service life and directs maintenance scheduling. For various types of corrosion, a multitude of corrosion sensor technologies have been developed based on various sensing principles. Corrosion sensors can be broadly divided into two categories: direct and indirect. The various corrosion causes and corrosive environments are directly monitored by the direct corrosion sensors, which track the corrosion process and rates. Through corrosion causes (such as low pH, water, or CO2) or consequences (such as changes in steel section thickness, leak vibration, or strain changes), the indirect corrosion sensors monitor corrosion. Corrosion coupons, electrical resistance (ER) probes, electrochemical sensors, ultrasonic testing (UT) sensors, and magnetic flux leakage (MFL) sensors are examples of conventional corrosion sensors.

The corrosion coupon measures corrosion rates by measuring weight loss. The corrosion coupon method's ease of use, adaptability to different materials and shapes, and straightforward working principle make it a popular choice (Figure 13.9). However, the coupons' installation, removal, and post-corrosion lab analysis take a lot of time. Without real-time data, the corrosion coupons only offer an average corrosion rate over a specific time period.

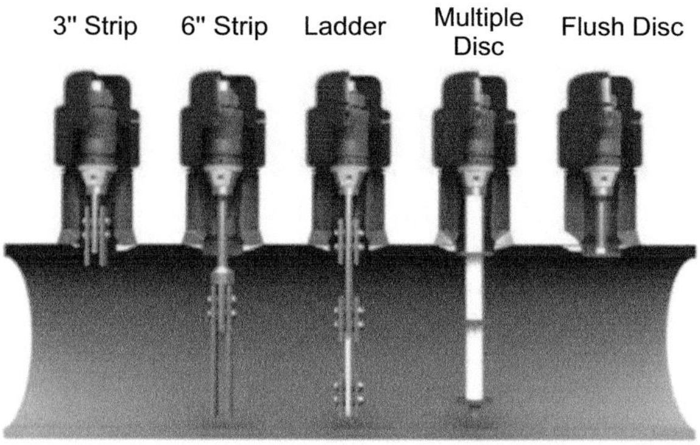

Fig. 13.9 Corrosion coupon installed in a steel pipe (Courtesy, Cosasco systems)

An electrical resistance (ER) probe is a real-time corrosion rate monitoring online tool that uses electrical resistance to track corrosion in real-time. In certain more advanced versions, it can log data automatically and remotely. An increase in electrical resistance is caused by mass loss in the metallic materials that are exposed. For every unique application, the material and shape of the exposed sensing element can be altered.

Electrochemical sensors make use of electrochemical methods like electrochemical noise (EN), electrochemical impedance spectroscopy (EIS), linear polarization resistance (LPR), and galvanic current measurement. Direct measurement of electrochemical corrosion rates and the ability to investigate the corrosion mechanism in situ using a range of electrochemical techniques are two benefits of electrochemical sensors. The disadvantage of electrochemical sensors is that an externally applied potential or current may cause an accelerated rate of corrosion relative to the true value.

Since corrosion detection is linked to the durability of structures, it is generally a very important problem in structural engineering. In the Northern US, for instance, where chlorides from road deicing salts can seep into concrete, identifying the corrosion of steel rebars in concrete roadway decks is of utmost importance. Over time, chlorides can lower the pH of concrete, which can lead to the breakdown of the passivation layer between concrete and steel rebar if the local pH falls below 11.5. As deck replacement is a major problem for long-span bridges that are supported by cables, corrosion cells should find many more uses in the concrete of composite decks on cable-stayed bridges as well as in the cable stays.

13.3 HEALTH MONITORING TECHNIQUES

SHM is typically employed to evaluate performance of the structure and to detect anomalies due to deterioration and damage during regular operation and/or after an extreme event. A comprehensive

SHM system can be achieved by continuously tracking the response of the bridge or its components to different loading environments and provide necessary information for health and performance about the structural system. Health monitoring techniques may be classified as global or local. Global methods attempt to simultaneously assess the condition of the whole structure whereas local methods focus non-destructive (ND) tools on specific structural components. Global methods utilizing the vibration characteristics of a structure are effective in evaluating the condition of the entire structure due to its large size. On the other hand, local methods are utilized upon identification of a damaged location, to detect, locate, and/or characterize defects more precisely. The global nondestructive examination technique utilizing vibration characteristics is accomplished in two stages: (1) dynamic testing for the evaluation of structural dynamic characteristics such as natural frequencies, mode shapes, and damping ratios; and (2) identification of structural damage through a damage detection algorithm.

13.3.1 Global Structural Health Monitoring

Global SHM utilizes the dynamic characteristics of the structure to evaluate the damage. The dynamic characteristics of the bridge include the natural frequencies of vibration, and their associated mode shapes and damping ratios. Data acquisition and extraction of vibration properties are associated with the first stage of global SHM. Dynamic characteristics of the structure are evaluated through dynamic testing. This approach requires a source of excitation to vibrate the structure, the use of sensors to acquire time history data, and experimental modal analysis procedures to extract modal parameters such as frequencies and mode shapes.

13.3.1.1 Dynamic Testing Procedures

The dynamic properties of a bridge structure can be measured and identified using two popular experimental techniques: (1) input-output methods of excitation and (2) free vibration output only methods. The structure's vibration responses to different kinds of external dynamic excitation are measured using the input-output method. One type of mechanical excitation apparatus that can provide the structure with external dynamic excitation is a linear or eccentric mass shaker. The ability to introduce a known input into the structure and lessen the influence of unrelated noise on the measured structural response are two advantages of input-output excitation techniques. However, the primary limitation of an input-output modal test is its inability to enable ongoing structural monitoring.

When a structure is in service, output-only ambient vibration methods are used to assess how it will respond to typical operating conditions. The technique is especially effective because it does not require traffic closures, and most importantly, it permits continuous bridge structure monitoring. Vehicle traffic and wind were the excitation sources for earlier ambient vibration tests. Other sources of excitation include foot traffic, waves in the ocean, and low seismic activity. Additionally, this approach assumes that modal parameters can be obtained solely from the recorded response of the structure. The uncontrollable amplitude of the excitation, which depends on the type of exciting agent—such as the amount of traffic or the strength of the wind—is one of the limitations of ambient vibration techniques. Furthermore, high frequency excitations are frequently unattainable with ambient vibration techniques. However, these limitations have been lessened over time by the development and advancement of modal extraction techniques and data acquisition systems. The ambient vibration method is the most widely used and practically approachable technique for determining the dynamic properties of long-span bridge structures. Operational modal analysis is another term used to describe this experimental methodology.

13.3.1.2 Testing Execution and Evaluation of Data

The discussion herein is limited to the ambient vibration method. Since acceleration is the main concern in the measurements, therefore selecting the types of sensors to be used, the location where the sensors should be placed, the number of sensors to be used, and the data-acquisition/storage/transmittal hardware is very significant. Another consideration is how often the data should be collected, whether continuously, at periodic intervals, or after a severe event. The locations of the sensors on the bridge must facilitate measuring the vibration responses in the vertical, torsional, longitudinal, and lateral directions.

To conduct the test through an SHM program a period of the data collection must be selected and the data acquisition system is operated automatically for data collection through this period. Next, the length of data sets to be processed is selected, which can be 30 minutes to one hour long. Finally, the sampling rate is set. The Nyquist Criterion states that the sampling rate must be at least twice the maximum frequency that is to be measured to prevent aliasing of the recorded signal. Aliasing is a phenomenon that can lead to erroneous frequency content. As an example, if the estimated maximum frequency of interest for the structure is 10 Hz, then the maximum theoretical frequency for this case would be 100 Hz or half of the sampling rate which must be for this case 200 Hz. Data acquisition is fundamentally consisting of 4 components: (1) the sensors (accelerometers), which convert the acceleration to an electric signal (voltage); (2) signal conditioners to pick up the signal and convert it into a higher level of electrical signal; (3) analog-to-digital converter to convert conditioned sensor signals to digital values. The signal conditioners also conduct signal amplification, which is the process of amplifying the signal for processing or digitization. There are two ways that signal amplification can be performed; by increasing the resolution of the input signal, or by increasing the signal-to-noise ratio. Another important function of a signal conditioner is filtering, and this is where the signal frequency spectrum is filtered to only include the valid data and block any noise.

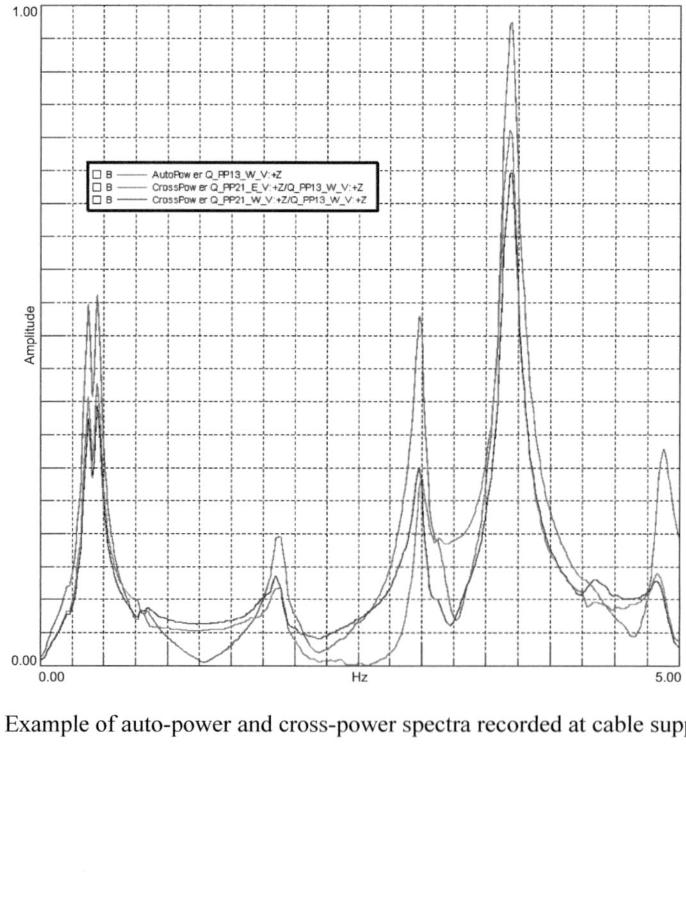

Fig. 13.10 Example of auto-power and cross-power spectra recorded at cable supported bridge

There are several filtering options including both low and high pass filters. Bessel, Butterworth, or anti-alias type filters also can be used. The filter is usually applied as a ratio of the total sampling rate, therefore, if for example the sampling rate is 200 Hz and the ratio is 10%, then the filter will limit the usable frequency range to 1/10th of the sampling frequency. Therefore, the maximum frequency that could be identified in this case would be 20 Hz.

The locations of accelerometer stations are optimized and classified to two categories: reference stations and roving stations for the purpose of evaluating the mode shapes. The accelerometers are used to measure acceleration time histories due to the vibration of the bridge in vertical, lateral, and longitudinal directions (triaxial accelerometers can capture the three directions in one channel).

Experimental modal analysis techniques are either frequency domain-based or time domain-based techniques. Examples of frequency domain modal identification techniques include peak picking of the auto- and cross-powers of the measured response, in which peaks are selected in the spectra to acquire estimates of the natural frequencies, which are the frequencies corresponding to high amplitudes in the spectrum and subsequently operational deflection shapes are obtained. An example of the auto-power and cross-power spectra recorded at a cable supported bridge is illustrated in Figure 13.10. This method by far is the most popular method employed for ambient vibration tests for long span structures.

The stochastic subspace identification (SSI) approach is a time domain method that was recently applied for modal parameter identification of several complex bridges based on ambient vibration data (Ren et al. 2004). The SSI algorithm extracts a system model in state-space using output-only measurement data directly. An SSI algorithm does not require any data pre-processing to calculate auto/cross-correlation functions or auto/cross-spectra of output data. In addition, robust numerical techniques such as QR factorization, singular value decomposition (SVD) and least squares are involved in this method.

Karbhari and Lee (2009) provided a complete list of other frequencies as well as time domain methods that can be employed for experimental modal analysis. Most of these methods are still in their infancy and never applied practically to a cable-stayed bridge.

13.3.1.3 Damage Detection Methods

Structural damage detection is typically carried out using vibration-based analysis. The basic idea behind this method is that modal parameters such as frequencies, mode shapes, and modal damping are functions of the physical properties of the structure (mass, damping, and stiffness). Therefore, changes in these properties will cause detectable changes in the modal properties. These properties should be dynamically obtained from a bridge before its initial opening, if possible, through an early ambient vibration test. A good example is the Alfred Zampa Memorial Bridge (AZMB), also known as the New Carquinez Bridge, which is located 32 km northeast of San Francisco on interstate Highway I-80. The AZMB, opened to traffic in November 2003, is a signature suspension bridge with an orthotropic steel deck, reinforced concrete towers, supported by large diameter drilled shaft foundations. Ambient forced vibration tests were carried out just before the bridge opened to traffic. They were mainly wind-induced, and vehicle-induced impact loads. These dynamic field tests provided valuable information on the dynamic characteristics of the bridge in its as-built (baseline) condition with no previous traffic loads or seismic excitation. Such properties are used as baselines for further health monitoring studies of the bridge (Conte et al. 2008).

A suite of vibration-based damage detection methods has been tried during the past few decades. Most of these approaches have been found to perform well in controlled laboratory or theoretical conditions but are extremely difficult to implement in the field. The modal-based damage detection method is the most utilized technique for vibration-based damage identification in recent decades (Lydon et al. 2004). This technique is classified into two categories: frequency-based methods and mode shape-based methods. Frequency-based methods are developed on the basis that the natural frequency shifts because of structural softening due to damage. Frequency shift methods indicate damage occurrence but cannot provide spatial information regarding the exact location of damage detection. Mode shape-based methods provide an advantage over frequency-based methods since they are spatially defined, which implies that damaged locations can be indicated. Nevertheless, many measurement locations may be necessary to precisely evaluate mode shape vectors. Mode shape-based methods are suitable for continuous monitoring of cable-stayed bridges as they compare measured mode shapes or their derivatives such as curvature or modal strain energy to enhance sensitivity to damage detection and localization.

Neural Networks is one of the damage identification approaches that have been receiving growing attention recently. Not only do neural networks not require information concerning the phenomenological nature of the system being investigated, but also, they are effective and robust in coping with uncertainty, insufficient information, and noise; they have fault tolerance, which makes them a robust means for representing model-unknown systems encountered in the real world. Nevertheless, these approaches are still in the research phase. Studies by Zang and Imgruen (2001), Chou and Ghaboussi (2001), and Hung and Kao (2002), showed that these methods are capable of identifying and locating damage. Nevertheless, they are incapable of severity estimation. Moreover, these methods were not applied yet to cable-stayed bridges.

Structural damage detection using wavelet transforms is considered one of the very simple and promising approaches. This time domain method is not based on modal parameters, rather it evaluates the damage directly based on evaluation of the pre-damaged and damaged sensor signals at a specific location in the time domain. A wavelet is a waveform with finite energy and limited duration that has an average value of zero and has its energy localized in both frequency and space. A wavelet still has the oscillating wavelike characteristics as its Fourier transform is concentrated around a specific frequency. Wavelet analysis is carried out by breaking a signal into shifted and scaled versions of the original (mother) wavelet through continuous or discrete wavelet transforms. Continuous Wavelet Transform (CWT) has been employed in the past for structural damage detection as it uses a wide range of frequencies in the analysis (Moyo and Brownjohn, 2002; Law et al., 2005; Bayissa et al., 2008; and Shama, 2014). CWT is first conducted to damaged and undamaged states and wavelet power is used as an indicator of a signal time history energy content. A damage index is then established by comparing total energy of the pre-damaged state to the damaged state.

13.3.2 Local Structural Health Monitoring

Global methods utilizing the vibration characteristics of the structure can identify whether a damage occurred or not from the shift in fundamental frequencies of the bridge. While the vibration based global methods can indicate the presence of damage in a structure, local methods are necessary for finding the exact location of the damage. Several non-destructive techniques are available for local structural health monitoring. Most nondestructive techniques are based on the use of mechanical waves (ultrasonic and acoustic), electromagnetic waves (magnetic testing, eddy current testing, and radiographic testing) and fiber optics. Application of each method is described in the following subsections.

13.3.2.1 Acoustic Emission

Acoustic monitoring (AE) is a continuous non-destructive monitoring method that is used to detect and locate failures in steel wires, strands and stay cables. The response of the structural element is caused by energy released when a bridge wire fails, or another event of interest occurs. This technology has been used recently for different applications such as main cables of suspension bridges, stay cables on cable-stayed bridges, prestressing strands in concrete structures, and reinforcement strands in post-tensioned concrete and steel structures. Acoustic monitoring typically requires installing an array of sensors along the main cable that can detect the energy release. The sensors are usually installed on a permanent basis as part of a global SHM strategy to continually monitor the main cables for wire breaks related to corrosion, fatigue, or other deterioration mechanisms. The sensors contain a piezoelectric transducer, which creates a voltage due to the energy released once a wire breaks. The voltage is sent to a data acquisition system (DAS), often located on or near the bridge site for analysis. Insignificant noises are filtered using screening software, and wire failures or other events of interest can be identified by date, time, and location/coordinates on the bridge's stay cable. Finally, a website with details of the structure can be created to show where wires break on the structure in near real-time.

Acoustic monitoring has been performed on a number of bridges, including but not limited to main cables of the Bear Mountain and Bronx White Stone suspension bridges in New York, eye bars of Oakland Bay Bridge in California, and stay cables of the Varina-Enon Bridge near Richmond Virginia. This study used acoustic emission (AE) to assess the condition of strands by examining for active defects (such as corrosion, cracks expansion and rubbing, wire breaks, and similar active defects) on a single stay-cable, from anchorage point to anchorage point, of the Varina-Enon Bridge (Figure 13.11).

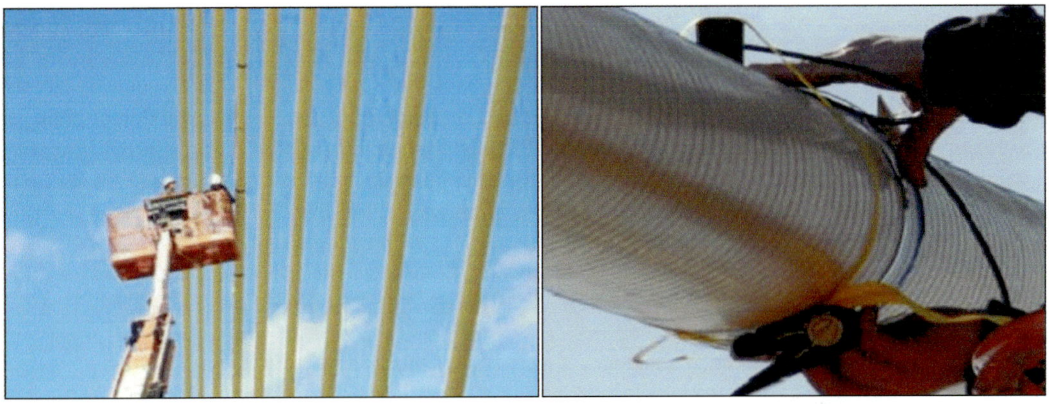

Fig. 13.11 Installation of AE sensors on Varina-Enon Bridge (left) Overview and (right) Close-Up of single AE sensor (Courtesy, Virginia Transportation Research Council

It is important to note that since the system is acoustic based, the sensors will detect any sounds being transmitted through the main cable and a corresponding voltage (or signal) will be created. Therefore, data on acoustic events associated with the everyday use of the structure will be sent to the data acquisition system. Accordingly, the acoustic monitoring system was set to filter the ambient noise on a structure to distinguish the events of interest (i.e., wire breaks). An AE signal amplitude threshold of 45 dB was set to discriminate the reliable damage-related emissions from the background noise signals. The advent of an AE event (when the AE signal amplitude is over the threshold of 45 dB), such as generation of a crack, a break, or weather-related effects resulting in AE, used to activate the system automatically. The AE DAQ system records the AE data consisting of the AE response parameters such as the date and time of the event, sensor response to the event, frequency of the AE, amplitude of AE, and a number of hits during the event. An active broadband wireless connection then transmits the data to the server, which can be remotely located, for downloading. Finally, a software is used for data analysis and plotting. Figure 13.12 shows AE activity across the center of the saddle region. Sensor 6 is located near the center of the saddle on its west face. The AE activity is attributed to initiation of a new crack or growth of a pre-existing crack.

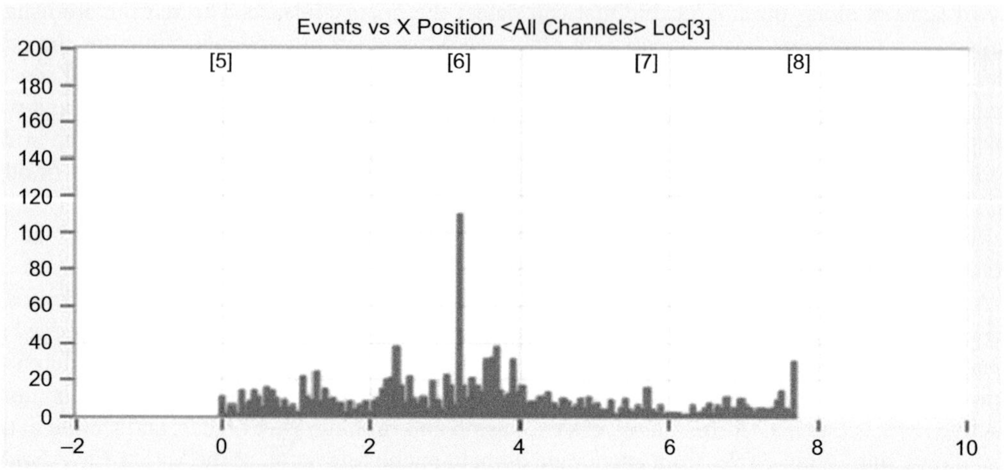

Fig. 13.12 Sample post processing of data for AE sensor (Courtesy, Virginia Transportation Research Council)

13.3.2.2 Fiber Optics

Fiber optics is typically used to sense temperature and displacement, but it can also detect changes in a variety of other parameters in bridges. The basis for sensing is the light waves' intensity, wavelength, and interference. Because of their geometric adaptability, optical fibers can be set up to sense a wide range of perturbations besides strains, as shown in Figure 13.13.

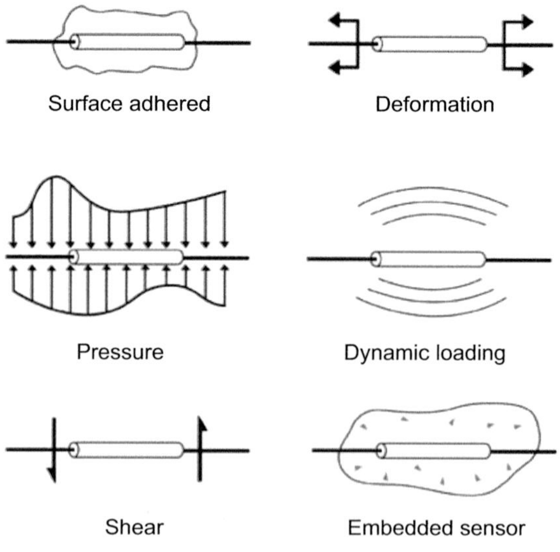

Fig. 13.13 Applications of Fiber optic sensors in structural engineering (Ansari, 2007)

Adhesives and surface protective coatings can be used to attach fiber optic sensors to the surface of structural materials, which is useful in situations where the structure is not subjected to extreme weather conditions or heavy loads. Additionally, during the production process, they can be embedded into materials like concrete. In this instance, the sensing body of the sensor is made of a metallic tube, and its gauge length is defined by two end flanges. It has been possible to attach fiber optic displacement sensors to structures. Certain displacement measuring devices, like linear variable differential transformers (LVDTs), are commonly used for monitoring cable stays in cable-stayed bridges or for measuring position and deflection. These sensors are held in place at two ends and run through stainless steel tubing, much like sensors that are embedded in structures.

Fiber optics has several benefits, such as geometric conformity, the ability to be used in a wide range of applications, and the absence of electric interference. Nevertheless, they are expensive and require highly skilled professionals to install and build the system. The pedestrian Lingotto Bridge in Turin, Italy, is presented as an example of how a fiber optic sensor is used to detect strains and deformations in the bridge cable (Talebinejad, 2007). Lingotto Bridge was built in 2005, to connect the general market plaza to the Lingotto exhibition center. The bridge goes over the railroad tracks over a total length of 365 meters with a main span of 150 meters. It is 69 meters above the ground, and it has a single arch that supports the suspended deck through cables (Figure 13.14).

Fig. 13.14 Aerial view of the Lingotto Bridge

The cable sensors were made up of individual FBG, which was pre-tensioned and protected inside a tube made up of Polyvinylidene fluoride. A fiber reinforced polymer composite cable was employed for strengthening the optical fiber and the FBG was attached to steel angle supports. The steel angle supports were then attached to the cables with a tension clamp (Figure 13.15).

13.3.2.3 Elasto-Magnetic Stress Sensor

The fundamental principle of EM sensors was presented earlier under section 13.2.6. These sensors can reliably monitor the true stress in tendons and cables, particularly applied to stay cables. The measuring principle as mentioned earlier is based on the permeability of ferromagnetic materials being a function of magnetic history and applied field (stress and temperature). This type of sensor was developed for future applications to the stay cables of Bridges in Japan. Three elasto-magnetic sensors were installed at the anchors of the cables on the Penobscot River Bridge (Maine, USA) to monitor cable forces during and after construction. Elasto-magnetic sensors have also been installed on the Stonecutters Bridge (Hong Kong) to monitor the tensions of post-tension concrete girders, at the interface between concrete and steel girders. A schematic of the application of EM sensors to bridge cables is displayed in Figure 13.16.

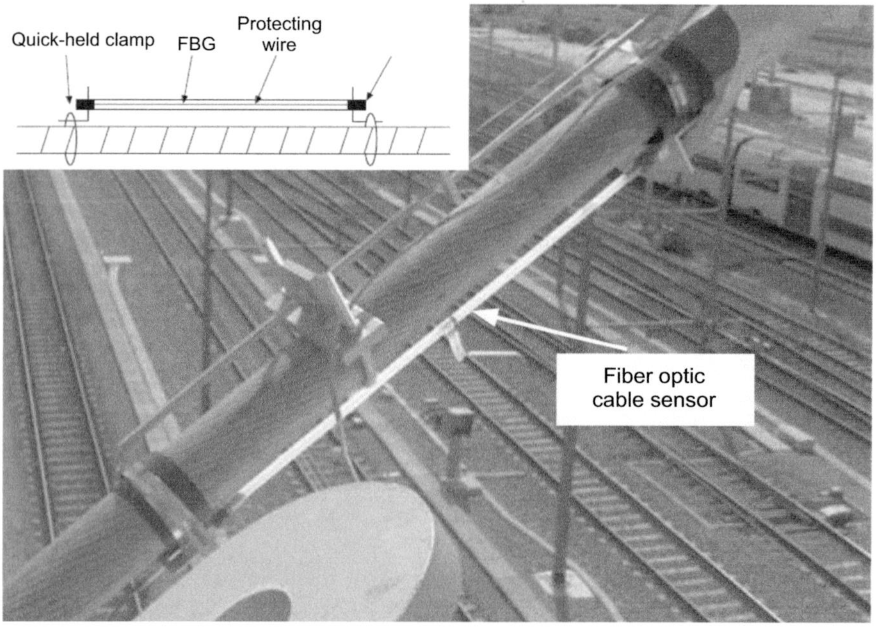

Fig. 13.15 Application of FOS ON a cable stayed bridge (Talebinejad, 2007)

13.3.2.4 Laser Doppler Vibrometers

Remote local techniques using laser readouts of vibrations are currently performed manually on bridge stay cables on cable-stay bridges. A laser Doppler vibrometer (LDV) is used to make non-contact vibration measurements of a surface. The device transmits a laser beam towards the surface of interest, and the vibration amplitude and frequency are extracted from the Doppler shift of the reflected laser beam frequency due to the motion of the surface. The output of an LDV is generally a continuous analog voltage that is directly proportional to the target velocity component along the direction of the laser beam. The main advantage of this method is its low cost and ease of use. A laser vibrometer can be used from a remote location relative to the measuring location and easy to use for areas that are difficult to access or that may be too small or too hot to attach a physical transducer. Hence, it does not involve the expense of installing and maintaining accelerometers at remote locations on a bridge that are frequently difficult to access. For cable-stayed bridge applications, the laser beam is targeted at the mid-length region of the cable from up to about 490 ft. Minute ambient cable vibrations created by normal wind speeds are measured with the laser sensor, and the cable's frequency spectra are analyzed. Then a software is used to calculate the tension in the cable based on a computation model that accounts for cable sag, bending stiffness, fixed boundary conditions, intermediate springs and dampers, cable mass, length, and external damping characteristics. Figure 13.17 displays application of the laser doppler vibrometer for stay cable vibrations. This method has been used successfully to measure the vibrations and compute the corresponding tension forces in all 84 stay cables of the Sunshine Skyway Bridge. Calculation of cable forces are based on the vibration frequency for the bridge.

13.3.2.5 Radiography

This method is based on either X-rays or gamma rays. X-rays are produced by X-ray tubes (Special high-voltage machines), and gamma rays are produced from radioactive isotopes. Radiography

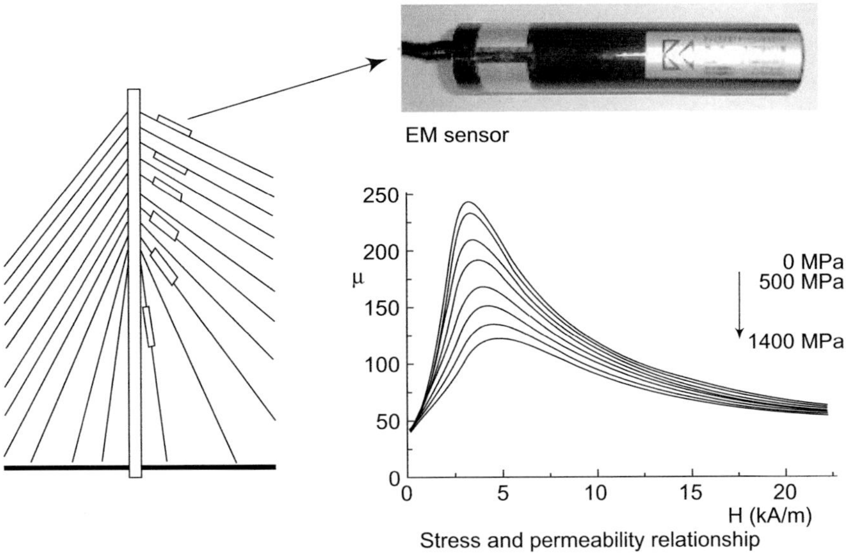

Fig. 13.16 Application of EM sensors on monitoring true stress in bridge cables

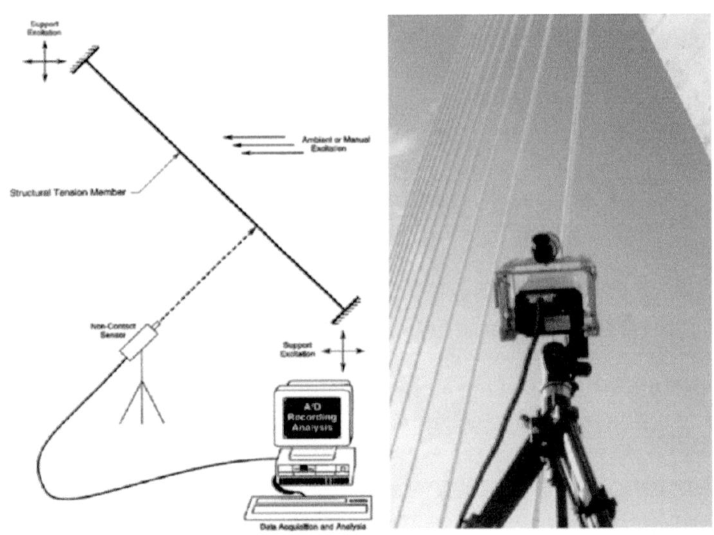

Fig. 13.17 Use of laser doppler vibrometer for stay cable vibration and force measurements

generally provides two-dimensional tomography for cross-sectional images of the three-dimensional object. (See Figure 13.18).

Nondestructive test methods based on radiography have been used to identify damage for the anchorage sockets of the Sacramento River Bridge. It was also used for a mock-up of a stay cable component for the C&D Canal Bridge in Delaware. This mock-up included wire and strand breaks and grout voids. The mock-up was tested by radiographic testing.

In general, application of radiography in bridge cable inspection is also limited due to the possibility of radioactive hazards to working personnel during inspection particularly at a high elevation of cable stay. Interpretation of the image would be difficult for grouted cables. Also, this method would only be able to detect gross section loss of the stay and is not precise enough to discern the onset or early stages of corrosion. Finally, the process may be considered lengthy, costly, and impractical.

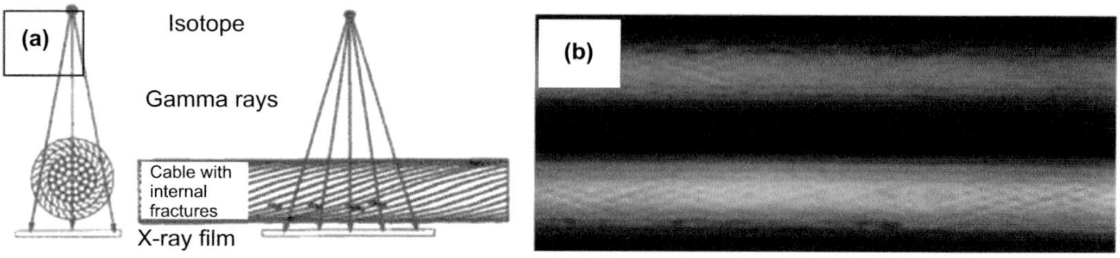

Fig. 13.18 Radiography inspection for bridge cable

13.3.2.6 Ultrasonic Testing

Ultrasonic testing (UT) measurement is a widely used nondestructive technique for tracking corrosion damage. High frequency (MHz) acoustic waves controlled by electric pulses are produced by a piezoelectric transducer and are emitted perpendicular to the object being tested. The transducer receives the waves after they are reflected back by the object's geometric inner and outer surfaces. The system determines the object thickness by timing the arrival of reflected echoes from both inner and outer surfaces. UT corrosion sensors can be integrated with in-line inspection devices and come in both fixed and portable forms. However, the geometry resolution is associated with ultrasonic frequencies and is frequently insufficiently sensitive to minute details like thin deposits or pitting corrosion.

Ultrasonic testing has proven effective in identifying concealed defects like wire breaks and grout voids in anchorage zones of several cable-stayed bridges such as the Cochrane Bridge in Mobile, Alabama., and the Talmadge Bridge in Savannah, Georgia. (Mehrabi, 2016).

Within five feet of the strand ends, UT can identify voids and breaks in the wire. However, because of the significant attenuation of pulses, the effectiveness of this method decreases with higher densities of filler material in anchorage sockets and thicker locking plates. Furthermore, it is strongly advised that the method and apparatus be calibrated on a cable anchor mockup with similar characteristics to be tested in order to ensure the accuracy of the results and interpretations.

13.3.2.7 Magnetic Flux Leakage (MFL)

The magnetic flux leakage (MFL) method is used to detect anomalies in steel elements. The sensing principle is based on the magnetic properties of steels. When the ferromagnetic material is magnetized close to saturation under the applied magnetic field, the magnetic flux lines will mostly pass through the inside of the material when there are no defects, whereas the defect or corrosion sites will result in bending and leakage of magnetic flux lines (Figure 13.19). The magnetic field is usually generated by an electromagnet, and a Hall-effect sensor is used to detect the magnetic flux leakage. A Hall effect sensor (or simply Hall sensor) is a type of sensor which detects the presence and magnitude of a magnetic field using the Hall effect. The output voltage of a Hall sensor is directly proportional to the strength of the field.

For application to cable stays, the cable under inspection is swamped by a homogeneous magnetic induction field, with the main axis of the field ideally parallel to the cable axis. The magnetic field is disturbed by the presence of a flaw in the cable such as loss of cross section or wire breakage. The field disturbance, the so-called MFL, is spread to the surface of the cable. The intensity of the MFL is related to parameters such as size and position of the flaw.

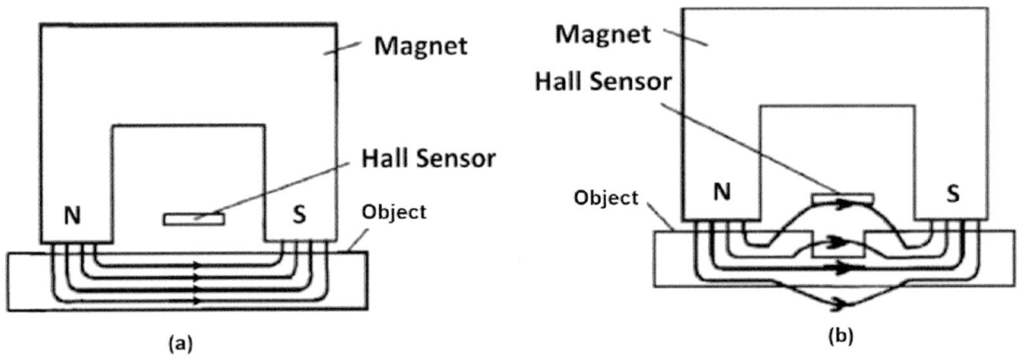

Fig. 13.19 Magnetic flux leakage method: (a) No defects; and (b) with defects

The testing unit must be mounted on the cable to investigate a specific location. To check the entire cable length the MFL unit must travel along the cable. In the past few years MFL found a wide application in cable inspection robotics. This topic will be discussed in Chapter 14.

13.3.2.8 Digital Video Cameras

Current uses for video cameras are for monitoring traffic flow in traffic centers, and for monitoring security of sensitive areas on bridges. Nevertheless, a digital camera can easily be used to make a video record of a range of motions and interactions of objects. The video record allows measurements of displacement and time, and hence calculation of velocities and other variables. Digital video cameras have multiple capabilities for use on cable-stayed bridges for monitoring structural health. Nowadays color, high resolution cameras are available with pan/tilt/zoom features as standard requirements. This enabled advancing Digital Image Correlation (DIC), which has been employed in SHM.

DIC is a vision-based technique that utilizes image detection and mapping technology to precisely measure variations of images in 2D and 3D. DIC is commonly used in many fields of science and engineering. It is frequently employed to analyze full-field displacement and strain, crack propagation, identifying material deflections, damage emergence, and progression in structural systems. DIC can be employed for routine or long-term surveillance with photographs of the targeted system taken at varying times. Images from the different periods can be examined optically with software, and the deformations can be measured from the data. The basic concept of Digital Image Correlation (DIC) compares the captured images at varying levels of deformation of a targeted system and analyzes them using correlation based matching algorithms. This technique can be used to calculate material deformation by monitoring a neighborhood of gray-level pixels and creating full-field deformations in 2D and 3D strain charts and vector fields. Photographs are obtainable across a broad spectrum of sources ranging from high-speed video recorders and portable commercial cameras to the traditional Charge-Coupled Device (CCD). Any structural difference can be accurately contrasted with the captured photographs, and thus, unforeseen changes that could induce irregularities would be effortlessly detected.

Video cameras have also been integrated with Weigh-in-Motion sensors to take short-term videos, to match heavy axle loads to corresponding strain gauge data. Digital cameras can be accessed and controlled manually via secure internet websites.

The design of a comprehensive structural health monitoring system can incorporate digital video cameras to document bridge movements over time or under extreme seismic and weather-related events.

13.3.2.9 Automated Motorized Total Stations

AMTS (Automated Motorized Total Stations) transmit high-precision total station measurements from project sites to the internet. They provide automated optical monitoring of displacements, deformations, and settlements. The system requires survey prisms, total stations, and controllers that schedule and direct AMTS measurements and then transmit the recorded data to the internet. The survey prisms are attached to predefined points on the structure, and the total station is used to record angles and distances to each prism. The measurements are offloaded to software that applies statistically weighted adjustments and formats data for a website database.

13.4 WEIGH-IN-MOTION METHODS

Weigh-in-motion (WIM) is a technique for monitoring various features of vehicles on roadways through an array of sensors. Data collected from WIM systems include the following: Wheel Load; Axle Load; Axle-Group Load; Gross-Vehicle Weight; Speed; Center-to-Center Spacing Between Axles; Vehicle Class (via axle arrangement); Site Identification Code; Lane and Direction of Travel; Date and Time of Passage; Sequential Vehicle Record Number; Wheelbase (frontmost to rearmost axle); Equivalent Single-Axle Loads (ESALs); and Violation Code.

ASTM E2328-09 Standard categorizes WIM systems into four distinct types, depending on the application and functionality. Type I must be designed for installation in one or more lanes at a traffic data-collection site and shall be capable of accommodating highway vehicles moving at speeds from 10 to 80 mph. The system must produce all data items listed above for each vehicle. Type II of the WIM system is identical to Type I except that the wheel load is not included in the produced items. Type III systems are identical for vehicles suspected of weight limit or load limit violations and have stricter functional performance requirements than Type I and Type II systems. Vehicle speed range is 10 to 80 mph. Type IV systems are not approved for use in the United States but intended for use at weight enforcement stations. Vehicle speed range is 2 to 10 mph.

Permanent WIM systems generally comprise the following components:

- WIM sensors designed to be installed once (surface or in-depth) or on/under a bridge deck to detect, weigh, and classify vehicles.
- Electronics to process sensor outputs and provide vehicle records.
- Support devices and infrastructure such as communication devices to transmit the collected data to a remote server, conduits, cabinets, poles, and junction boxes.
- Software installed in the WIM electronics to process sensor measurements and analyze data.

13.4.1 Sensors used in WIM

There are several commonly used WIM sensing devices currently in use. It is recognized that a higher level of data quality requires more precise WIM sensors and systems. In general, the required level of data quality and the available funding often dictate the type of sensor adequate for a project. The most frequently used sensors include bending plate, load cell, quartz piezoelectric, polymer piezoelectric, and strain gauge strip sensors.

The bending Plate sensor utilizes strain gauges bonded to the underside of the plate to collect loading data. The bending plate WIM sensor is usually 6 feet long, 2 ft wide, and 1 inch thick. The system measures the strain on the plate as axles pass over the bending plates and calculates the load required to induce that level of strain. This sensor is very accurate and durable. It also demonstrates very little speed dependency, especially in smooth pavements, and almost no temperature dependency. It is installed on the surface of pavement as illustrated in Figure 13.20 and recommended for Portland cement concrete (PCC) pavements.

Fig. 13.20 Bending plate installation (FHWA, 2018)

A Load Cell consists of two 6 ft × 6 ft platforms placed adjacently to cover the 12-ft monitored lane. A single hydraulic load cell is installed at the center of each platform to measure the tire-load-induced forces that are then transformed into tire loads. The installation of a load cell scale requires the use of a concrete vault. The size of the vault can be as large as 12 feet long, 5 feet wide, and nearly 3 feet deep.

The Polymer Piezoelectric Sensor can be embedded in the pavement. It produces a charge that is equivalent to the deformation induced by the tire loads on the pavement's surface. Each sensor is typically installed for half-lane coverage, with four sensors installed in each lane, 2-inch-deep slots, with at least 1 inch of grout cover. A chair is used to hold the sensor in place during installation (Figure 13.21). Each set of sensors in each wheel path is spaced 16 feet apart. A properly installed and calibrated piezoelectric WIM system yields GVW that are within 15% of the actual vehicle weight for 95% of the measured trucks.

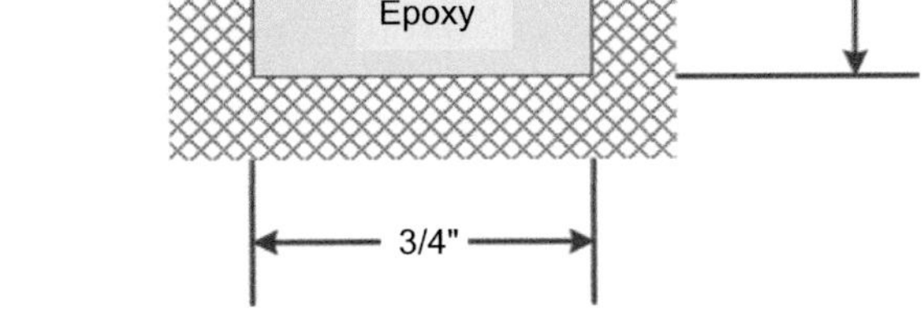

Fig. 13.21 Polymer Piezoelectric sensor installation (FHWA, 2018)

The quartz Piezoelectric Sensor is a force sensor based on quartz crystal technology. A wheel traveling on the sensor applies vertical forces which are distributed through the quartz crystals, which produce an electrical charge proportional to the applied forces. It is about 5 ft to 6.5 ft in length and can be combined in varying lengths to provide half-lane or full-lane width coverage. It can be installed in Asphalt concrete (AC), over 4 inches thick (Figure 13.22), or PCC, but the installation is more durable in PCC pavements.

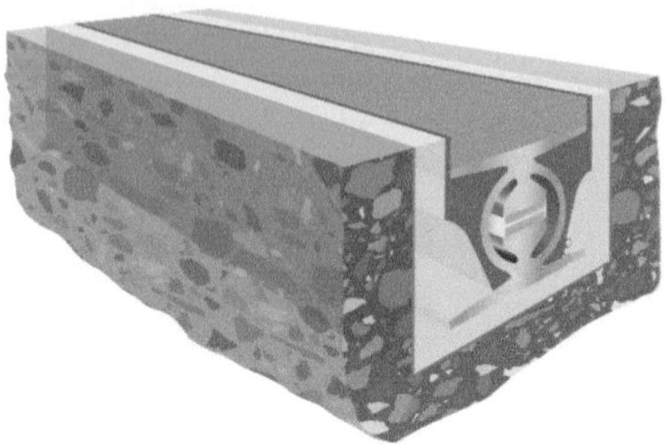

Fig. 13.22 Quartz Piezoelectric sensor installation (FHWA, 2018)

Strain Gauge Strip sensors come in 59-, 69-, or 79-inches lengths and are installed in sets of 1 to 4 pairs (2 to 8 strip sensors) that can be combined to cover different road widths, with one pair covering the width of a single road lane. Each strain gauge WIM strip sensor is approximately 3 inches wide, 3 inches tall. As a vehicle passes over the WIM sensor, the system measures the vertical strain placed on the sensor by the weight of the wheel. The resultant change in the electronic properties of the strain gauge load cells is then converted to the dynamic load by the WIM software and the wheel, axle, and vehicle weights are produced.

There are some factors that need to be taken into consideration in the selection of a specific sensor for a project:

1. *Cost:* Load cell sensors and bending plate sensors are in general more expensive than the other kinds of sensors. Polymer piezoelectric sensors are the cheapest. The initial cost of the bending plate sensor is slightly more than quartz piezo sensors or strain gauge strip sensors and less than load cell sensors.
2. *Accuracy:* Bending plate WIM systems using two 6-foot-long bending plates in each lane may be expected to provide GVW with an error of 10 percent or less of the actual vehicle weight. When properly installed and calibrated, load cell WIM systems should be expected to provide GVW that are within an error of 6 percent of the actual vehicle weight, while polymer piezoelectric sensors would provide the GVW within 15% error. Therefore, if the project requires high accuracy weight data, then either a load cell or bending plate sensor is a favorable solution.
3. *Calibration:* Some sensors such as polymer piezoelectric are affected by temperature fluctuations and by seasonal changes in pavement stiffness. Consequently, polymer piezo sensors must be calibrated every 6 to 12 months to maintain accuracy in weight measurements.
4. *Duration of project:* Polymer piezoelectric sensors are best suited for short-term studies of 1 to 3 years. Quartz piezoelectric and strain gauge strip sensors are considered most cost-effective for mid-term data collection (3 to 5 years). Load cell and bending plate sensors are recommended for long projects over a long period of time, say more than 7 years provided the PCC pavement requirement can be satisfied.

13.4.2 WIM Calibration

A WIM sensor produces a signal whose value is a function of the instantaneous dynamic wheel weight of a moving vehicle. WIM calibration is the process of adjusting the system so that measured items such as weight, axle spacing, overall vehicle length, and speed values reported by the WIM system match their known values. The initial calibration is achieved to confirm that the WIM system accuracy meets contract specifications after installation. Periodic routine WIM calibrations are performed to ensure that the data accuracy remains consistent with the selected performance specifications.

The following steps are included to perform initial calibration:

- Statically weigh and measure each axle and record each axle spacing and total bumper to bumper length of each of the calibration test trucks.
- Conduct enough test truck runs to establish an acceptable confidence in the accuracy of the WIM system.
- The influence of temperature changes on measurement errors beyond those observed during calibration cannot be addressed without vendor assistance in setting site-specific temperature compensation factors.
- Evaluate the influence of temperature changes on measurement with vendors who will set site-specific temperature compensation factors.
- Install the system error compensation factors in the WIM system and document them. This step is necessary to compensate for the measured system bias.
- Conduct additional test truck runs to gain higher confidence in the calibration process.
- Recalibrate, if necessary, to meet contract specifications.

13.4.3 Advantages and Disadvantages of WIM

Advantages of WIM Sensing Systems over Static Scales:

High Processing rate: When a vehicle needs to be weighed, weigh-in-motion systems don't require it to stop; instead, they just slow down while taking measurements and recording the data accurately and quickly.

Improved productivity: The process is far more efficient since there are no needless delays because cars don't have to stop to be weighed; instead, they continue to drive while their weight is verified. In crowded areas like freight terminals, where efficiency and speed are essential to avoiding lines and delays and maintaining smooth operations, WIM is the perfect solution.

Instant Identification of overloaded vehicles: Vehicle overloading is a major problem that can be expensive in a number of ways. Fuel prices may rise, and vehicles will experience premature and needless wear and tear. The roads may be harmed by overloaded cars. The WIM weighbridge is a useful tool for protecting vehicles with overloaded axles by instantly identifying them before they are allowed to drive on public roads.

Economical: It is a more economical choice because of the low installation costs. They typically require very little power and are reasonably priced to maintain, which allows for even more savings.

Safety: Vehicle accumulation at highway lanes leading to weight stations is greatly reduced when static weighing is minimized.

Disadvantages of WIM Sensing Systems over Static Scales:

Less precise: Static scales are more precise than WIM systems. Wheel load scales must maintain a 2% accuracy after certification, with an accuracy of 1% required during testing, according to the

National Bureau of Standards. For 95% of measured trucks, the best accuracy achieved with the priciest, most widely used WIM sensors is within 6% of the actual vehicle weights.

Reduced information: Conventional WIM systems are unable to retrieve truck information that is readily gathered at static weight stations, such as fuel type, state of registry, model year, loaded or unloaded status, origin, and destination.

Damage from electromagnetic transients: WIM systems are vulnerable to electromagnetic disturbances, which are primarily brought on by lightning strikes that occur close to the apparatus.

WIM systems can be used in a number of ways to enhance the design of cable-stayed bridges. Live-streaming WIM data can be utilized to always give the frequency of heavy loads and days of the week and hence can be used in estimates of fatigue life of bridge components. Because site-specific WIM data has more accurate loading and impact information, it may also have an impact on bridge load rating calculations. It may be used to provide accurate information regarding live loads on a particular bridge. Moreover, truck loads applicable to a specific bridge lane can be found using WIM data because of geometric characteristics like lane width, roadway curvature, superelevation, and the presence or absence of a median barrier. This is important for assessing loading and impact damage to specific bridge members.

13.5 EXAMPLES OF SHM IMPLEMENTATION TO CABLE-STAYED BRIDGES

Structural health monitoring (SHM) is advantageous in the sense that it provides continuous real-time data to evaluate the performance of the structure. Thus, many bridge owners are considering SHM as the supplement and alternative for improving bridge inspection and management practices. A comprehensive SHM program for a cable-stayed bridge may include measurement of environmental conditions (temperature and humidity), tilt and soil pressure sensors to monitor foundations, crack displacement sensors to track crack widths in structural elements, strain gauges to track stress cycles associated with steel fatigue; GPS to track structural motion; and accelerometers to capture the vibration and characterize bridge dynamics. The following subsections provide some examples of SHM systems that were installed on cable-stayed bridges in the US and worldwide.

13.5.1 Sunshine Skyway Bridge

The Sunshine Skyway Bridge, spanning the main entrance to Tampa Bay, is a significant transportation link in the State of Florida, with an average of over 50,000 vehicles daily traffic. The bridge consists of trestle, low-level approach spans, high-level approach spans, and 1200 ft of cable-stayed main span. The low-level approach spans were constructed with reinforced concrete twin box girders and the main span was constructed with a single-cell box girder. The pylons, constructed at a height of 435 ft above sea level, support 21 cable stays for each half of the main span. These cables are anchored along the centerline of the box sections. The bridge was opened to traffic in 1987. Refer to section 6.1.5 for a detailed description of different bridge elements. According to Rice and Davis (2016), the FDOT Surveying and Mapping Office installed five GPS receivers and established a website to enable continuous data monitoring early in the 1990s, (Figure 13.23) and the receivers were upgraded in 2011. Two Automatic Motorized Total Stations (AMTSs) were installed on the main dolphins (one at each side of the main span) for scanning targets (prisms) located on the bridge piers and beneath the deck. The AMTS were eliminated in 2016 due to high costs of maintenance and/or replacement.

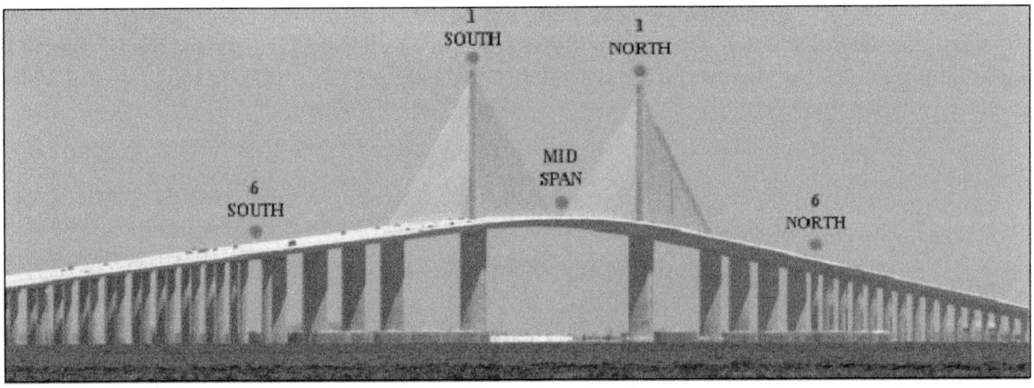

Fig. 13.23 Location of GPS receivers on Sunshine Skyway Bridge (Bridge and Davis, 2021)

In 2005, three weather stations were added, two at each tower and one at the center of the main span. Another weather station was added in 2015 at the beginning of the end of the north side span. These stations include ultrasonic anemometer, an ambient air temperature sensor, and a relative humidity sensor Figure (13.24).

Fig. 13.24 Weather station at the center of the main span of Sunshine Skyway Bridge (Bridge and Davis, 2021)

Six triaxial accelerometers, to provide cable force data, were installed in 2015 by Diagnosis Inc. contracted by FDOT District 7 Structures Maintenance Office Bridge (D7 DSMO) as shown in Figure 13.25. Four are on cables, one is at the top of the North Tower, and one is at deck level. The triaxial accelerations are oriented such that the X axis measures vibrations along the cable (longitudinal), the Y axis measures horizontal/transverse acceleration (out of plane), and the Z axis measures vibrations perpendicular to the cable and the horizontal plane.

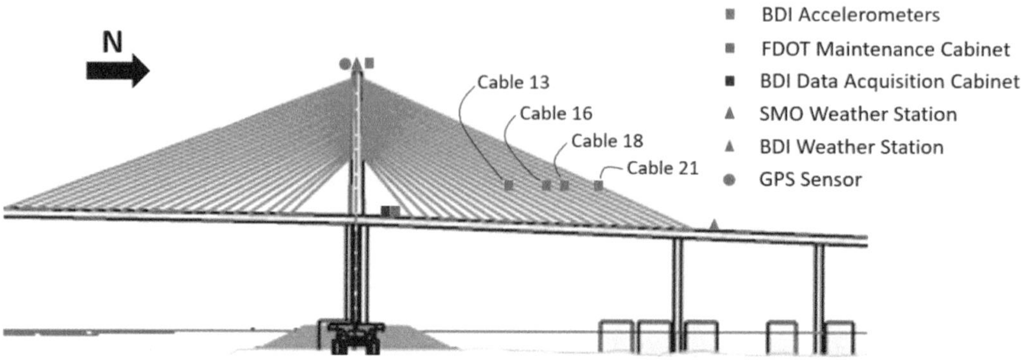

Fig. 13.25 Locations of sensors near the North Tower of Sunshine Skyway Bridge (Bridge and Davis, 2021)

D7 DSMO contracted University of Florida monitoring to lead a project for the purpose of integrating monitoring components on the Sunshine Skyway Bridge and creating a centralized system available via a web interface capable of providing alerts for anomalous bridge response data. This project included the design of a web interface, the development of algorithms to produce bridge response information, analysis of long-term data to establish bounds of typical bridge responses, and recommendations for a bridge alert response plan (Bridge and Davis, 2021).

13.5.2 The Tatara Bridge (Japan)

The Tatara Cable-Stayed Bridge is a part of the Nishiseto Expressway, commonly known as the Shimanami Kaidō. The bridge has a center span of 2,920 ft. The bridge was opened on May 1, 1999, and carries two lanes of traffic in each direction and has additional lanes for bicycles, motor bikes, and pedestrians. Refer to Chapter 7.2.11 for a detailed description of the bridge. The SHM system of the bridge is limited to verification of its seismic design. As shown in Figure 13.26. accelerometers were deployed on top of the pylon, foundation level and along the orthotropic deck to monitor all modes of vibration that were strongly excited by the 2001 Geiyo (near Hiroshima) Earthquake (moment magnitude Mw = 6.7). The maximum ground acceleration at the bridge site was 0.15g. Observation of the seismic responses revealed that the actual seismic load in terms of the response spectra calculated from the recorded ground motion was below the design specification (Fujino et al. 2019).

13.5.3 The US Grant Bridge

The US Grant Bridge is a two lane, three span cable-stayed structure spanning the Ohio River and connecting Portsmouth, Ohio and Fullerton, Kentucky. It was built to replace its predecessor, a suspension bridge that was built in 1927 but was closed and demolished. The new cable-stayed bridge has a total suspended span length of 1685 feet. The bridge employs steel support girders, floor beams as well as a post-tensioned concrete deck; it has two single shaft towers, which rise over three hundred feet above the riverbed, with two planes of stays anchored to each single tower head, and

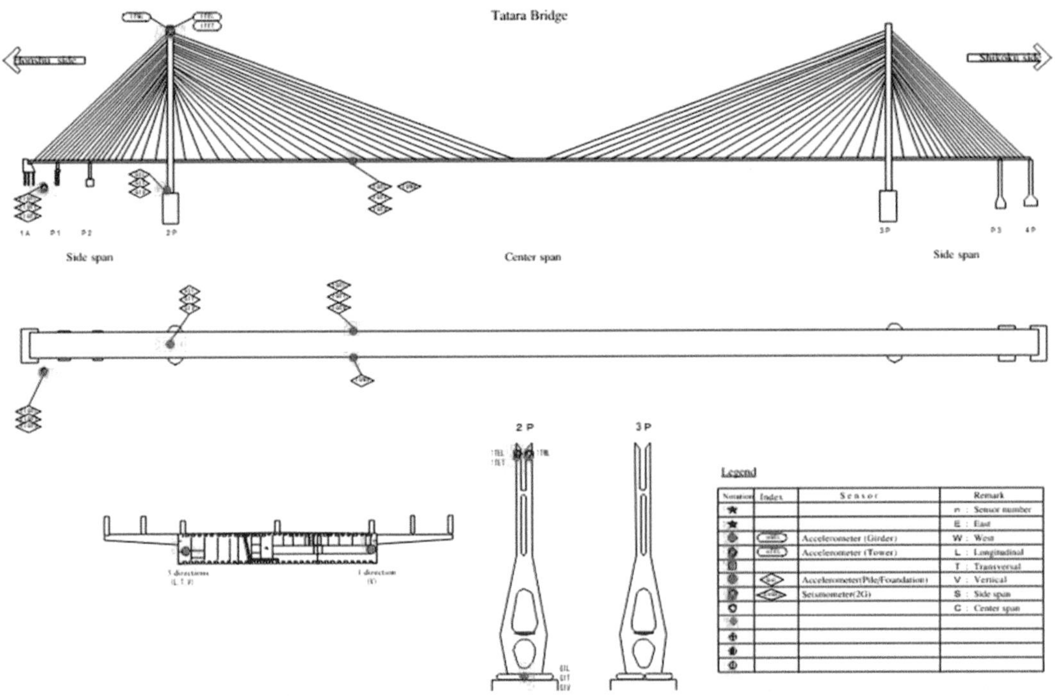

Fig. 13.26 Sensor configuration of Tatara Bridge

the superstructures are fixed at the towers. After 5 years of construction, the bridge was opened to traffic in October 2006. Refer to Chapter 6.1.12 for more details about the bridge.

Ohio Department of Transportation (OHDOT) sought to identify an appropriate instrumentation and field-testing program to support management of the US Grant Bridge and augment the traditional visual inspection program to provide objective, quantitative data in assessing the status of the structure. Therefore, a team from the University of Cincinnati (UC) (Helmicki and Hunt, 2013) conducted the project which came in several phases. The first phase was concerned with monitoring the structure during construction. Key parameters of interest included stress levels at pylons; deck displacement in response to thermal, wind, and dead loads; and cable acceleration and displacement during erection to ensure proper tensioning. Next, upon completion of construction, well-defined loads were applied to the bridge, and the UC team ran test sequences to confirm the structural behavior, proper operation of the instrumentation, and validatation of pre-built computer models. Finally, the team in collaboration with ODOT invested the instrumentation installed during erection for a long term SHM plan and some additional measurements were taken into consideration. The key variables that were considered for the SHM program included:

1. Weather conditions (i.e., temperature, precipitation wind direction and velocity).
2. Thermal cross-sections of pylons, exterior girders, and concrete-decking sections, over time, to resolve the environmental responses and internal strains of the structural components and separate them from traffic responses.
3. Stress levels (longitudinal and tangential) at selected pylon cable anchorages, to continually monitor settlement and live load effects (wind and traffic).
4. Acceleration of selected deck sections (particularly at abutments, pylons, and mid-spans) in response to thermal, wind, and traffic loads.
5. Longitudinal stress of selected exterior girders and deck sections, particularly at abutments, pylons, and mid-span (high moment regions), in response to thermal, wind, and traffic load.

These were to be used to monitor movement of the neutral axis of the deck section to obtain information on the integrity of the prestressing tendons in the deck sections.

6. Cable acceleration in response to thermal, wind and traffic loads to characterize vibration levels and mechanisms, as well as to calculate loads and stresses.

The US Grant Bridge monitoring system is an automated health monitoring system which is composed of multiple modules: sensor network, data acquisition system, data processing, data management system, warning system and a web application. A layout of the sensors is displayed in Figure (13.27)

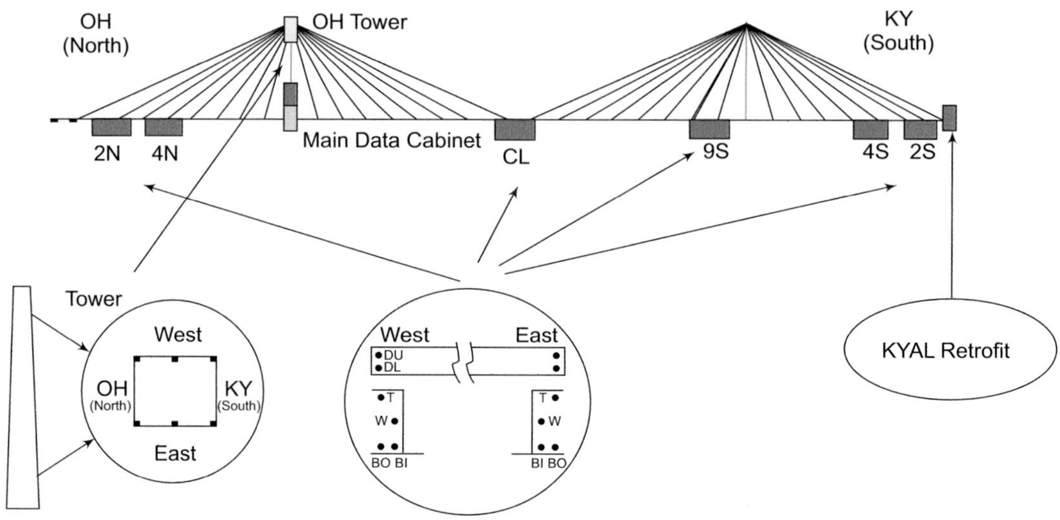

Fig. 13.27 SHM system for the US Grant Bridge (Helmicki and Hunt, 2013)

The sensor network consists of low-speed vibrating-wire strain gages, resistive (high-speed) strain gages, strand meters, inclinometers, and a wind sensor. The low-speed strain gauges are also capable of measuring temperature and are permanently (i.e., from construction to service life of the bridge) installed and monitored whereas the high-speed gauges are used only in special events such as truckload testing. The locations of the sensors are determined jointly by finite element analysis and inspection. The instrumented members include edge girders of the OH tower, and KY abutments. The data acquisition system included multiplexers, data loggers (sensor to receive the information and a computer chip to store it), and power supply.

13.5.4 The Stonecutters Bridge (Hong Kong)

Stonecutters Bridge is a high-level cable-stayed bridge spanning the Rambler Channel in Hong Kong, connecting Nam Wan Kok, Tsing Yi to Stonecutters Island. The bridge's main span is 3,340 ft. Refer to Chapter 7.1.4 for a more detailed description of different bridge components. The bridge's SHM was developed and put into operation to monitor the following parameters in real-time: operational loads (such as highway and rail traffic); environmental factors (such as wind, temperature, seismic activity, humidity, and corrosion status); bridge characteristics (such as influence coefficients and modal parameters); and bridge responses (such as geometrical profile, cable force, displacement/deflection, strain/stress histories, and cumulative fatigue damage). The SHM program's main goals are to: (i) gain a better understanding of the structural behavior of bridges in use; (ii) develop and validate bridge evaluation techniques based on measurement results; (iii) assess structural integrity immediately following uncommon occurrences like large earthquakes and collisions involving vehicles and vessels; (iv) provide data and analytical tools for the purpose of organizing, scheduling, assessing, and designing efficient long-term bridge inspection and maintenance strategies; and (v) reduce the amount of time that lane closures are necessary for bridge inspection and maintenance activities.

The SHM for the Stonecutters Bridge comprises six components: Sensor System (SS); Data Acquisition and Transmission System (DATS); Data Processing and Control System (DPCS); Data Management System (DMS); Structural Health Evaluation System (SHES); and Inspection and Maintenance System (IMS). Figure 13.28 illustrates the SHMS deployed on the bridge.

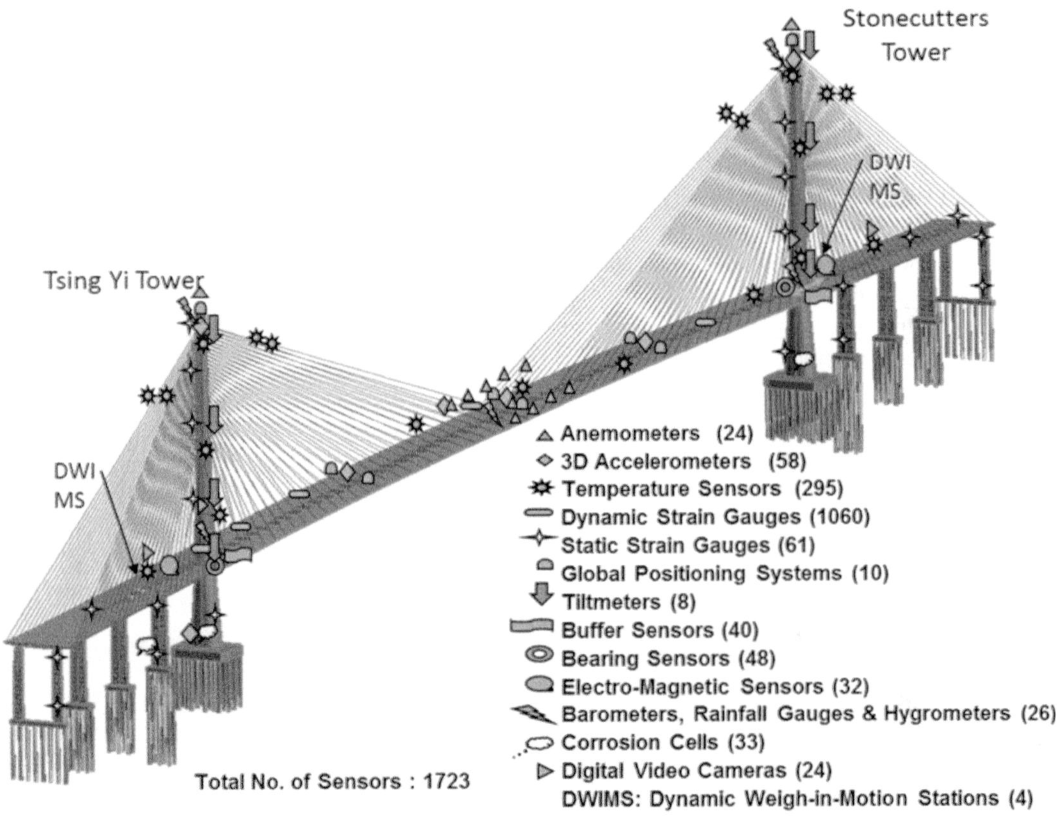

Fig. 13.28 Sensor configuration of Stonecutters Bridge (Dascotte, 2011)

For the purpose of gathering, processing, and transmitting signals, the SS and DATS sensors, on-structure data acquisition units (DAUs), and cabling networks are deployed. The DPCS is a computer system for the execution of system control, system operation display, and processing and analysis of data. The monitoring data and analysis findings are stored and retrieved using the DMS, a data warehouse system. For the purposes of performing finite element analysis, sensitivity analysis and model updating, bridge feature and response analysis, diagnostic and prognostic analysis, and visualization of analyzed results, the SHES is a high-performance computer system outfitted with the necessary software and sophisticated analysis tools. It consists of both an offline structural health and safety assessment system and an online structural condition evaluation system. The IMS is a laptop-computer-aided portable system for the inspection and maintenance of the SHM itself.

13.5.5 The Jindo Bridge (South Korea)

The Jindo Bridges are twin cable-stayed bridges, which connect Jindo Island and the southwestern tip of the Korean Peninsula near the town of Haenam. The 1st Jindo Bridge, constructed in 1984, is the first cable-stayed bridge in Korean bridge history. The 2nd Jindo Bridge, which is the subject of the SHM plan described herein, was constructed to accommodate increasing traffic loads in 2006. The bridge is a three-span steel-box girder cable-stayed bridge composed of a 1129 ft main span and 230 m of side spans. The deck system is a streamlined steel-box girder supported by the sixty stay cables connecting the two A-shaped steel pylons on concrete piers.

The SHM system was deployed on the bridge in 2009. The system employed the wireless smart sensor network system (WSSN) for vibration and environmental monitoring. It was installed on the bridge (Jang et al. 2010, and Rice et al. 2010). A total of 113 wireless sensor nodes resulted in 330 channels of acceleration measurements, nine channels of wind measurements, and 113 channels each of temperature, light and humidity measurements. The 330 acceleration channels measured three axes of acceleration on each of the stay cables, along both sides of the underside of the main steel box girder, and at three locations on each of the pylons, as shown in Figure 13.29. Each wireless sensor node was solar powered with its own microprocessor and radio and supported multiple data channels. The microprocessor of each sensor carries out much of the data processing at the sensor locations without having to communicate large amounts of raw data wirelessly. The processed results are then sent to one of two base stations located on the bridge (one on each tower). The in-network data processing included cable tension estimation and dynamic analysis of the bridge (natural frequencies, mode shapes and damping ratios).

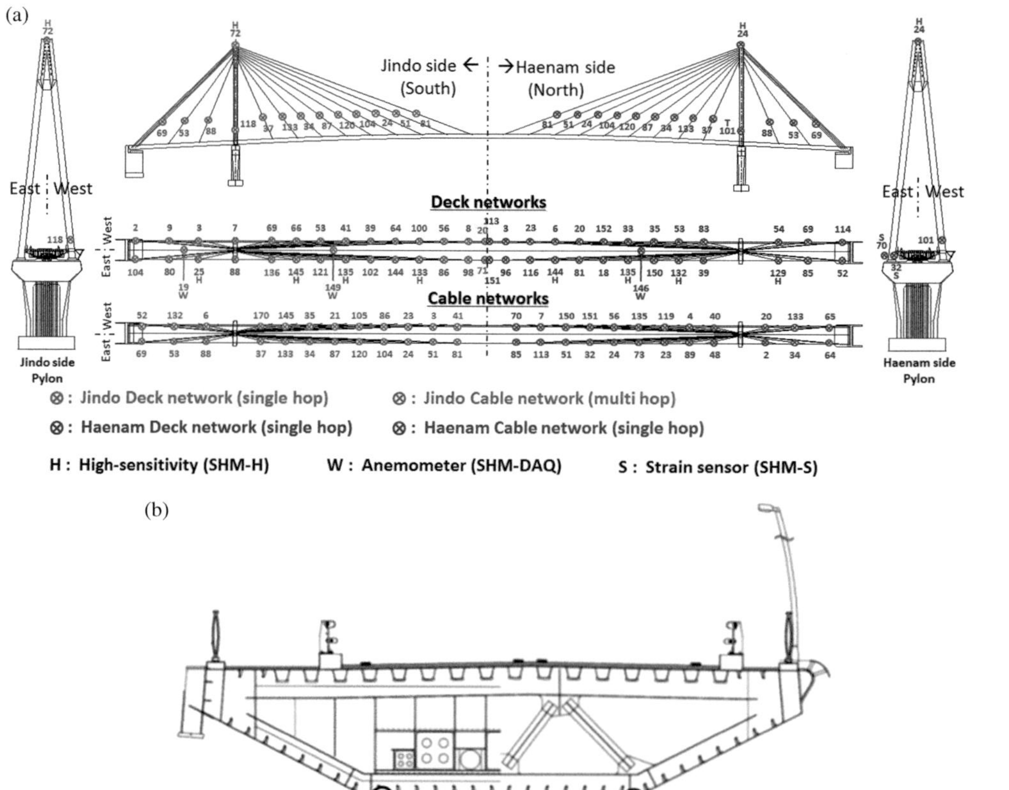

Fig. 13.29 SHM system for the 2nd Jindo Bridge: (a) Layout of the wireless sensor nodes; and (b) sensor placement on either side of the box girder (Jang et al., 2010)

Once the pre-processed data was sent to the base stations, the data was stored locally on PCs which can be accessed via a cable modem. The SHM network operated autonomously, collecting data on a set schedule, unless triggered by the exceedance of predetermined threshold values. Both vibration and wind level thresholds were set so that data could be collected under higher than usual loading conditions. The data was also used to calibrate an FE model of the bridge (Jang et al., 2010).

13.5.6 The Charles W. Cullen Bridge

The 1,749-foot Charles W. Cullen Bridge at the Indian River Inlet, also known as the Indian River Inlet Bridge (IRIB), is a cable-stayed structure with two 397-foot back spans and a main span that is 948 feet long. A combination of precast and cast-in-place reinforced concrete was used in the bridge's design. The bridge has two traffic lanes, a shoulder in each direction, and is 105 feet wide. In May 2012, the bridge was made available for traffic.

The Delaware Department of Transportation (DelDOT) views the bridge as a signature structure in the state, so the owner was very interested in obtaining accurate, quantitative data about the bridge's condition and how it responded to external factors and live loads. As a result, installing an SHM system for the bridge was decided (Shenton et al. 2017). A decision was made to track the bridge's performance over time and gather information. This was done for making well -informed decisions about bridge upkeep and repairs. In addition to the usual visual inspection data, such quantitative information will guarantee that the bridge can be maintained effectively for the duration of its service life. The SHM system was installed throughout the whole structure and was intended to deliver ongoing information about the primary performance metrics of the bridge. Strain, acceleration, tilt, displacement, chlorides, temperature, wind direction, and speed are among the parameters that are observed. The fiber-optic based system was chosen due to its noncorrosive components and immunity to electrical noise.

According to Shenton et al. (2017), the three primary parts of the SHM system are sensors, conduit and cabling, and a data acquisition and control system. Seventy locations on the bridge are used to measure strains, including the deck, pylons, and edge girders. Micron Optics OS3600 strain sensors are used for all strain measurements. An integrated temperature sensor at the sensor location provides relative temperature changes.

Accelerations are measured with a combination of uniaxial and biaxial Micron Optics model OS7100 at 27 distinct locations on the bridge. There are forty-four distinct acceleration measurements in all. Accelerations are measured on the top of pylon 5's east leg as well as the tops of pylon 6's west and east legs. A biaxial sensor to measure motion in both the longitudinal and transverse directions is situated in each of these locations. On the west side of pylon 6, at the top of the east leg, is another sensor that measures motion in a longitudinal direction. When combined with the other accelerometer on top of the same leg, this single sensor can provide some insight into the torsional movement of the pylon. There are therefore seven different ways to measure pylon acceleration. Twelve distinct locations are used to measure accelerations on the deck: nine on the east side and three on the west. The acceleration is measured in the vertical direction at each of these locations and in the transverse direction at three additional locations. Therefore, a total of 15 distinct deck acceleration measurements are obtained from the combination of 9 uniaxial accelerometers and 3 biaxial accelerometers. The quarter points, midpoint, and back spans of the main span and the back spans are where the sensors are located. This arrangement will guarantee that the bridge's dynamic properties—natural frequencies, mode shapes, and damping ratios—are recorded at any point in its service life, allowing for the identification of any damage worldwide. Eleven distinct stay cables—nine on the east side and two on the west—are also used to measure accelerations. A biaxial accelerometer measures acceleration in both the vertical and transverse directions at each of these locations. The sensors are all situated about 35 feet above the deck. These accelerometers are important because they can be used to calculate the forces acting on the stay cables and the installed stay dampers' damping ratio. Monitoring these two factors will guarantee that the stays operate as intended. Using tiltmeters of the FBG Tech model FBG-TI-310, the east edge girder's tilt, or inclination, is measured at nine different points along the girder. These places are midway between each of these locations, the pylons, the ends of the bridge, and the middle of the span. The tilt meters are on the east side of the bridge. The tilts are measured with respect to the Y axis, or the girder's "pitch".

In order to measure only static movements, Micro Optics SM130 Interrogator sensors were employed. These sensors have a wavelength stability of 2 pm, yielding a resolution of 0.0044 degrees and a range of +/– 3 degrees. In order to estimate the global deflection of the bridge at strategic points, edge girder tilts are measured under controlled vehicle loads. This information is then integrated to provide the bridge's stiffness distribution, which can be monitored over time. The tilts can also be used to track the effects of long-term shrinkage and creep.

Displacement transducers from Cleveland Electric Labs are used to measure the displacements at the bearing of pylon 5's east leg as well as at each of the two expansion joints. The resolution of the transducers is 0.1 mm. The transducers, which measure longitudinal movement in the direction of traffic, are always located on the east side of the bridge. By correlating these measurements with temperature and time, it will be possible to verify that the expansion joints and bearings are operating and moving as they should.

Measurements of wind direction and speed are taken at the deck level on the east side, just south of the mid-span, and at the top of the west leg of pylon 6. This is accomplished by using R.M. Young model 85106 ultrasonic anemometers, which have a resolution speed of 0.33 ft/s at 1 degree. Since anemometers are analog devices, they are integrated into the optical system by means of an analog-to-optic converter, which is placed at the sensor and converts the analog output to an optical signal.

Ten distinct locations in the deck are used to measure the chlorides using a combination of six fiber-optic and ten conventional galvanic ladder sensors. Cosasco Systems, Inc. model 900 sensors are used at ten different locations to measure the amount of chloride ingress in the deck. Measurements are obtained at four different depths, usually 3, 4, 5, and 6.25 inches above the bottom face of the deck, using the Cosasco multi-depth ladder type sensor. Additionally, in six distinct locations, Innovative fiber-optic chloride sensors are installed concurrently with the Cosasco sensors. The fiber optic sensors are produced by QPS Photonics and utilize a coating that changes reflective properties as it is subjected to chloride. Chloride measurements are used to track the bridge's durability by indicating the degree of chloride penetration into the deck.

In conclusion, 144 distinct sensors have been placed at various points along the bridge to track a range of structural response scenarios under varied environmental loads and live load circumstances: Seventy strain sensors, situated in the deck, pylons, and edge girders; nine tiltmeters positioned along the east edge girder; forty-four accelerometers fixed to the deck, pylons, and stay cables; Two anemometers that measure wind speed and direction, one at deck level and one at the top of one pylon; three displacement sensors, one at each of the bridge expansion bearings (the two abutments and the south-east pylon); 16 chloride sensors spread across 10 locations on the deck.

With the exception of the anemometer and ten chloride sensors, which are traditional analog devices, every sensor is optical in nature. Figure 13.30 shows the sensor layout. Two 4-channel Micro Optics SM130 Interrogators handle data acquisition. With the exception of accelerometers and a few strain sensors, the first one processes all of the sensors and operates at a maximum scan rate of 500 Hz, typically operating at 125 Hz. The second one can scan at up to 1 kHz, but it typically operates at 250 Hz. It manages all of the remaining strain sensors and accelerometers. Each interrogator has a 16-channel multiplexer attached to it, making the total effective number of main fibers in the system 32. The software programs "Enlight" and "Intellioptics" are used for data management. The former manages the instruments, while the latter is a graphical user interface program that handles the database and offers general control over the SHM system. In addition, the

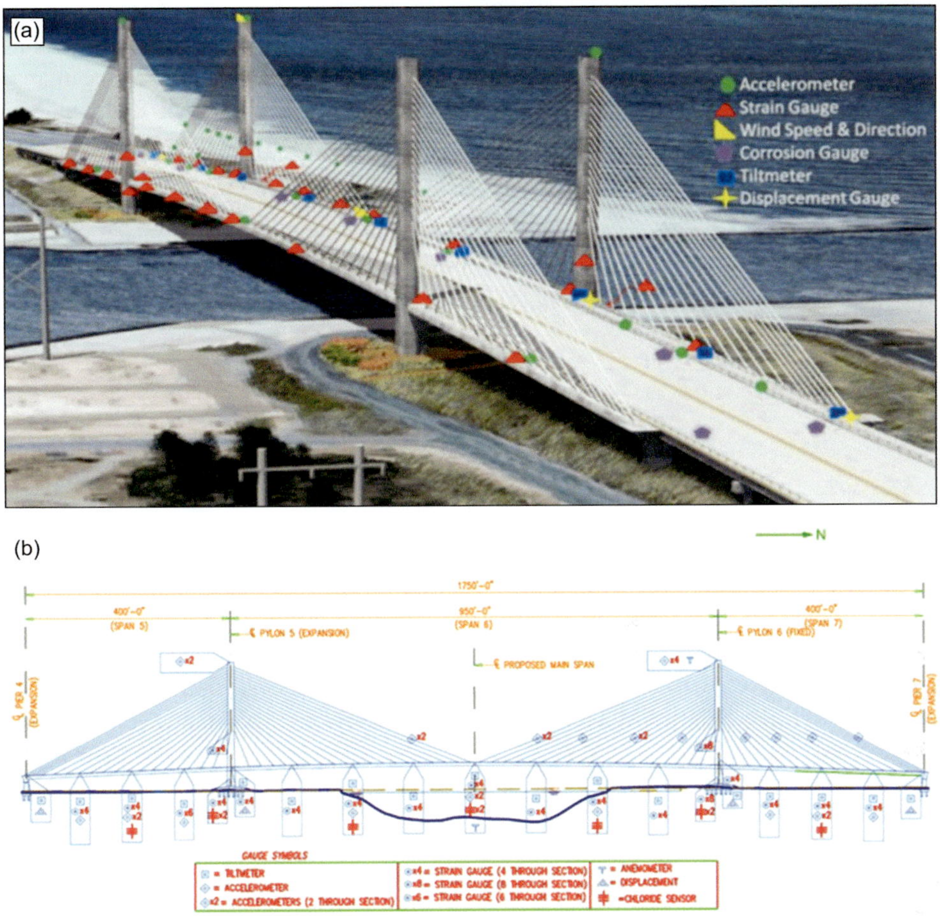

Fig. 13.30 SHM system for the Charles W. Cullen Bridge: (a) 3D Illustration; and (b) schematic of overall layout (Shenton et al. 2017)

user can view real-time sensor data that is broken down by zones, generate reports from archived data using custom time windows, set thresholds for sensors that, when exceeded, can send important users text or email alerts, and have the system automatically generate weekly reports with a plot of the data and important sensor statistics.

13.5.7 The ZhiJiang Bridge (China)

Built in 2013, Zhijiang Bridge spans the Hangzhou Qiantang River, the largest river in Zhejiang Province, and is a cable-stayed bridge. Zhijiang Bridge is 1568 feet long, with a main span that is 807 feet long. The bridge is a steel box girder cable-stayed dual-pylon dual-cable plane bridge with an arched twin-tower space and a twin-cable plane structure.

The SHM plan aims to address the following: provide for the needs of operational management; raise the bar for pre-alarm security; boost maintenance management effectiveness; and enable efficient, effective, and scientific operational management. In order to ensure structural safety and cost-effective operational decision-making to provide full technical support, the primary responsibilities of the SHM system are to ascertain the environmental load, structural response, partial damage, and other information. Additionally, based on a thorough assessment of this information, the system obtains security state information for the traffic and structure.

The information acquisition system (IAS), data management system (DMS), evaluation and decision-making system (EDS), and application service system (ASS) are the four functional subsystems that make up the bridge's SHM. Data monitoring and maintenance management subsystems are part of the lower-level IAS system. The middle-level systems DMS and EDS each have a subsystem for data management; the latter also has subsystems for maintenance management, security pre-alarming, and structural state evaluation. The upper-level system that has a user interface subsystem is called ASS. The overall workflow of Zhijiang Bridge's integrated system is depicted in Figure 13.31 (Chen et al. 2014). Through direct inputs and wired fiber communication, the IAS and DMS exchange data. All data is filed, stored, and managed by the DMS, which also gives the ASS access to data queries and the essential data support for the EDS.

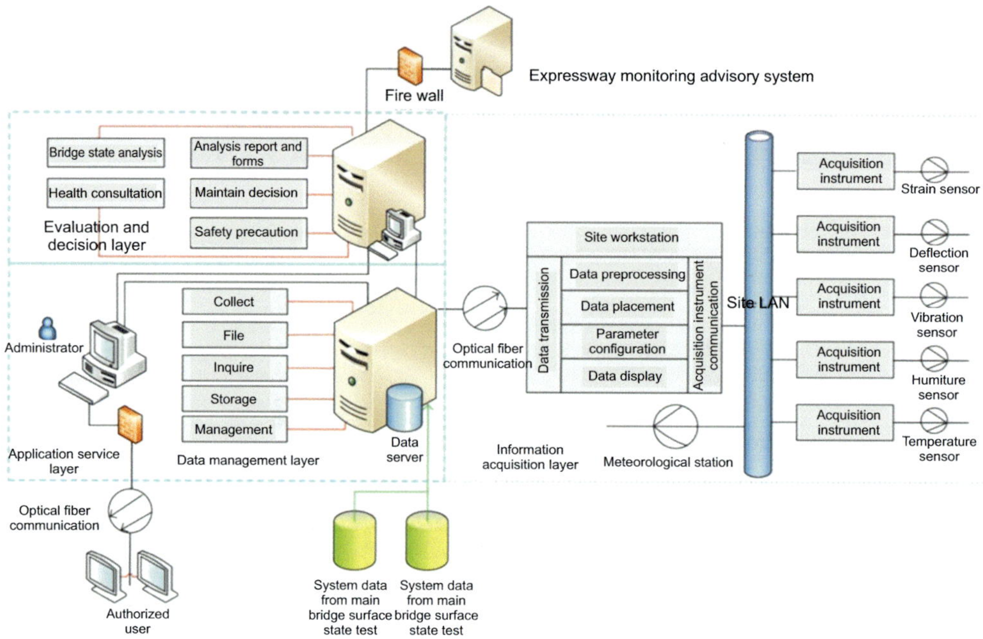

Fig. 13.31 Overall workflow of SHM system at Zhijiang Bridge (Chen et al., 2014)

In order to support decision-making and Pre-alarming maintenance management—where the security Pre-alarming subsystem feeds back to the ASS—EDS offers analytical results. The Expressway Monitoring Advisory System (EMAS) of Hangzhou City is interfaced with the DMS and EDS. The monitoring data and decisions produced by the system are shared between the Zhijiang Bridge monitoring center and the Hangzhou City Traffic Emergency Command Center by means of simultaneous transfers of monitoring information, evaluation results, maintenance decisions, and warning information pertaining to the bridge.

The Sensor System (SS) monitors the following aspects of the work environment: wind load, temperature, humidity, rainfall, visibility, temperature and humidity of the steel arch tower and steel box girder, and vehicle load; structural spatial deformation; bridge alignment; section stress; fatigue and welding cracks; vibration; impact force; bridge pier monitoring; earthquake response; cable force monitoring; and monitoring of the anchor force in the steel-concrete joint segment. Figure 13.32 shows the number of sensors and their arrangement.

The meteorological station on board the SS is used to track rainfall, wind direction and speed, visibility, and atmospheric temperature and humidity. Within the steel structure of the bridge, temperature and humidity sensors are also installed along with a dehumidification system. The dehumidification system regulates the temperature and relative humidity inside the steel box girder and steel arch tower. The temperature and humidity sensors track variations in these parameters. In order to track the volume of traffic, SS also has a Weigh-in-Motion system. This system features a 0–50 mph speed range, 1.5% average speed accuracy, and 3% gross vehicle weight accuracy. Additionally, SS has GPS sensors to track the spatial deformation at various tower and superstructure locations. There are two GPS sensors at the top of the steel towers in the east and west, and there are two GPS sensors upstream and downstream of the steel box girder in the middle main span. Strain sensors are positioned throughout the bridge, as seen in Figure 13.33, to keep an eye on any possible signs of fatigue or crack initiation.

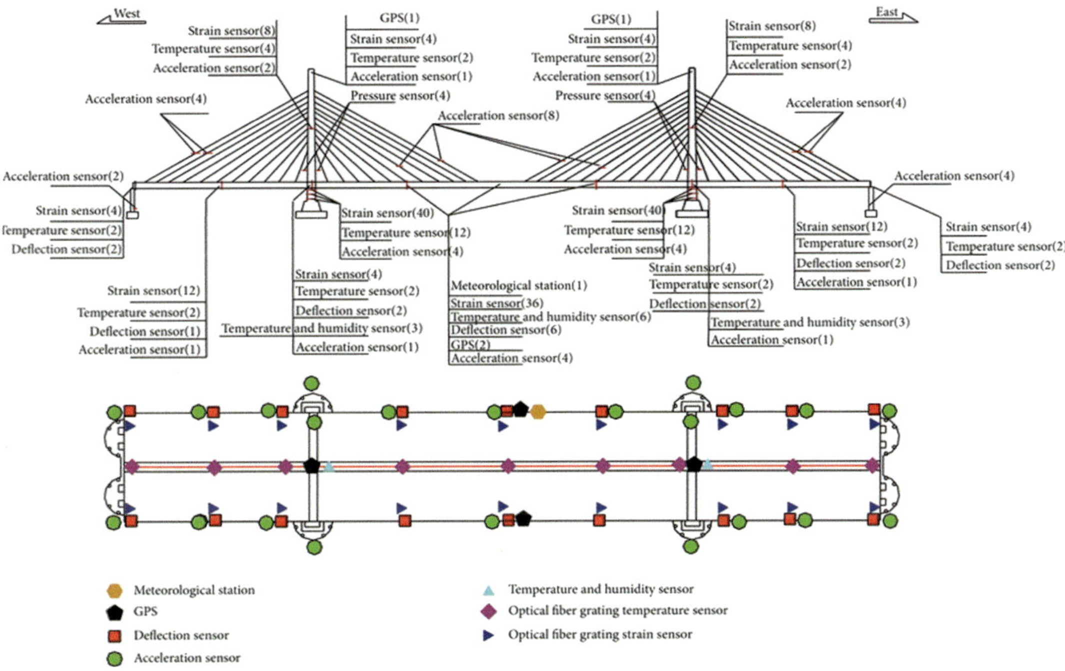

Fig. 13.32 Layout of sensors on Zhijiang Bridge (Chen et al., 2014)

Acceleration sensors are used in structural vibration monitoring to track vibrations in the steel box girder and arch tower, as well as the impact force on the bridge pier and seismic reactions. The steel box girder's vibration monitoring points are designed to guarantee the structural displacement's maximum modal matrix type. As a result, the steel box's vibration sensors are placed in the center of each span, at the main span's quarter-point, and on either side of the main span's pier top.

The tension of stayed cables is monitored using acceleration sensors, which have a measurement range of ±10 g, a frequency response range of zero to 100 Hz, a dynamic range >80 dB, and a work temperature range of −68°F to 176°F. Additionally, acceleration sensors are used to track earthquake responses, the impact force on the bridge pier, vibrations of the steel arch tower and steel box girder, and a measurement range of ±2 g, zero to 100 Hz for the frequency response, >120 dB for the dynamic range, and a working temperature range of −68°F to 176°F for the frequency response.

Preprocessing, transmission, acquisition, and temporary storage facilities are all part of the data acquisition and transmission system.. Wired data transmission is used between the sensor and the acquisition facility as well as between the acquisition facility and the instrument to increase the stability of the data transmission in the SHM system. Since there are monitoring points located all over the main bridge and the approach area next to it, using a single low-level instrument would result in a very long transmission distance from some far-end sensors to the low-level instrument, which would reduce transmission efficiency and quality. As a result, two low-level instruments are placed in the steel box girder section at the intersection of the main bridge's steel box girders and two steel arch towers. An integrated optical communication cable transports data between the monitoring center and the low-level instruments.

13.5.8 The Governor Mario M. Cuomo Bridge

The New Tappan Zee Bridge, officially named the Governor Mario M. Cuomo Bridge is a twin cable-stayed bridge over New York's Hudson River that is owned and maintained by the NY State

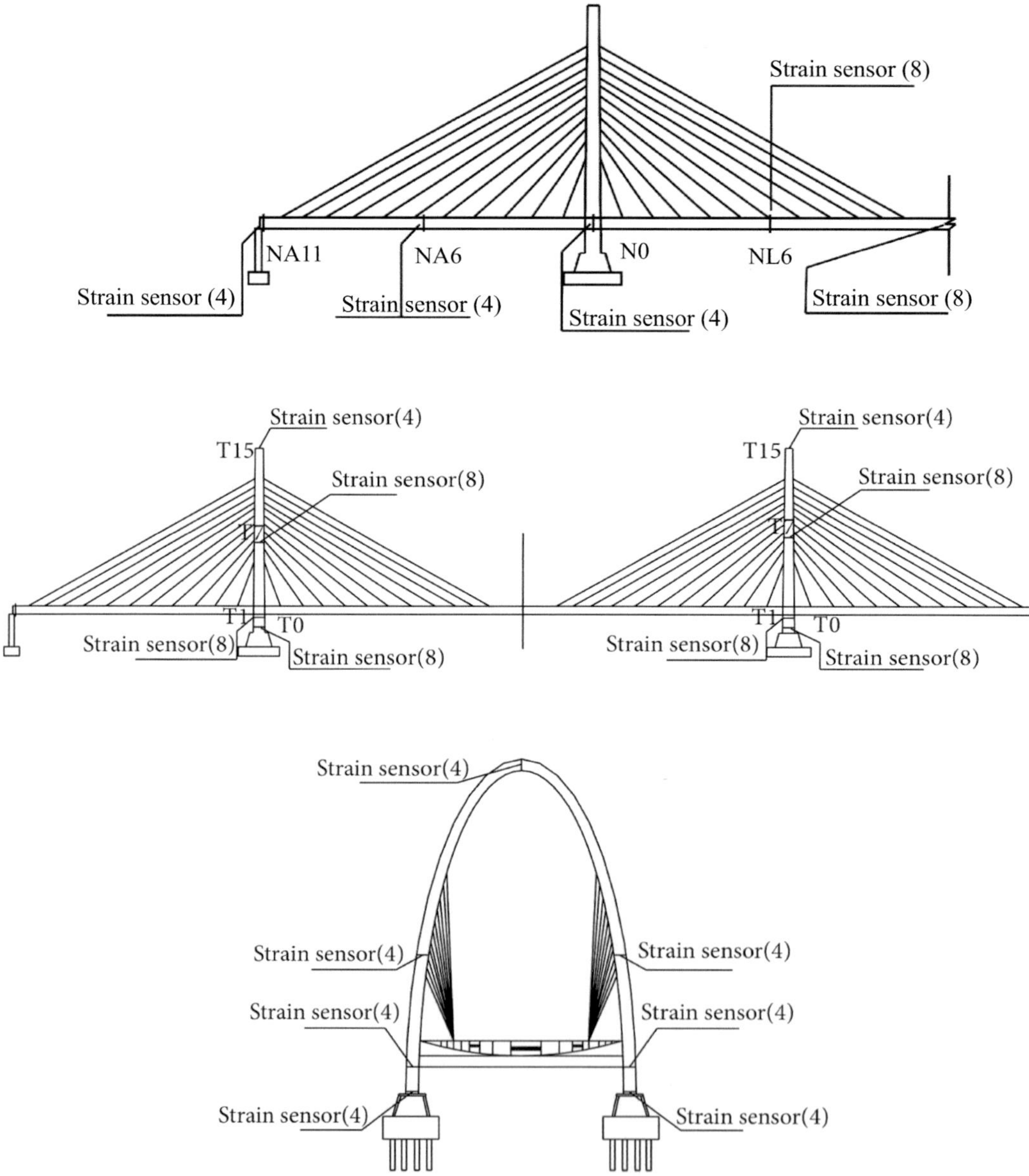

Fig. 13.33 Stress and fatigue monitoring points: (a) superstructure; and (b) tower

Thruway Authority (NYSTA). The Bridge includes two parallel structures 16370 ft long with a center span length of 1214 ft. The Bridge was completed in September 2018. The SHM of the Cuomo Bridge is considered one of the largest and most comprehensive systems of its kind in the United States that includes more than 450 sensors measuring temperature; strain, and fatigue on the cables and joint movements; and normal corrosion in the concrete and on the deck. It was designed for the following objectives: measure wind speed and movement; collect load demands to gauge strains on critical elements of the bridge; capture the behavior of the bridge due to daily or seasonal temperature changes; trigger maintenance alters to inspect components; inform the bridge's command center for efficient operations of traffic. As a result, operators can better understand and track how the bridge is aging and reacting as loads are applied and seasonal temperature changes occur.

The SHM was installed during construction for monitoring operations over the lifespan of the structure and included: corrosion measurements, weather measurements, expansion measurements and effects, acceleration measurements, strain measurements, and the data acquisition network. Figure 13.34 displays the layout of the SHM system of the Cuomo Bridge.

Fig. 13.34 SHM system for the Governor Mario M. Cuomo Bridge (Courtesy, New York State Bridge Authority)

The SHM includes four different types of data collection systems (Pietrobelli et al., 2019): static data loggers, high speed data Loggers, GPS Receivers, fiber optic interrogators. All sensors are sampled continuously with statistical values (min/max/avg/STD dev) over displayed and stored ten minute intervals. This method greatly reduces data storage requirements and provides the expected response for setting response values. All the loggers are configured into seventy-five (75) Data Acquisition Units (DAU) positioned across the length of both bridge structures. DAU are integrated into a Fiber Optic Local Area Network (LAN) that terminates in the control room, which includes mainframe Server and multiple monitors for real-time data review, event log and reporting. All the measurements are initially processed in the Data Acquisition Units (DAU) for data quality, statistical summaries over defined time periods. Filtered data are then forwarded to the Mainframe Server where they are further analyzed for event triggers, data archival and display.

Response values can be established that trigger action for traffic control or maintenance decisions based on seasonal variations. These are collected and reviewed. The SHMS provides automated reporting of Event Logs that capture any exceedance of a limit value. This could be the result of sensor malfunction requiring replacement, listing of all sensors related to other sensors' exceedance or specific criteria established by the owner. The schedule for reporting can be daily, weekly, monthly or combinations of time periods, conditions, or planned schedules for maintenance. The focus of automated reporting is on eliminating the need for dissemination of data later, but rather a proactive reporting schedule of needed information for decision making. Some other featured reporting includes weather conditions and wind speeds during high windstorms as they are important parameters for traffic control decisions. This kind of information is fed directly to the Intelligent Traffic Management System (ITS). Reporting also includes bearings and expansion joints displacements since the manufacturer provides an expected service life based on cumulative displacement for these elements; this data can provide notifications on potential replacement or time of detailed inspection.

In summary, the SHM system for the Cuomo Bridge consists of measuring the response of individual components and their effect on the structure when changes in expected responses occur. The system will autonomously provide alerts to these changes based on predetermined response values. These alerts can be based on event driven conditions, anomalies in behavior or long-term degradation of components to provide notifications for focused inspections.

References

Ansari, F., Practical Implementation of Optical Fiber Sensors in Civil Structural Health Monitoring, *Journal of Intelligent Material Systems and Structures*, Volume 18, pp 879–889, 2007.

Bayissa, W.L., Haritos, N., Thelandersson, S., Vibration based structural damage identification using wavelet transform, *Mechanical Systems and Signal Processing*, Volume 22, pp 1194–1215, 2008.

Bridge, J.A. and Davis, J.R., *Sunshine Skyway Bridge Monitoring Phase II: System Development*, Technical Report, Department of Civil and Coastal Engineering, University of Florida, Sponsored by FDOT, 2021.

Chen, B., Wang, X., Sun, D. and Xie, X., Integrated System of Structural Health Monitoring and Intelligent Management for a Cable-Stayed Bridge, *The Scientific World Journal*, Volume 2014, Article ID 689471, 12 pages, 2014.

Chou, J.-H. and Ghaboussi, J., Genetic Algorithm in Structural Damage Detection, *Computers and Structures*, 79(14), 1335–1353, 2001.

Conte, J.P., Xianfei, H., Babak, M., Masri, S., Caffrey, J. et al., Dynamic Testing of Alfred Zampa Memorial Bridge, *Journal of Structural Engineering*, Volume 134 (6): 1006–1015, 2008.

Dascotte, E., Vibration Monitoring of the Hong Kong Stonecutters Bridge, *EVACES 2011 – Experimental Vibration Analysis for Civil Engineering Structures*, 2011.

FHWA, *Weigh-in-Motion Pocket Guide Part 1 WIM Technology, Data Acquisition, and Procurement Guide*, Publication No. FHWA-PL-18-015, 2018.

Fujino, Y., Siringoringo, D.M., Ikeda,Y., Nagayama, T., and Mizutani, T. Research and Implementations of Structural Monitoring for Bridges and Buildings in Japan, Engineering 5 PP 1093–1119, 2019

Helmicki, A. and Hunt, V., *Instrumentation of the US Grant Bridge for Monitoring of Fabrication, Erection, In-Service Behavior, and to Support Management, Maintenance, and Inspection*, Report No, FHWA/OH-2013/23 by University of Cincinnati Infrastructure Institute, Sponsored by Ohio DOT, 2013.

Hung, S-L. and Kao, C.Y., Structural Damage Detection Using the Optimal Weights of the Approximating Artificial Neural Networks, *Earthquake Engineering and Structural Dynamics*, 31, 217–234, 2002.

Jang, S., Jo, H., Cho, S. and Bridge, J., Structural Health Monitoring of a Cable-Stayed Bridge Using Smart Sensor Technology: Deployment and Evaluation, *Smart Structures and Systems*, Vol. 6, No. 5-6, pp 439–459, 2010.

Karbhari, V.M. and Lee, L.S-W, Vibration-based damage detection techniques for structural health monitoring of civil infrastructure systems, *Structural Health Monitoring of Civil Infrastructure Systems, Edited by Karbhari and Ansari*, pp 178–212, 2009.

Law, S. S., Li, X.Y., Zhu, X.Q., Chan, S.L., Structural damage detection from wavelet packet sensitivity, *Engineering Structures*, Volume 27, 2005, pp 1339–1348, 2005.

Lydon, D., Lydon, M., Kromanis, R., Dong, C-Z., Catbas, N. et al., Bridge Damage Detection Approach Using a Roving Camera Technique, *Sensors*, Volume 21, 21 p, 2021.

Mehrabi, A., Performance of Cable-Stayed Bridges: Evaluation Methods, Observations, and a Rehabilitation Case, Journal of Performance of Constructed Facilities, Volume 30, No.1, 2016.

Moyo, P., Brownjohn, J.M., Detection of anomalous structural behavior using wavelet analyses, *Mechanical Systems and Signal Processing*, Volume 16, pp 429–445, 2002.

Pietrobelli, M., Celik,O. and Weinmann, T.L., Structural Health Monitoring System for the Governor Mario M. Cuomo Bridge, Proceedings 9th International Conference on Structural Health Monitoring of Intelligent Infrastructure August 4-7, 2019 – St. Louis, Missouri.

Ren, W-X, Zhao, T., and Harik, I.E., Experimental and Analytical Modal Analysis of Steel Arch Bridge, *Journal of Structural Engineering*, Volume 130 (7): 1022–1031, 2004.

Rice, J.A., Mechitov, K., Sim, S.H., Nagayama, T., Jang, S., Kim, R., Spencer, Jr., B.F., Agha, G. and Fujino, Y., Flexible smart sensor framework for autonomous structural health monitoring, Smart Structural Systems., 6(5-6), 423–438, 2010.

Rice, J.A., and Davis, J.R., Sunshine Skyway Bridge Monitoring Phase I: System Assessment and Integration Recommendations, Technical Report, Department, of Civil and Coastal Engineering, University of Florida, Sponsored by FDOT, 2016

Shama, A., Seismic Damage Assessment using Wavelet Analysis, *Tenth US National Conference on Earthquake Engineering*, Anchorage, Alaska, 2014.

Shenton, H.W., Al-Khateeb, H.T., Chajes, M.J. and Wenczel, G. et al., Indian river inlet bridge (part A): Description of the bridge and the structural health monitoring system, Bridge Structures, Volume 13, pp 3–13, 2017.

Talebinejad, I.F., Structural Health Monitoring of Lingoto Cable-Stayed Pedestrian Bridge, *SHMII 3 A, Mufti (Ed.)*, Vancouver, 2007.

Zang, C. and Imregun, M., Combined Neural Network and Reduced FRF Techniques for Slight Damage Detection Using Measured Response Data, *Archive of Applied Mechanics*, 71, 525–536, 2001.

CHAPTER **14**

Inspection and Maintenance of Cable-Stayed Bridges

14.1 INTRODUCTION

Inspection of cable-stayed bridges is a crucial part of maintenance and ensuring safety. Inspections allow engineers to identify small defects and potential problem areas before they develop into major issues. Structural health monitoring (SHM) of cable-stayed bridges facilitate the inspection and maintenance process, and optimize inspection budgets. The level of inspection is guided by the Comprehensiveness of the SHM. Level 1 SHM, which is basically global, will warn the bridge operators that there is degradation occurring in the structure without locating and evaluating the damage. In this case, it is a very comprehensive inspection program for all the bridge elements to determine the location and severity of damage. On the other hand, other levels of SHM will ease up the inspector's mission as it will provide guidance regarding the location and severity of the damage. Local methods of SHM such as the Acoustic monitoring method of the cable using a piezoelectric transducer, which creates a voltage due to the energy released once a wire breaks, is an excellent tool for detection of cable damage and hence implementation of a comprehensive inspection of the defected cable. The rest of this chapter will start by providing general guidelines on the inspection of cable-stayed bridges and the rest of the chapter will focus on methods of inspection and maintenance of cables.

14.1.1 Inspection Rating Scale

The primary goal of the inspection is to document the state of each and every component of the bridge. This scale can be used to rate elements:

Rate 1: Completely degraded or in a failing state
Rate 2: Used to conceal between ratings of 1 and 3
Rate 3: Severe degradation or failure to perform as originally designed
Rate 4: Used to conceal between ratings of 3 and 5
Rating 5: Slight wear and tear, but still operating as intended
Rating 6: Used to shade between ratings of 5 and 7
Rating 7: Excellent state – no deterioration

Items with a rating of three or lower might need extensive rehabilitation. Items with ratings of four or higher could use maintenance work to make improvements. Along with the rating procedure, some states, like NEWSDOT, also employ a flagging procedure. This flagging procedure is a standardized way to promptly notify the relevant responsible parties of significant bridge deficiencies that need to be addressed. Additionally, it specifies the conditions that must be met in order to attest to the timely and proper implementation of corrective or preventive measures. Conditions that present a clear and present danger or that could become dangerous if ignored for an extended period of time should be reported using this procedure. Three different kinds of flags exist: (NYSDOT, 2008):

Red Structural Flag: Employed to report an important primary structural component's failure or impending failure. Potentially forthcoming denotes the likelihood of a failure occurring prior to the next inspection date.

Yellow Structural Flag: Used to report a potentially dangerous condition that, if ignored past the next scheduled inspection, will likely develop into a clear and present danger.

Safety Flag: Used to report an issue that does not involve a structural failure or collapse but instead poses an obvious and present risk to vehicle or pedestrian traffic.

14.1.2 Types of Inspection

There are five types of inspection, each requiring different levels of intensity and are conducted under certain structural conditions (AASHTO-2011). Items such as the extent of access to structural elements, the level of detail required for the physical inspection, and the degree of testing will vary considerably for each type of inspection.

14.1.3 Initial (Inventory) Inspections

An Initial Inspection is the first inspection of a bridge as it becomes a part of the bridge file. It is conducted for the purpose of determination of baseline structural conditions and the identification and listing of any existing problems or locations in the structure that may have potential problems. The Initial Inspection is a fully documented investigation, and it must be accompanied by an analytical determination of load capacity. The inspector will note any fracture-critical members or details during this Initial Inspection, aided by a prior detailed review of plans.

14.1.4 Routine (periodic) Inspections

Routine inspections are usually biennial. Special equipment, such as under-bridge inspection equipment, rigging, or staging, is necessary for Routine Inspection in circumstances where its use provides for the only practical means of access to areas of the structure being monitored. If the cable-stayed bridge is provided with a mechanical traveler, then it will facilitate inspection of the under deck. Areas to be closely monitored are those determined by previous inspections, load rating calculations, or both to be critical to load-carrying capacity. The results of a Routine Inspection should be fully documented with appropriate photographs and a written report that includes all recommendations for maintenance or repair and for scheduling of follow-up In-Depth or Special Inspections, if necessary.

14.1.5 Damage Inspection

This is an unscheduled inspection to assess structural damage resulting from environmental factors or human actions. Examples of environmental factors include earthquakes and hurricanes. Examples of human actions include barge collision with a pylon foundation, an aircraft hitting the top of pylons, or terrorist explosion activities. The scope of inspection should be sufficient to determine the need for emergency load restrictions or closure of the bridge to traffic, and to assess the level of effort necessary to affect a repair. If significant damage has occurred, inspectors must determine the extent of section loss of damaged elements, quantify misalignment of members, and evaluate foundations for a potential loss of support.

14.1.6 In-Depth Inspections

An In-Depth Inspection concentrates on one or more members above or below the water level to identify any deficiencies not detected by the Routine Inspection. Cable stays and their anchorages to the tower and deck are potential elements for this inspection. This type of inspection can be scheduled independent of a routine inspection, though generally at a longer interval, or it may be a follow-up for damage or initial inspections. To determine the extent of any deficiencies or damage, performance of nondestructive field tests may be required. A class of these tests was discussed in Chapter 13. The inspection may include a load rating to assess the residual capacity of the member or members, depending on the extent of the deterioration or damage. The activities, procedures, and findings of In-Depth Inspections should be completely and carefully documented.

14.1.7 Special (Interim) Inspection

A Special Inspection is one scheduled at the discretion of the bridge owner according to the Bridge Manual. It is usually performed during the calendar year between the required biennial inspections; however, inspection frequency should consider the severity of the known deficiency. It is required if some conditions exist such as condition rating of 3 or less, or presence of an active or inactive red flag, or yellow flag for a certain member. Examples include but are not limited to foundation settlement or scour; member condition; or bulging of a cable sheathing that indicates damage. Special inspections usually are not sufficiently comprehensive to meet the requirements for biennial inspections.

14.2 RESPONSIBILITIES AND QUALIFICATIONS OF INSPECTION TEAM

Bridge inspection teams usually include a Team Leader (TL), who must be a professional engineer and an Assistant Team Leader (ATL). There are five main points that the TL must consider in any inspection program:

- *The bridge is a public asset*: *The* ATL must be on guard for minor problems that can be corrected before they lead to major costly repairs. The inspector must also be able to recognize bridge elements that need repair to maintain bridge safety and avoid replacement costs.
- *Provide accurate bridge records*: Accurate bridge records are required to establish and maintain a structure history file; to identify and assess bridge deficiencies and to identify and assess bridge repair requirements; to identify and assess minor bridge deficiencies and to identify and assess bridge maintenance needs in a similar manner to the repair requirements.
- *Provide an integral report*: The ATL must ensure the bridge is inspected completely, and that the inspection report conforms with all requirements of the Bridge Manual.
- *Fulfill legal responsibilities*: In this regard, descriptions in the inspection report must be specific, detailed, quantitative (where possible), and complete.

The ATL may inspect and measure components, if working under direct supervision of the TL. Other personnel may be assigned as needed, such as Laborers and ATL Trainees. All field work must be reviewed by a Quality Control Engineer (QCE).

14.3 ASPECTS ASSOCIATED WITH INSPECTION OF CABLE-STAYED BRIDGES

Inspection and maintenance of cable-stayed bridges should be considered from the beginning of the preliminary design. Design should consider identifying areas of the bridge where access for inspection and maintenance are critical and the inspection-maintenance manual should describe viable methods for inspection and maintenance access. An optimal design of the bridge is one that requires minimal maintenance and designed with convenient access and flexibility in replacing components at the end of their useful life. Some significant issues that need to be considered for an easy inspection process are discussed in what follows:

- *Avoid Tie-downs*: Some critical elements need special attention because their failure may lead to bridge failure such as tie-downs as shown in Figure 14.1. They are uplift restraint elements and in most cases are non-redundant. Therefore, design should avoid using tie-downs. If unavoidable, then they should be easily accessible for inspection and maintenance. They should also design for enabling replaceability. If cables are used as tie-downs, they should be sufficiently long so that the change in length due to thermal creep and shrinkage movements is very small.

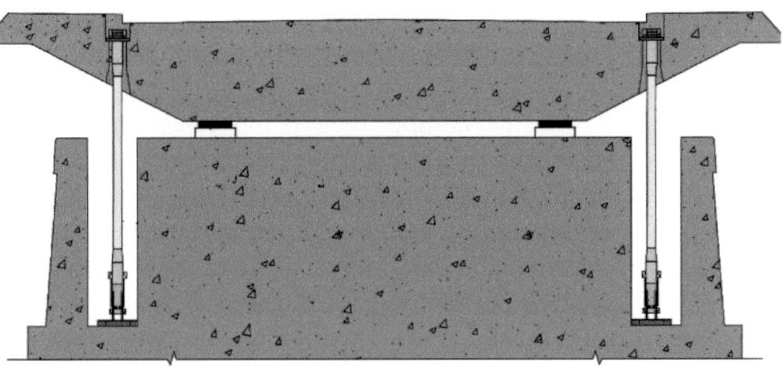

Fig. 14.1 Schematic of a tie-down to a cable-stayed bridge auxiliary pier to prevent uplift

- *Limitations of Bucket Truck*: Under-deck inspection may require a bucket truck which is the cost-effective option particularly for a roadway that is not too wide and traffic conditions are tolerant to partial lane closure. It should be noted that working around stays with shallower angles could be time consuming when access is provided from the roadway by a bucket truck, which puts inspectors at risk and may hit cables.
- *Stay cables*: Inspectors must keep in mind the following common defects that can occur on the cable members: failure of the paint system; pitting, surface rust; section loss, fatigue cracking; collision, overload, or heat damage.
- *Inspection traveler*: For a very wide roadway, the use of an inspection and maintenance traveler is a must for the inspection of under-deck and edge girders (Figure 14.2). Nevertheless, the traveler and the mechanical system used to operate it will require inspection and maintenance.
- *Routine Maintenance*: Access to the underside of the superstructure, lightning arresters, as well as lighting (aviation, navigation, and aesthetic) should all be considered during the design phase.
- *Innovative methods of inspection*: Visual inspection has its own limitations as it can only report the surface conditions based relatively on the inspector's judgment. Therefore, advanced inspection techniques may be required to evaluate the underlying condition. Some of the advanced nondestructive methods are discussed in the next section.

Fig. 14.2 Use of Inspection traveler to inspect under-deck (courtesy, SISSCO)

- *Accessing the stay cables*: As the number of existing cable-stayed bridges increase, one problem that continues to arise is on how to inspect the cable stays for corrosion and deterioration. An inherent difficulty in developing an inspection procedure for cable-stayed bridges is that each cable-stayed bridge is unique in its cable configuration. Therefore, innovative systems to comfort crews with visual inspection and maintenance of the cable stays may be necessary. Some of these innovative systems are discussed in the next section.

14.4 ADVANCED NONDESTRUCTIVE INSPECTION TECHNIQUES

Simple nondestructive testing (NDT) has evolved during the past few decades. Different NDT methods are adopted to identify damage and cracking in both concrete and steel bridge elements. A method has proved to be a noble solution to this issue. It is widely used nowadays for periodical inspection of bridges. NDT results can help to evaluate the performance of certain components of the cable-stayed bridge. Advanced testing methods can aid decision makers in the determination of the best plan for repairs and maintenance or perhaps complete replacement.

14.4.1 Advanced NDT for RC Members

The following methods can be used for detecting delamination, honeycombing, voids, and cracks in reinforced concrete bridge members:

- *Infrared Thermography*: This technique uses an infrared camera to detect temperature differentials in a concrete surface. A thermal infrared camera obtains an image showing subsurface defects as a higher rate of radiant temperature change than the surrounding concrete thus producing some spots within the thermal infrared image. A "cold spot", as displayed in Figure 14.3, indicates a delamination. Although subject to weather conditions, this technique is efficient for large surfaces.
- *Ground Penetrating Radar (GPR)*: This technique uses a low-power, high-frequency pulsed radar. Ultra-high frequency radio waves are transmitted using a portable control unit and recorder. Some of the penetrating waves are reflected by interfaces or objects with dielectric

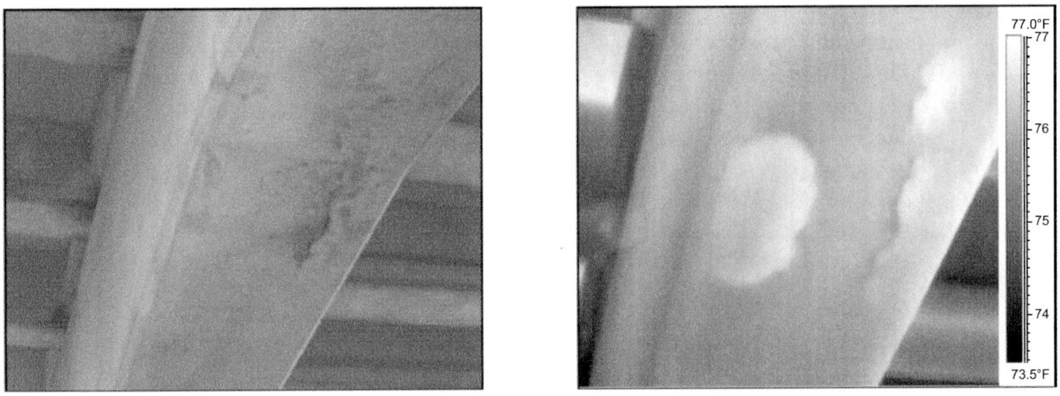

Fig. 14.3 Use of infrared thermography to evaluate defects in a bridge member (Bhandari et al., 2016)

properties differing with those of the general medium. This method can also be used to examine the condition of some members that may otherwise be inaccessible such as the top flange of box beams.

- *Electromagnetic methods*: Advancements in ground penetrating radar motivated researchers in the Lawrence Livermore National Lab to develop the High-Speed Electromagnetic Roadway Measurement and Evaluation System (HERMES) Bridge Inspector. HERMES technology was designed for the evaluation of concrete bridge deck deterioration, with a particular emphasis on the detection of corrosion-induced delamination. The system is set up in a trailer mounted towing vehicle that travels over the structure and sends high frequency electromagnetic pulses into the bridge deck from radar antennas and other components of the system to analyze the reflected waves. At speeds of around 20 mph, the system can sample the concrete deck every 9/16" in the direction of travel (FHWA,2002). Output information can be reconstructed to show cross-sections of the deck being inspected. The depth of penetration depends on time and the material type.
- *Impact-Echo Testing*: Testing involves introducing a stress pulse into the concrete by mechanical impact. The pulse is reflected by internal flaws such as cracks and voids. A transducer placed near the impact point, as shown in Figure 14.4, monitors surface displacements caused by the reflections. The response can then be interpreted to detect flaws within the concrete. This

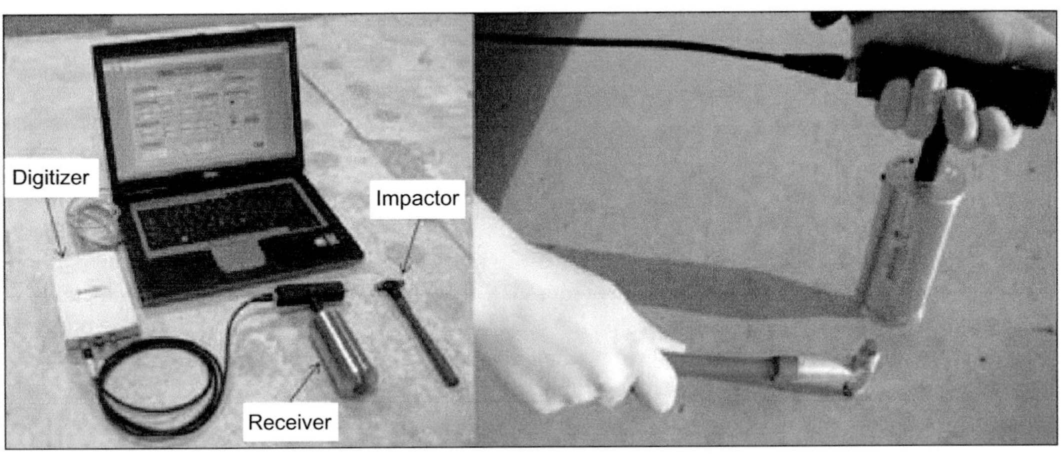

Fig. 14.4 Illustrating Impact Echo testing (Giannini et al., 2012)

method is effective in detecting flaws in slabs and pavements and to assess the condition of concrete beams and columns.

- *Pulse Velocity*: UPV of concrete is a simple tool for testing the uniformity of concrete quality. It is an ASTM designated method (ASTM C597). The travel time of an ultrasonic pulse between two transducers is measured, and the velocity of the compression wave calculated by simply dividing the distance traveled by the travel time. Figure 14.5 illustrates a typical UPV test setup.

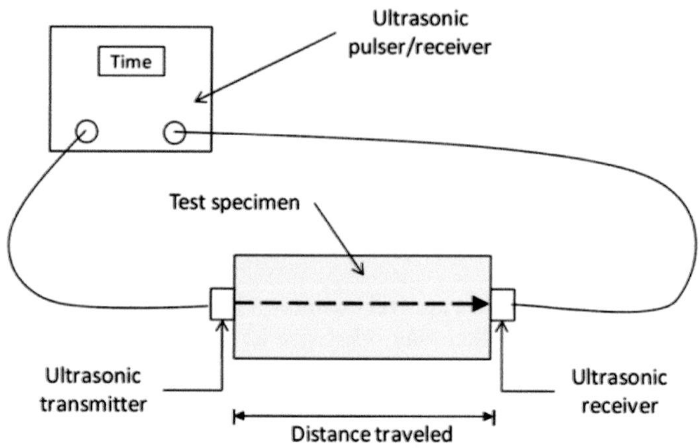

Fig. 14.5 Typical UPV test setup (Kreitman, 2011)

- *Ultrasonic Testing (UT)*: Ultrasonic testing consists of a sending transducer, which produces high frequency sound waves. Discontinuities in the element interrupt the sound wave and deflect it toward a receiving transducer. The magnitude of the return signal allows a measurement of the flaw size. The distance to the flaw can be estimated from the known properties of the sound wave and the material being tested. It can be used to detect discontinuities, surface damage, internal flaws, and medium size cracks. It can provide valuable information regarding the condition of cable-stayed bridge concrete members. However, the method can be difficult to use with reinforced concrete members because the presence of steel parallel to the line of transmission may cause misleading results. Therefore, some skill is required to obtain usable results for reinforced concrete.
- *Copper Sulphate Electrode/Half Cell Potentials*: This technique is used to evaluate the corrosion activity of reinforcing steel embedded in concrete. As the copper sulfate electrode (CSE) contacts the concrete over an actively corroding rebar, voltage is induced. Measured potential values indicate corrosion activity.

14.4.2 Advanced NDT for Steel Members

The following methods can be used for cracks and damage in steel bridge members:

- *Dye Penetrant (DP)*: Dye Penetrant Inspection (DPI) is widely used to detect surface breaking flaws. This non-destructive testing technique is a cost-effective method used to locate surface breaking flaws such as cracks, porosity, laps, seams, and other surface discontinuities. A dye is applied to the steel surface after being cleaned to the bare metal (Figure 14.6). The penetrant is allowed to penetrate the surface, and any excess material is removed. A developer is then applied, which draws the dye out of the irregularities and defines the extent and size of surface flaws. This method is a preliminary fast tool to detect cracks, but it reveals neither the depth of cracks nor any subsurface flaws.

- *Magnetic Flux Leakage (MFL)*: The MFL method was discussed in section 12.3.2.7. As mentioned, a magnetic field is induced into the member, and cracks or other irregularities in the surface of the member cause disturbance in the magnetic field. MFL is useful in detecting surface defects, such as voids, inclusions, cracks, and holes in ferromagnetic materials. It can also detect subsurface gouges and cracks located close to the surface.

Fig. 14.6 Illustrating detection of a crack using Dye Penetrant (courtesy, Bob Hildebranski)

- *Radiographic Testing*: Radiographic testing is very effective in detecting surface cracks, voids, and inclusions. Through this test, X-rays are applied to the member. The absorption of the rays for a defected region is different from that for a non-defective region.
- *Ultrasonic Testing (UT)*: UT has broad application in the inspection of steel members, detecting voids, corrosions, inclusions, and cracks. It is also used in the inspection of welds.

14.5 INSPECTION OF SUPERSTRUCTURE

As discussed in Chapter 4, there are several types of superstructure cross-sections for cable-stayed bridge depending on the length and cable configuration. They include an orthotropic steel box girder, concrete box girders, open concrete cross-section, and composite cross-section. Inspection of each of these systems is discussed herein.

The orthotropic steel box girder section comprises a thin steel plate stiffened in the longitudinal direction by a series of closely spaced longitudinal ribs and the deck acts integrally with the steel superstructure overlaid with a layer of wearing surface. The box is strengthened in the transverse direction with plate diaphragms. Examples of steel orthotropic box girder bridges include but are not limited to the Saint-Nazaire Bridge, France; Kohlbrand Bridge, Germany; Luling Bridge, Louisiana; Faro-Folster Bridge, Denmark; Wye River Bridge, England; Millau, Bridge, France, SuTong Bridge, China; and Stonecutters Bridge, Honk Kong. It is very imperative that inspection is accounted for during the design of the box girder and manholes are arranged either in the top plate or in the web to facilitate inspection. The inspector can enter through these manholes and climb through the box from one end to the other and move in the longitudinal directions through an opening in the transverse diaphragm as illustrated in Figure 14.7 (Stonecutters cable-stayed Bridge in Honk Kong). This can facilitate the inspection of the rib stiffeners on the interior of the box girder. Common steel defects of concern include bent, buckled, or damaged members; corrosion; section loss; and fatigue cracks. In addition to visual inspection, other NDE testing methods employed may include dye penetrant, ultrasonic testing, radiographic testing, and MFL methods. Since an orthotropic box girder comprises numerous welded connections, it is prudent that a sampling number/location of representative orthotropic weld details, be determined as part of the inventory inspection, and receive inspections on a biennial inspection. These predetermined details are then monitored over time to determine whether the weld is exhibiting any cracks due to fatigue.

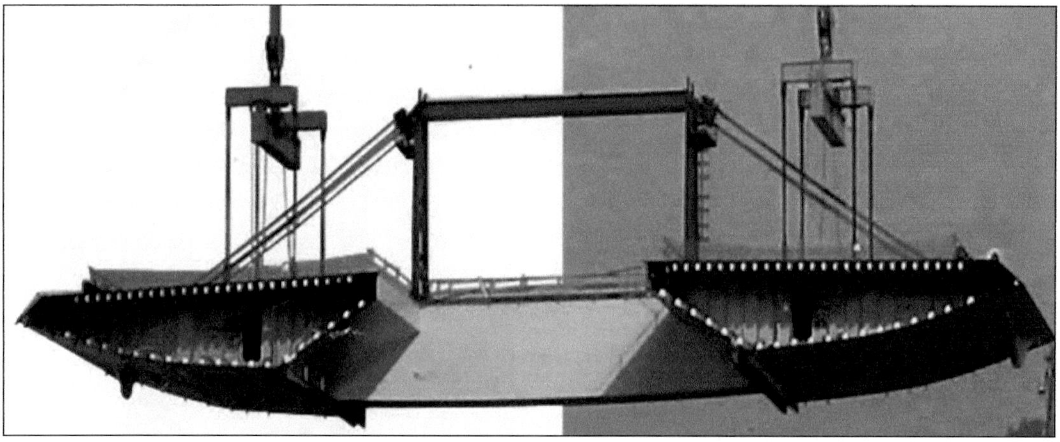

Fig. 14.7 Openings in the transverse diaphragms for inspection of Stonecutters Bridge

Orthotropic decks typically have a layer of asphalt, ranging from 1 inch to 2.5 inches in thickness, as the wearing surface. An epoxy asphalt polymer concrete also is used for orthotropic bridge deck wearing surfaces. Common defects of wearing surfaces include cracking and debonding. A loss of bonding can occur in the wearing surface due to a lack of blast cleaning, corrosion-protective painting, and bonding layer application on the upper steel plate's top surface.

Cable-stayed Concrete Box Girder Bridges are either single-cell such as the Wadi Kuf Bridge in Libya or multi-cell such as the Maracaibo Bridge, Venezuela; the Wadi Kuf Bridge, Libya; Tiel Bridge, Holland; River Foyle Bridge, Northern Ireland; Brotonne Bridge, France; Sunshine Skyway, Florida, US; Suez Canal (Al Salam) Bridge, Egypt; Aswan Bridge, Egypt; Veterans Glass City Skyway Bridge, Ohio, US; Kota Chambal Bridge, India; and Atlantic Bridge, Panama. Open concrete cross section cable-stayed bridges consist of two edge girders and an intermediate cross girder to carry the transverse bending. The concrete roadway slab is supported on top of this system. Examples of cable-stayed bridges include but are not limited to, Pasco-Kennewick Bridge, Washington, US; Reinbridge Diepoldsau Bridge, Switzerland; East Huntington Bridge, Ohio, US; Dame Point Bridge, Florida, US; and Sidney Lainer, Georgia, US.

The construction of concrete box girders is either cast-in-place or uses the segmental method. The inspection of a box girder bridge requires an awareness of the girder function. Therefore, a comprehensive review and understanding of design or as-built drawings prior to the inspection is required. This will entitle the inspector to get an insight of the regions that may exhibit high stress under different loading conditions. The bridge manual should shed some light on regions of interest for inspection. In general, the inventory inspection is the most important inspection a box girder will receive. It will serve as a benchmark for all future inspections. An engineering survey can be performed when the construction is over and a schedule for future surveys can be established. The results of these surveys will aid in evaluating the behavior and performance of the bridge. Permanent survey points can be set at several locations along the superstructure in the transverse directions to aid the inspector to check the girder for the proper camber.

Some of the significant locations for inspection in box girder bridges are discussed. Figure 14.8 displays, wind tongue at the abutment of a cast-in-place cable-stayed bridge box girder bridge. This region is susceptible to undesirable conditions due to effects of restraint, temperature, creep, and concrete shrinkage. This may lead to improper movements. Therefore, bearing areas need to be inspected thoroughly. It is also important to note that cable-stayed bridge box girders will be subjected to torsion under severe wind loads. Therefore, one of the basic functions of the box girder in general is to increase the torsional stiffness of the bridge. Torsion combined with shear can induce significant cracking in both the web and flange as indicated in Figure 14.9.

Fig. 14.8 Bearing area of cast-in-place box girder (courtesy, FHWA)

This cracking will produce a helical configuration if torsion alone was present. Tension cracks can appear as a series of parallel cracks running transverse to the longitudinal axis of the bridge and spaced at approximately 1 to 2 times the minimum thickness of the girder component. The cracks can potentially propagate through the whole depth of the box girder section. The girder should be also inspected throughout for cracks due to prestressing or post-tensioning. Some shrinkage cracks are to be expected. All cracks should be measured carefully with an optical crack gauge or crack comparator and its location, length, width, and crack spacing should be documented. Areas such as joints, scuppers and curb lines exposed to drainage should receive special attention. The condition of the drainage holes must be checked to verify they are clear and functioning properly. The inspector must also concentrate on the roadway surface for delamination, cracking, spalling, and deformation as the presence of these defects may exaggerate the impact effect of traffic particularly if the top flange does not have an added wearing.

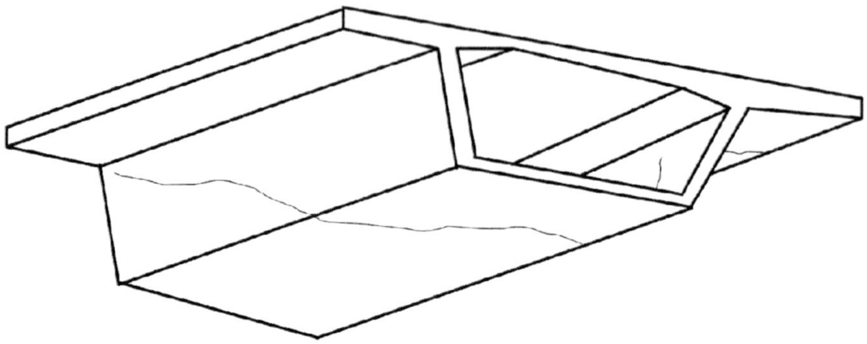

Fig. 14.9 Box girder cracks due to combined torsion and shear (courtesy, FHWA)

All repairs that have been previously made need to be carefully inspected to determine if repaired areas are functioning appropriately. Other potential cracking areas include the anchor blocks. Anchor blocks contain the termination of the post-tensioning tendons. Hence, very large, concentrated loads are developed within these blocks. They tend to crack if not sufficiently reinforced or if there are voids adjacent to the post-tensioning tendons. For cast-in-place box girders, the cracking would be oriented in the direction of the post-tensioning tendon and may be more of a splitting failure in the web (Figure 14.10). For segmental construction, the inspection needs to focus on the box girder webs adjacent to the anchor blocks and determine if vertical cracks on either side of the anchors are developed. The condition of the tendons adjacent to the anchor blocks needs to be examined. The flange or web on which the anchor block is located will require attention concerning the potential for transverse cracking in the vicinity of the anchor (Figure 14.11).

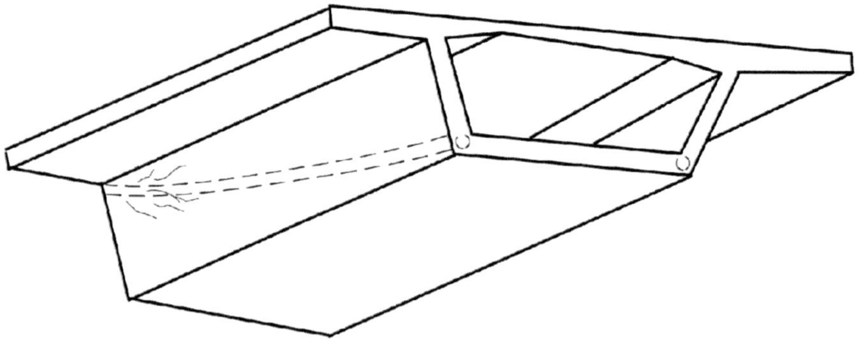

Fig. 14.10 Web splitting near anchorage block (courtesy, FHWA)

Joints should be inspected for any potential damage or movement of the shear keys. Any open or loose joints need to be documented. Areas where the type of construction requires closure joints or segments to be poured in place will need close attention. For externally post-tensioned box girders (Figure 14.12), deviation blocks and blister blocks should be carefully examined for spalling and/or cracking distress. These are points of very high stress concentrations, and their integrity is essential to the integrity of the span, and post-tensioning continuity. Locating and mapping areas of delamination, spalling and delamination on the top flange is essential because of the structural importance of this component.

Common defects that occur on reinforced concrete boxes and open girder bridges include (FHWA, 2002): cracking, scaling, delamination, spalling, efflorescence, abrasion, and reinforced

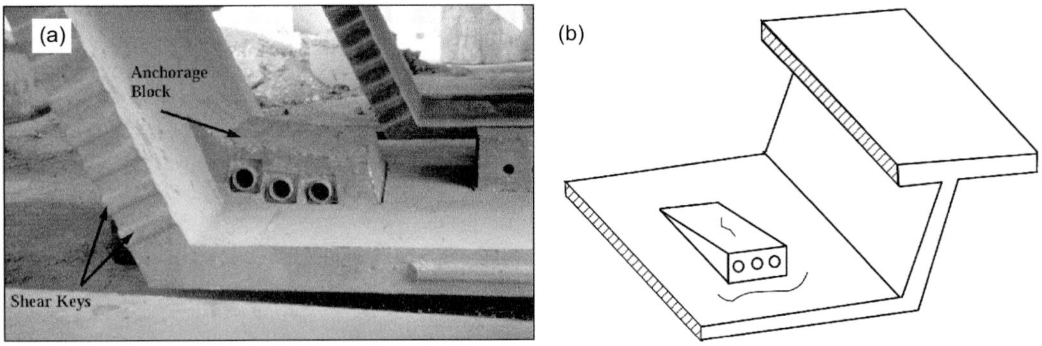

Fig. 14.11 Segmental anchor block cracking pattern: (a) View of the box girder anchor block; and (b) cracks adjacent to Anchorage block (courtesy, FHWA)

Fig. 14.12 Illustrating an externally post-tensioned box girder

steel corrosion. The inspection of concrete girders for cracks, spalls, and other defects is mainly a visual activity. Therefore, hammers are primarily used to detect areas of delamination. In general, delaminated areas will exhibit a hollow "clicking" sound when knocked with a hammer. Nevertheless, sound concrete will result in a solid "pinging" type sound. Depending on the results of this simple test other testing levels can be pursued such as core sampling, carbonation, concrete permeability, concrete strength, endoscopes and videoscopes, and reinforcing steel strength. In addition, advanced nondestructive methods can also be employed for concrete inspection. These methods include but are not limited to pulse velocity measurements, (ASTM, 2023) electromagnetic method, ground-penetrating radar, impact-echo testing, infrared thermography, and ultrasonic testing.

A composite superstructure is widely used in the US. It comprises two steel edge girders connected transversely with floor beams that are spaced uniformly. This system supports a concrete roadway deck. Inspection methods of steel and concrete had been discussed above for steel orthotopic box girders and concrete girders along with potential defects.

14.6 INSPECTION OF PYLONS AND FOUNDATIONS

The majority of cable-stayed bridges in the United States and other countries are made of reinforced concrete, but some, like the Luling Bridge in Louisiana and the Strelasund Bridge in Germany, are constructed with steel for their pylons. Ladders, protected climbing techniques, and a motorized suspended rigged basket are among the equipment used for pylon inspection.

For steel pylon members, visual inspection of the welded connections—which consist of the surrounding base metal and the weld material itself—is necessary. In particular, any weld distress, discontinuities, or deterioration (such as pinholes/porosity, undercut, slag inclusion, and lack of fusion) should be graphically documented and digital color photos should be added. The tower's steel members need to be inspected for flaws like paint system failure, pitting, deterioration, missing, damaged, or defective components, and misalignment. To ascertain the effective dimensions of the remaining structure and the extent of section loss, the deteriorated areas need to be cleaned. Light grinding can be used by the inspector if hand cleaning doesn't work in certain areas. Ultrasonic measuring devices can be used to verify thickness measurements obtained with a caliper and ruler, if needed. When physical properties are required, members' measurements must be sufficient to determine the structural section's as-inspected configuration. In addition to the rust formation on the weathering steel, inspection needs to take care of any suspected steel cracks and poor weld conditions. When applicable, current circumstances can be compared to those from earlier reports. An ultrasonic thickness gauge can be used to measure the amount of thickness lost and to pinpoint areas where rusting, pitting, or flaking has occurred.

It is necessary to inspect RC pylons both inside and out for indications of deterioration. It is necessary to examine every surface for areas that are "honeycombed," unsound, or split. It is particularly crucial to check for cracks. A pylon's cracking may indicate that a particular area is weak or overstressed. The bottom of the pylon's leg where it interacts with the foundation is one example of such a region. All unusual circumstances need to be reported right away. If there are underwater sections of the foundation, such as piles or drilled shafts, an underwater inspection program must be put in place. To ascertain the structural condition of every underwater member, a thorough inspection must be conducted. It is essential that certified engineers assess the inspection results. The individuals conducting underwater inspections with diving equipment need to be certified as both divers and inspectors. The qualifications and inspection protocols for bridge inspectors are outlined in detail in the National Bridge Inspection Standards (NBIS). The U.S. Occupational Safety and Health Administration (OSHA) regulations, 29 Code of Federal Regulations, Part 1910 (29 CFR 1910), Subpart T-Commercial Diving Operations, with relevant updates and directives, at the very least, govern all diving operations. Information on diving equipment, dive hazard analyses, and diving inspection procedures and techniques are all included in this handbook. Doing underwater bridge inspections safely requires the right experience and commercial diving training. When it is feasible, piles should be inspected spirally (FHWA, 2010). Starting at the top of one pile, the diver descends to inspect it; moving on to the next pile, he or she inspects it while ascending (Figure 14.13). In low water visibility, the inspector might have to climb and descend the same pile. If a defect is discovered during an inspection without communication between the crew below and the inspector, the diver should come to the surface right away, report the specifics of the defect, and then go back to the defective site to complete the inspection.

14.7 INSPECTION OF STAY CABLES

Inspection and maintenance of cable-stayed bridges is a very challenging process. The main steel cable is hidden from the view of inspectors. Access to cables for visual inspections or non-destructive testing (NDT) is generally difficult, and in the case of the anchorage zones, nearly impossible. Those who are tasked with inspection and maintenance of stay cables are faced with challenges for which

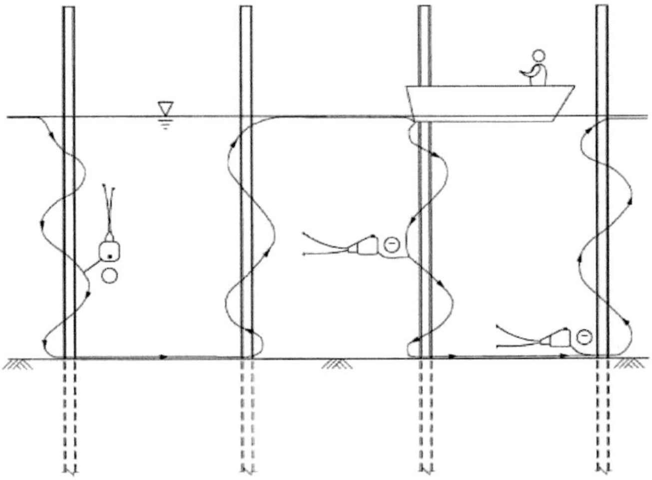

Fig. 14.13 Schematic representation of a diver inspecting a pile foundation (FHWA, 2010)

proven and accepted methodologies and tools are limited and, in many cases, very costly (Tabatabai, 2006). This section outlines effective inspection and maintenance techniques for stay cables in cable-stayed bridges. Inspection and monitoring methods are categorized to two classes: short term methods and long-term methods (NCHRP, 2005).

14.7.1 Visual Inspection

The most popular method for inspecting stay cables is visual inspection. The following are usually included in general visual inspections of stay cables:

- Find any evidence of excessive bulging or longitudinal or transverse cracking in the sheathing, along with any damage at the joints with the dampers or cross cables.
- Checking for abnormalities in the cable alignment, such as waviness or excessive sag. It is possible to estimate (measure) cable sag with optical devices or by processing images from photos or videos. An inclinometer can be used at certain locations to measure the cable angle.
- Inspect sheathing damage, particularly in cases where PVF tape is not utilized. Cracking in the sheathing needs to be monitored, particularly in high-stress areas.
- Damage to the connections between cable sheathing and anchorage pipes should be identified.
- Examining the neoprene boots and band clamps for wear and tear, damage, loosening, and degradation in order to ensure they are waterproof.
- Examining the neoprene rings and, if applicable, the keeper rings for wear or displacement.
- Finding the spaces that exist between the sheathing and the neoprene rings.
- Using a borescope or another tool, examine the sheathing surface inside the guide pipe to check for any damage or deformation to the sheathing close to the anchorage..
- Examine any evidence of guide pipe cracking or damage, as well as any proof that cable components have an effect on the pipes.
- Surface conditions of the visible anchorage components, such as bearing plates, end caps, and ring nuts, are examined.
- Check visible areas of saddles for corrosion, damage, and cracking, if any.
- Look for signs of moisture or fillers leaving the anchorage parts, like grease.
- Inspection of any dampers, if any, in accordance with the manufacturer's recommendations.

Lifts, basket-equipped cranes, or bucket trucks (Figure 14.14) can all be used to gain access to cable components. The visual inspection of cable-stayed bridges has adopted some creative ideas.

Fig. 14.14 Use of Bucket Truck for inspection of under-deck (courtesy, ODOT)

The Luling Bridge, located in St. Charles Parish, close to New Orleans, Louisiana, spans the Mississippi River and serves as one example (Elliot and Heymsfield, 2003). To help inspectors with the visual inspection and maintenance of the bridge, the Louisiana Department of Transportation and Development (LADOTD) has constructed two distinct trolleys. Figure 14.15 shows the trolley, which is a steel frame carriage with a detachable basket. With the help of a wire rope to pull the trolley up the cables, the system can support two inspectors and 400 lbs. of equipment. The two trolleys allow LADOTD bridge inspectors to conduct close-quarters inspections of the cable stays. This system has little effect on bridge traffic and is both time and cost effective.

In order to facilitate access to the stay cables of the 1,300-foot main span of the Dames Point Bridge, an additional novel system was created. The climbing system was designed using site visits and bridge plans; it is primarily a customized rolling device (Figure 14.16). Because inspectors would need to carry equipment up ladders to the tops of the 300-foot towers, the device weight was a significant consideration. In order to provide anchorage points at the top of the towers for load and safety lines and ropes to descend to the deck, the system needed specific rigging plates and connection hardware.

Inspection of the anchor zone includes removal of anchorage caps and visual inspection of strand ends, sockets, and locking plates. Visual inspection has been used efficiently for verifying in many cases the presence of water in and around anchorages and severity of corrosion activity because of this presence. It has also identified locations along the cable free length where the extent of deterioration has damaged the corrosion and ultraviolet (UV) protection barriers including UV protection tape, cover pipe, and filler grout visible. Nevertheless, the effectiveness of this method is limited to visible and accessible areas (Mehrabi, 2006).

14.7.2 Robotic Cable Scan

This technology is considered the most up to date. It does not require lane closures, man-lifts, bucket trucks, or night-time inspections. This system utilizes Magnetic Flux Leakage (MFL) technology to

Fig. 14.15 Trolley employed for inspection of Luling Bridge cable stays (Elliot and Heymsfield, 2003)

locate section loss and corrosion through the HDPE sheathing on the exterior strands of steel inside the steel cables. The robotic cable-stay inspection system utilizes high-definition video technology. The scanner robotically climbs the bridge taking a video of 360° of the exterior while conducting (MFL) of the interior steel locating the loss of metallic area and any broken wire within 360° of the cable (Figure 14.17). The self-propelled robotic system is controlled by a base station that receives High Definition (HD) videos and MFL transmissions from the robotic crawler. By utilizing this system all imperfections along the cable stay are recorded as well as the length traveled along with the stay, and the corresponding length traveled along the deck. It provides the length and size of cracks and other abnormalities on the cable stay.

14.8 MAINTENANCE OF CABLE-STAYED BRIDGE ELEMENTS

The concept of preventive bridge maintenance suggests that many relatively small repairs and activities are performed to keep the bridge in good condition and thereby avoid large expenses in

Fig. 14.16 Use of climbing devices for cable inspection (Courtesy, Burgess and Niple)

Fig. 14.17 Use of robotic cable scan (Courtesy, Infraspect)

major rehabilitation or replacement. Preventive maintenance activities can be classified into two groups: Scheduled and response (NYSDOT). Preventive maintenance is typically applied to bridge elements on structures with a significant remaining service life.

Scheduled: These are activities that are conducted on a scheduled interval basis including:

- cleaning decks, seats, caps, and salt splash zones.
- cleaning bridge deck drainage systems.
- cleaning and lubricating expansion-bearing assemblies.

- sealing concrete decks or substructure elements

Response: These are activities that are performed on an as-needed basis including:

- resealing expansion joints
- painting structural steel members
- repair of cable sheathing
- replacement of cables
- replacing deck wearing surfaces
- extending or enlarging deck drains.

Most of these activities can be readily performed by in-house forces and are cost-effective investments.

However, the engineer must ensure that preventive maintenance procedures contemplated for use are consistent with environmental standards and safety codes within the jurisdiction and obtain any required permits before commencing work.

14.9 INSPECTION AND MAINTENANCE MANUAL

Maintaining as-built documentation and an extensive maintenance handbook are essential for comprehending the significance of different parts and the prompt maintenance required to increase the bridge's longevity. The maintenance handbook is unique to each project. It must include the following information (NCHRP, 2005):

- A description of the bridge.
- The frequency of in-depth, twice-yearly inspections.
- Significant design information including:
  - the estimated (calculated) cable frequencies, sag, and inclination angles (at a point) as determined by the designer, both with and without the use of dampers or cross cables.
  - The estimated (calculated) bending stiffness and damping of the cable in the anchorage zones and free length by the designer.
  - The designer's estimated (calculated) stiffness of neoprene rings and/or proprietary dampers in contact with the cable.
  - The designer's estimated (calculated) wind speeds at which vibrations owing to vortex shedding would be expected.
- recommendations for baseline measurements of cable sag, cable inclination angles, cable frequency, and damping ratio should be done (at specific points accessible by inspectors).
- methods for determining whether viscous dampers or other dampers are functioning as intended, including damper maintenance methods.
- loads for vital components like wind tongues and bearings.
- details about the stay cable include identification numbers, the quantity of strands and wires, the diameter of the cable, mass per unit length, inclination angles, length, and an estimate of the cable tension both at the conclusion of construction and after creep and shrinkage effects have been taken into account.
- stay cable shop drawings including as-built anchorage design, materials used, any repairs done during construction, and history of problems during construction.
- frequency of maintenance and inspections, as well as the qualifications of inspection teams.
- information regarding platforms, ladders, and snooper trucks for access.
- techniques for retensioning cables.
- specific replacement protocols and precautions, along with traffic patterns and cable replacement procedures.

- bridge component inspection protocols for pylons, superstructure, cable stays, anchorages, guide pipes, neoprene boots, neoprene washers, sheathing, dampers, and other components; these include identifying critical areas, detecting techniques, and possible locations for moisture and corrosion.
- listings of vendors and contractors offering stay cables and related items.
- a summary of the entire stay cable system's qualification test results.
- performance standards and shop drawings.
- surveys of deck elevation.
- repair techniques including welding connections, guide pipe damage repair, PVF tape repair, and sheathing repair.
- traffic management and safety throughout inspections.
- techniques for calculating cable forces.
- survey forms for deck elevation and inspection forms.

References

American Association of State Highway and Transportation Officials, *Manual for Condition Evaluation of Bridges*, Washington D.C., 2011.

ASTM, *Standard Test Method for Pulse Velocity Through Concrete*, West Conshohocken, PA, 2023.

Bhandari, P., Kanawade, B., and Sengupta, A., Advanced NDT Methods for Evaluation of Bridges, *International Journal of Advance Research in Science and Engineering*, Vol. 5, # 09. 2016.

Elliot, M. and Heymsfield, E., Inspection of Luling Bridge Cable Stays: Case Study, *Journal of Construction Engineering and Management,* Vol. 129, No. 2, April 1, 2003.

Federal Highway Administration, *Bridge Inspector's Reference Manual, 2002.*

Federal Highway Administration, *Underwater Bridge Inspection, Underwater Bridge Inspection* Publication Number: FHWA-NHI-10-027, Washington, D.C., 2010.

Giannini, E.R., Folliard, K.J., Zhu, J., Bayrak, O., Kreitman, K., Webb, Z., and Hanson, B., Non-Destructive Evaluation of In-Service Concrete Structures Affected by Alkali-Silica Reaction (ASR) or Delayed Ettringite Formation (DEF)—Final Report, Part I

Kreitman, K. *"Nondestructive Evaluation of Reinforced Concrete Structures Affected by Alkali-Silica Reaction and Delayed Ettringite Formation."* MS Thesis, The University of Texas at Austin, Austin, Texas, 2011.

Mehrabi, A., In-Service Evaluation of Cable-Stayed Bridges, Overview of Available Methods, and Findings, *Journal of Bridge Engineering*, Vol. 11, No. 6, November 1, 2006.

National Cooperative Highway Research Program, *Inspection and Maintenance of Bridge Stay Cable Systems,* Transportation Research Board,, Washington, D.C., 2005.

New York State Department of Transportation, *Fundamental of Bridge Maintenance and Inspection of Bridge,* Albany, New York, 2008.

Occupational Safety and Health Administration, *Regulations (Standards-29CFR),* Washington, DC,1974.

Tabatabai, H., Maintenance and Inspection of Bridge Stay Cable Systems, *Wind-Induced Vibration of Cable Stayed Bridges Workshop,* St. Louis, Missouri, 2006.

Index